Practical Hatchery Management of Warmwater Fishes

In Memory of Jack Snow

This volume is made possible by the work of the late Jack Snow. Jack graduated from Auburn University earning a B.S. Degree in 1942 and a M.S. Degree in 1948. He started his fisheries career in 1950, when he joined the U.S. Bureau of Sport Fisheries and Wildlife. His assignment was at the Marion, Alabama National Fish Hatchery and Training School, where he served as Director until 1974 when he retired. He joined the Auburn University Fisheries Department serving until 1983. During his career, he published more than 55 scientific articles and papers on warm-water fish culture. Jack R. Snow is widely known among fish culturists and others for his excellent training program at the Marion school. During his career at Marion, Jack Snow trained more than 121 culturists using a unique "hands on and classroom" program. His service at Auburn University involved training, extension work and consulting with assignments in at least seven foreign countries. His capability as a dedicated teacher instilled high standards of knowledge and excellence into warm-water fish culturists nationally and internationally. He is recognized worldwide as a leader in warm-water fish culture.

Enshrined into the "NATIONAL FISH CULTURE HALL OF FAME,"
Spearfish, South Dakota, 1986.
Executive Committee, Fish Culture Section, American Fisheries Society

Practical Hatchery Management of Warmwater Fishes

Jack R. Snow
School of Fisheries, Aquaculture and Aquatic Sciences
Auburn University
203 Swingle Hall, Auburn, Alabama 36849, USA
(retired)

and

Ronald P. Phelps
School of Fisheries, Aquaculture and Aquatic Sciences
Auburn University
203 Swingle Hall, Auburn, Alabama 36849, USA
(retired)

American Fisheries Society
Bethesda, Maryland
2020

A suggested citation format for this book follows.

Snow, J., R. and R. P. Phelps. 2020. Practical hatchery management of warmwater fishes. American Fisheries Society, Bethesda, Maryland.

Front cover photos: (Clockwise from top-left) a collected egg sample; a Channel Catfish *Ictalurus punctatus*; Goldfish *Carassius auratus* spawning on a spawning mat made of Spanish moss; Common Carp *Cyprinus carpio* fry before stocking into primary nursery pond; a battery of MacDonald hatching jars, each with its own water supply; Largemouth Bass *Micropterus salmoides* fingerlings harvested from a primary nursery pond; seining of a pond; Largemouth Bass brooder sexually mature at one year of age.

Any use of trade, firm, or product names is for descriptive purposes only and does not imply endorsement by the authors.

Printed in the United States of America on acid-free paper.

Library of Congress Control Number 2020949960
ISBN 978-1-934874-59-2

American Fisheries Society Web site address: *www.fisheries.org*

American Fisheries Society
425 Barlow Place, Suite 110
Bethesda, Maryland 20814
USA

Table of Contents

Preface....ix
List of Figures....xi
List of Tables....xix
Symbols and Abbreviations....xxii

Section I: Hatchery Management Techniques and Procedures

1 **Introduction**....3

2 **Hatchery Facilities**
2.1 Water Supply....9
2.2 Earthen Ponds....13
2.3 Troughs, Tanks, and Raceways....17
2.4 Service Buildings and Infrastructure....19

3 **Production Alternatives and Brood Management**
3.1 Production Alternatives....25
3.2 Broodstock Selection and Management....28
3.3 References....42
3.4 Recommended Reading....43

4 **Control of the Spawning Process**
4.1 Environmental Stimuli....45
4.2 Hormone-induced Spawning....46
4.3 Hormone Administration....50
4.4 Spawning and Egg Handling....52
4.5 References....56
4.6 Recommended Reading....56

5 **Incubation and Post-Hatch Care**
5.1 Introduction....59
5.2 Water Considerations for Incubation....60
5.3 Incubation Equipment....61
5.4 Egg Management....65
5.5 Hatching....69
5.6 Larval Management Planning Considerations....70
5.7 Recommended Reading....71

6 Larval Management in the Hatchery

6.1 Post-hatch Management....73
6.2 Harvesting and Counting Larvae....74
6.3 Management Alternatives....76
6.4 Use of Live Food Organisms for Fish Larvae....77
6.5 Environmental Setting for Indoor Larval Rearing....80
6.6 Swim Bladder Inflation....81
6.7 Considerations Before Outdoor Stocking....81
6.8 Recommended Reading....82

7 Extensive Pond Production: Nursery Pond Preparation

7.1 Introduction....83
7.2 Pond Preparation....84
7.3 Fertilizers....87
7.4 Factors Affecting Fertilization Success....92
7.5 Recommended Reading....94

8 Extensive Pond Production: Pest Control

8.1 Weed Control....95
8.2 Animal Pest and Predator Control....97
8.3 Lower Vertebrates as Pests and Predators....100
8.4 Chemical Control of Invertebrates and Lower Invertebrates....102
8.5 Higher Vertebrates as Pests and Predators....103
8.6 Recommended Reading....108

9 Extensive Production: Pond Management

9.1 Primary Nursing Pond Stocking....109
9.2 Primary Pond Management....110
9.3 Secondary Nursery Pond Management....112
9.4 Production of Large Fingerlings....114
9.5 Recommended Reading....115

10 Feeds and Feeding

10.1 Formulated Feeds in the Hatchery....117
10.2 Feed Use in Pond Settling....119
10.3 Determination of How Much to Feed....120
10.4 How to Feed in a Pond Setting....122
10.5 Recommended Reading....126

11 Harvesting and Handling

11.1 Harvesting Techniques....127
11.2 Harvesting Considerations....131
11.3 Holding and Cleaning the Harvest....132
11.4 Grading....135
11.5 Sampling, Disease Control, and Hardening....138
11.6 Recommended Reading....139

12 Distribution

12.1 Modes of Transportation....141
12.2 Aeration....141
12.3 Open Container Systems....145
12.4 Closed Container Distribution....149
12.5 Factors Affecting Transportation....156
12.6 Distribution Problems....157
12.7 Recommended Reading....158

Section II: Specific Hatchery Techniques by Species

13 Catfishes

13.1 Channel Catfish and Its Hybrids....163
13.2 *Clarias* Catfish....184
13.3 *Pangasius* Catfish....195

14 Carps-Cyprinidae

14.1 Common Carp *Cyprinus carpio*....207
14.2 Chinese Carps....224
14.3 Bait Minnows....244

15 Tilapias

15.1 Introduction....267
15.2 Brood Management....270
15.3 Seed Production Alternatives....271
15.4 Fry Production Procedures for Use in Sex Reversal....276
15.5 Sex Reversal Procedures....282
15.6 Production of Sex-reversed Fingerlings to Stocker Size....285
15.7 Within Species Selective Breeding....286
15.8 Harvest and Distribution....287
15.9 References....287
15.10 Recommended Reading....288

16 Sportfish Species

16.1 Largemouth Bass *Micropterus salmoides*, Lacepede....291
16.2 Sunfish *Lepomis* sp. and Their Hybrids....320
16.3 Striped Bass *Morone saxatilis* (Walbaum) and White Bass *Morone chrysops*....331

Section III: Toolbox with More Details and Other Hatchery Topics

17 Toolbox with Recommended References

17.1 Aquaculture/Fisheries Management Planning....357
17.2 Aquatic Weed Control....361
17.3 Chemical Use and Dose Calculations....364
17.4 Control of Egg Adhesiveness....368
17.5 Counting Techniques....372

17.6 Disease Management....378
17.7 Effluent Management....380
17.8 Facilities and Equipment....381
17.9 Feeds and Fertilizers....391
17.10 Fish Transport....392
17.11 Genetics....393
17.12 Grading....394
17.13 Induced Spawning....395
17.14 Live Food Production....396
17.15 Nursery Pond Preparation....397
17.16 Predator Control....398
17.17 Water Filtration....399
17.18 Water Quality....400

Preface

This book has its foundations as the Training Manual for Warm-water Fish Culture used as part of the U.S. Fish and Wildlife Service training program directed by the senior author, Jack Snow, at Marion, Alabama for 28 years. During his career at Marion, he trained more than 121 culturists using a combination of "hands on" and classroom activities. Following his retirement in 1974 he joined the faculty of the Fisheries Department of Auburn University where he continued to teach and conduct hatchery-related research as well as being active in the Department's International program until 1983 and his second retirement.

The book was expanded and updated by the junior author during his 27 years leading the fish hatchery management program at Auburn University. During his career, Dr. Phelps has worked with hatchery aspects of more than 30 warmwater freshwater and marine fishes in the Americas, Africa, and Southeast Asia. The species covered in this book are ones with which the authors have first-hand experience.

Ron Phelps
November 2020

List of Figures

Figure 2.1. Monk-style concrete catch basin for harvesting, with a water supply to help maintain good water quality, and stairs to facilitate safe harvesting and movement of fish.

Figure 2.2. Monk-style drain with adjustable valve over the drain pipe, and grooves in the sidewalls to hold screens in front of drain.

Figure 2.3. Catch basin with a standpipe designed to pivot into the catch basin, a water supply to the basin and a pier to provide access.

Figure 2.4. Circular tanks can be used for brood holding, spawning, egg incubation, or fry or fingerling production depending on the fish species.

Figure 2.5. Raceway designs similar to those used for trout can be appropriate for several warmwater specie where gravity supplied water of an appropriate temperature is available.

Figure 2.6. Drive-under outdoor feed storage bin (Photo courtesy of Michael Luthi, Farm and Commercial Storage Mfg. Inc.).

Figure 2.7. Indoor feed storage room.

Figure 3.1. Egg sample taken by catheter of a marginal quality brooder based on having numerous small globules and the lack of transparency in the eggs, and the presence of primary oocytes in the sample.

Figure 3.2. Egg sample taken by catheter of an acceptable quality brooder based on enough globules have fused increasing the transparency of the eggs.

Figure 3.3. Egg sample taken by catheter of a hormone-induced female where a majority of eggs are advancing toward ovulation as oil globules have become larger and fewer in number and the eggs are more transparent.

Figure 3.4. Egg sample taken by catheter of a hormone-induced female nearing ovulation where the eggs have hydrated to the point they are starting to push against each other and loosing their round appearance.

Figure 4.1. Common Carp *Cyprinus carpio* pituitary exposed after brain is moved forward.

Figure 4.2. Intramuscular injections are given above the lateral line into the flesh in an area near the base of the dorsal fin.

Figure 4.3. A temporary tag made of yarn used to identify individual fish for hormone-induced spawning.

Figure 4.4. Stripping Grass Carp *Ctenopharyngodon idella* eggs into a dry bowl for "dry fertilization."

Figure 4.5. Sperm collection from a Grass Carp male.

Figure 5.1. Heath tray incubator, often referred to as a vertical flow incubator, shown as: (a) a single tray, and (b) a set of trays where the water flows down through the series of trays.

Figure 5.2. Adhesive *Clarias* eggs attached to a screen for incubation. Brown eggs are alive and white eggs are dead.

Figure 5.3. A battery of MacDonald hatching jars, each with its own water supply.

Figure 5.4. A MacDonald jar loaded with striped bass eggs, effluent of jar flows into an aquarium to collect the larvae.

Figure 5.5. Zug jar incubation where water supply enters from the bottom.

Figure 5.6. Fiberglass cones for egg incubation.

Figure 5.7. Cloth cone incubator suspended in a tank of water with the water supply for the cone entering at the tip of the cone.

Figure 5.8. Paddlewheel troughs for Channel Catfish *Ictalurus punctatus* eggs.

Figure 5.9. A paddlewheel trough with eggs in baskets set between paddles attached to a shaft running the length of the trough.

Figure 5.10. Whitish-colored dead eggs float higher in an incubator than live eggs.

Figure 5.11. Dead eggs close-up.

Figure 6.1. Hapa used for hatching of adhesive eggs or holding fry.

Figure 6.2. Volumetric determination of the number of Largemouth Bass *Micropterus salmoides* sac fry.

Figure 6.3. *Artemia* cyst, recently hatched *nauplii* and a 24-h *nauplii*

Figure 6.4. *Brachionus* rotifers with eggs.

Figure 7.1. Ponds should be dried thoroughly to the point where the clay soils cracks (a), with no water remaining in any nests or the catch basin (b).

Figure 7.2. Tilapia nests in a spawning pond, which were treated with quick lime to eliminate any fry or other potential pests before the pond is refilled.

Figure 8.1. Fish trapped inside a chewed-into filter sock.

Figure 8.2. Propane powered gas cannon used to scare birds.

Figure 11.1. A mud line added to the bottom lead line of the seine to minimize the seine from digging into the mud.

Figure 11.2. Glass V-trap with a wing to guide fish to the trap.

Figure 11.3. A small transport tank for on-farm use and local deliveries (Photo courtesy of Charles Weibel, Kentucky State University).

Figure 11.4. Pathway from harvest to distribution of fish.

Figure 11.5. a) Small tank bar grader; b) large tank bar grader.

Figure 11.6. Floating bar grader.

Figure 12.1. 12-volt (DC) electric agitator. Minnow-Saver #3. used in fish transport (Photo courtesy of Brian Smith and Joshua Senkevech, Minosaver, MinoTech Inc.).

Figure 12.2. A transport tank with two 1.2 m^3 compartments set on a pickup truck along with associated equipment used with fish transport.

Figure 12.3. A dedicated truck with four side-draining tanks for off-station fish transport (Photo courtesy of Robby Mays, American Sport Fish).

Figure 12.4. A gooseneck trailer with a compartmentalized tank and aeration systems set up specifically for fish transport (Photo courtesy of Michael Matthews, Texas Parks and Wildlife).

Figure 12.5. A 3- to 5-mil polyethylene plastic bag placed in a Styrofoam box, protected by an outer cardboard box used as a closed container for fish transport.

Figure 12.6. Transport bags made by cutting off sections from a tubular roll of polyethylene plastic and tying off each end.

Figure 13.1.1. A male Channel Catfish *Ictalurus punctatus* in breeding season is typically dark in color and has enlarged head pads.

Figure 13.1.2. Papilla of a mature male Channel Catfish during breeding season.

Figure 13.1.3. A mature female brood Channel Catfish is typically gray in color and should have an abdomen that spreads laterally when the female is placed on a flat surface.

Figure 13.1.4. Female Channel Catfish genital opening becomes distinct during the spawning season, appearing swollen and reddish in color.

Figure 13.1.5. A large Channel Catfish egg mass where the eggs are held in a gelatinous matrix.

Figure 13.1.6. Change in color of Channel Catfish eggs over the incubation period.

Figure 13.1.7. Catfish egg mass collection from a spawning pond.

Figure 13.1.8. Spawning pens for spawning specific pairs of catfish.

Figure 13.1.9. Manually stripping eggs following exogenous hormone induced ovulation.

Figure 13.1.10. Channel Catfish eggs in paddlewheel trough.

Figure 13.2.1. Manually fertilized, adhesive *Clarias gariepinus* eggs scattered over window screen and incubated indoors.

Figure 13.2.2. Size variation among hatchery-reared *Clarias* contributes to cannibalism.

Figure 13.2.3. Manual stripping of *Clarias gariepinus* following induced spawning.

Figure 13.3.1. Manual stripping of HCG–induced female *Pangasianodon hypophthalmus*.

Figure 13.3.2. *Pangasianodon hypophthalmus* eggs incubated in an up-flowing incubator.

Figure 13.3.3. *Pangasianodon hypophthalmus* larvae soon after hatch, dependent on water currents to stay up in the water column.

Figure 13.3.4. Feeding response of pangasius fingerlings given a formulated feed.

Figure 14.1.1. Natural spawning of Common Carp *Cyprinus carpio* in a pond setting, scattering eggs over vegetation.

Figure 14.1.2. Acceptable common carp brood female with soft abdomen and enlarged genital papilla.

Figure 14.1.3. Common Carp fry before stocking into primary nursery pond.

Figure 14.1.4. Common Carp fingerling after a 30-day primary nursery.

Figure 14.2.1. Male Grass Carp *Ctenopharyngodon idella* anus and papillae.

Figure 14.2.2. Female Grass Carp *Ctenopharyngodon idella* anus and papillae.

Figure 14.2.3. Male Grass Carp with pearl organs on pectoral fin evident by touch.

Figure 14.2.4. Female Grass Carp with extended pelvic fins, and starting to quiver, just prior to ovulation.

Figure 14.2.5. Ovulated female Grass Carp with a free flow of eggs.

Figure 14.2.6. Female Grass Carp with an inverted ovary limiting egg flow.

Figure 14.2.7. Un-water-hardened grass carp eggs added to a fiberglass cone for incubation.

Figure 14.2.8. (a). Un-water-hardened Grass Carp eggs in a MacDonald jar, (b) egg after water-hardening, (c) water-hardened eggs filling approximately 2/3rds of jar volume.

Figure 14.2.9. Grass Carp larvae ready to be stocked into nursery ponds.

Figure 14.3.1. Adult male Fathead Minnow *Pimephales promelas* in breeding colors (Photo courtesy of Barry W. Smith, American Sport Fish).

Figure 14.3.2. Fathead Minnow eggs deposited on a fibrous substrate. (Photo courtesy of Barry W. Smith, American Sport Fish).

Figure 14.3.3. Goldfish *Carassius auratus* spawning on a spawning mat made of Spanish moss.

Figure 14.3.4. Adult Golden Shiner *Notemigonus crysoleucas*. (Photo courtesy of Brian Zimmerman, Zimmermans Fish).

Figure 15.1. Adult *Tilapia zilli/Coptodon zillii* (common name Redbelly Tilapia).

Figure 15.2 *Tilapia zilli* nest dug into the pond bank.

Figure 15.3. Nests prepared by male *Oreochromis niloticus*.

Figure 15.4. Female *Oreochromis niloticus* holding eggs in mouth.

Figure 15.5. Genital papilla highlighted with carmine red of a male (a) and a female (b) tilapia used as a key for separating sexes. Note the slit of the oviduct across the papilla of a female.

Figure 15.6. Tilapia spawning hapas.

Figure 15.7. Harvesting a tilapia spawning hapa.

Figure 15.8. Down flow incubator (a) for tilapia eggs providing (b) a vigorous role of eggs.

Figure 15.9. Once tilapia larvae have developed enough they flow out of the incubator and are harvested daily from a mesh collecting box.

Figure 15.10. Screen grader for separating tilapia fry for sex reversal.

Figure 15.11. Tilapia held in a hapa set up with a feeding ring and a belt feeder.

Figure 16.1.1. Largemouth brooder can be sexually mature at one year of age and are often suitable as brooders.

Figure 16.1.2. Largemouth Bass tapeworm *Proteocephalus ambloplites* capable of causing significant damage to the ovaries of Largemouth Bass.

Figure 16.1.3. Male (a) and female (b) Largemouth brooders.

Figure 16.1.4. Visual estimation of the quantity of swim-up largemouth bass collected.

Figure 16.1.5. Placing spawning mats around the shoreline of the Largemouth Bass spawning ponds.

Figure 16.1.6. Largemouth Bass eggs on mat.

Figure 16.1.7. Indoor raceway for spawning of Largemouth Bass.

Figure 16.1.8. Indoor spawning with brooder on nest.

Figure 16.1.9. Incubation of mats containing Largemouth Bass eggs.

Figure 16.1.10. Largemouth Bass with functional mouthparts, ready for stocking into a primary nursery pond.

Figure 16.1.11. Largemouth Bass fingerlings harvested from a primary nursery pond.

Figure 16.2.1. Male Bluegill *Lepomis macrochirus* (above) and female Bluegill (below).

Figure 16.2.2. Several sunfishes such as Bluegill *Lepomis macrochirus* illustrated here are colonial spawners where a male will build and guard a nest in close vicinity to other males and their nests.

Figure 16.3.1. Sampling female Striped Bass *Morone saxatilis* with a catheter to collect an egg sample to determine egg maturity.

Figure 16.3.2. Wet fertilization of Striped Bass eggs.

List of Tables

Table 2.1. Water quality criteria for a freshwater hatchery.

Table 3.1. Species of warmwater fishes covered in this book where viruses have been reported.

Table 3.2. An example of a broodstock replacement plan covering a three-year period.

Table 4.1. Time after ovulation when approximately 50% of eggs are over-ripe (Woynarovich and Horvath, 1980).

Table 7.1. Amount of N-P-K nutrients contained in various organic fertilizer materials (dry weight).

Table 7.2. Composition of some common inorganic fertilizer source materials.

Table 10.1. Guide for feeding Oregon Moist Pellet to Largemouth Bass *Micropterus salmoides* at water temperatures in the range of 26–31°C.

Table 10.2. Feeding guide for Channel Catfish *Ictalurus punctatus* according to temperature and size.

Table 10.3. Feeding schedule for Channel Catfish *Ictalurus punctatus* fingerlings.

Table 11.1. Estimated time requirements for harvesting fingerlings of Largemouth Bass *Micropterus salmoides* or Bluegill *Lepomis macrochirus* from a 0.32 ha earthen pond by draining.

Table 11.2. A guide for loading holding tanks with fry or fingerlings.

Table 11.3. Sizes of Bluegill *Lepomis macrochirus* fingerlings that can be separated by grader mesh size.

Table 12.1. Relationship between truck size and quantity of fish hauled.

Table 13.1.1. Stocking rates for Channel Catfish fry stocked into earthen ponds, fed a commercial diet and cultured for 90–120 d.

Table 13.1.2. Suggested plan for feeding Channel Catfish fry to fingerling size in earthen ponds on formulated feeds when the water temperature is above 26°C.

Table 13.1.3. A feeding schedule for Channel Catfish fry and fingerlings based on anticipated biomass and water quality.

Table 13.1.4. Bar grader widths to separate Channel Catfish fingerlings.

Table 13.2.1. Hormones and dose rates used to induce spawn and manually strip female *Clarias* or *Heterobrachus*.

Table 13.2.2. Number of eggs/ kg of female body weight obtained from different Clariidae when hormone induced to spawn and manually stripped.

Table 13.2.3. Approximate hatching time, hours post-fertilization, by species and temperature.

Table 13.3.1. Hormone protocols used successfully to induce ovulation in pangasius.

Table 14.1.1. Hormones and dose rates used to induce spawn and manually strip female Common Carp.

Table 14.2.1. Hormones and dose rates used to induce spawn and manually strip female Grass Carp.

Table 14.2.2. Stocking rates appropriate to produce advanced fingerlings of Bighead Carp *Hypophthalmichthys nobilis*.

Table 14.2.3. Number of Chinese carp fry that can be transported in a container of 18.5 L capacity containing 8.5 L of free oxygen (from Hora and Pillay 1962).

Table 14.2.4. Guide for transporting small fingerling carps in a closed container supplied with oxygen (from Antalfi and Tolg 1968).

Table 14.3.1. Growth of Fathead Minnows on a natural diet.

Table 14.3.2. Fathead Minnow length and the number expected per kg or liter.

Table 15.1. Daily ration per 1,000 tilapia fry of known or assumed average length for the first month of feeding.

Table 16.1.1. Characteristics of spawns from Largemouth Bass *Micropterus salmoides* given artificial or natural diets.

Table 16.1.2. Forage sizes able to be consumed by various size classes of Largemouth Bass (from Lawrence, 1958).

Table 16.1.3. Results of growing brood bass in Bluegill *Lepomis macrochirus* spawning-rearing ponds.

Table 16.1.4. Time schedule for rearing 40 mm bass fingerlings at a water temperature of 19 °C.

Table 16.1.5. Extensive production of large Largemouth Bass fingerlings in ponds.

Table 16.1.6. Production of Largemouth Bass in ponds using Oregon Moist Pellets.

Table 16.3.1. Water quality parameters for Striped Bass egg incubation.

Table 16.3.2. Recommended feed sizes and feed rates for various sizes of Phase-2 Striped Bass and hybrid Striped Bass.

Symbols and Abbreviations

The following symbols and abbreviations may be found in this book without definition. Also undefined are standard mathematical and statistical symbols given in most dictionaries.

A	ampere
AC	alternating current
Bq	becquerel
C	coulomb
°C	degrees Celsius
cal	calorie
cd	candela
cm	centimeter
Co.	Company
Corp.	Corporation
cov	covariance
DC	direct current; District of Columbia
D	dextro (as a prefix)
d	day
d	dextrorotatory
df	degrees of freedom
dL	deciliter
E	east
E	expected value
e	base of natural logarithm (2.71828…)
e.g.	(exempli gratia) for example
eq	equivalent
et al.	(et alii) and others
etc.	et cetera
eV	electron volt
F	filial generation; Farad
°F	degrees Fahrenheit
fc	footcandle (0.0929 lx)
ft	foot (30.5 cm)
ft^3/s	cubic feet per second (0.0283 m^3/s)
g	gram
G	giga (10^9, as a prefix)
gal	gallon (3.79 L)
Gy	gray
h	hour
ha	hectare (2.47 acres)
hp	horsepower (746 W)
Hz	hertz
in	inch (2.54 cm)
Inc.	Incorporated
i.e.	(id est) that is
IU	international unit
J	joule
K	Kelvin (degrees above absolute zero)
k	kilo (10^3, as a prefix)
kg	kilogram
km	kilometer
l	levorotatory
L	levo (as a prefix)
L	liter (0.264 gal, 1.06 qt)
lb	pound (0.454 kg, 454g)
lm	lumen
log	logarithm
Ltd.	Limited
M	mega (10^6, as a prefix); molar (as a suffix or by itself)
m	meter (as a suffix or by itself); milli (10^{-3}, as a prefix)
mi	mile (1.61 km)
min	minute
mol	mole
N	normal (for chemistry); north (for geography); newton
N	sample size

NS not significant
n ploidy; nanno (10^{-9}, as a prefix)
o ortho (as a chemical prefix)
oz ounce (28.4 g)
P probability
p para (as a chemical prefix)
p pico (10^{-12}, as a prefix)
Pa pascal
pH negative log of hydrogen ion activity
ppm parts per million
qt quart (0.946 L)
R multiple correlation or regression coefficient
r simple correlation or regression coefficient
rad radian
S siemens (for electrical conductance); south (for geography)
SD standard deviation
SE standard error
s second
T tesla
tris tris(hydroxymethyl)-aminomethane (a buffer)
UK United Kingdom
U.S. United States (adjective)
USA United States of America (noun)
V volt
V, Var variance (population)
var variance (sample)
W watt (for power); west (for geography)
Wb weber
yd yard (0.914 m, 91.4 cm)
α probability of type I error (false rejection of null hypothesis)
β probability of type II error (false acceptance of null hypothesis)
Ω ohm
μ micro (10^{-6}, as a prefix)
′ minute (angular)
″ second (angular)
° degree (temperature as a prefix, angular as a suffix)
% per cent (per hundred)
‰ per mille (per thousand)

Section I

Hatchery Management Techniques and Procedures

Chapter 1

Introduction

To produce an agricultural crop, you need seed that can be planted, grown to a larger size, and then harvested. To culture fish you also need seed to begin the production. A role of a fish hatchery is to produce the seed. How the final product is used may vary. The fish may be produced as a food for human consumption, as a sportfish for recreation, as a means for restoring natural populations, or in the case of ornamental fishes, as a companion animal. In each case the fish are cultured under controlled conditions (aquaculture) and seed is needed to begin production.

Finfish aquaculture had its beginnings in the method of collecting seed from the wild and culturing them under controlled conditions. However, this had numerous disadvantages, including seasonal changes in availability as well as year-to-year variability in supply, the lack of genetic control, and the potential for overfishing of the natural seed supply. A more controlled supply of seed was needed before the culture of a given species could become widespread. As appropriate hatchery techniques were developed adult fish could be brought into maturation, stimulated to spawn, eggs and larvae managed under suitable conditions for growth and survival, and then reared to a suitable size to provide the seed needed to begin the production phase.

Often under hatchery conditions, solving a few critical steps in the life history of a fish has resulted in major changes in the quantities of fish being produced. In the case of the Chinese carps, they were cultured for thousands of years but the extent of their culture was limited geographically by how far wild-caught larvae could be safely transported from the collection site to the culture site. Chinese carp such as Bighead (*Hypophthalmichthys nobilis*), Silver (*Hypophthalmichthys molitrix*) and Grass Carp (*Ctenopharyngodon idella*) are riverine spawners that will mature in ponds but will not spawn in static water. In the 1950s and 1960s, hormone induced spawning methods were developed that allowed static water raised brood fish to be artificially spawned. This allowed Chinese carp larvae and fingerlings to be shipped to other parts of the world where they could come to maturity in static water ponds, successfully be induced to spawn, and produce fingerlings.

Seed availability was a major factor limiting pangasius production throughout the Mekong river basin until induced spawning techniques were developed. Pangasius migrate up river to spawn during the rainy season, the larvae carried downstream into the flooded floodplain where they develop, and are carried back into the rivers as the water level drops. Fingerlings were captured from the wild for culture at such rates to the extent that their capture in Cambodia was prohibited because of overfishing. Up until the 1990s, the annual production of pangasius, which was dependent on wild-caught seed, rarely reached 20,000 tons a year. However, with the increased availability of seed following the development of hatchery-based induced spawning techniques annual production levels of over 140,000 tons were reached by 2010. Solving this bottleneck and other culture problems has resulted in pangasius becoming a worldwide commodity.

Often there are a series of hatchery related problems that must be solved before adequate quantities of seed can be available. This was the case with Gilthead Sea Bream *Sparus aurata* in Europe. Its culture is another example of where a traditional culture of a species based on wild-caught seed expanded rapidly once hatchery techniques were developed. In this case a progression of hatchery issues were solved over a ten-year period, enough to give a 15% survival from eggs to fingerlings out of the hatchery. Achieving this threshold made commercial culture of sea bream practical, and production rose from less than 1,000 tons in 1990 to over 100,000 tons by 2000. Since crossing that threshold hatchery techniques have continued to improve and fingerling production has become more efficient.

Successful seed production is not just one activity but includes: obtaining seed, rearing, harvesting, holding, and distribution. These topics will be covered in this book as they apply to many of the warmwater freshwater fishes cultured in the world. In this book hatchery management is defined as "the implementation of appropriate production systems to achieve predetermined seed production goals." There are several key concepts in this definition. "Systems" is a key word, as a hatchery is a series of activities linked together that must be accomplished successfully if the goals are to be met. This includes the linking of biological activities such as broodstock management, fish spawning and egg incubation, as well as larval and juvenile rearing to the resources available. The resources needed include adequate facilities (tanks, ponds, buildings etc.), supplies, staff, and budget. Of the resources needed a well-trained motivated staff is key. A well-trained motivated staff can make an underfunded hatchery with out-of-date facilities productive, but a well-funded hatchery with state-of-the-art infrastructure will not be productive unless the staff is well trained and motivated.

"Predetermined goals" is another key concept in the definition of Hatchery Management in that we set goals before we start taking steps to meet those goals. To accomplish our goals efficiently and profitably we need to have them clearly defined. A goal common for any fish produced in a hatchery is that they be of good quality and are healthy. This is key in that the fish produced by a hatchery are not the end product but are the seed for producing a larger animal. In addition, specific hatchery goals are typically defined in terms of:

1. Specified species—single or multiple species
2. Specified number of fish of each species
3. Specific sizes of each species
4. Specified culture activity (tagged, fed trained, disease free, etc.)
5. Specific genetic history
6. Available at specified time
7. Delivery to designated locations

Once the goals are defined you can develop a strategy to accomplish the goals. You can evaluate the appropriateness of the resources available, select methodologies for implementation, identify what additional resources that will have to be obtained and budgeted for, set up schedules for each phase of production, and develop backup plans in case certain resources are not available when you need them.

"Management" is another key word in the definition of Hatchery Management. This book will focus on the technical aspects of fish production and the tools available but the management is often what makes a hatchery successful. A successful hatchery manager must juggle a finite quantity of resources to make a facility productive. The manager should recognize the resources available and the constraints they impose, the biological constraints of the animals to be produced, and the production options available for the animals to be produced. From there the manager should develop a plan of work that is practical, is realistic, and will come as close as possible to meeting the goals, while keeping in mind the constraints. The risks associated with the production methodology need to be reviewed. Some production methods can result in large yields under ideal conditions but the yields might be highly variable due to variations in weather or some other hard-to-control factor. Other production methodologies may give a lower mean yield but with more consistency from one production unit to another resulting in a greater overall yield from a block of production units.

Goals are meant to be adjusted. By carefully reviewing the constraints and the corresponding plan of work, it may be apparent the goals cannot be met. In that case the goals need to be reviewed. Who is depending on the animals you are producing? Will they accept a lesser quantity, a smaller animal, or a different delivery date? This discussion needs to be held before you go into production. It may be that in reviewing the goals and the constraints to meeting those goals the manager may find that the goals can be easily met and resources are available to produce more. In that case you need to ask yourself is there a need to produce more than asked for, or can those resources be used to produce another product?

Who sets the goals for a hatchery? First you need to know who the clientele is. In the public sector, government hatcheries often support services to other programs. In that case your clientele is biologists that need fish of certain characteristic for use in their programs. Often the biologists are asked once a year by their supervisor what their needs are in the coming year. The supervisor would compile the requests, maybe making readjusts, and then sets a hatchery's goals based on those requests. The hatchery staff may have little input in setting such goals but should provide feedback to the supervisor as what will be needed to accomplish those goals.

In the private sector, hatchery goals depends on who owns the facility and how much freedom the hatchery manager has to set the goals. An established hatchery should keep a good set of records as to: who has bought what in the past, are they repeat customers, how many new customers did you gain last year, and how many did you lose? The records should also be reviewed for trends over time as to: the demand for specific sizes of a given species, the number sold and when, and changes in demand by geographic area. For a new hatchery, goals should have been stated as part of the business plan along with a clear identification of the market opportunities.

How well did you accomplish your goals? Good record keeping will let you see what quantity and quality of fish were produced and when such fish were available. In a hatchery we commonly evaluate the production based on biological criteria such as growth rates, feed conversion ratios, and fish survival. From a management view we can evaluate how efficiently a hatchery met its goals in terms of: number or weight of fish produced/money spent; number or weight of fish produced/m^2 of pond or m^3 of raceways available; and number or weight of fish produced/person day spent. More information on setting goals, budgeting, and evaluation is given in Chapter 17.1.

The following chapters will familiarize the reader with some of the components of a hatchery and the basic techniques used in a freshwater fish hatchery, illustrated by some commonly cultured fishes. The goal of this book is to ensure that if the reader knows the basic biology of any fish species then the reader should be able to select an appropriate combination of techniques covered in this book to successfully produce that species in a hatchery setting. The book is organized into three major sections. Chapters 2 to 12 discuss basic hatchery infrastructure, techniques, and procedures available for the production of a variety of fish species. Many basic procedures can be applied across species when there is an understanding of why the techniques are used and how they can be modified to address specific problems unique to a given species. Chapters 13 to 16 give details on the hatchery production of commonly cultured warmwater food fish and sport fishes. A final section entitled Toolbox (Chapter 17) covers specific techniques that have application at most hatcheries and to a number of species of fish. The focus is on providing references, most available through the Internet, that provide detail on specific issues and techniques where the reader's personal experience may be lacking or needs to be reinforced.

For more information on fish hatchery contributions, components, planning and goal setting, see Chapter 17.1. Aquaculture Planning, as well as the recommended reading listed below.

Recommended Reading

Allan, G., and G. Burnell, editors. 2013. Advances in aquaculture hatchery technology. Woodhead Publishing Series in Food Science, Technology and Nutrition, Elsevier, Amsterdam, Netherlands.

Araki, H., and C. Schmid. 2010. Is hatchery stocking a help or harm? Evidence, limitations and future directions in ecological and genetic surveys. Aquaculture 308:52–59.

Avery, J. L., and J. A. Steeby. 2004. Chapter 7. Hatchery management. *In* C. S. Tucker and J. A. Hargreaves, editors. Biology and culture of Channel Catfish, Developments in Aquaculture and Fisheries Science 34:145–165.

Baluyut, E. A. 1989. Aquaculture systems and practices: a selected review. United Nations Development Programme. Food and Agriculture Organization of the United Nations, Rome. Available: http://www.fao.org/3/t8598e/t8598e00.htm#Contents. (May 2020).

Bondad-Reantaso, M. G. editor. 2007. Assessment of freshwater fish seed resources for sustainable aquaculture. FAO Technical Paper. No. 501. FAO, Rome.
Available: http://www.fao.org/3/a1495e/a1495e00.htm. (May 2020).

Boyd, C. E., C. S. Tucker, A. McNevin, K. Bostick, and J. Clay. 2007. Indicators of resource use efficiency and environmental performance in fish and crustacean aquaculture. Reviews in Fisheries Science 15:327–360.

De Silva, S. S., and N. T. Phuong. 2011. Striped catfish farming in the Mekong Delta, Vietnam: a tumultuous path to a global success. Reviews in Aquaculture 3(2):45–73.

Diana, J. S., H. S. Egna, T. Chopin, M. S. Peterson, L. Cao, R. Pomeroy, M. Verdegem, W. T. Slack, M. G. Bondad-Reantaso, and F. Cabello. 2013. Responsible Aquaculture in 2050: Valuing local conditions and human innovations will be key to success. BioScience 63(4):255–262. Available: https://academic.oup.com/bioscience/article/63/4/255/253239. (May 2020).

Gui, J., Q. Tang, Z. Li, J. Liu, and S. S. De Silva, editors. 2018. Aquaculture in China: success stories and modern trends. Wiley Blackwell, Somerset, New Jersey.

Halls, A. S., and M. Johns. 2013. Assessment of the vulnerability of the Mekong Delta Pangasius catfish industry to development and climate change in the Lower Mekong Basin. Report prepared for the Sustainable Fisheries Partnership, January 2013, 95 pp.

Hargreaves, J. A. 2002. Channel Catfish farming in ponds: lessons from a maturing industry. Reviews in Fisheries Science 10:499–528.

Joffre, O. M., L. Klerkx, M. Dickson, and M. Verdegem. 2017. How is innovation in aquaculture conceptualized and managed? A systematic literature review and reflection framework to inform analysis and action. Aquaculture 470:129–148.

Kumar, G., C. Engle, and C. Tucker. 2018. Factors driving aquaculture technology adoption, Journal of the World Aquaculture Society 49(3):447–476. Available: https://onlinelibrary.wiley.com/doi/epdf/10.1111/jwas.12514. (May 2020).

Moretti, A., M. Pedini Fernandez-Criado, G. Cittolin, and R. Guidastri, 1999. Manual on hatchery production of seabass and gilthead seabream, Volume 1. Part 1. Historical background and present situation. FAO, Rome. Available: http://www.fao.org/3/x3980e/x3980e04.htm#bm04.1.3. (May 2020).

Patrick, W. S., O. Bin, K. A. Schwabe, and P. W. Schuhmann. 2006. Hatchery programs, stock enhancement, and cost effectiveness: a case study of the Albemarle Sound/Roanoke River stocking program 1981–1996. Marine Policy 30:299–307.

Wang, Q., L. Cheng, J. Liu, Z. Li, S. Xie, and S.S. De Silva. 2015. Freshwater aquaculture in PR China: trends and prospects. Reviews in Aquaculture 7:283–312.

Weber, G. M., and C. S, Lee. 2014. Current and future assisted reproductive technologies for fish species. Pages 33–76 *in* G. Lamb and N. DiLorenzo, editors. Current and future reproductive technologies and world food production. Advances in Experimental Medicine and Biology, vol 752. Springer, New York.

Chapter 2

Hatchery Facilities

There are several aspects of hatchery facilities that are unique to a fish hatchery often not considered in standard construction practices. A hatchery biologist does not have to be an architect or an engineer but needs to be able to explain to an architect or an engineer what is needed for a fish hatchery to function effectively and why.

2.1. Water Supply

A good water supply is an essential requirement for consistent production of high-quality fish seed. Each species of fish has an acceptable range of physical and chemical characteristics of water conducive to growth and reproduction. The water quality requirements for successful fish reproduction are more demanding than those for growth. Often the first indication of poor environmental conditions at a hatchery is a depression in fish reproduction. Acceptable water quality conditions for freshwater fish spawning are shown in Table 2.1. The chemical profile of the water supply will impact pond fertilization success, efficacy of chemical weed control, levels of photosynthesis, organic matter decay rates, and other processes. Physical characteristics such as temperature and suspended materials are also highly important in culturing some species.

The quantity of water available should be ample enough to fill all the ponds on a hatchery within 5 to 10 days. After the ponds are filled, a similar volume of water may be required to replace seepage and evaporation losses during the culture period. When the fish crop is being harvested some new water may be needed for trapping or draining operations. Additional volumes of water may be needed to flush ponds in the event of low oxygen levels, or dilute the pond water following weed or disease treatments. A substantial amount of water is also needed for hatching, holding, processing, and shipping. At a minimum, a flow of 0.4 m^3/min/ha will be needed for all these purposes. The yearly quality of water needed for holding facilities is often equal or greater than that needed for filling ponds.

In intensive culture units the quantity of water available should permit 1 or 2 water changes or more per hour, depending upon the type unit and species involved. Supply and drain lines should be of a size to quickly admit and remove the quantity of water required. Generally, the younger the fish, the greater the requirement for high quality water with a minimum fluctuation in any parameter.

Sources of water for a fish hatchery include springs, wells, lakes, reservoirs and streams, each with advantages and disadvantages. Two or more sources may be used to supply a given hatchery. The best quality water should be reserved for the hatchery buildings where fish are to be reproduced and eggs or larvae are to be held. The source providing the largest volume should be available for the ponds.

Table 2.1. Water quality criteria for a freshwater hatchery.

1. Correct temperature for species being cultured.
2. Low CO_2 – less than 5 ppm, no more than 20 ppm.
3. High O_2 – 100% saturation, more than 5 ppm.
   1–5 ppm slow growth.
   less than 1 ppm lethal.
4. Nitrogen (gas).
   100% saturation acceptable.
   supersaturation causes problems.
5. Suspended solids < 2000 ppm.
6. pH – 6.5–9.0.
7. Alkalinity and hardness.
   minimum 25 ppm.
   ideal 80–150 ppm.
8. Salinity < 5 ppt, species specific.
9. Iron – less than 0.1 ppm.
10. Pesticides/pollution – not acceptable.

Groundwater is often anoxic when tapped by a well or emerging out of the ground as a spring and should be aerated before being used in a hatchery. Minerals such as iron and manganese can go into solution under such anoxic conditions. Provisions need to be made to aerate such groundwater to bring the oxygen concentration up to saturation and allow enough time for any dissolved iron or manganese to become oxidized, and precipitate out, or the precipitant be filtered out before the water is used in the hatchery (see 17.17. Water Filtration). Manganese precipitates out slower than iron and water supplies high in this mineral need to be held longer before being used. Precipitated minerals will clog values reducing water flow to egg incubators or where larval are being held as well as clogging fine mesh drain screens. If such precipitants are not removed prior to egg incubation or larval holding they can settle on the eggs and larvae and suffocate them.

Groundwater is also often supersaturated with other gases (CO_2, N_2) due to the pressure of the soil above the water table. These gases will begin to come out of solution when the pressure is removed as the water reaches the surface. Such supersaturations should be stripped out before use, otherwise gas bubble problems may occur in the fish being held in such water. As the pressure is removed and the gas comes out of solution the bubbles may cause eggs to float out of the incubators or result in over-inflation of the swim bladder or bubbles in the intestine of larval fish. A small reservoir pond can be constructed near the hatchery building where groundwater can be held to allow the water to warm up, iron and manganese to precipitate out, and any supersaturation of gases to be reduced before it is used with fish eggs or larvae.

Groundwater may have other water quality issues that should not be overlooked. In many parts of the southeastern USA there are strata of groundwater high in alkalinity but with little to no calcium hardness, but such water sources can be used if calcium (agricultural lime $CaCO_3$, or Calcium Chloride $CaCL_2$) is added before use to increase the hardness.

Depending on the depth at which the water source originates, the water is often too cool for direct use in reproducing warmwater fishes (see Chapter 5). In such cases the water can be warmed using a water heater and mixing valve, a heat exchanger, a heat pump or a combination of these approaches. However, warming ground water can result in gases such as nitrogen and carbon dioxide coming out of solution resulting in gas bubbles in the water supply that need to be stripped from the water prior to its use. This is often done using a packed column or other degassing device. Such devices also help to aerate oxygen deficient groundwater.

Although several negatives have been described for groundwater, most problems can be easily corrected, and it is the preferred water source for a hatchery. Its main advantages are that it is less contaminated (diseases and other undesirable organisms) than surface water sources, there are less season variations in quantity and quality, and there is more on-site control of the source. As the volume of water needed for egg incubation and larval rearing is typically much less that needed for pond culture, it is often justified to drill a well to provide water specifically for the egg and larval phases of a hatchery operation.

Surface waters can be collected as runoff from within the local watershed or be diverted from a nearby stream, lake, or reservoir. The fundamental problem with surface water is contamination. Wild fish found in surface waters can contaminate hatchery ponds, preying on or competing with desired fishes, and also be a reservoir of infectious diseases. Undesirable plants associated with surface water sources can be potential weed problems in production ponds. Such contamination can be controlled by screening and filtration before the water enters the hatchery building or is used to fill a pond.

Often the watershed providing the surface water for a hatchery is not completely owned by the hatchery. As such, there are risks of the surface water being contaminated on occasions with sediments or pollutants following heavy rains or accidents occurring in the watershed that cannot be controlled or predicted. To minimize the risks of such problems, a reservoir should be built on hatchery property that will provide a minimum of a week's water storage. This will allow the hatchery to continue to operate if there is an issue with the surface water coming down the watershed. A reservoir with such a retention time will also allow any sediment in the water source to settle out in the reservoir before the water is used elsewhere on the hatchery.

The reservoir should be designed so that there is an inlet control structure so water coming into the reservoir can be controlled, adding water only as needed. This will help minimize the risk of storm-event contamination or known pollution accidents that may have occurred in the watershed. In addition, a diversion canal around the reservoir should be built so the water supply to the reservoir can be diverted around the reservoir in case of pollution events or excessive rainfall in the watershed.

The reservoir should be designed so that water can be drained off at two or more depths within the reservoir as well as a drain low enough so the reservoir can be drained dry for maintenance and repair. Water reservoirs often stratify resulting in a warmer layer of water above the thermocline and cooler layer below. There may be occasions on a hatchery located in a temperate climate where you may want to delay the spawning season of a portion of the broodfish by adding cooler water taken from the reservoir below the thermocline in an attempt to keep the brood pond temperature below what is appropriate for spawning of that species. This can be an effective technique, however, such water may be low in oxygen and must be aerated before being added to a brood holding pond. In locations where there are notable rainy and dry seasons, the drawdown of the reservoir by the end of the dry season may be significant. In such locations, the reservoir should be designed to be large enough to hold enough water to supply gravity-fed water to all hatchery locations where fish are held even at the end of the dry season.

A gravity flow water supply is highly desirable. This can be achieved from any of the listed water sources provided that the site lends itself to this feature. Not only does the gravity flow supply minimize cost of delivering the water where it is needed, but also it is less likely to be interrupted at critical times. Even when pumping is necessary, provision for a reservoir pond of adequate capacity can reduce danger of the water supply being cut off because of power failure or mechanical breakdown.

The appropriateness of a given water supply should take several factors into consideration. Chemical analysis of any water source should be made and the results compared to Table 2.1 to determine if the water quality meets the basic requirements appropriate for fish spawning and rearing. Biological tests evaluating any effects of a water source on successful spawning fish and egg hatching should be conducted. Holding small fish for several months and evaluating their growth and survival can help identify any chronic problems associated with the water supply. Being able to obtain good phytoplankton and zooplankton production in response to organic and inorganic fertilizers is another important indicator of a good water supply.

Supply lines large enough to carry the volume of water needed to operate the facilities at full capacity should be provided. Details on supply line and pump selection are given in Chapter 17.8 Facilities and Equipment. Ideally the water supply system is adequate to simultaneously fill several hatchery ponds within a 48 h period. Keep in mind that the holding facilities will seasonally require large volumes of water that may divert water needed for pond filling if the supply lines are not of adequate size.

In locations where the quantity of water is limited or where restrictions against disposal of hatchery effluent into public waters exist, a reuse system may be necessary. Such a system might include a provision for treatment after use, mechanical filtration, biofiltration or a combination of these treatments. Following treatment, the water is pumped to the reservoir pond for reuse. The use of ultra-violet light treatment might be necessary to control chronic disease problems particularly if such water is to be used with fish eggs and larvae. When good quality ground or surface water is plentiful, the increased cost of a water reuse system (equipment, maintenance, electricity etc.) is often not justified. However, where fish escapement or the water quality of a hatchery's effluent is an issue then a water reuse system may be justified or additional treatment of the effluent may be necessary.

As ponds are drained the last water leaving a pond may be very high in suspended solids and a settling pond with enough retention time for suspended solids to settle out might be appropriate. Pond water temperatures are often higher than the temperature of the receiving stream and cause native fauna to be displaced when ponds are drained. A large enough holding pond to receive pond drain water can store the effluent, allowing a slow release of hatchery water into the receiving stream will minimize temperature effects and also dilute any BOD (Biological Oxygen Demand) or nutrient impact on the receiving stream.

Often a hatchery may be called on to produce a species or genetic strain of a species that is not native to the watershed in which a hatchery is located. If so, a risk assessment needs to be conducted. A no-discharge hatchery can be set up so that all discharges flow by gravity into a holding pond and then the water pumped back up into a reservoir pond where it can be re-used. Another approach is to design the holding pond so that it does not hold water well and all water entering the pond will seep out of the pond allowing the pond walls and bottom to act as a filter. For more detail on effluent management see Chapter 17.7.

2.2. Earthen Ponds

Earthen ponds are used for the reproduction and nursery of many important fish species. A hatchery pond should be of an appropriate size to provide the number of fish needed at that time, hold water well, have its own individual supply and drain, be deep enough to permit weeds to be controlled, slope evenly from shallow to deep end into a harvest basin, and be equipped with a water supply to the harvest basin to aid in safely harvesting the fish. The optimum size pond may range from 0.1 to 0.5 ha depending on the fish to be produced and the quantity of fish that need to be available at a given point in time. Ponds larger than 1 ha are more difficult to manage and may produce more fish than can be readily harvested and distributed in a short period of time. Ponds as small as 0.02 to 0.04 ha are often needed and most hatcheries could use at least 5 to 10 of this size for temporarily holding small numbers of large fish or greater numbers of the smaller sizes as well as spawning of certain species.

Ponds should be built on soils with good water holding capacity. Construction of the dikes should include clay core walls extending into the subsoil below the pond bottom. Compaction of the fill is essential and adequate slope provided for the dikes to minimize sloughing and lateral seepage. The pond dikes should be planted with a suitable vegetative cover as soon as construction is complete to reduce the effect of wave action and to minimize erosion. Excessive seepage will result in nutrients being lost from the pond, reducing the production of natural food organisms, and in turn reducing production of fish dependent on such food organisms. The loss of nutrients from a pond makes it more difficult to establish phytoplankton blooms and favors the establishment of aquatic weeds. If water must be pumped to replace excessive seepage, production costs also are increased. Trying to save money during the pond construction phase can increase the cost of fish production in the long term. For more detail on pond construction see the references in Chapter 17.8.

For hatchery ponds $\leq$1,000 m^2 a rubberized liner may be a cost-effective solution where soils do not have adequate clay content to minimize seepage. The pond needs to be shaped to the desired configuration prior to adding the liner making sure that the pond bottom and sides do not have any exposed rock, tree roots or other material that can damage the pond liner. Several types of liner material are available with EPDM (Ethylene Propylene Diene Terpolymer) and HDPE (High Density Polyethylene) being commonly used. EPDM liner material is more flexible, easier to install, and repair than HDPE liner material but initially more expensive/m2 than HDPE. Both materials have a life expectancy of approximately 20 years.

When small fish are to be produced in ponds they are best harvested by draining the pond into a catch basin. The pond should have a uniform slope from the shallow to deep end, sloped about 1 percent from the shallow to deep end and about 0.5 percent from the toe of the side dikes to the center of the pond. Such bottom slopes are needed to assure downward movement of small fish as the pond is being drained. A minimum depth of 75 cm in the shallow end sloping to 165 cm at the drain has proven to be satisfactory for ponds of 0.2 to 1.0 ha size. When high weights of fish (>500 kg/ha) are anticipated to be harvested, a partial harvest by seining may be needed before the pond is completely drained. A seining basin equal to 5–10 percent of the pond area should be prepared in the deepest part of the pond to have a firm bottom and be free of objects that might interfere with seining.

Nursery ponds are often used to culture larval fish for 30–45 days until they have begun to taken on juvenile characteristics. However, when the fish are harvested they are still small and often sensitive to handling and poor water quality. Such fish are best harvested in water using a catch basin similar to Figure 2.1. The pond should be carefully constructed so the pond will drain completely into the catch basin. Water draining into the catch basin warms up during the day and carries suspended solids, in turn reducing the availability of oxygen. A supply of well aerated freshwater should be available at the catch basin so that pond water entering the catch basin can be diluted, helping to control the water temperature while adding oxygen.

To harvest small fish from a catch basin, the fish should be crowded together in the catch basin and scooped into a bucket of water. Moving the buckets of fish safely out of the pond can be an issue. To avoid sliding on the side of the pond dike and spilling a bucket of fish, stairs should be built that extend from the catch basin up the side of the dike (Figure 2.1). The extra costs of having a catch basin, a source of fresh water at the catch basin and stairs up from the catch basin will pay for itself over time resulting from the improved survival and quality of the fish being harvested.

The optimum shape of such pond units is rectangular with a ratio of length to width of about 2.5:1. Ponds with a length: width ratio of 4:1 or more have been built but a ratio this width increases construction cost and may cause a problem in establishing and maintaining an adequate slope to the pond bottom of smaller ponds. If possible, a water supply at both the deep and shallow ends of the pond should be provided. Where this provision is not made, a water supply at the deep end (Figure 2.1) is preferred over a shallow-end water inlet. Size of the supply line needs to be sufficient to fill the pond in 48 to 60 hours or less.

Drain lines should permit a pond (<1,000 m) to be emptied in 24 to 48 hours with main lines handling the effluent of least three ponds simultaneously. Larger ponds may take 3 to 5

Figure 2.1. Concrete catch basin for harvesting, with a water supply to help maintain good water quality, and stairs to facilitate safe harvesting and movement of fish.

days to drain. The longer the drain time the longer the fish might be exposed to poor water quality and are subject to predation by a variety of animals. Ponds should be designed so each pond drains independently without one pond draining into another.

A commonly used drain structure is a "monk," a three-sided structure that is connected at its base through its back wall to the drainpipe. The monk is attached to the floor of the catch basin and extends upward to above the maximum desired water level of the pond. Each sidewall has two groves down the inside to hold a screen or dam board. Most monks have a threaded rod running down the inside back-wall that connects to a flap valve that covers the drain opening, allowing the valve to be cranked up or down to cover the mouth of the drain (Figure 2.2). A screen frame is designed to fit in the groves on the sidewalls. This screen extended from the base of the monk to above the full water level of the pond. Such a screen provides a large surface area so when the pond is being drained the screen does not get clogged too rapidly with debris, and spreads out the water pressure on the screen, reducing the chance of impinging fish on the screen. A second screen fits into the other set of groves and when the first screen starts to get clogged the second screen is added, and the clogged screen is removed and cleaned to later replace the second screen.

Figure 2.2. Monk-style drain with adjustable valve over the drain pipe, and grooves in the sidewalls to hold screens in front of drain.

Where larger fish are being cultured and the pond does not need to be drained so often, boards can be placed across the open side of the monk into the slots on each side. The space between the boards is filled with clay and packed to prevent any seepage. Such an approach eliminates the need for a valve over the drain, however, adding or removing the boards and the clay is labor intensive. The advantage of monks is that if properly constructed they will last for years.

Another widely used drain structure is a pivotal standpipe (Figure 2.3). The chief advantages of the standpipe drain are simplicity of design, economical construction, and ease of operation. A pivotal standpipe consists of the end of the drainpipe being connected to an elbow set horizontally. This elbow is connected via a short section of pipe to another elbow that contains the standpipe. The second elbow, which is set into the horizontally-set elbow, is not glued so it can pivot 90° into the catch basin. The length of the standpipe should be long enough so that when positioned upright it will determine the desired water level of the pond when full. The catch basin should be long enough so the standpipe and its screen can lay flat in the basin when draining the final water from the pond.

A rope or chain should be attached to the top of the standpipe so that the pipe can be lowered to the desired depth and tied off to a post to prevent the standpipe from dropping down farther and accidently draining the pond. If the standpipe is made of PVC pipe, which is not very heavy, after the pipe is lowered and the water level drops the pipe is closer to the surface and air might be sucked into the pipe causing the pipe to rise to the surface preventing

Figure 2.3. Catch basin with a standpipe designed to pivot into the catch basin, a water supply to the basin and a pier to provide access.

the pond from drained more. To prevent that, a detachable weight can be added to the mouth of the pipe to prevent the pipe from rising back to the surface. Figure 2.3 also shows a small pier extending from the pond bank to the catch basin and standpipe. This makes the standpipe more accessible and easier to control the standpipe without having to get into the pond. Such a pier also facilitates monitoring water quality at various pond depths and zooplankton sampling.

In nursery ponds that have a standpipe and catch basin where small fish need to be harvested, the screen area of the standpipe needs to be increased to avoid impinging the fish on the drain screen. A 0.5 × 0.5 × 1.0 m screen box, appropriate for a 200–400 m^2 pond, can be made to slip over the drainpipe. The box width should be narrow enough to fit inside the catch basin. One side of the box should be a solid piece with a hole cut to match the outside diameter of the drainpipe. The hole should be located along the side in the center approximately 2.5 cm from the edge. The side that will touch the bottom of the catch basin can be one solid piece of wood. The other four sides should be covered with screen of a mesh size that will allow a free-flow of water through the mesh but not allow any fish to past through.

Knowing the life history of the species to be produced and the quantity of fish to be produced is helpful in designing the pond layout. However, such information is often lacking during the planning and construction stages, and often production goals change over the years. Building the best hatchery pond one can afford is often a wise procedure as demand may change over time and the pond should be suitable for a variety of production practices.

2.3. Troughs, Tanks and Raceways

Tanks, troughs, and raceways may be used for a variety of purposes at a hatchery. Tanks might be used for holding brood fish before, during, and after spawning. When ponds are harvested, fish can be temporarily held in tanks and prepared for distribution. Tanks are needed for holding the fish for cleaning, grading, enumeration and disease treatment before the fish are distributed. Broodfish can be held for selection and prophylactic disease treatment. Depending on the species and the propagation method, tanks and troughs may be used for controlled spawning, egg incubation, fry holding, or artificial feeding.

A rectangular holding tank 3 m long, 76 cm wide and 81 cm deep with a water volume of about 1.5 m^3 is a useful general-purpose unit. A water flow of 20 to 100 L/min will be adequate for most situations. With good quality water, up to 100 kg of fish can be held in a tank of this size if water changes are adequate for maintaining the oxygen level. For a medium size hatchery, 10 to 12 tanks of this dimension and type will be needed for temporary fish holding.

Troughs should have a sufficient water supply to give 1 to 2 changes of water per hour. A commonly used, general purpose trough has dimensions of 2.44 m × 46 cm × 30 cm, fitted with a standpipe drain 3.54 cm in diameter and 20 cm in length, with a volume of 0.22 m^3. It should have a water flow of 7.5 to 20 L/min for rearing fry. Such a trough can be used for holding larvae through larval development and early feeding stages. Having a number of smaller troughs gives a hatchery more flexibility than just a few larger ones.

Circular tanks (Figure 2.4) can be used instead of rectangular ones particularly for species of fish that prefer to swim up in the water column. Circular tanks can be self-cleaning particularly when a circular pipe is placed over the standpipe that extends above the top of the standpipe and to within a few centimeters of the tank bottom. Creating circular flow of water in the tank will cause wastes to accumulate near the base of the standpipe when the standpipe

Figure 2.4. Circular tanks can be used for brood holding, spawning, egg incubation, or fry or fingerling production depending on the fish species.

is located in the center of the tank. The pipe around the standpipe, being off the bottom and higher that the top of the standpipe, will cause water inflowing into the tank to drain by flowing up between the two pipes caring waste out in the effluent.

Circular fiberglass or metal tanks are commercially available in several diameters. Standard sizes of 1.2, 1.5, and 1.8 m diameter with a water depth of 76 cm can be obtained. Units such as these are adequate for rearing but are less suitable for temporary holding of small fish where large quantities must be handled for cleaning, grading, and other preparations for shipment. Circular tanks can also be used for holding brooders. Circular tanks work well for spawning of fish that have non-adhesive floating or semi-buoyant eggs. The circular flow helps to concentrate the eggs and the eggs can be collected in the tank effluent.

Rectangular raceways are commonly used for rearing cold-water species (i.e. Rainbow Trout *Oncorhynchus mykiss*) but less so for warmwater species. Most warmwater fish hatcheries do not have a year-around supply of water at a proper temperature to allow raceways to be an appropriate culture unit. A typical raceway is 30.5 m long, 2.44 m wide, and 1 m deep. Two or more units may be located side by side with the water flow draining into other raceways set at lower levels (Figure 2.5). As many as 4 to 6 raceways may be placed in succession but reuse of water more than 3 to 4 times is of questionable value. Trout fingerlings are started in troughs or tanks and reared to a size of 10 to 15 cm before being placed in the rearing raceways. The tanks are actually small raceways, however, being 10 to 15 m long, 90 cm wide and 80 cm deep. The issue at most warmwater fish hatcheries is that they do not have enough volume of consistent temperature water to make raceways a practical production unit. Floating raceways in static water ponds where pond water is flushed through the raceway are an effective way to produce food-size catfish as long as the producer if in a temperate climate is willing to accept that fish growth will not be at its optimum year-around.

Figure 2.5. Raceway designs similar to those used for trout can be appropriate for several warmwater species where gravity supplied water of an appropriate temperature is available.

2.4. Service Buildings and Infrastructure

2.4.1. Administrative/Educational Center

A fish hatchery needs a variety of buildings to help support the fish production. An administrative area is needed either as a stand-alone building or combined with other activities in a central building. Record keeping is an essential part of a productive hatchery. An administrative area provides a central location for record keeping as to how each pond is being used, how it is being managed, what inputs had been used, what fish production was obtained, etc. Likewise, records need to be kept as to how tanks and troughs are used, how successful was a given culture technique was, etc. Good record keeping is essential for budgeting. Typically, someone at the hatchery has the primary responsibility for making purchases, receiving deliveries, documenting such expenditures, as well as keeping hatchery production information up to date and organized. Likewise, personnel management topics such as work schedule, hours worked, vacation and leave time needs to be managed. Space needs to be provided for such administrative functions. Typically, this includes computer space, access to the internet and other modes of communication, copy machines, and room for staff meetings. Many hatcheries are being designed to allow remote monitoring of ponds, and indoor facilities. The central administrative area may need to be designed to accommodate such monitoring as well.

Hatcheries often receive a variety of visitors and need to be designed to funnel any visitor to a single first contact area. This might be a reception area as part of the administrative area or another area that limits unsupervised access to other parts of the hatchery. This can help improve visitor safety, and reduce accidents and theft, and improve biosecurity. A role of increasing importance at public hatcheries is natural resources education, explaining how hatcheries contribute to natural resources management. Fish hatcheries have always attracted the attention of the general public, and can be a valuable public relations tool particularly as the general public becomes more urban. An aquarium room or display pool area will draw many visitors who can be provided elements of a fish conservation message as they view the fish displays. Such a facility requires careful planning along with daily care and maintenance to assure that the quality of the exhibit does not deteriorate into a liability instead of an asset. Such a display is particularly attractive to school groups, and a guided tour of the hatchery can do much to build good will in the community and acquaint the public with the mission of the facility.

Hatcheries are often built in rural areas and can be hard to find without signage at key turning points on the highways. A large sign at the entrance to the hatchery should include the hatchery name, a contact phone number, and the hours and days when the hatchery is open to the public. This signage is the first impression visitors will have of the hatchery and as such should be neat, attractive, and well maintained.

2.4.2. Holding Building

One of the most important buildings at a hatchery is the fish holding building. This building can have multiple functions including holding fish after harvest, preparing them for distribution, and also be used for fish spawning, egg incubation and fry rearing. In addition, such a building may have a lab for checking fish for diseases, conducting water quality analyses, and preparation of specialty feeds. The building should be as centrally located as possible on the hatchery for easy access to the most actively used areas of the hatchery.

The holding building should be designed to allow the largest truck commonly used at the hatchery to drive in one end and exit the other. Space should be allocated near this driveway for storing nets and other equipment needed for harvesting fish from the ponds. A water supply is needed in this area so the truck with the transport tank can drive in and fill the tank. If the nets and other equipment are stored in this area they can be cleaned after being used in pond harvests. At least 10 to 12 permanent tanks should be located adjacent to the driveway to facilitate loading and unloading of fish on and off the vehicles. The demand for tank space can be seasonal and enough tanks need to be installed to accommodate peak seasonal demand.

The holding building will have a high demand for water. The yearly volume used is often more than the yearly requirements needed to fill and maintain water levels in the production ponds. The holding building should be plumbed to allow two water exchanges/h/holding tank assuming that all holding tanks are in use. An adequate reservoir of gravity fed water should be available to supply at least 24 h of water as insurance in case of a power outage or if a pump breaks down.

Attention should be given to effluent management from the holding building. Fish are routinely given prophylactic chemical treatments (formalin, potassium permanganate, etc.) when held in the holding building to help control disease outbreaks. These chemicals may be applied at high concentrations but for short durations to the fish holding tanks then flushed rapidly from the tank. The effluent from such tanks needs to be significantly diluted before such water is released into any receiving stream (formalin <1 mg/L), or be held in a holding pond long enough so such chemicals can detoxify naturally before the water is released. The half-life of formalin in water is estimated to be 36 h.

Another consideration related to effluent management is how to deal with any chemical waste associated with water quality analysis or used in any other laboratory activity. Typically, hatcheries are built in rural areas not connected to municipal sewage treatment systems but have to rely on septic tank systems to handle domestic wastes. An additional separate septic tank system may be needed for the laboratory effluent as the chemicals in that effluent may kill the bacteria that are needed to decompose the organic wastes in a normal septic tank system. The additional system should have a large enough tank volume to give sufficient retention time so any chemical can degrade naturally before entering the field line.

The holding building should have more than enough aeration available to handle all tanks when loaded to their capacity. At least two low-pressure blowers should be available to minimize the risk of a mechanical failure. Blowers should be installed in an accessible area not in direct sunlight with a pressure relief valve to avoid excessive backpressure and overheating. Having two blowers plumbed into a common distribution system will allow you to operate more efficiently using just one blower at times when only a small portion of the tanks are being used at a time. Diffusers should be selected with a large enough pore size to avoid excessive backpressure on the low-pressure air blowers. Depending on the reliability of the electrical supply in the area a back-up generator with an automatic transfer switch may be needed to power the air blowers and other critical equipment.

A portion of the holding building may be set aside for fish spawning and larval rearing. This can be an area of open floor space where water supplies, aeration, and drainage are available. Spawning is often a seasonal activity and it may be more convenient to move in the necessary incubators and larval rearing tanks for a few weeks of use. However, many hatcheries spawn a number of species over an extended period of time. Therefore, it may be more appropriate to permanently dedicate a portion of the holding building to fish reproduction and larval rearing or have a separate building for such activities.

2.4.3. Shop and Storage Buildings

A hatchery has a variety of supplies and equipment that need to be stored as well as equipment that needs to be maintained and kept out of the weather. Depending on the size of the hatchery this can be done in one or more buildings.

A hatchery often uses a variety of formulated feeds that differ in feed size, feed quality, and quantity to be used. For a hatchery that produces a large quantity of food fish in ponds or raceways using tons of one type of feed in a production cycle, bulk storage may be practical. Upright hopper bottom bins with a center draw should be mounted on legs high enough that a small tractor pulling a feed blower trailer can drive under and fill up the trailer (Figure 2.6). The bin size should be able to provide 1–2 months of feed.

In most cases a hatchery will be using a variety of feeds and organic fertilizers purchased by the bag. A building specifically built for feed and fertilizer storage, or a room is included as part of the holding building, is needed for the storage of these bagged materials (Figure 2.7). The storage area needs to be readily accessible so bags of feed can be moved in and out with ease. It needs to be well ventilated with a stable air temperature and low humidity. The area needs to be screened to prevent insect infestations and rodent damage. The area also needs to be adequately lit so the feed and feed bags can be inspected for mold growth, oil migration, and animal damage. The area should have a smooth hard floor so that the area can be cleaned easily. Bagged feed should be set on pallets to avoid direct contact with the floor with the pallets set about 10 cm away from the walls. This will reduce the contact with any moisture condensing on the floor or walls that could damage the feed. A walk-in cooler is desirable for storage of feeds rich in oils and fats.

Figure 2.6. Drive-under outdoor feed storage bin (Photo courtesy of Michael Luthi, Farm and Commercial Storage Mfg. Inc.)

Figure 2.7. Indoor feed storage room.

Hatcheries have a wide variety of equipment and facilities that need to be maintained, repaired or replaced, much of which is done in-house by the hatchery staff. A shop building should be large enough for vehicles to enter with enough open floor space to accommodate other activities. Covered sheds located near the shop building are also needed for vehicle and equipment storage. Both areas should be located near the administration area to facilitate easy access as well as security.

2.4.4. Housing

Hatcheries are often located in rural areas where there may be few housing options. An alternative is to provide on-site housing. This can be a significant incentive in recruiting good staff, as well as a way to help provide security for the fish and the facility. A few minutes spent checking essential details during off-duty hours such as weekends and holidays can do much toward preventing losses of fish and other valuables. When personnel live on the station these observations can be made conveniently and in a minimum amount of time. At least two residences are needed and for larger hatcheries three or four are desirable.

2.4.5. Hatchery Access and Location

Fish hatcheries are in operation year-around and need an internal road system that is passable under all weather conditions. Other considerations are how centrally located the hatchery is to where fish need to be delivered and the quality of the road network to reach there. If air shipment of fish is part of the hatchery plan then the distance to a major airport hub must be considered.

Detailed infrastructure considerations are given as part of the United States Department of the Interior, Fish and Wildlife Service, Bureau of Sport Fisheries and Wildlife: Standards for Fish Hatchery Development; Warmwater Hatcheries. August 1969 and excerpts are given in Chapter 17.8. The following references provide additional information, and other useful references regarding hatchery infrastructure are included in 17.7, 17.8, and 17.17.

Recommended Reading

Clay, C. H., editor. 1985. Inland Aquaculture Engineering, ADCP/REP/84/21, FAO, Rome. Available: http://www.fao.org/3/x5744e/x5744e00.htm. (May 2020).

Steeby, J., and J. Avery. 2002. Construction of Levee Ponds for Commercial Catfish Production, SRAC Publication No. 101. Southern Regional Aquaculture Center, Stoneville, Mississippi. Available: https://wkrec.ca.uky.edu/files/pondconstruction.pdf. (May 2020).

Tidwell, J. H., editor. 2012. Aquaculture production systems. Wiley-Blackwell, Oxford, UK.

Whitis, G. N. 2002. Watershed fish production ponds guide to site selection and construction, SRAC Publication No. 102. Southern Regional Aquaculture Center, Stoneville, Mississippi. Available: https://wkrec.ca.uky.edu/files/watershedponds.pdf. (May 2020).

Chapter 3

Production Alternatives and Brood Management

3.1 Production Alternatives

The mode of reproduction and feeding habits of a given species of fish needs to be understood before they can be effectively produced in a hatchery setting. This includes knowing the basic biology of the species as to: under what environmental conditions the species mature and reproduce, what are the natural food habits at the different life stages, what factors stimulate reproduction, what are the characteristics of the eggs and larvae, and what type of care is given by the parents to the eggs and larvae? Based on that, an appropriate set of hatchery techniques can be selected to produce a given species. What techniques are applied also depends on:

1. Difficulty of spawning and rearing the species to a distributable size
2. Number of seed required
3. Delivery schedules
4. Distribution requirements such as size of seed, distance of shipment, hardiness of the species, etc.
5. Labor available and their knowledge and skills
6. Facilities available
7. Disease problems

3.1.1 Spawning-Rearing

There are a number of techniques and combinations of techniques that can be used to meet production goals, varying in the level of intervention required by the hatchery staff, the input requirements, and the consistency in yields. The most basic approach is a Spawning-Rearing approach where much of the operation takes place in the same production setting with little management intervention. Broodfish are stocked into a spawning pond, allowed to spawn naturally, and the progeny and brooders are harvested at a later date. This can be an acceptable approach when working with species that will spawn naturally in a static water pond setting as little human intervention is needed. The disadvantages include the uncontrolled spawning results in an unknown number of fish being produced, and because the fish may be spawning over an extended period of time there will be variation in sizes of fish produced. This size variation will contribute to competition for food and possible cannibalism, resulting in poor yields from the later spawns. However, because of the simplicity of the method it is still used with some species such as Bluegill *Lepomis macrochirus*.

3.1.2 Spawning-Egg/Fry Transfer

The Spawning-Egg/Fry Transfer approach is where spawning takes place in one locale where eggs or fry are collected and transferred for development in another location. This approach works well for species that will spawn naturally in ponds or tanks using an artificial substrate where the substrate with eggs attached can be collected and transferred to the hatchery or nursery ponds where the eggs will hatch. A variation in this approach is where fry instead of eggs are collected from a spawning pond and transferred. There a spawning pond is filled with clear water, stocked with brooders, and the brooders allowed to spawn naturally. Once the eggs hatch and the fry start swimming, the fry are collected and transferred to a nursery pond. This technique works well for species such as Largemouth Bass *Micropterus salmoides* where the male guards the school of fry, keeping the school compact, which greatly facilitates the collection of the fry. The fry transfer technique also works well for species that spawn on a substrate where the substrate holding eggs cannot be easily transferred to the hatchery or a nursery pond.

The egg or fry transfer method demands more technical skill, closer adherence to time schedules, and better facilities and equipment. However, by being able to control the timing of when brood fish are stocked, being able to collect eggs or fry of similar age, a much more predictable production schedule can be developed to obtain specific sizes and numbers of fish. Moreover, yields are more consistent, higher production per unit of effort is possible, losses to cannibalism are lower, the rearing unit can be stocked with a more predictable number of fish for rearing to distributable size, and by separating the young from parents at an earlier stage helps to minimize disease transmission. Largemouth Bass, Channel Catfish *Ictalurus punctatus* and bait fishes such as Golden Shiner *Notemigonus crysoleucas* are often produced using an egg/fry transfer system.

3.1.3 Controlled Systems

Controlled systems are where the hatchery staff intervenes in all steps in the reproductive and culture process and assumes control over all life stages of a species to ensure its successful propagation. Broodfish are induced to spawn and eggs are taken by hatchery staff, fertilized and incubated in a controlled environment. The resultant fry are reared to a marketable size in a controlled production environment on a nutritionally complete feed. The following requirements should be met for controlled propagation of a species on a production scale:

1. The demand for seed should be great enough to justify developing a controlled system of propagation
2. The species must be tolerant of the unavoidable stresses associated with controlled propagation
3. Broodstock must mature and produce quality gametes under hatchery conditions
4. A means of collecting eggs or fry must be available
5. Eggs must be adaptable to artificial incubation, and appropriate incubating devices be available

6. Larvae should be tolerant to intensive culture
7. It must be practical to provide the appropriate feeds for all life stages of the fish being cultured
8. Effective disease control must be possible

Controlled systems may be practical for highly domesticated species with high market value, or necessary with a species with other unique traits such as being specific pathogen free. Complete control over the life cycle requires active human intervention at all phases and few fish justify this level of intervention.

3.1.4 Combined System

Complete control over the life history of a species may be technically feasible, but it may not be practical to manage the entire life history so intensively. Often it is more practical to have complete control over critical phases of the life history and less control over others, a Combined System. Such critical areas include the act of spawning and egg incubation. Typical solutions for gaining control over egg production and incubation of warmwater species include:

1. Manual stripping of wild broodstock collected during a seasonal spawning run and ovulated with hormone treatment. Eggs are incubated in hatching jars. Post-larvae are reared in earthen ponds on diet of natural food supplemented by artificial rations. This approach is widely employed to propagate Striped Bass (*Morone saxatilis*) and Striped Bass hybrids (*Morone saxatilis* × *M. chrysops*). Also applicable to Paddlefish (*Polyodon spathula*) and other species that make spawning runs, respond to hormone treatment, and release eggs with manual stripping.

2. Manual stripping of domesticated broodstock that have been induced to ovulate with hormone injection. Eggs are incubated in jars, troughs or baskets. Swim-up fry are stocked in earthen ponds for growth to distributable size on a natural diet. This approach is applicable to common carp, Chinese carps, Indian carps and Goldfish.

3. Natural spawning of domesticated broodstock ovulated with hormone injection; eggs deposited on spawning mats or similar artificial substrate. Eggs are transferred on the collection substrate to the hatchery or nursery ponds for hatching and growth to a size large enough to collect and transfer to rearing pond. Utilized for spawning Goldfish, Channel Catfish in aquaria, and other species on occasions.

4. Natural spawning of domesticated broodstock on spawning mats or other media; eggs are moved from the spawning pond to an incubating pond or separated from the collecting device for incubation in a jar, tray, or trough. Used to spawn Golden Shiner minnows, Goldfish, Channel Catfish, Largemouth and Smallmouth Bass.

These approaches have been developed because there are stages in the reproductive process where hatchery intervention can increase spawning rates, and/or survival, resulting in more progeny compared to what would happen naturally. These levels of intervention need to be matched to the biology of the fish being worked with to determine what level of intervention may be appropriate and what tools can be applied.

The challenge to the hatchery manager is to select and apply the appropriate tools and techniques that will improve on the success that would occur naturally. One of the most important tools for a successful hatchery is to have quality broodstock.

3.2 Broodstock Selection and Management

3.2.1 Genetic Considerations

Genetic management of wild-caught or domesticated brooders is a major factor in broodstock management. Goals often differ considerably between hatchery programs using wild-caught brooders to produce fish for release into natural setting and where domesticated brooders are used for aquaculture. When fish are to be released into the wild they should reflect the genetic signature of that species found in that locale. This can be a challenge to a hatchery as there are often unique genetic sub-sets of a species over its natural range, then the question becomes where within that range should broodstock be collected and where within that range should the progeny be released? Such questions should be answered at a higher administrative level with the role of the hatchery manager being to remind administrators of these questions.

Using domesticated brooders presents another set of challenges. The initial acquisition of broodstock often determines the future genetic diversity within a population of fish (founder effect). The limited genetic pool resulting from the founder effect can result in inbreeding depression within a few generations unless steps are taken to begin with a large founder population or periodically introduce more of that species with a different genetic history. Natural selection of fish that successfully survive and reproduce in a hatchery setting results in a more domesticated population over the generations that is suitable for aquaculture but not for restoring a wild population. Active selection of fish by hatchery staff for a particular genetic trait may further limit the gene pool resulting in the increased frequency of an undesirable trait as well.

The genealogy of any domestic stock is important to know when establishing a new population or adding to an existing population. The stock being considered for introduction may have a common genetic heritage as the existing stock and thus would contribute little to diversifying the genetic pool. Often fish are shared among hatcheries with intent of improving genetic diversity, only to find out that the stock of fish to be introduced has a common genetic history with your existing stock. Hatchery staff move on and memories fade; good written record keeping is essential to know the genetic history of a given stock of fish.

Another issue with any domesticated stock of fish is that there may be a reduction in the genetic diversity in the population from one generation to the next. This reduction, often referred to as a bottleneck, can be the result of mortality from one generation to the next of a portion of the fish being reared to be future brood stock. Maintaining genetic diversity in highly fecund fishes can also be a challenge. Spawning a few females of a highly fecund species may give enough eggs to meet the hatchery's production goal, but the genetic diversity in the progeny is limited to that inherited from those few parents. If such progeny become the broodstock for the next generation then their progeny inherits the limited genetic diversity inherited from their parents.

Introduction of new species faces both the issues of founder effect and genetic bottlenecks. The original source of fish is often founded from a limited number of fish, and a limited number of progeny may have been available for transport, with a lesser number of fish surviving the transportation, and once arriving at the hatchery only a portion of those survive, reach maturity and spawn successfully. Knowing the genetic history is key in selecting brooders. Founding populations should be from as a diverse genetic population as possible, and in each spawning season as many fish as possible should be spawned to reduce future genetic bottlenecks and the rate of inbreeding. For more background on fish genetics and how it can be manipulated see Chapter 17.11.

3.2.2 Sources of Brooders and Their Acquisition

A source of broodstock is essential for any successful hatchery, whether it be from the wild or a domesticated source. A clear understanding of what a hatchery's objectives are is needed as part of broodstock acquisition and use. If the hatchery's role is to support a native population of fish then the hatchery's broodstock should be developed from fish representative of the genetic diversity of the watershed where the hatchery-produced fish will be stocked. If the hatchery is to support an aquaculture program then broodstock should be selected with genetic traits that favor the intensive production of that species in an aquaculture setting. Records should be maintained on the number and origin of the initial broodstock and any subsequent importations, as well as how well the introductions had survived and performed.

Rearing future broodstock from eggs is recommended particularly for species used in aquaculture. Going from egg to brooder under hatchery conditions facilitates domestication, resulting in brooders that have grown, survived and matured under hatchery conditions. It allows specific brooders of known genetic linage to be spawned to provide eggs of a specific genetic makeup. It can help reduce the rate of inbreeding by making sure eggs from each female are divided into subsets and each subset fertilized by a different male. The best approach is to rear a sufficient number of fry or fingerlings from known genetic families, keeping track of growth rates and survival for each family. When such fish reach sexual maturity, crosses between families can be made, selecting families that have desirable aquaculture traits while trying to minimize inbreeding. Enough lead-time should be allowed for such fish to become sexually mature. Some fish species become sexually mature at less than one year of age while other species take four years or more.

This approach requires good record keeping to be able to track the fish from each year class and genetic family. This initially requires enough space to keep the groups separate. Once the fish are large enough they can be individually tagged, identifying their year class and genetic group, and then reared communally.

Knowing the disease history of any potential introduction to a hatchery is very important. The stock being considered for introduction may be a carrier of diseases that currently do not exist at the hatchery. Similarly, knowing the disease history of the fishes in the area where the hatchery-produced fish are to be released is also important. Are the hatchery-produced fish introducing a new disease into an area where it did not previously occur or will the hatchery-produced fish be uniquely susceptible to diseases that might occur in their new locale? In the case of Channel Catfish production in Alabama, the bacteria *Edwardsiella ictaluri* is a reoccurring problem in the major commercial area, but in some areas of the state this bacterial problem did not occur. Producers in the main commercial area learned that purchasing fingerlings from the areas that did not have an *E. ictaluri* problem was not profitable as such fingerlings often died from *E. ictaluri* once introduced into the commercial area.

Viruses pose several challenges when selecting brooders for a hatchery whether brooders are collected from the wild, or obtained as eggs, fingerlings or adults from other hatcheries and developed as brooders. Knowing the history of the fish and any related issues is a starting point. Viral problems are often not evident unless there is an active infection. Brooders may be carriers of the virus and not exhibit any signs of disease, and pass it on to their progeny without there being an active infection. The onset of a virus outbreak is often associated with some type of stress. This might be sub-optimum water quality conditions, poor handing or hatchery practices, or the onset of bacterial or parasitic diseases. But the magnitude of such stressors may not be enough to trigger an active infection in a given year. As such, broodstock or their progeny may be carriers of a potentially lethal virus that is not routinely expressed.

Virus-free broodstock can be difficult to obtain. All the fish species covered in this book are known to be potential carriers of viruses (Table 3.1). Virus induced mortalities can be severe to chronic with virus particles evident only during an active infection. However, diagnostic techniques based on the use of DNA or RNA primers (Office International des Epizooties 2000) continue to be refined that are able to detect fish that have been exposed to a given virus even when there is no history of a previous exposure.

Some viruses are not very host specific. The viral hemorrhagic septicemia virus, VHS IVb variant, found in the Great Lakes drainages, has been found in 28 fish species (Bain et al. 2010). Some viruses are more host specific. Channel Catfish virus is found in only a few species of catfish. However, Channel Catfish is a popular food fish and sportfish and has been widely introduced outside its native range, and in turn spreading the range of the virus as well. Recent efforts to improve genetic quality of tilapia have resulted in improved strains being introduced around the world but in several cases a virus has been also introduced. Tilapia lake virus (TiLV) has been identified as causing tilapia mortalities in Colombia, Ecuador, Egypt, Israel, and Thailand.

There always will be a disease risk when bringing in new fish to a hatchery. A hatchery manager should know as much as possible about the history of the fish being introduced, always treat the fish prophylactically before they are released into ponds, and use protected quarantine ponds where no birds or other animals could carry a sick or dead fish to other locations. The effluents from quarantine ponds or tanks holding new introductions should not be allowed to enter natural bodies of water without first being sterilized. Diagnostic techniques needed for virus detection are not typically available at most hatcheries. A hatchery manager needs to plan ahead as to what is needed to have a given lot of fish examined for viruses. This should include locating qualified facilities to perform the evaluation, determining the size of samples needed, how samples should be taken, and what is the lead-time needed before examination results will be known and fish can be available for introduction to the hatchery.

Some fish species may require more than 5 years to reach maturity. Bringing such species to maturity under hatchery conditions requires a considerable allocation of space and labor. In addition, not all species produce the quality of eggs under hatchery conditions that might be obtained from wild-caught brooders. Using wild-caught brooders should be the foundation of any program when the intent is to release the progeny into the wild. Brooders should be collected from over the range where the progeny are to be released.

Wild-caught brooders often can be collected on the natural spawning grounds as they move to such areas to spawn. In that case what is collected are mature fish with well-developed gametes. However, there are issues associated with using wild-caught brooders. Their prior background as to nutrition, disease exposure, pollution stress, and other influences is

Table 3.1. Species of warmwater fishes covered in this book where viruses have been reported.

Fish Species	Virus	Reference
Catfishes		
Channel Catfish	Channel catfish virus (CCV) Viral Hemorrhagic Septicemia virus	Camus, 2004. SRAC Publication No. 4702 Bain et al. 2010
Channel × Blue catfish hybrids	Channel Catfish virus (CCV)	Camus, 2004. SRAC Publication No. 4702
Clarias batrachus	EUS rhabdovirus (ERV)	Lio-Po 1998.
Pangasius hypophthalmus	Channel Catfish virus (CCV)	Siti-Zahrah et al. 2014
Carps		
Common Carp & Koi	Spring viraemia of carp (VHC)	pp 46–52, in Diagnostic Manual for Aquatic Animal Diseases 2000 https://www.oie.int/doc/ged/D9568.PDF
	Cyprinid herpesvirus 3 (CyHV-3 Carp picornavirus 1 (CPV-1) Fish pox, Cyprinid herpesvirus-1 (CyHV-1) Carp oedema virus	Hendrick et al. 2000. Lange et al. 2014. Radosavljević. et al. 2017 Soliman, H., and M. El-Matbouli, 2018.
Grass carp (*C, idellus*)	Grass carp reovirus Spring viraemia of carp	Chen Y.X. and Y. Jiang, 1984 https://www.dnr.state.mn.us/fish_diseases/svc.html
Bighead carp (*A. nobilis*)	Spring viraemia of carp	https://www.dnr.state.mn.us/fish_diseases/svc.html
Silver carp (*H. molitrix*)	Spring viraemia of carp	https://www.dnr.state.mn.us/fish_diseases/svc.html
		The Center for Food Security and Public Health, Iowa State University 2007

Table 3.1. Continued.

Fish Species	Virus	Reference
Carps		
Goldfish (*C. auratus*)	Carp pox, Cyprinid herpesvirus-1(CyHV-1)	Goodwin. 2012.
	Goldfish herpesvirus (GHV)	" "
	Cyprinid herpesvirus-2 (CyHV-2)	" "
	Koi herpesvirus (KHV)	" "
	Spring viraemia of carp	https://www.dnr.state.mn.us/fish_diseases/svc.html
Golden Shiner	Golden Shiner/Grass Carp reovirus (Aquareovirus C)	McEntire et al. 2003.
	Fathead Minnow nidovirus	Baird and Faisal M. 2016.
Fathead Minnow	Fathead Minnow nidovirus	Iwanowicz and Goodwin. 2002.
	Fathead Minnow picornavirus (FHMPV)	Phelps et al. 2014.
Tilapias		
	Tilapia larvae encephalitis virus (TLEV)	Shlapobersky et al. 2010.
	Infectious spleen & kidney necrosis virus (ISKNV)	Subramaniam et al. 2016.
	Tilapia lake virus (TiLV)	Jansen and Mohan. 2017.
	Infectious pancreatic necrosis virus	Mulei et al. 2018.

Table 3.1. Continued.

Fish Species	Virus	Reference
Sportfish Species		
Largemouth Bass	Ranavirus LMBV	Plumb et al. 1996
	Largemouth Bass reovirus	Sibley et al. 2016
	Viral encephalopathy and retinopathy (VER)	Bovo et al. 2011
	Viral hemorrhagic septicemia	Kim and Faisal. 2011
Bluegill	*Viral* hemorrhagic septicemia	Goodwin and Merry. 2011
	Bluegill picornavirus 1 (BGPV-1)	Barbknecht. 2013
		http://www.picornaviridae.com/limnipivirus/limnipivirus_a/limnipivirus_a.htm
	Bluegill hepatitis B virus [TAX:2169918]	https://www.genome.jp/virushostdb/2169918
		https://nas.er.usgs.gov/queries/GreatLakes/FactSheet.aspx?NoCache=9%2F16%2F2011+8%3A53%3A17+PM&SpeciesID=2657&State=&HUCNumber=
Striped Bass	Infectious pancreatic necrosis virus (IPNV)	Schutz et al. 1984
	Striped Bass aquareovirus	Baya et al. 1990
	Ranavirus LMBV	Plumb and Zilbero. 1999
	Viral haemorrhagic septicaemia virus (VHSV)	Gagné et al 2007
Striped × White bass	Viral encephalopathy and retinopathy (VER)	Bovo et al. 2011

unknown. Often the availability of such fish is seasonal, resulting in a limited window during the year in which they can be collected and reproduced at a hatchery. The collecting site may be remote with limited access and located many hours from the hatchery. Temporary holding facilities may be needed at the collecting site until enough brooders are collected to justify their transportation back to the hatchery.

The act of collecting the brooders and their transport can be very stressful on the brooders to the point where the fish may not spawn once they reach the hatchery. When possible, it is best for hatchery personnel to personally collect wild broodstock or to be present when others do this. Even when collected by experienced commercial fishermen, they usually have little understanding of the needs of the spawning operation. Hatchery staff assuming responsibility for care of potential spawners as soon as they are captured is the best assurance possible that the wild fish will be properly handled during transport to the hatchery. If a species is hard to keep alive or in spawning condition following capture, special treatments may be required to reduce stress and its effects. This can include transportation in a special bath solution to reduce osmotic stress (e.g. salt or a tris buffer), being anesthetized to reduce metabolic rate (e.g. quinaldine, MS 222), or using special containers (e.g. mesh bags or burlap bags to facilitate holding, or a circular tank for a species that needs to be able to swim continuously), and/or injection with antibiotics (e.g. terramycin) to prevent bacterial infections.

Some species of catfish, cichlids, and sunfish can be collected from the wild out of the natural spawning season, transported to the hatchery and if well cared for will have adequate gonadal development under the hatchery conditions to spawn the next season. Such fish can serve as the founding stock of a domesticated population, or spawned for one or two seasons and be replaced if the progeny are to be used in stocking natural waters.

3.2.3 Brood Replacements

Each spawning season, a hatchery manager should be prepared to replace a portion of the brood fish. This would include 1) replacing any mortality that occurred from one spawning season to the next, 2) culling fish at the start of each spawning season that are not in adequate condition for spawning, and 3) culling fish at the end of the spawning season are un-healthy or fish that will be too large to be conveniently managed in the next spawning season. In addition, fish may be replaced as part of a genetic management plan. In the following example (Table 3.2) mature 2 yr. old Largemouth Bass are introduced to a new hatchery and spawned that year and for the next two years before being completely replaced. Each year progeny are set aside (Yr. classes A, B, and C) and cultured to become replacement broodstock when they are two years old. In this example, it is assumed that there is a 20% loss of brooders/yr. (mortality and culling), and a 20% mortality/yr. in each replacement year class. For example, if Largemouth Bass are 2-yr old when used for the first time for spawning and a brood loss of 20%/year each year thereafter is expected, and if that year-class is used for several more years, then new fish will be needed each year to replace mortality. This will require that a number of 2-year-old fish equal to 20% of the brood population be available every year. A plan must then be implemented every year that will provide young-of-the-year, 1-yr. and 2-yr. old fish.

If replacements are reared at the hatchery, the rearing system can be a monoculture or polyculture. Each system has advantages and disadvantages which relate to the species involved and the local conditions under which propagation is being done. Regardless of the system followed, a planned, systematic approach consistent with sound hatchery management is needed.

Table 3.2. An example of a broodstock replacement plan covering a three-year period.

Goal: have 250 2 yr+ old brooders available each year

Year-0 (intro class)
Introduce 350 2-yr old brooders
Use 250 bass for spawning
Progeny to be used as replacement brooders
Culture progeny
Save 500 as replacements (Yr-class A)
Lose 20% of brooders before Yr-1

Year-1
280 brooders of intro class available
Use 250 bass for spawning
Lose 20% of brooders before Yr-2
Culture Yr-class A (400 fish)
Progeny to be used as replacement brooders
Culture progeny
Save 500 as replacements (Yr-class B)

Year-2
224 brooders of intro class available
320 brooders of Yr- class A available
400 Yr-class B available (need 1 more yr)
Use 26 brooders Yr-class A
Total for spawning 250
Culture progeny as Yr-class C
Save 500 as replacements (Yr-class C)
Culture Yr-class B (400 fish)

Year-3
180 brooders of intro class available
256 brooders of Yr- class A available
Use 70 brooders Yr-class A
320 brooders of Yr- class B available
400 Yr-class C available (need 1 more yr)
Total for spawning 250
Discard Intro yr class brooders after spawning season
Culture progeny
Save 500 progeny as replacements (Yr-class D)

3.2.4 Brood Maintenance

Broodfish are a valuable commodity that needs special care to ensure that they are in sufficient condition to spawn successfully. This includes being held in a non-stressful environment, being properly fed, with diseases kept under control. Quality and quantity of food, fish density, temperature, light, and water quality are factors that can be managed to ensure successful brood maturation and spawning.

At freshwater warmwater fish hatcheries, fish are typically held in earthen ponds from one spawning season to the next. The stocking rate should be low enough to allow brooders to increase in average weight 50 to 100% from one spawning season to the next with the maximum standing crop of brooders not exceeding about 50% of the maximum standing crop typical for that species of fish when produced in a production settling. The greater the weight of fish being held the greater the quantity of feed required, the greater the quantity of feed the more the risk of water quality issues. Both acute and chronic water quality problems will stress the brood fish and in turn reduce gonadal development. Ideally at least four ponds should be available for brood holding, two for males and two for females. This helps to minimize the risk of losing all your brooders to disease, predators or other problems. If enough ponds are available, brooders should be held separately by year-class, as older fish often are ready to spawn sooner in the spawning season than younger fish. This will allow you to collect mature brooders of one age group without having to handle and stress less-mature fish of another year class. For species that will spawn in static water ponds it is best to hold males and female brooders in separate ponds until the hatchery manager is ready to spawn them. This is particularly true for temperate climates where spring temperatures may temporarily rise into an appropriate spawning temperature range only to drop again a week later. If males and females are held together they may spawn when the temperature rises but the when the temperature drops any eggs spawned will be slow to hatch and may get fungus and die. When a limited number of ponds are available, brooders can be identified by year-class using a distinct tag or brand and held in a common pond for growth and gonadal development, then separated prior to the start of the spawning season. Likewise, brooders of compatible species can be maintained as a polyculture in a common pond and separated prior to the start of the spawning season.

Adequate feed must be available for brood growth and gonadal development. Brooders of some predacious species are best maintained on live forage or fresh natural feeds, however, brooders of the majority of the freshwater fish species will accept formulated diets and have adequate gonadal development. Often commercial diets used for food fish production are also used for brood fish but in some cases specific brood feeds are commercially available. The quality of the ration is important and may need to be supplemented with fatty acids and vitamins for a month or more prior to the start of the spawning season to help improve egg quality. Diet compositions commonly used for several species are given in Chapter 17.9. Feeds and Fertilizers.

Brooders are fed at approximately 1% body weight/feeding when given a commercial ration. Frequency of feeding depends on the species and water temperature. For Channel Catfish, brooders might be fed daily at temperatures >24°C, three times/week at 18–24°C, once or twice a week at 14–17°C and those days when the temperature rises to >12°C even though the mean weekly temperature is <12°C. It is important to observe the brooders during feeding as to the intensity of the feeding response, the general health of the brooders and any signs of mortality. Overfeeding will result in excessive visceral fat and a lower quality brooder, underfeeding will result in poor egg quality and a limited percentage of the brooders spawning successfully.

The success of a hatchery is dependent on the quality of broodstock available. Attention to nutrition, disease outbreaks, environmental conditions, and handling during transfers minimizes losses and maximizes reproduction of these valuable fish. As a routine, brood fish should be treated prophylactically every time they are handled in the post-spawning period. For fish that are going to be pair spawned for several years, each fish should be marked individually so that the reproductive performance and the quality of the offspring can be monitored.

3.2.5 Selection of Brood Fish for Spawning

Selection of brood fish for spawning can be somewhat subjective based on a number of characteristics. Brooder age is one consideration. In temperate climates males are often mature a year earlier than females. It is not uncommon that a portion of the females may be mature one year before the majority of the females are mature. At first maturity, females will have smaller eggs and lower fecundity than older females. Older females are typically fully mature earlier in the spawning season than younger brooders. Younger females examined early in the spawning season may not be acceptable but if set aside and examined again a few weeks later they may be adequately mature for use. Brooders should be identified as to their age or be held separately by age and sex. This will minimize the amount of handling needed to find mature brooders.

When working with a new fish or in a new environmental setting, one way to determine when the spawning season is or at what size a fish might be expected to be mature is to sacrifice some fish and examine their gonadal development. The ratio of gonad weight to total body weight expressed as a percentage is referred to as the Gonadosomatic index (GSI). The GSI of a mature fish can be an indicator of the frequency of spawning. In species that spawn only once a year the GSI of a gravid female may range between 10 and 25% and males having GSI's around 5 to 10% with the female GSI dropping to <2% post-spawn. Such females typically will have two sets of eggs, mature eggs to be released that season, and immature eggs that will become mature later and released in the next spawning season. This pattern is common in species that respond to distinct environmental changes to stimulate final maturation and spawning.

In species where the environment is more stable, gravid females may spawn multiple times a year and this is reflected in the gonadal development and GSI. The gonad will have multiple sizes of eggs present often ranging from primary oocytes to mature eggs ready to be ovulated. The GSI of a mature female that spawns multiple times within a year is often 5 to 10% with mature male GSI's often <5%.

Fish often display secondary sexual characteristics that facilitate separating males and females as well as judging the degree of sexual maturity. When working with a new species the significance of the external appearance of such fish can be verified using GSI if enough fish are available for sacrifice. If there is a commercial fishery for the fish of interest, there might be an opportunity to examine the catch to determine seasonal changes in GSI and its relationship to the expression of secondary sexual characteristics.

Some external characteristics such as body shape, fin size and shape, and genital papilla shape are permanent traits in mature fish usable to separate the sexes whether in or out of the spawning season. Some of these traits are more prominent in the spawning season. For example, mature Channel Catfish males have a broader head than the females and in the spawning season the males exhibit muscular pads on the head making the head more distinct from that of a female. Female Channel Catfish have a distinct genital papilla that is different than that of a male. In the spawning season the papilla of a very mature female is enlarged and less so in a lesser mature female.

Other sexual differences are evident in some species only during the spawning season. Many cyprinid and catostomid males display breeding tubercles on the head, body and/or fins that can be seen or felt during the spawning season but are not evident at other times of the year. During the spawning season males of some species show very distinctive color pattern making it easy to separate mature males from females and immature males. However, such colors can be ephemeral, particularly when the fish are stressed. Males may show a bright color pattern when held in brood ponds but the colors may fade when the fish are handled, transported or held in holding tanks.

Final selection of brooders should begin with a plan as to how to minimize stress on the brooders when they are handled. Brooders should not be fed for several days before being examined to avoid mistaking a female with a stomach full of feed as being one with well-developed ovaries. Brooders are often seined from brood ponds but often not all the brooders are needed at one time. Seine selection should keep in mind that the larger the mesh the easier it is to pull the seine and the less chance the seine will fill with mud that otherwise could damage the brooders. However, large mesh seines are made of knotted material that can be abrasive to the fish resulting in hemorrhaging and scale loss. A brood seine should have a bag in the center where brooders can be concentrated and individual fish examined. Not all the brooders in the pond are going to be equally mature at the same time. Seining the brood pond allows the best quality brooders to be selected at the pond and the other brooders returned back to the pond where they can be examined at a later date to see if they have been improved in quality. Draining a pond to harvest a portion of the brooders, and having to hold the remaining fish in holding tanks typically results in a decrease in brood quality in those being held if they are held for a week or more.

Individual brooders should be carefully examined, selecting fish that are in good health with no external hemorrhaging, broken or eroded fins, or deformities. Fish showing these signs may not spawn successfully, and could introduce a disease to the locale where the fish are to be spawned. Deformities may be the result of poor handling or diet during the earlier history of the fish or be an inherited genetic trait. The cause of a deformity can be difficult to confirm and the safest thing is to avoid using such fish as brooders.

In the spawning season the eggs of a gravid female have become larger, in turn enlarging the size of the ovaries, giving the abdomen a swollen appearance. Depending on the species, when a gravid female is viewed ventral side up, the two lobes of the ovary will shape the abdomen and their outline should be visible. The abdomen of a gravid female should be soft to touch. The genital papilla is often enlarged and depending on the species may be pink to a light red in color.

The papilla of males will be of a different shape from that of the female and also may be enlarged in the breeding season. Males with lobular testes should release semen when light pressure is applied to the abdomen just above the genital papilla. The color and flow of the semen depends on the fish species ranging from a strong flow of white-colored semen with striped bass to a limited volume of a more transparent semen from a Largemouth Bass. If working with a cyprinid or catostomid species where the males are known to display breeding tubercles, such tubercles should be evident on any male selected. When working with species such as channel catfish that have villiform testes, the testes are small and there is not enough tissue mass that semen can be expelled from the genital papilla by applying pressure to the abdomen. In that case male brooders are selected based on head shape, body color, and prominence of the genital papilla.

How mature a given female is can be difficult to know in some cases based on just external appearance, particularly with species that may spawn several times in a year such as pangasius. Holding hatchery brooders without feed for several days before final brood selection will help reduce mistaking a female full of feed as being a gravid female. However, among the gravid females there may be differences in how close the eggs are to final maturation and how that female will respond to induced spawning. This variation is particularly true when working with wild-caught brooders that have an extended spawning season, or species such as Striped Bass that make a spawning migration. In the case of Striped Bass, brooders may be collected on the spawning grounds but some of the fish collected may have arrived more recently than others that had arrived earlier and are closer to final egg maturation and ovulation. In such cases, it may be necessary to use a catheter to take an egg sample from the ovary and base brood selection on egg diameter, egg maturity, and their relative abundance.

A catheter may be a rigid glass tube, or a flexible plastic or silicone rubber tube that is long enough to enter the oviduct and enter into the ovary. The end of the tube should be smooth so the tube does not damage the oviduct as the catheter is inserted. The outside diameter of the catheter used will vary with the species of fish and its size. Commonly used catheters have an outside diameter of 1.5 to 3.0 mm with an inside diameter large enough to accommodate the size of eggs being sampled (typically ≥ 1.0 mm). Potential brooders should be anaesthetized and held upside down in water but with the genital papilla extending out of the water. The catheter should be gently inserted into the genital papilla, up the oviduct and into the ovary to collect an egg sample. Upon entering the ovary, the catheter is often given a slight twist and a slight suction applied to help collect an egg sample.

Once the catheter is withdrawn the quality of the eggs can be ranked subjectively by eye based on apparent egg diameter and color as in the case with pangasius (See Chapter 13.3.3.). In Striped Bass, White Bass, and many marine fishes a catheter sample may be taken and examined microscopically for a more accurate measurement of egg diameter, relative abundance of mature eggs, and degree of oil globule collation. A female with eggs appearing like Figure 3.1 should be considered to be of marginal quality for induced spawning. In this case there are a mixture of eggs some approaching maturation along with some primary oocytes visible. Such females may or may not response to hormone induction, the response time can be highly variable, and if eggs are released the quantity and quality may be limited.

Eggs from acceptable brooders for induced spawning should show signs of advancement toward final maturation. There should be some evidence of oil globule fusion and the egg becoming more transparent (Figure 3.2). How fast a hormone-induced female progresses from that point to ovulation depends on the fish species, how she has been handled or otherwise stressed, the hormone and the protocol used, and the water temperature at which the fish is held. The response time may vary from hatchery to hatchery. Keeping detailed records will provide the basis for developing a successful protocol for future induced spawning efforts. Females may be sampled to see if the eggs have advanced from stage seen in Figure 3.2 to

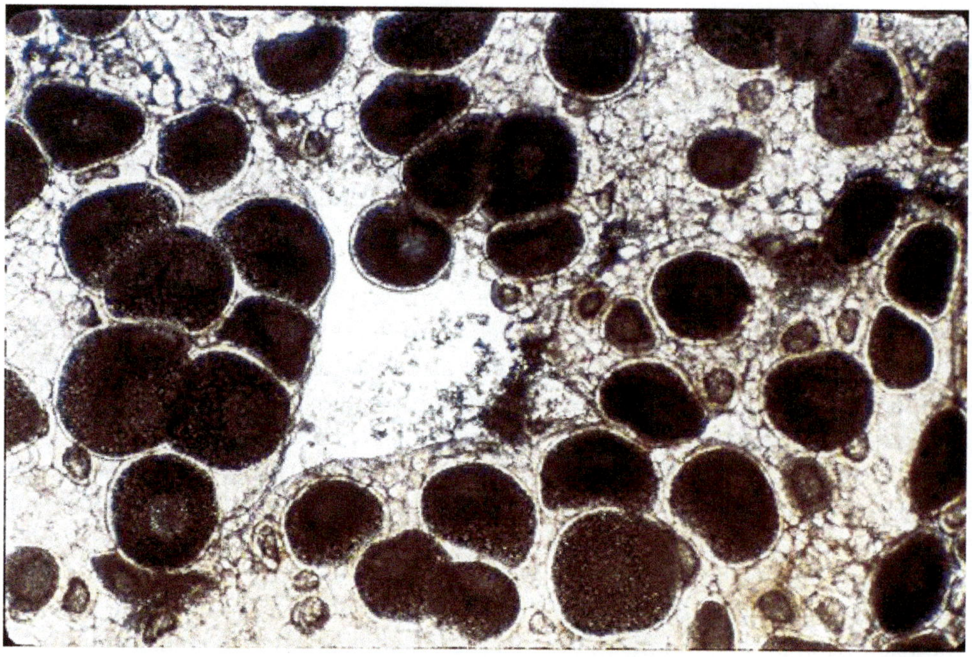

Figure 3.1. Egg sample taken by catheter of a marginal quality brooder based on the numerous small globules and the lack of transparency in the eggs, and the presence of primary oocytes in the sample.

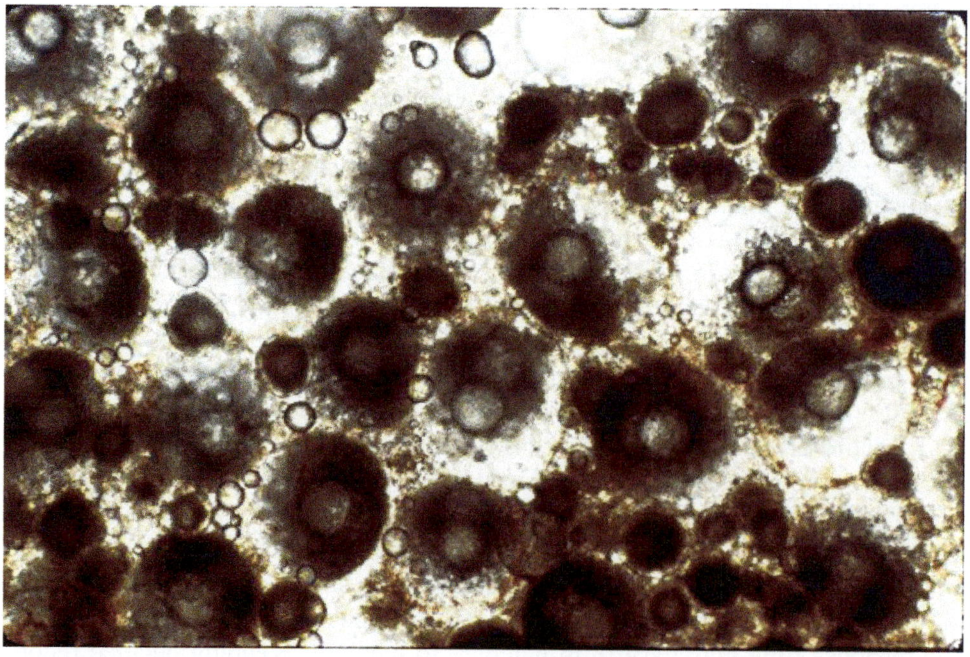

Figure 3.2. Egg sample taken by catheter of an acceptable quality brooder based on enough globules have fused increasing the transparency of the eggs.

one where there are only a few large oil globules seen and the egg is much more transparent (Figure 3.3). However, keep in mind that each time the fish is sampled the handling may delay or inhibit ovulation. As ovulation approaches, the eggs have increased in diameter and start to press against each other and may lose their round appearance and start to appear more hexagonal with some being loose in the lumen of the ovary (Figure 3.4).

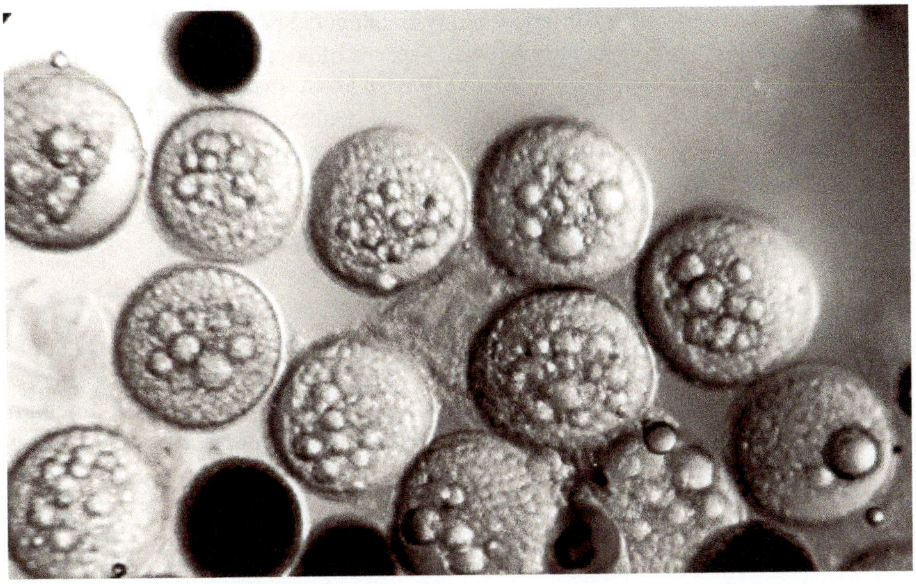

Figure 3.3. Egg sample taken by catheter of a hormone-induced female where a majority of eggs are advancing toward ovulation as oil globules have become larger and fewer in number and the eggs are more transparent.

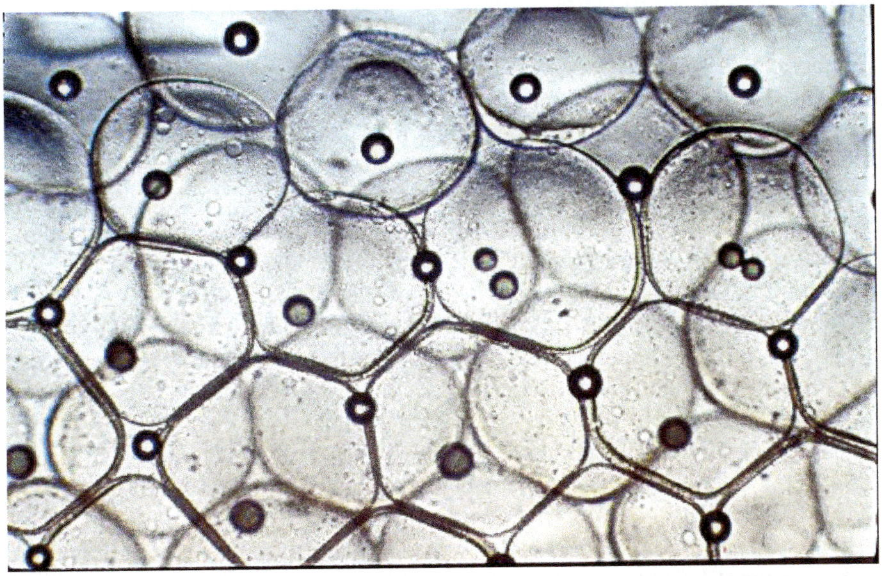

Figure 3.4. Egg sample taken by catheter of a hormone-induced female nearing ovulation where the eggs have hydrated to the point they are starting to push against each other and losing their round appearance.

3.3 References

Bain, M. B., E. R. Cornwell, K. M. Hope, G. E. Eckerlin, R. N. Casey, G. H. Groocock, R. G. Getchell, P. R. Bowser, J. R. Winton, W. N. Batts, A. Cangellosi, and J. W. Casy. 2010. Distribution of an invasive aquatic pathogen (viral hemorrhagic septicemia virus) in the Great Lakes and its relationship to shipping. PLoS One. 2010 Apr 13;5(4):e10156. doi: 10.1371/journal.pone.0010156.

Baird A., and M. Faisal. 2016. Fathead minnow nidovirus infects Spotfin Shiner *Cyprinella spiloptera* and Golden Shiner *Notemigonus crysoleucas*. Diseases of Aquatic Organisms 119:37–44. Available: https://www.int-res.com/abstracts/dao/v119/n1/p37-44. (May 2020).

Barbknecht, M., S. Sepsenwol, E. Leis, M. Tuttle-Lau, M. Gaikowski, N. J. Knowles, B. Lasee, and M. A. Hoffman. 2013. Characterization of a new picornavirus isolated from the freshwater fish *Lepomis macrochirus*. Journal of General Virology 95:601–613.

Baya, A., A. E. Toranzo, S. Nunez, J. L. Barja, and F. M. Hetrick. 1990. Association of a Moraxella sp. and a reo-like virus with mortalities of Striped Bass, *Morone saxatilis*. Pages 91–99 *in* F. O. Perkins and T. C. Cheng, editors, Pathology in marine science. Academic Press, New York.

Bovo, G., A. Gustinelli, F. Quaglio, F. Gobbo1, V. Panzarin, A. Fusaro1, F. Mutinelli, M. Caffara, and M. L. Fioravanti. 2011. Viral encephalopathy and retinopathy outbreak in freshwater fish farmed in Italy. Diseases of Aquatic Organisms 96:45–54.

Camus, A. C. 2004. Channel Catfish virus disease. SRAC Publication No. 4702. Southern Regional Aquaculture Center, Stoneville Mississippi. Available: https://agrilifecdn.tamu.edu/fisheries/files/2013/09/SRAC-Publication-No.-4702-Channel-Catfish-Virus-Disease.pdf. (May 2020).

Chen Y. X., and Y. Jiang. 1984. Morphological and physico-chemical characterization of the hemorrhagic virus of Grass Carp. Chinese Science Bulletin 29:832–835.

Gagné, N., A-M. Mackinnon, L. Boston, B. Souter, M. Cook-Verslot, S. Griffiths, and G. Olivier. 2007. Isolation of viral haemorrhagic septicaemia virus from Mummichog, Stickleback, Striped Bass and Brown Trout in eastern Canada. Journal of Fish Diseases 30(4):213–223.

Goodwin, A. E. 2012. Herpesviruses in fish. SRAC Publication No. 4710, Southern Regional Aquaculture Center, Stoneville Mississippi. Available: https://agrilifecdn.tamu.edu/fisheries/files/2013/09/SRAC-Publication-No.-4710-Herpesviruses-in-Fish.pdf. (May 2020).

Goodwin, A. E., and G. Merry. 2011. Mortality and Carrier Status of Bluegills Exposed to Viral Hemorrhagic Septicemia Virus Genotype IVb at Different Temperatures. Journal of Aquatic Animal Health 23(2): 85–91.

Hedrick R. P., O. Gilad, S. Yun, J. V. Spangenberg, G. D. Marty, R. W. Nordhausen, M. J. Kebus, H. Bercovier, and A. Eldar. 2000. A herpesvirus associated with mass mortality of juvenile and adult koi, a strain of Common Carp. Journal of Aquatic Animal Health 12(1):44–57.

Iwanowicz, L., and A. E. Goodwin. 2002. A new bacilliform fathead minnow rhabdovirus that produces syncytia in tissue culture. Archives of Virology 147(5):899–915.

Jansen, M. D., and C. V. Mohan. 2017. Tilapia lake virus (TiLV): Literature review. Penang, Malaysia: CGIAR Research Program on Fish Agri-Food Systems. Working Paper: FISH-2017-04.

Kim, R., and M. Faisal. 2010. Comparative susceptibility of representative Great Lakes fish species to the North American viral hemorrhagic septicemia virus sublineage IVb. Diseases of Aquatic Organisms 91:23–34.

Lange, J., M. Groth, D. Fichtner, H. Granzow, B. Keller, M. Walther, M. Platzer, A. Sauerbrei, and R. Zell. 2014. Virus isolate from carp: genetic characterization reveals a novel picornavirus with two aphthovirus 2A-like sequences. Journal of General Virology (2014), 95:80–90.

Lio-Po, G. 1998. Studies on several virus isolates, bacteria and a fungus associated with Epizootic Ulcerative Syndrome (EUS) of several fishes in the Philippines. Doctoral dissertation, Simon Fraser University, Burnaby, British Columbia, Canada.

McEntire, M. E., L. R. Iwanowicz, and A. E. Goodwin. 2003. Molecular, physical, and clinical evidence that Golden Shiner virus and Grass Carp Reovirus are variants of the same virus. Journal of Aquatic Animal Health 15(4):35–38.

Mulei, I. R., P. N. Nyaga, P. G. Mbuthia, R. M. Waruiru, L. W. Niagai, E. W. Mwihia, A. A. A. Gamil, Ø. Evensen, and S. Mutoloki. 2018. Infectious pancreatic necrosis virus isolated from farmed Rainbow Trout and tilapia in Kenya is identical to European isolates. Journal of Fish Diseases 41(8):1191–1200.

Office International des Epizooties. 2000. Office International Des Epizooties. Diagnostic manual for aquatic animal diseases. Third edition. Office International Des Epizooties, 2000, Paris, France.

Phelps, N. B. D., S. K. Mor, A. G. Armien, W. Batts, A. E. Goodwin, L. Hopper, R. McCann, T. Fei Fan Ng, C. Puzach, T. B. Waltzek, E. Delwart, J. Winton, and S. M. Goyal. 2014. Isolation and molecular characterization of a novel Picornavirus from Baitfish in the USA. PLoS One. 2014 Feb 21;9(2):e87593. doi: 10.1371/journal.pone.0087593.

Plumb, J. A., J. M. Grizzle, H. E. Young, A. Noyes, and S. Lamprecht. 1996. An iridovirus isolated from wild largemouth bass. Journal of Aquatic Animal Health 8(4):265–270.

Radosavljević V., J. Maksimović-Zorić, L. Veljović, D. Ljubojević, M. Ćirković, Z. Marković, and V. Milićević. 2017. CYPRINID HERPESVIRUS DISEASES. Arhiv Veterinarske Medicine 10(1):51–60.

Schutz, M., E. B. May, J. N. Kraeuter, and F. M. Hetrick. 1984. Isolation of infectious pancreatic necrosis virus from an epitzootic occurring in cultured Striped Bass, *Morone saxitilis* (Walbaum). Journal of Fish Diseases 7:505–507.

Shlapobersky, M., M. S. Sinyakov, M. Katzenellenbogen, R. Sand, J. Don, and R. R. Avfalion. 2010. Viral encephalitis of tilapia larvae: Primary characterization of a novel herpes-like virus. Virology 399(2):239–247.

Sibley, S. D., C. Puzach, M. A. Finley, A. G Armien, D. Giehtbrock, T. L Goldberg, and B. B. Baker. 2016. Novel reovirus associated with epidemic mortality in wild Largemouth Bass (*Micropterus salmoides*). Journal of General Virology 97:2482–2487.

Siti-Zahrah, A., M. Zamri_Sad, M. Firdaus-Nawi, M. K. Hazreen-Nita, and M. Nur-Nazifah. 2014. Detection of Channel Catfish virus in cage-cultured *Pangasius hypophthalmus* (Sauvage, 1878) in Malaysia. Journal of Fish Diseases 37(11):981–983.

Soliman, H., and M. El-Matbouli, 2018. Rapid detection and differentiation of carp oedema virus and cyprinid herpes virus3 in koi and Common Carp. Journal of Fish Diseases 41(5):761–772.

Subramaniam, K., M. Gotesman, C. E. Smith, N. K. Steckler, K. L. Kelley, J. M. Groff, and T. B. Waltzek. 2016. Megalocytivirus infection in cultured Nile Tilapia *Oreochromis niloticus*. Diseases of Aquatic Organisms 119(3):253–258.

3.4 Recommended Reading

Arthur, J. R., M. G. Bondad-Reantaso, and R. P. Subasinghe. 2008. Procedures for the quarantine of live aquatic animals: a manual. FAO Fisheries Technical Paper 502. Food and Agriculture Organization of the United Nations, Rome. Available: http://www.fao.org/3/a-i0095e.pdf. (May 2020).

Bobe, J. 2015. Egg quality in fish: present and future challenges. Animal Frontiers 5(1):66–72.

Bromage, N. R., and R. J. Roberts, editors. 1995. Broodstock management and egg and larval Quality. Wiley-Blackwell, New York.

Brooks, S., C. R. Tyler, and J. P. Sumpter. 1997. Egg quality in fish: what makes a good egg? Reviews in Fish Biology and Fisheries 7:387–416.

Izquierdo, M. S., H. Fernandez-Palacios, and A. G. J. Tacon. 2001. Effect of broodstock nutrition on reproductive performance of fish. Aquaculture 197:25–42.

Lubzens, E., G. Young, J. Bobe, and J. Cerdà. 2010. Oogenesis in teleosts: how eggs are formed. General and Comparative Endocrinology 165:367–389.

Migaud, H., G. Bell, E. Cabrita, B. McAndrew, A. Davie, J. Bobe, M. Herráez, and M. Carrillo. 2013. Gamete quality and broodstock management in temperate fish. Reviews in Aquaculture 5(1):194–223.

Ross, R. M. 1984. Catheterization: a non-harmful method of sex identification for sexually monomorphic fishes. The Progressive Fish-Culturist 46(2):151–152.

Rottmann, R. W., J. V. Shireman, and F. A. Chapman. 1991. Determining sexual maturity of broodstock for induced spawning of fish. SRAC Publication No. 423. Southern Regional Aquaculture Center, Stoneville, Mississippi. Available: https://www.ncrac.org/files/biblio/SRAC0423.pdf. (November 2020).

For more specifics on the topics of this chapter as they apply to a given species of fish see: Chapters 13 Catfishes, 14 Carps, 15 Tilapias, and 16 Sportfish Species. Additional references regarding genetics are given in Chapter 17:11.

Chapter 4

Control of the Spawning Process

4.1 Environmental Stimuli

The natural spawning by fish is a complex interaction of factors often dependent on a number of environmental factors, some of which we can provide in a hatchery setting, others we cannot. In the proper range and combination, these environmental factors can impact the fish's physiology, causing a cascade of hormone production resulting in ovulation.

Temperature is perhaps the most important environmental factor affecting fish reproduction, as it impacts growth rate, time when first maturity is reached, and how quickly new eggs develop from one spawning cycle to the next. In a north temperate region, over a species natural range there may be a year or more difference in the times it takes for a fish to first reach maturity. Similarly, the length of the growing season is affected by temperature. A given species may be able have enough nutrient intake in a long growing season to spawn every year, but at the northern part of its range the growing season is shorter, and it may take two seasons to take in enough nutrients to complete a spawning cycle.

Each species has an optimum range of temperature, often no wider than 5°C, where it can successfully reproduce. How long the temperature stays within that temperature range influences the length of the spawning season. In a north temperate climate, the temperature rises in the spring into the appropriate range for spawning. The temperature rise helps to synchronize brood maturity, resulting in the majority of the brooders beginning mature at the start of the spawning season. In a more tropical setting there is less seasonal change in temperature. In such a setting, the same species will have some mature brooders throughout a wider spawning season, but at any one point in time only a portion of the brooders are mature enough for spawning.

In temperate climates not only does the temperature rise in the spring but the hours of daylight increase as well. In a natural setting it is hard to separate the effects of temperature and photoperiod on spawning. Under controlled conditions both temperature and photoperiod can be manipulated to impact spawning. At warmwater hatcheries fish spawning can be delayed in the spring if cool water from a well or other source can be added to the brood holding pond. Similarly, moving the fish into a warm-water setting can advance the spawning season. Under conditions where temperature and photoperiod can be controlled, spawning seasons can be significantly advanced or delayed, or the spawning cycle condensed from a one spawning period/yr. to two or more periods/yr. Lunar phase is another factor influencing natural spawning of a number of marine species and some species of freshwater sunfish. The lunar cycle is regulated under controlled hatchery conditions for a number of marine species, some species spawn more frequently around the dark of the moon and other species around a full moon. The lunar cycle is not commonly regulated in freshwater hatcheries.

Chronic stressors such as high ammonia or low oxygen will limit spawning. In Channel Catfish *Ictalurus punctatus*, spawning is depressed where morning oxygen values are consistently below 4 mg/L during the spawning season. Pheromones are thought to suppress spawning of many cyprinid species when the fish are held at high densities in static water ponds. Adding enough new water to dilute such pheromones can stimulate spawning in that setting. Moving the brooders into new water will also stimulate spawning.

Access to spawning sites is another technique to control the timing of fish spawning in warmwater freshwater hatcheries. In species that spawn by scattering their eggs over fibrous substrate, controlling when the fish have access to such material can control when spawning occurs. For species that build nests in shallow water, spawning can be controlled by holding the brooders in steep-sided deep ponds and transferring the brooders into shallow ponds when spawning is desired. Other environmental stimuli such as specific water flow velocities are harder to provide in a hatchery setting.

4.2. Hormone-induced Spawning

When in the right range and combination, the above-mentioned environmental factors will stimulate sexually mature fish to produce a cascade of hormones resulting in successful spawning. However, it may not be practical to provide all the needed environmental stimuli in a hatchery setting. It is then more practical to provide an exogenous source of hormone critical to the reproductive process. In nature the fish's hypothalamus serves as a translator of environmental stimuli to reproductive hormones, including gonadotropin-releasing hormone. The pituitary in response to gonadotropin-releasing hormones produces gonadotropins that in turn stimulate gonad maturation. In a hatchery setting it is possible to intervene and provide the above hormones or equivalents to stimulate the release of eggs or sperm.

4.2.1 Fish Pituitary

Fish pituitary was the first material used as a source of gonadotropins for induced spawning and continues to be widely used. Pituitaries collected from the same species being spawned, or more often, pituitaries from common carp are used. The tissue can be used fresh, preserved in alcohol, or after a series of alcohol and acetone rinses, desiccated and stored for later use. While the choice of a donor species for pituitary glands is strongly influenced by its availability at a reasonable cost, some attention should be given to the phylogenetic relationship between the donor and the recipient. The closer the phylogenetic relationship of the donor to the recipient, the better the response that can be anticipated.

Pituitaries should be collected from mature adults at the start of the spawning season. If used fresh, the size of the donor should approximate that of the recipient or two or more donors whose collective weight equals that of the recipient can be sacrificed to provide the necessary quantity of pituitary. The pituitary, which is located in a cavity beneath the brain, can be collected in several ways. Most commonly the head is cut from the body and held vertically to allow blood to drain from brain cavity. The top of the skull is cut off to expose the brain, and the brain is then pulled forward to expose the pituitary (Figure 4.1).

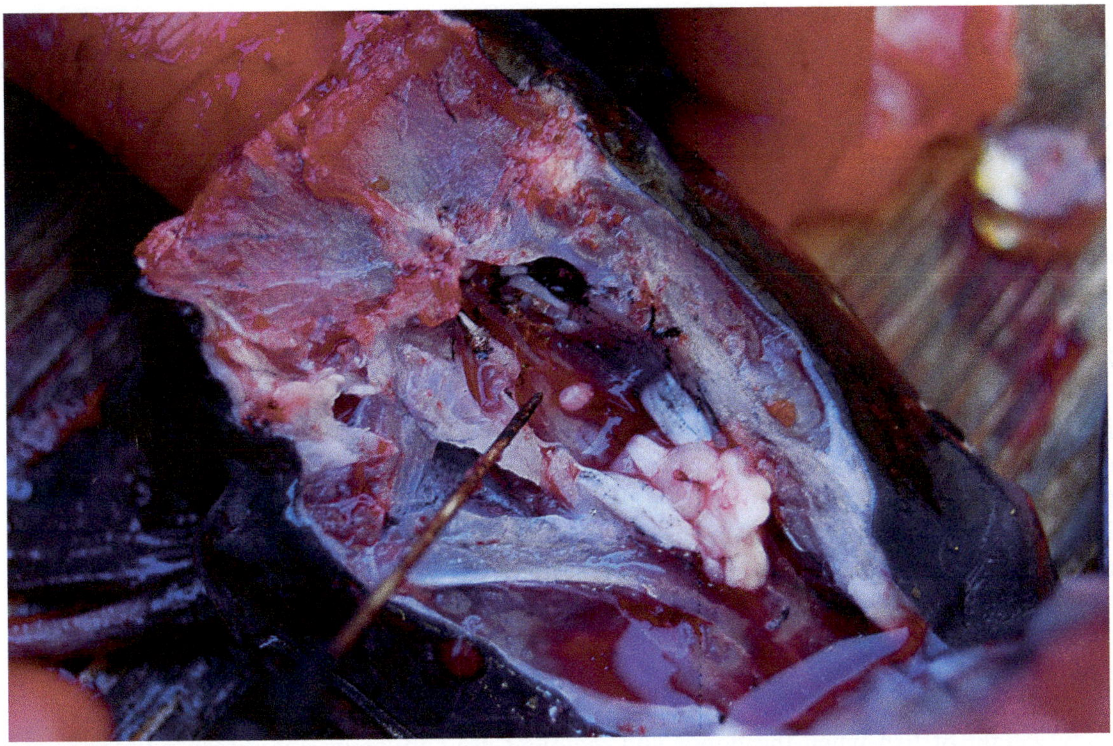

Figure 4.1. Common Carp *Cyprinus carpio* pituitary exposed after brain is moved forward.

The pituitary is removed from its cavity with a pair of forceps. The pituitary can also, after a little practice, be collected with a hollow drill bit that will permit a plug of tissue containing the pituitary to be cut out of the skull. It is also possible to collect pituitary from fresh market fish while doing minimal damage to the product by cutting laterally from the ventral side of the head through the gill arch into the skull just off the midline and exposing the pituitary from underneath.

Fresh pituitary can be used by macerating or homogenizing it into small particles in sterile water or a sterile 0.85% saline solution and injected using a hypodermic syringe having a 16 to 20-gauge needle. Smaller diameter needles may be used if the particle sizes of the pituitary are fine enough to pass through the needle. To minimize the risk of the needle clogging, it is best to allow the suspension to stand for a few minutes so any pituitary particles can settle out, then withdraw the supernatant for use.

Pituitaries can also be preserved for later use. Once collected from the fish, they are first placed in a 70–100% ethyl alcohol solution, then after 6–12 hours the solution is replaced with 100% ethyl alcohol. The solution is replaced again after another 6–12 hours with 100% ethyl alcohol. The pituitaries can be held in the final alcohol solution for up to a year before use. Before use pituitaries are removed from the alcohol and held until the alcohol evaporates and then prepared for injection.

Pituitaries can also be stored dry for later use. This is done by first replacing the final bath of 100% ethyl alcohol with 100% acetone and holding the pituitaries for 12–24 hours. The acetone is then poured off and the pituitaries allowed to air dry. The dried pituitaries should be placed in a clean vial with an airtight cap for storage. Dried pituitaries can be stored at room temperature. In humid climates the dried pituitaries should be stored in a desiccator. If carefully prepared and stored, dried pituitaries can be held for an extended period of time. For example, one lot of acetone-dried whole Common Carp pituitaries was stored 8 years in a laboratory desk drawer and still retained enough potency to ovulate Goldfish using the customary dose.

Ground, dried pituitary is given in one or more injections in proportion to the weight of the recipient fish, most often expressed as milligrams of pituitary per kilogram of fish. A balance capable of weighing pituitary to the nearest milligram will be required. Dose rates range from 2 mg/kg female where the female is allowed to spawn naturally with a male, to 4 mg/kg female where the eggs are manually stripped from the female. Higher dose rates (10 mg/kg) have been used for manually stripping of Channel Catfish.

A common injection protocol is to give the females 10% of the total dose in the first injection and the remaining 90% 6 to 12 h later. Dividing the dose into a preparatory dose and a resolving dose allows a more gradual final maturation of the eggs and helps to avoid a premature hydration of the ovary that might result from injecting the total dose in a single injection. The time interval between preparatory and resolving injection is a function of water temperature. For example, at a water temperature of 21°C an injection interval of 12 h might be used but at 28°C, where the fish's metabolic rate is higher, an interval between injections of 6 h would be more appropriate. Males are given half the female total dose rate at the time when the females are given their second injection.

The dosage of whole dried pituitary is sometimes expressed as number of pituitaries per body weight of fish when the donor and recipient are similar in size. Fresh pituitaries are given as a ratio of weight of donor to weight of recipient (equivalent weights). For example, if the protocol was to give 2 equivalent weighs per kg of female and the female to be injected weighed 1 kilo then pituitaries need to be collected from 2 kg of donor fish for injection.

Where ground dried pituitary is to be given based on fish weight, a maximum concentration of 25 mg of pituitary per mL of sterile water or 0.85% sterile saline solution carrier is recommended to ensure a good dispersal of pituitary within the carrier. Efforts should be made to calculate the total quantity of pituitary needed and the dose required for each individual fish so that a hormone solution of the proper strength can be prepared to give the required dose in a single intramuscular injection of 1 mL or less. On occasions more than one intramuscular 1 mL injection may be needed for large fish to give the required volume of injection. In such cases the total required volume can also be given as a single intraperitoneal injection.

The process of preparing the dose and its injection will result in a loss of some of the hormone solution. To account for this, an additional 10–15% of the calculated required quantity should be added to give a final quantity prepared of 100% of the calculated required dose + 10–15 % anticipated loss. Once the quantity of pituitary and carrier solution is calculated, the pituitary is ground in the carrier solution for a few minutes. This can be done using a mortar and pestle or a tissue grinder. Once ground the pituitary solution should be allowed to stand a few minutes or be centrifuged so the particulate matter can settle out. The supernatant is then drawn off and used to inject the fish. This solution can be stored at least 12 hours under refrigeration. This allows one initial solution to be prepared from which the initial and resolving dose can be given when the time interval between injections is ≤12 h. It is best to prepare new solutions for each new lot of fish so that the concentrations/mL is more appropriate for the average weight of the fish in that lot.

4.2.2 Chorionic gonadotropin

Chorionic gonadotropin (CG) is a gonadotropin typically isolated from mammalian sources that is used either by itself or in conjunction with fish pituitary to induce spawn fish. Human chorionic gonadotropin (hCG) is widely available, as well as veterinary grade CG is also suitable. CG is commonly sold as a lyophilized powder of a known strength sealed in a glass vial with an accompanying vial of diluent. This convenient packaging facilitates its preparation and administration. Having a product of a known strength also facilitates its use. The strength of CG is generally expressed in international units (IU) and occasionally as rat units (RU) or frog units (FU). One international unit is equal to 0.001279 mg of the 2nd International Standard Chorionic Gonadotropin. It can be commonly purchased in vials containing 5,000 or 10,000 IU per vial. Less common are vials of 50,000 IU per vial. Dose rates used have ranged widely depending on the species, the degree of maturity and the environmental conditions. When mature striped bass females are collected on a spawning migration, a single injection at 250 IU/kg is effective in inducing ovulation. More typical dose rates for most species when using mature females range from 1,000 to 1,500 IU/kg giving 10% of the dose as the initial injection and anywhere from 6 to 24 h later the remaining 90% as the resolving injection. Males are typically given half the female dose rate as a single injection at the time when the resolving injection is given to the females.

The lyophilized powder dissolves readily in water or saline solution, and concentrations up to 5,000 IU per mL can be prepared. Once dissolved, the CG solution can be stored several days under refrigeration while retaining its potency.

4.2.3 Gonadotropin-Releasing Hormone (GnRH)

Both pituitary solutions and CG provide gonadotropin that acts directly on the ovary. A more subtle approach is to stimulate the fish to produce its own gonadotropins by administering a gonadotropin-releasing hormone. The synthetic analogue of Luteinizing Hormone-Releasing Hormone: D-Ala^{6}, des-bly^{10}. LHRH ethylamide (LHRH$_a$) has been successfully used to induce spawn over 20 species. Another formulation [Des-Gly^{10}, D-Arg^{6}, Trp^{7}, Leu^{8}] – LHRH, ethyl amide (sGnRH$_a$) is the analogue of salmon GnRH. Doses are generally expressed as micrograms per kilo of fish. Effective dose rates to induce ovulation maybe as low as 20 μg/kg for Chinese carps to as much as 100 μg/kg for Channel Catfish or Paddlefish *Polyodon spathula*, giving females 10% in a preparatory dose and the remaining 90% as a resolving dose.

LHRH$_a$ and sGnRH$_a$ are commercially available as a powder in milligram quantities. They dissolve readily and are most easily prepared by first making a strong stock solution (1–5 mg/mL) using sterile water or a 0.85% sterile saline solution, then making one or more dilutions to make a concentration appropriate for injection. The stock solution and dilutions can be frozen and held for later use.

4.2.4 Dopamine Antagonists

Just as the hypothalamus translates favorable environmental conditions for fish reproduction by producing GnRH, it can produce dopamine in response to unsuitable conditions.

This can include stress resulting from brood collection and holding. Dopamine is a chemical that inhibits the release of hormones from the pituitary and blocks the pituitary's response to GnRH. The effects of dopamine can be reduced using a dopamine antagonist. Two of the most common dopamine antagonists used along with GnRH for induced spawning are domperidone (1–5 mg/kg) (see Figure 4.4) and pimozide (10–20 mg/kg). These two products are not very water-soluble but domperidone will dissolve in dimethyl sulfoxide (DMSO) (48 mg/mL) and ethanol (1 mg/mL), and pimozide in DMSO (18 mg/mL), and ethanol (3 mg/mL). Other dopamine antagonists have been used along with a GnRH include: 0.5 mg/kg haloperidol and 10 mg/kg metoclopramide. The addition of a dopamine antagonist along with a GnRH is particularly helpful in increasing the spawning success of fish that are easily stressed. They can be injected separately or mixed in a common solution and injected together. Commercial products such as Ovaprim are sold as a liquid containing salmon $sGnRH_a$ at 20 µg/mL and domperidone at 10 mg/mL.

4.2.5 $GnRH_a$ pellets

The above-mentioned hormones are typically powders with an appropriate quantity of solvent added to give the desired concentration then are given as an injectable solution, but GnRH is also given as an implant to obtain a slower hormone release. A pellet can be prepared using cholesterol or cholesterol-cellulose (See Lee et al. 1986 reference in Chapter 17.13.) or purchased commercially, to give a slow $GnRH_a$ release over several days or weeks. Such pellets can be injected intramuscularly to advance maturation, induce natural spawns or induce ovulation for manual stripping. Typical $GnRH_a$ concentrations/pellet may range from 25 to 150 µg/pellet depending on the anticipated size of the brooders and the release period. Actual dose/fish is not as exact as when an injectable solution is given, however, once a brooder is weighted, an approximate dose/kg fish can be obtained by dividing pellets or giving multiple pellets of different $GnRH_a$ concentrations.

4.3 Hormone Administration

Injection is the most common method to administer hormones for induced spawning. Injections can be given intramuscularly or intraperitoneal depending on the volume to be injected. Intramuscular injections are given above the lateral line into the flesh in an area near the base of the dorsal fin (Figure 4.2). The point of the needle is slipped bevel side down underneath a scale and the angle of the needle increased more perpendicular to penetrate the muscle. The needle should penetrate approximately halfway from the skin to the backbone or about 1 cm. A maximum of 1 mL per injection site is recommended for intramuscular injections to minimize fluid loss when the needle is withdrawn. Fluid loss can also be minimized using a finger to put pressure over the needle insertion point and slowly withdrawing the needle after the injection. Intraperitoneal injections are made into the body cavity with the needle being inserted at a low angle into the scale-less area at the base of the pelvic or pectoral fin. One can feel the needle break through the muscle wall and into the body cavity. At that point stop and inject the hormone. Care must be given to not penetrate any internal organ. Intraperitoneal injections permit larger amounts of fluid to be injected with a minimum of fluid loss into large fish.

Figure 4.2. Intramuscular injection given above the lateral line into the flesh in an area near the base of the dorsal fin.

Intramuscular injections are best given using 1 mL syringes and 18 to 22 gauge needles. Using 1 mL syringes allows an accurate measurement of the dose/fish and helps reduce accidental hormone loss more common when a larger volume syringe is used to inject multiple fish. Needle diameter is expressed in terms of its gauge, which is a measure of the outer diameter (OD) of the wire used to make the needle. However, this is an inverse relationship. The smaller the gauge number, the greater the needle diameter. For example, a 16 gauge needle has an outside diameter of 1.65 mm while a 22 gauge needle has an outside diameter of 0.72 mm. Higher gauge needles (20–22 gauge) are suitable for hCG and LHRHa as these hormones go into solution easily and will not clog the needle. Higher gauge needles also make a smaller opening in the tissue. Lower gauge needles (16–18 gauge) are appropriate for use when injecting pituitary suspensions that otherwise might clog higher gauge needles. Intraperitoneal injections are typically given with larger syringes (>1 to 5 mL) using a longer needle length to ensure penetration into the body cavity.

As most hormones are given based on the weight of the individual fish, the first step is to have a way to identify each fish and record its weight. In some cases, it is practical to hold each fish in its own container. This allows handling of one fish at a time without stressing other fish. However, in most cases it is necessary to hold several fish in the same container but with each individually identified. In addition to the more classical types of fish tags, for induced spawning a temporary tag that will identify fish for no more than 48 hrs can be adequate. An example of a temporary tag might be using a different colored piece of yarn per fish, tying the yarn around the hard spine of the dorsal fin (Figure 4.3).

A key to successful induced spawning is minimizing stress on the brooders. This begins with brood handling. Brooders are often large and hard to handle however, handling stress can be reduced using a cradle to wrap around the fish or a soft mesh bag where the fish can be inserted. Cradles and bags can be used to hold individual fish for weighing and holding the fish while they are being injected. Calmly collecting the fish avoiding quick movements will reduce

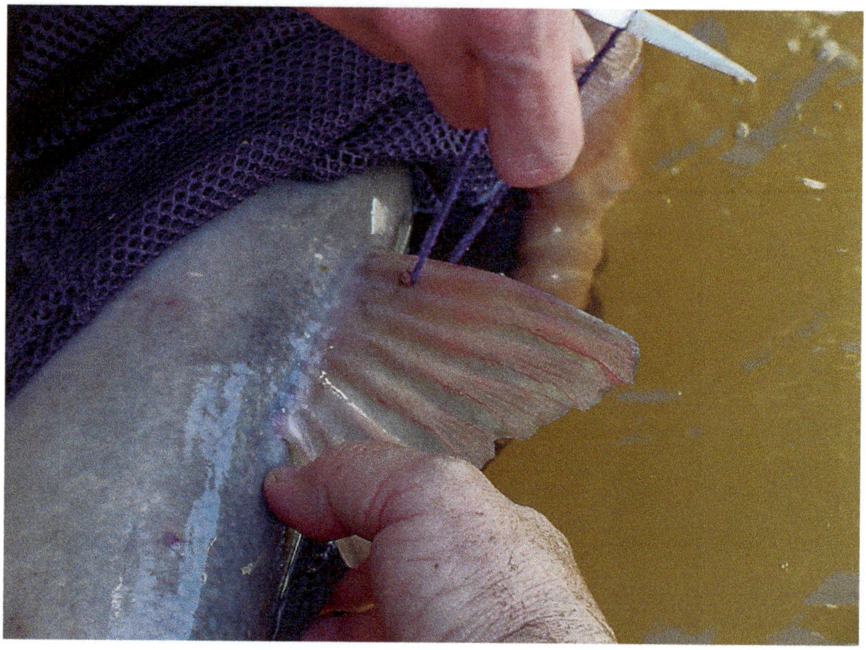

Figure 4.3. A temporary tag made of yarn used to identify individual fish for hormone-induced spawning.

fish damaging themselves and minimize stress. When working with delicate fish it may be necessary to anesthetize the fish prior to handling however for most fish this is not necessary. Covering the eyes with a cloth can also help to calm down many fish when they are being handled.

The holding container used to hold the brooders can also be a source of stress. Rectangular tanks where the water enters one end and exits the other are adequate for species that do not need to swim continuously. Circular tanks are better for holding most brooders as it allows the fish to continuously swim around the tank and the curved wall reduce frequency of head-on collisions with the wall. An appropriate water flow is needed to maintain water quality keeping oxygen near saturation and ammonia near zero. The flow rate should be enough to create a current for the fish to orient into but not so strong that the fish have to swim hard to maintain position. Brood holding tanks should be located in a building where there is a minimum amount of noise and other activity that might disturb the brooders.

4.4 Spawning and Egg Handling

When fish are induced spawned the fish can be allowed to release and fertilize the eggs naturally, or the eggs can be manually stripped from the female and artificially fertilized with sperm collected from the males. If the eggs are to be manually stripped, there is an optimum time following the hormone injection to collect the eggs. Once the eggs are ovulated into the lumen of the ovary they are no longer receiving oxygen from the blood stream as when they were attached and the egg quality starts to deteriorate. The rate of deterioration is greater the higher the water temperature. Table 4.1 gives the time window after ovulation as to when there can be significant loss in egg quality for several fish species.

Table 4.1. Time after ovulation when approximately 50% of eggs are over-ripe (Woynarovich and Horvath 1980).

Species	minutes
Common Carp	50–80
Grass Carp	30–40
Silver Carp	30–40
Bighead Carp	50–80
Prochilodus sp.	20–30

Knowing when to collect the eggs is key to successful induced spawning and artificial fertilization. As most fish are considered to be "cold blooded," rates of physiological activities can be described as a function of water temperature, including how soon a female will ovulate following hormone injection. The response time can be expressed as degree hours (water temperature°C × time in hours) from the time of the resolving injection to ovulation. When induced spawned with carp pituitary the response time for Common Carp *Cyprinus carpio* is between 240–260 degree hours and 200–220 degree hours for bighead and silver carp for fish. Careful use of the degree-hour concept can predict the time of ovulation to within a relatively short window of time. This will help reduce excessive handling of brood fish, allow hatchery personnel to have all the needed resources available prior to fish spawning, as well as to do other work while waiting for the fish to ovulate.

Degree-hour response times apply best when developed around a specific set of circumstances (hormone protocol, fish condition, etc.), and other indicators of possible ovulation should also be used. As a female nears ovulation, the eggs in the ovary are hydrating and this may affect the fish's equilibrium. Such females may be seen moving up and down in the water column, readjusting to the new weight. If males are held in the same tank as the female, there may be an active courtship as ovulation nears. Males may pursue the female around the tank, nudging the vent area, occasionally pushing the female to the surface. For some species a noticeable change in body conformation is visible as the fish ovulates. This might include a swelling of the lower body near the vent area.

One of the easiest ways to recognize that a fish has ovulated is that many species of fish will release a few eggs into the water as they begin to ovulate. As the lumen of the ovary fills some eggs are forced out the oviduct into the water and these eggs can be seen in the water column or on the tank bottom and collected with a fine-mesh dip net. This reduces the need to handle the brooders but does not identify which female in a tank has ovulated. How recently a Chinese carp ovulated can be estimated by the egg diameter of the eggs collected in the water column, as once the eggs are released into the water they hydrate rapidly often reaching their maximum diameter within about an hour. If fully-hydrated eggs are found in the tank when the tank is first sampled for the presence of ovulated eggs, a female may have ovulated earlier than anticipated and poor quality eggs should expected from that female.

The clearest indicator that a female has ovulated is the free flow of eggs from the vent when the fish is handled, particularly when pressure is put on the abdomen. Although this is the surest sign of ovulation, frequent handling of brood fish before ovulation may delay or prevent ovulation from occurring. Sampling eggs while still attached in the ovary can be used to tell how mature the eggs are based on egg diameter, egg clarity, number of oil globules formation and location of the germinal disc (see Chapter 17.13). However, egg sampling can be stressful to the brooder with a risk of damage to the ovary or oviduct.

Once an ovulated female has been identified she should be carefully handled. When handled the vent should be covered and pressure applied to prevent egg loss. The head and anterior portion of the brooder should be slipped into a pillowcase or other soft mesh bag to facilitate handling. Covering the eyes with such a bag helps to calm the fish and covering the body makes it easier to hold the fish. An ovulated fish should be anesthetized prior to egg collection. In many cases more than one female is held in the same tank. Adding an anesthetic to such a tank may facilitate brood examination and capture of the female that has ovulated, however, the anesthetic may delay the ovulation of other females in the tank. It is often more practical to calmly collect and transfer the ovulated female to a smaller container containing the anesthetic and once anesthetized prepare her for egg collection. Tricane methane sulfonate (MS-222) at 50–100 mg/L is effective in putting most fish into a deep sleep. However, MS-222 can change the pH of poorly buffered holding water and to avoid any resultant damage to the gills an equal weight of sodium bicarbonate can be added to the water to increase its buffering capacity. Some fish species are extremely sensitive to handling may not survive after stripping. With such species, once the fish ovulates it is best to euthanize the fish before stripping of eggs.

Eggs can be collected by allowing them to flow into a dry bowl (Figure 4.4) or a container of water. If the eggs are to be fertilized after being collected in a dry bowl the area around the vent as well as the hands of the person holding the fish should be dried very carefully to avoid the eggs coming in contact with water as they are stripped into the bowl. The bowl should be

Figure 4.4. Stripping Grass Carp *Ctenopharyngodon idella* eggs into a dry bowl for "dry fertilization."

held near the vent to minimize the distance that eggs drop into the bowl. Eggs should flow freely from the vent when only a minimum of pressure is applied to the abdomen. Little or no blood or fluid should be expelled with the eggs. Such fluids would suggest poor egg quality. A clear fluid suggests that some eggs have become overripe and started to breakdown. Bloody eggs suggest that the female has not fully ovulated and the pressure applied to the abdomen was breaking un-ovulated eggs away from the ovary.

Sperm can be collected from males before the anticipated ovulation or stripped directly from the males and mixed with the eggs (Figure 4.5). Collecting sperm ahead of time facilitates a smooth flow of activities during the spawning process. Males should be held with the vent up and the area around the vent dried carefully. Sperm can be expelled with a slight pressure on the abdomen and collected using a syringe or an aspirator. Care must be given in the collection and storage of sperm to avoid contact with water, once in contact with water sperm become motile and soon die. If sperm is collected and held dry it can remain viable it can at room temperature for several hours and even longer under refrigeration. Before a given batch of sperm is used to fertilize a set of eggs the sperm should be checked microscopically for motility. This can be done by adding a drop of semen and a drop of water to a microscope slide then mixing the two and examining the slide under the microscope for sperm movement.

Collecting the eggs into a dry bowl and mixing in un-diluted sperm before adding water is often called the “dry method.” That approach is used with most freshwater fish. With this approach it is especially important to prevent eggs or sperm from coming in contact with water before they are mixed together. Undiluted sperm is added to eggs held in a smooth-sided container at a ratio of approximately 5 mL semen per liter of eggs and gently mixed. Once well mixed, water or a fertilizing solution (a dilute physiological saline solution for extending sperm life) is to activate the sperm. In most cases, 1–2 min of gentile stirring of the egg/sperm mixture is adequate for egg fertilization. After that time period, additional fresh water or fertilizing solution should be added and the eggs rinsed. The eggs will begin to adsorb water and

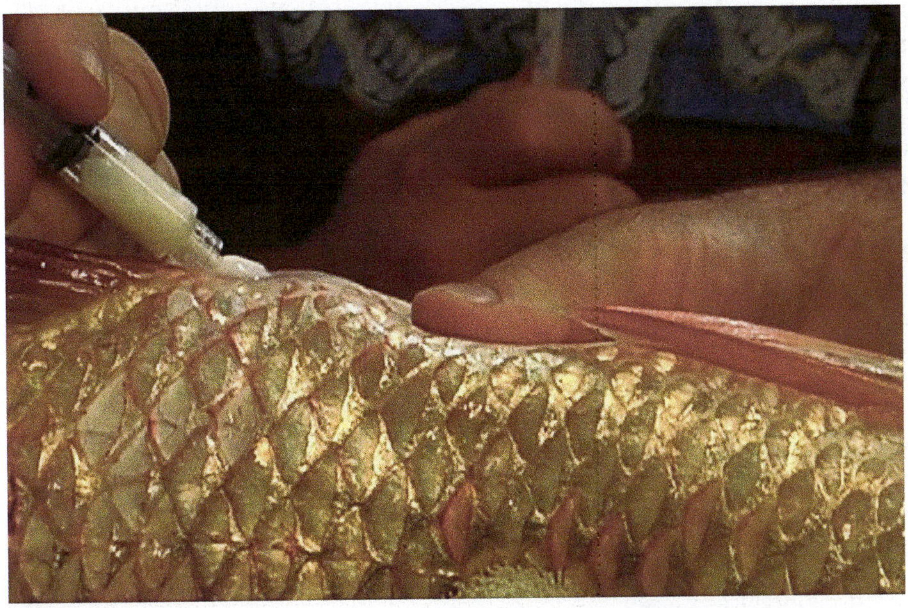

Figure 4.5. Sperm collection from a Grass Carp *Ctenopharyngodon idella* male.

increase in diameter and additional water needs to be added as the eggs hydrate. This process of absorbing water, often takes about an hour, is called water hardening, and may result in a 10-fold increase in egg diameter for fish with floating or semi-buoyant eggs. Once the eggs have been rinsed with water several times to remove excess sperm and organic matter they can be added to incubators or held and given several more rinses until they water harden then loaded into the incubator.

For Striped Bass and many marine fish a "wet method" is used. It is used with species where the male releases large volumes of sperm when stripped and the female releases eggs that may be delicate to handle. With this approach, eggs are stripped into a pan or bowl containing fresh water or a saline solution and sperm added simultaneously or within a minute or two after adding the eggs. Careful coordination of activities is needed to ensure successful fertilization. Once the eggs are in contact with water the micropyle of the egg becomes constricted, making it more difficult for sperm to enter and fertilize the egg. Water also stimulates the sperm to become motile and the sperm is only viable for a few minutes from that point. Once the eggs have had several minutes of opportunity to become fertilized, they can be rinsed and moved to the incubators.

A number of fishes have eggs that are adhesive within a few minutes after they come in contact with water. Such eggs can be incubated by spreading the eggs over a substrate within a few minutes after coming in contact with water, trying to avoid individual eggs coming in contact and sticking to each other. Such eggs can also be treated to reduce the adhesiveness and collectively incubated in conventional incubators. Techniques to reduce the adhesiveness are discussed in Chapter 17.4.

An alternative approach to obtain fertilized eggs, especially working with delicate brooders, or fish where egg quality deteriorates rapidly after ovulation, is to inject the brood fish and hold both males and females in a tank or hapa (a fine mesh cage) where they are allowed to spawn freely. Substrate is provided for collecting adhesive eggs. Non-adhesive sinking or semi-buoyant eggs can be collected from the bottom of the tank by siphon or with a net. Floating eggs can be collected with a net or allowed to flow out of the tank into another container. Issues with this approach include the female may not release all of her eggs into the water. Males may release their semen into the water before the female has completely ovulated and the short life of the sperm may result in a low rate of successful fertilization. Similarly, the female may release eggs into the water but the male does not, resulting in unfertilized eggs.

4.5 References

Woynarovich, E., and L. Horvath. 1980. Artificial propagation of warmwater fin fishes—a manual for extension. FAO Technical Paper 201. FAO, Rome. Available: http://www.fao.org/3/ac742e/AC742E00.htm. (May 2020).

4.6 Recommended Reading

Chattopadhyay, N. R. 2016. Induced Fish Breeding: A Practical Guide for Hatcheries. 341 pp. Academic Press, Cambridge, Massachusetts.

Dufour, S., M. E. Sebert, F. A. Weltzien, K. Rousseau, and C. Pasqualini. 2010. Neuroendocrine control by dopamine of teleost reproduction. Journal of Fish Biology 76(1):129–160.

Lahnsteiner, F., F. Soares, L. Ribeiro, and M. T. Dinis. 2009. Egg quality determination in teleost fish, Pages149–180 *in* E. Cabrita, V. Robles, and P. Herraez, editors. Methods in reproductive aquaculture marine and freshwater species, CRC Press, Boca Raton, Florida.

Mañanos, E., N. Duncan, and C. C. Mylonas. 2008. Reproduction and control of ovulation, spermiation and spawning in cultured fish. Pages 3–80 *in* E. Cabrita, V. Robles, and M. P. Herráez, editors. Methods in reproductive aquaculture: marine and freshwater species. CRC Press, Boca Raton, Florida.

Mylonas, C. C., and Y. Zohar. 2001. The use of GnRH delivery systems for the control of reproduction in fish. Reviews in Fish Biology and Fisheries 10(4):463–491.

Mylonas, C. C., A. Fostier, and S. Zanuy. 2010. Broodstock management and hormonal manipulations of fish reproduction. General and Comparative Endocrinology 165:516–534. Available: https://www.researchgate.net/profile/Constantinos_Mylonas/publication/229885378_Reproduction_and_Broodstock_Management/links/5a6a6b9ba6fdcc2aedee0999/Reproduction-and-Broodstock-Management.pdf. (May 2020).

Ohs, C. L., and J. S. Broach. 2005. Hormone Preparation, Dosage Calculation, and Injection Techniques for Induced Spawning: Bait fish and Ornamental Fish. SRAC Publication No. 0428. Southern Regional Aquaculture Center, Stoneville Mississippi. Available: https://fisheries.tamu.edu/files/2019/01/SRAC_0425.pdf. (May 2020).

Urbanyi, B., A. Horvath, and Z. Bokor. 2008. Artificial fertilization in aquaculture species: from normal practice to chromosome manipulation. Pages 183–216 *in* E. Cabrita, V. Robles and M. P. Herráez, editors. Methods in reproductive aquaculture: marine and freshwater species. CRC Press, Boca Raton, Florida.

Yanong, R. P. E., C. Martinez, and C. A. Watson. 2017. Use of ovaprim in ornamental fish aquaculture. UF/ IFAS Fact Sheet FA-161. Available: http://edis.ifas.ufl.edu/fa161. (May 2020).

For additional references regarding induced spawning see Chapter 17.13.

Chapter 5

Incubation and Post Hatching Care of Fry

5.1 Introduction

Better hatching success can be expected when it is possible to incubate the eggs in a hatchery settling, that is, if the eggs can be collected safely. Some fish build nests that are not practical to transfer to the hatchery or attach their eggs so firmly to a substrate that it is not practical to remove the eggs. In such cases, it may be best to leave the eggs in their natural settling, allow the parents to guard the eggs, then collect the fry after hatching. Another alternative would be to provide a substrate for egg deposition, then transfer the substrate to a better environment for incubation (See Chapter 3 on egg/fry transfer). Eggs and the associated substrate can be transferred to a prepared nursery pond where the eggs will hatch and the fry can develop. This approach is disadvantageous in that the magnitude of disease problems during hatching are hard to quantify and even harder to control. This results in an unknown quantity of fry in the nursery pond. Past production records from ponds where similar production practices were used can be used as a guide as to the quantity of fish that can be expected.

An alternative is to transfer the substrate holding eggs to the hatchery where the substrate can be rinsed to remove any sediment or debris and the eggs treated with formalin to control fungus. The substrate with eggs attached can be held in aerated aquaria or tanks that receive a 2 times/h water exchange. A given aquaria or tank should be stocked with egg-bearing substrates holding eggs of similar age. This will facilitate the harvesting of similar age fry. If the eggs take ≥48 h to hatch they should be treated twice a day with formalin at 100–250 mg/L for an hour until 12 h before the anticipated hatching. In some cases where the eggs can be removed from the substrate, i.e. Channel Catfish *Ictalurus punctatus*, the eggs can be collected from the spawning pond and transferred to the hatchery for incubation.

A cleaner way for a natural spawn is to bring the brooders into the hatchery, provide the appropriate substrate and water quality, and then allow the fish to spawn naturally, or stimulate the fish to spawn naturally following the administration of an exogenous hormone. In either case, the substrate and its eggs will be cleaner than natural spawns collected from a pond and the eggs should have less disease problems and a higher hatch rate. Once the egg-bearing substrate is collected, it can be incubated as described in the previous paragraph.

Many species of fish with adhesive eggs have fry that initially will remain in the nest or substrate until more of the yolk sac is absorbed, then the fry will start swimming up into the water column. Where the substrate cannot be easily transferred to the hatchery for incubation it is often easier to collect the fry after they have begun to swim-up versus trying to collect recently hatched fry often hidden in the nest or substrate where the eggs were attached. Where the substrate can be transferred to the hatchery and suspended vertically in a clean aquarium, the sac fry will drop to the bottom of the aquarium and can be siphoned out daily and stocked into new tanks along with other sac fry of a common age.

For fish that are to be induced to spawn and the eggs artificially fertilized, there are a variety of ways to incubate the eggs depending on the characteristics of the eggs. Fish eggs can be grouped into a few basic categories: adhesive sinking, non-adhesive sinking, non-adhesive semi-buoyant, or non-adhesive buoyant. Incubators need to be provided that take into consideration these characteristics.

5.2. Water Considerations for Incubation

One thing common to any incubation system to help ensure success is a clean environment with good water quality. The water supply used for incubation and fry holding should be as clean as possible, free of contaminates, high in oxygen, little to no ammonia, and be stable in temperature (See Chapter 2). The total volume of water required for egg incubation and larval rearing is often much less than for other hatchery uses and the installation of a well, or special water filtration or treatment procedures are justified to obtain the water quality needed.

Incubation success and fry development is temperature dependent. Every fish species has an appropriate water temperature for egg incubation and fry development. Temperatures at the lower end of the range slow egg and fry development. This longer development period increases the opportunity for fungus development on eggs, and by extending the time until mouthparts are formed results in larvae with less nutrient reserves by the time of first feeding. At higher temperatures the rate of development is accelerated, and when it is too high can result in pre-mature hatching and a greater rate of deformed larvae. Each species has a range of appropriate temperatures, and from a hatchery perspective, incubation near the higher end of the optimum range can be beneficial. A one-degree temperature increase can speed up development time by several hours and in turn speed up when fry are ready to leave the hatchery. This can reduce the opportunity for human or mechanical hatchery errors, while freeing up hatchery space sooner.

Groundwater is typically too cool for the incubation of warmwater fish eggs and needs to be warmed up before use. This can be done by holding such water in a small reservoir, allowing it to warm up naturally before use in the hatchery. When well water is held in a small reservoir in a temperate climate the water will warm seasonally and its temperature will be optimum for egg incubation of a particular fish species for only a few weeks of the year. However, that window often corresponds to the temperature appropriate for spawning that particular species if the brooders have been held in static water ponds on the hatchery.

Groundwater is relatively stable temperature-wise and can be heated to provide the desired temperature needed for egg incubation of the particular fish being spawned. However, heating the water can produce gas bubble issues as discussed in Chapter 2.1. Where gas supersaturation is an issue the water should be agitated after being heated but before contact with the eggs or larvae. Aeration of anoxic groundwater may result in precipitation of minerals such as iron or manganese (See Chapter 2.1).

The water temperature of surface water can cycle diurnally 5°C or more, with more significant changes occurring over a few days as related to weather changes. A stable water temperature of the right temperature will help make egg incubation and larval rearing more successful and predictable. To stabilize water temperature a steady flow of surface water can be added to an insulated indoor water reservoir linked with a heating/cooling unit to provide a flow-through supply of water for. Another approach is to use a mixing valve to blend cool groundwater with a smaller volume of heated ground water to achieve a stable desired water temperature.

Eggs can be incubated up until hatching approaches on recirculating systems equipped with biological filters. During incubation ammonia level may build due to dead eggs decomposing but as the eggs begin to hatch the ammonia level in the water rapidly rises, faster than a biological filter can convert it. At that point a clean source of water at the same temperature as the eggs are held needs to be flushed through the system to dilute the ammonia concentration. Larvae have a lower tolerance to ammonia than un-hatched embryos making it more difficult to hold recently hatched larvae on recirculating systems.

Surface water will have a variety of organisms that can be harmful to fish eggs and larvae. These might range from pathogenic bacteria to predaceous copepods. Sand filters can remove the larger organisms while UV filters may be necessary to reduce bacterial problems (Chapter 17.17). The turbidity of surface water can vary daily and such turbidity should be filtered out prior to its use in the hatchery. Materials contributing to the turbidity, if not filtered out, can settle out, suffocating fish eggs or larvae, clogging supply line valves, and also reduce the efficacy of any UV filter. Such problems can develop rapidly and filters need to be checked frequently to ensure they are not clogged and are functioning properly.

5.3. Incubation Equipment

There are a number of types of incubators and variations thereof. The type of incubator to use depends on the nature of the eggs and how they are obtained. Non-adhesive sinking eggs and adhesive sinking eggs that have or have not been treated to eliminate the adhesiveness can be incubated as one layer spread evenly over a horizontal screen or netting with a water flow coming up through, flowing down through the eggs, or across the eggs. A traditional incubator used with trout and salmon is a Heath tray incubator (Figure 5.1). This incubator has a series of trays set one above the next. Each tray is designed to hold a layer of eggs set over a screen, with water flowing up through the eggs, then draining out and down to the next tray. This type of incubator has been used successfully with several warmwater fish species as well. Its main disadvantage is the limited quantities of eggs that can be incubated per incubator space.

A more rustic approach is to make a square or rectangular wooden frame where window screen is tautly attached. The frame should be held horizontally in a tank of water with a current of water across the eggs. Eggs should be scattered uniformly over the mesh. Adhesive eggs should be scattered over the screen within a few minutes post-fertilization (Figure 5.2). Adhesive eggs can also be incubated by scattering the eggs over fibrous material, and that material suspended in a tank or a fine-mesh hapa.

Perhaps the most common incubator used in a fish hatchery is a MacDonald jar (Figures 5.3. and 5.4). It is suitable for sinking or semi-buoyant non-adhesive eggs. The jar has a hemispherical bottom with cylindrical sides made of clear plastic, and a channeled lip at the top for water discharge. Typical volumes range between 6.5 and 7.5 L. A tube, having a fluted base, located in the center of the jar, extends down from above the lid of the jar to the center of the hemispherical bottom. Water flows via a flexible hose down through the tube, spreading out at the bottom of the jar, and then flows up and out the jar. Adding non-adhesive sinking or semi-buoyant eggs to the jar will result in an up-flow of water through the eggs. A screen at the top of the jar prevents eggs from floating out. The flow rate needs to be adjustable to

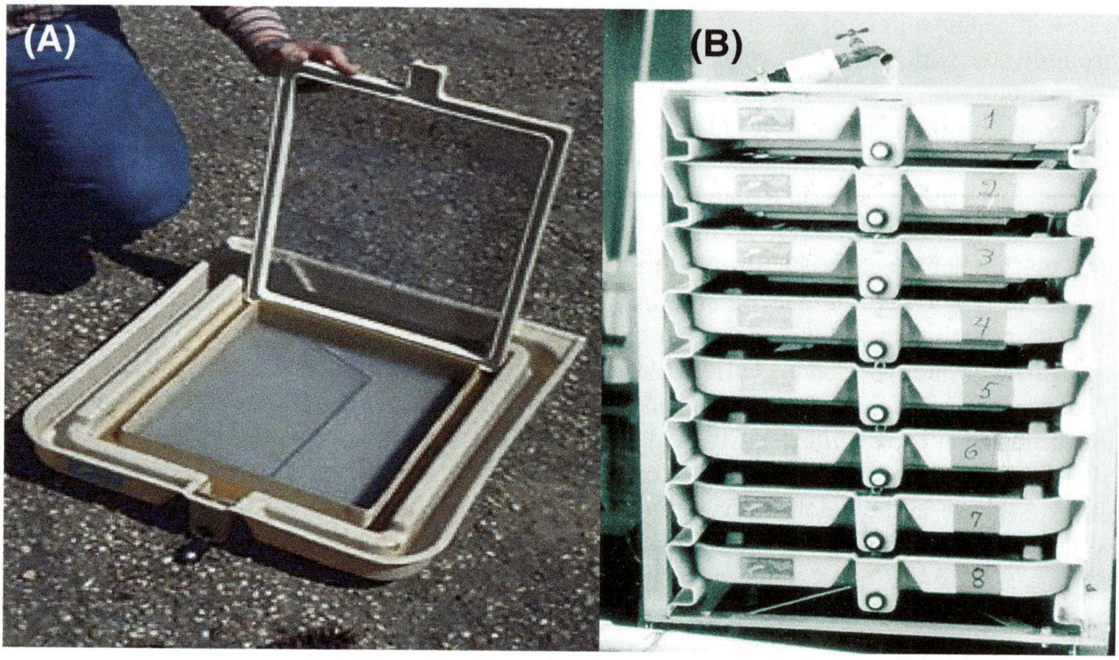

Figure 5.1. Heath tray incubator, often referred to as a vertical flow incubator shown as (a) a single tray and (b) a set of trays where the water flows down through the series of trays.

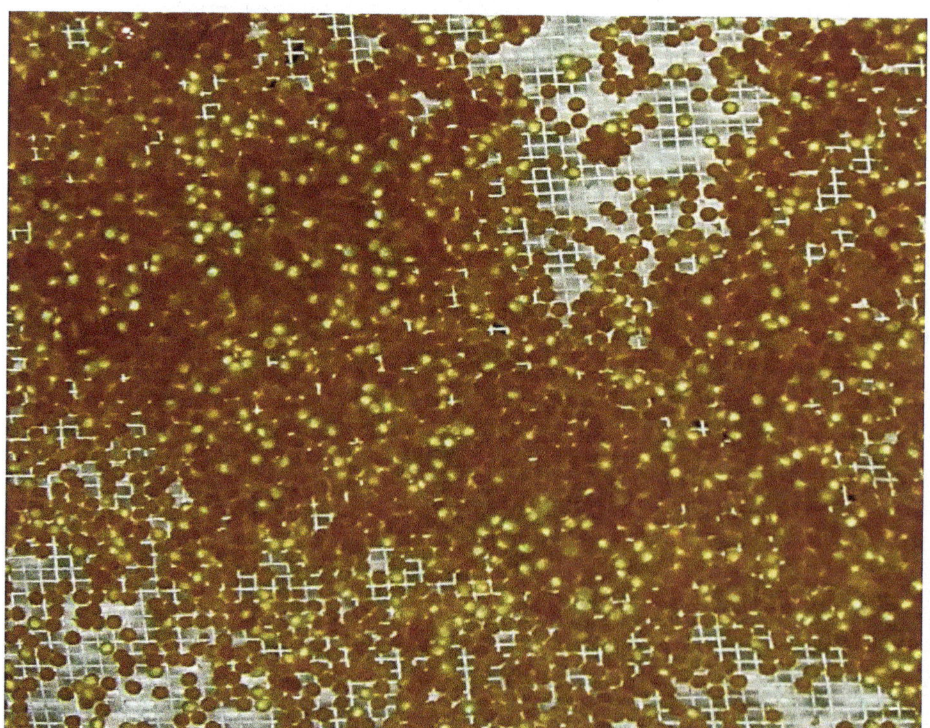

Figure 5.2. Adhesive *Clarias* eggs attached to a screen for incubation. Brown eggs are alive and white eggs are dead.

Figure 5.3. A battery of MacDonald hatching jars each with its own water supply.

Figure 5.4. A MacDonald jar loaded with Striped Bass *Morone saxatilis* eggs, effluent of jar flows into an aquaria to collect the larvae.

accommodate eggs of different weight. A flow rate of <0.5 L/min is needed for semi-buoyant eggs while large sinking eggs such as that of Paddlefish need a flow of >1 L/min. The initial flow rate of water should be just enough to give the eggs a slow gentle roll, increasing the flow rate as hatching approaches to help meet the oxygen demand of the developing embryo.

A cone-shaped incubator is suitable for sinking or semi-buoyant non-adhesive eggs, and even floating eggs if the water exchange is slow enough. The entire incubator may be conical in shape or have a conical base attached to a cylinder (Figures 5.6) In either case the water enters from the bottom of the cone creating a plume of water toward the surface (Figure 5.5). The slope of the cone should be 45° or more to prevent eggs from settling on the slopes of the cone and not being able to roll to the bottom where they can be re-suspended in the water column. Most cone-shaped incubators are hard-sided made from fiberglass or plastic, however, the upper sides can be made of cloth if the incubator is to be suspended in water (Figure 5.6.). Some hard-sided cones may have a screen inside the cone to prevent egg loss in the effluent; taller cones may not need screens if the water flow rate can be controlled to prevent egg loss. Cloth cones (Figure 5.7) should be of a large enough mesh to allow lateral water flow but not allowing egg or larvae to pass through. A cloth cone should be set with enough freeboard so if the mesh of the incubator in the water clogs and the water level rises there is new clean mesh that will allow water flow.

Each incubator design has advantages and disadvantages. The disadvantage of a standard size MacDonald jar is that one jar is often not large enough to hold all the eggs of one fish. A fish that has semi-buoyant eggs may require 10–20 MacDonald jars to accommodate the eggs from a 5–10 kg female. The more incubators needed per female means more labor and

Figure 5.5. Zug jar incubation where water supply enters from the bottom.

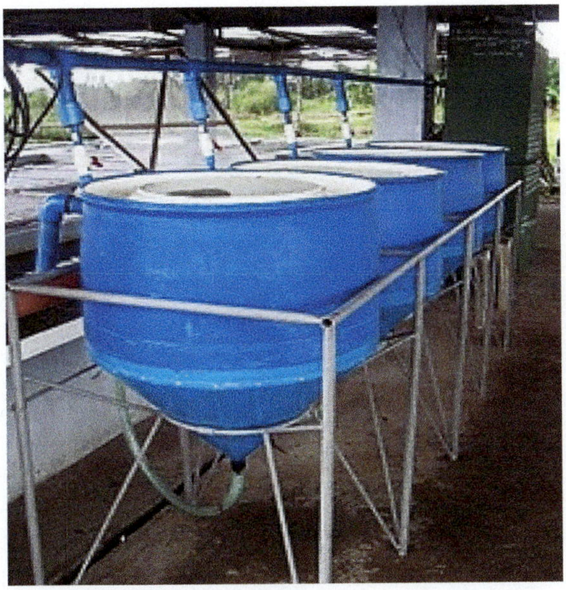

Figure 5.6. Fiberglass cones for egg incubation.

more physical space devoted to egg incubation but it does split the risks of a poor hatch or management issues among several jars. MacDonald jars and other designs where the water flows down through an open tube then up through the eggs has the advantage in that if there is a break in the water supply line, the eggs will still be covered in water in the jar. However, if the portion of the supply line is submerged in water located at a lower elevation than the incubator and a break in that portion of the supply line, eggs may be siphoned out of the incubator. In hard-sided cones where the water comes into the cone from the bottom, a break in the water supply line may drain the incubator often with eggs flowing into the supply line. When the water supply is surface water, eggs in cloth cones set in a large tank of water may be less stressed due to diurnal temperature variations than eggs incubated in MacDonald jars or hard-sided cones. A paddlewheel trough is a more specialized incubator commonly used for Channel Catfish eggs. This species deposited its eggs in layers making a mass held together in a gelatinous matrix. The male incubates the eggs by fanning the eggs causing current of water around the egg mass bringing in oxygen-bearing water while washing away any sediment accumulation. At Channel Catfish hatcheries egg masses are collected from the ponds, and the eggs incubated in paddlewheel troughs (Figure 5.8) that simulate this action. A paddlewheel trough consists of a rectangular trough approximately 0.5 m wide and 20–25 cm deep, and commonly 2.5–3 m long. A shaft that has blades attached every 25–30 cm is centered down the length of the trough. The blades extend down into the water to about half the water's depth. The shaft is rotated at 30 rpm and the blades create a current that pulsate the egg masses held in baskets located between the blades (See Chapter 13.1 for more details).

5.4. Egg Management

When fish are induced spawned and the eggs manually fertilized, the eggs can be loaded into incubators within a few minutes after they were fertilized. However, at that point the eggs are delicate as they are undergoing the first cell divisions, and they also are undergoing water-

Figure 5.7. Cloth cone incubator suspended in a tank of water with the water supply for the cone entering at the tip of the cone.

Figure 5.8. Paddlewheel troughs for incubation of Channel Catfish *Ictalurus punctatus* eggs.

hardening. During water hardening the eggs are expanding in diameter as they absorb water until they reach an osmotic equilibrium with the surrounding water. Common Carp *Cyprinus carpio* produce sinking eggs with a diameter of 1.0–1.5 mm prior to water-hardening but reach 2.0–2.5 mm post-water hardening. Grass Carp *Ctenopharyngodon idella* have semi-buoyant eggs with a diameter of 0.9–1.2 mm prior to water-hardening and reach 3.7–5.3 mm post water-hardening. How much egg diameters increase is related to the total hardness of the water that the eggs are incubated in. The lower the total hardness, more water is absorbed into the eggs until an osmotic balance is reached resulting in a greater egg diameter. In low water hardness setting around 20 mg/L, Grass Carp eggs greatly expand, stretching the egg shell to the point that the eggs are much more delicate and subject to being ruptured. In such cases, egg loading rates need to be reduced and water flow rates restricted and controlled carefully, otherwise eggs may be damaged and washed out of the incubator.

Management of water flow rate is critical for successful egg incubation and a valve or clamp on the water supply to each incubator that will allow small adjustments in water flow as needed. When a small pressurized water line is used to supply a series of incubators changing the flow rate to one incubator may accidentally change the flow in other incubators. This can be avoided by using a "header tank." A header tank is a non-pressurized water storage tank located above the incubators. Individually-valved supply lines to the incubators are connected to the header tank to provide gravity-fed water. Typically, the header tank receives a continuous flow of water so the header tank has a constant volume with any excess draining out the standpipe. This helps to ensure a uniform head-pressure and flow to each incubator.

A common problem with supplying water to egg incubators and larval rearing tanks, particularly when the water supply is surface water, is sediment and other material collecting on the intake side of the value or clamp supplying water to the incubator, resulting in a reduced flow rate. Filtering the water supply will help control the sediment but some accumulation behind the valve should be expected. Snails clogging the valves are a common problem when using header tanks or recirculating systems. Checking flow rates and flushing the line connected to an incubator every 3–12 h should be part of the hatchery management routine.

Dead eggs are to be expected when incubating eggs in the hatchery. Unfertilized eggs will be apparent within a few hours after the eggs are loaded into the incubator. Another period when mortality can be high is when the eggs undergo gastrulation. Dead eggs will contribute to reduced water quality, and increased risks of disease problems, and should be removed. Dead eggs are often lighter than live eggs and will float higher in the water column (Figures 5.9 and 5.10). In MacDonald jars and other incubators of similar shape, dead eggs can be removed from the incubator by reducing the water flow rate, allowing the eggs to layer out, then slowly siphoning off the dead eggs which form the upper layer, then restoring the water flow back to normal.

Fungus (typically Saprolegnia) is a common problem with fish egg incubation. When eggs are collected from an outdoor setting, you must assume that the eggs have been exposed to fungus and must be treated to control fungus before they are introduced into the hatchery. Chemical baths with hydrogen peroxide (750–1000 mg/L, 15 min), iodine (100 mg/L, 10 min), formalin (100–250 mg/L, 15–30 min) or potassium permanganate (10 mg/L, 5–10 min) are used. When fish are spawned in the hatchery, recently fertilized eggs are cleaner and begin with less fungus problems. If the eggs will hatch within 48 h post-fertilization, fungal growth is slow enough that few eggs will be damaged in that time period and such eggs do not need to be treated.

Figure 5.9. Whitish-colored dead eggs floating higher in an incubator than live eggs.

Figure 5.10. Dead eggs close-up.

Eggs from species that take longer than 48 h to hatch need to be treated within the first 24 h post-fertilization. Typically, treatments need to be applied at least twice a day and continued until the embryos reach the eyed stage. Treatments can be given as a bath or as a flush using hydrogen peroxide (750–1000 mg/L, 15 min), or formalin (100–250 mg/L, 30 min). Bath treatment of eggs can be given when eggs are attached to a substrate and the substrate can be moved in and out of the treatment solution, or for eggs held in troughs where the water flow is stopped for the treatment period then the chemical rapidly flushed out. For MacDonald jars and similar incubators, bath treatments can be given by moving the jars with eggs to a small recirculating system where the chemical can be added to the water and pumped through the jars for the treatment period. In some cases, it may be possible to temporally re-plumb the existing incubation system into a recirculating system for when treatments need to be applied. Flush treatments can be used in continuous flow systems using higher dose rates and more frequent applications. Keep in mind that both formalin and hydrogen peroxide can kill non-target organisms. When eggs are being treated the effluent should be diluted at least 75X as it is discharged.

Other than removing dead eggs and giving prophylactic treatments, eggs need little care except making sure water flow rates are adequate to provide ample oxygen, not being too slow causing eggs to pile up and smother; or being too strong, harming the eggs or washing them out of the incubator.

5.5. Hatching

Hatching is not an instantaneous occurrence; eggs from the same female may differ 6–12 h or more as to when they start to hatch and finish hatching. In the case of species that are induced to spawn where large quantities of eggs are fertilized in a short period of time, and incubated under similar conditions, there is more uniformity in the time to hatch. In the case of Chinese carp eggs incubated in MacDonald jars, once hatching starts the rate of hatching can be sped up by temporarily reducing the water flow rate for a minute or two, which creates a mild oxygen stress. The larvae within the eggs can be seen to become more active and start breaking out of the egg shell sooner. As soon as this is seen, the previous water flow rate through the jar is restored, and the larvae as well as shell debris are washed out of the incubator. The egg shell also can be weakened using a 0.1 % protease solution making it easier for the larvae to break through the eggs shell. Narrowing the window when hatching will occur will result in a pulse of egg shell debris, making it more efficient for hatchery staff to keep things clean. When hatching should begin can be anticipated by periodically sampling the eggs to see how fast the embryos have developed from one point in time to another. Careful record keeping will allow the development of a degree-hour response (= water temperature × hours to hatch) that can help predict hatch time with a range of water temperatures. Prior to hatching any dead eggs present should be removed. During hatching, egg shells and other wastes must be removed out of the incubator and the container where larvae are held, as such organic material contributes to a spike in ammonia during hatching.

Larvae can be removed from an incubator in several ways. Larvae of species with large yolk sacs are often not developed enough and weigh too much to swim up into the water column soon after hatching. Larvae of Paddlefish and Tilapia, which have large yolk sacs, may remain several days in MacDonald jar-like incubators after hatching. They need time to have adsorbed enough of their yolk sac and have developed enough, so they can swim up into the water column, and be carried out in the effluent from the incubator. Larvae of Chinese carps are much lighter and depending on the water current will be washed out of the hatching jar soon after hatching. It is not uncommon to find some larvae remaining in the incubator after the others have left the incubator. These larvae are often of poor quantity, are not strong enough to swim up into the water column, and should be discarded.

When adhesive eggs are incubated on a substrate the larvae may remain attached to the substrate or the slides of the holding tank for a day or so then swim up, or drop to the bottom of the holding tank after hatch. For species where the sac fry drop to the bottom of the tank, good quality sac fry will wiggle to the edges of the holding tank and can be siphoned out of the tank. The sac fry are then rinsed free of debris, counted based on number/unit volume, and transferred to a clean tank where they remain until the larvae have developed mouthparts and swim up into the water column. The poor quality larvae will drop off the spawning substrate and remain on the tank bottom near where they fell and should be siphoned out and discarded.

5.6 Larval Management Planning Considerations

How long to hold the larvae in the hatchery should take into consideration the species characteristics, the resources available and how the fish are to be used once they leave the hatchery. There are several alternatives including: holding the larvae until mouthparts are functional then stocking the larvae into nursery ponds; hold the larvae for several days after mouthparts are functional and intestinal tract is complete, feeding an appropriate feed, then stock nursery ponds; or holding the larvae in hatchery until the larvae have developed enough to take on juvenile characteristics, feeding appropriate feeds during this transition, then restocking the fish at lower densities into cages, tanks or ponds.

Deciding what alternative to take is best based on what alternative has worked best at that hatchery in the past. Most hatcheries produce several species of fish and for each species several batches may be produced to meet the production goals. How to optimally use the resources available can be a challenge. By keeping good records, using degree-hour/degree-day responses, and monitoring water temperature, it is possible with many fish species to predict when a female should ovulate, when eggs should hatch and when larvae are ready to feed. This will help identify the lead-time needed to have the appropriate hatchery facilities available, suitable food organisms produced, and nursery facilities prepared. The window of time between hatching and first feeding is also a good time to ship larvae from one hatchery to another.

Monitoring larval development is needed to confirm when mouthparts will be functional and the larvae are ready to feed. At hatch fish larvae, often called sac fry, are not fully formed and are not ready to feed. They rely on the yolk sac to provide the energy and building blocks for growth and development. How long the larvae rely on the yolk sac depends on the species of fish and the water temperature. In general, when comparing egg size effects, smaller eggs yield smaller larvae at hatch with less nutrient reserves available when the larvae is ready to feed.

A key to larval survival is to have an appropriate food available when the larvae have functional mouthparts and a complete digestive tract. How long larvae survive depends on how much of the yolk sac remains when the larvae are ready to feed and how much time passes before the larvae find an acceptable food. Many larvae have little yolk reserves remaining by the time they are ready to feed and will have begun to starve to death within 72 h. How fast yolk reserves are used and the larvae develop is water temperature dependent. In the case of Largemouth Bass larvae, they will swim-up and be ready to feed 5 days sooner when held at 23°C vs. when held at 17°C.

Providing the first food can be approached in several ways. One approach is to prepare an earthen pond to have an abundance of natural foods, and stock larvae into the pond as soon as they are ready to feed. Another approach is to hold the larvae in the hatchery and feed them an appropriate feed (typically a live food vs. a formulated feed) until they reach a life stage that is hardier before being stocked elsewhere. These approaches are discussed in more detail in the following chapters.

5.7 Recommended Reading

Aral, F., E. Şahınöz, and Z. Doğu. 2011. Embryonic and larval development of freshwater fish *in* F. Aral, editor. Recent advances in fish farms. InTech Europe, Rijeka, Croatia. Available: https://cdn.intechopen.com/pdfs/24074/InTech-Embryonic_and_larval_development_of_freshwater_fish.pdf. (May 2020).

Chapman, D. C., W. A. Hubert, and U. T. Jackson. 1988. Influence of access to air and of salinity on gas bladder in Striped Bass, The Progressive Fish-Culturist 50(1):23–27.

Conceicao, L. E. C., M. Yufera, P. Makridis, S. Morais, and M. T. Dinis. 2010. Live feeds for early stages of fish rearing. Aquaculture Research 41:613–640.

Dhert, P. and P. Sorgeloos.1994. Live feeds in aquaculture. Pages 99–109 *in* Aquaculture Towards the 21st Century: Proceedings of INFOFISH-AQUATECH '94, International Conference on Aquaculture. 29–31 Aug. 1994. Colombo, Sri Lanka

Dhont, J., K. Dierckens, J. Støttrup, G. Van Stappen, M. Wille, and P. Sorgeloos. 2013. Chapter 5, Rotifers, Artemia and copepods as live feeds for fish larvae in aquaculture. *In* G. Allan, and G. Burnell, editors. Advances in aquaculture hatchery technology. Woodhead Publishing Series in Food Science, Technology and Nutrition 242.

Friedmann, B. R., and K. M. Shutty. 1999. Effect of timing of oil film removal and first feeding on swim bladder inflation success among intensively cultured Striped Bass larvae. North American Journal of Aquaculture 61(1):43–46.

Govoni, J., editor. 2004. The development of form and function in fishes and the question of larval adaptation. American Fisheries Society, Symposium 40, American Fisheries Society, Bethesda, Maryland.

Heidinger, R. C., R. C. Brooks, D. Leitner, and I. Soderstrom.1997. Prediction of Walleye egg and embryo survival at two stages of development. The Progressive Fish-Culturist 59:64–67.

Park, J. M., S. J. Mun, H. S. Yim, and K. H. Han. 2017. Egg development and larvae and juveniles morphology of Carp, *Cyprinus carpio* in Korean. Development and Reproduction 21:287–295. Available: https://www.ncbi.nlm.nih.gov/pmc/articles/PMC5651695/. (May 2020).

Shields, R. J., N. P. Brown, and N. R. Bromage. 1997. Blastomere morphology as a predictive measure of fish egg viability. Aquaculture 155:1–12.

Woolley, L. D., and J. G. Qin. 2010. Swimbladder inflation and its implication to the culture of marine finfish larvae. Reviews in Aquaculture 2(4):181–190.

For additional references see Chapters 17.14. Live Food Production and 17.5. Counting Techniques.

Chapter 6

Larval Management in the Hatchery

6.1 Post-Hatch Management

After hatching, larvae should be held in aquaria, tanks or hapas, until the larvae have developed enough to be able to swim horizontally and have functional mouthparts. A continuous water flow should be provided when using aquaria or tanks to help maintain water quality and wash wastes away. Aquaria have the advantage in that the larvae are easier to observe. Things to observe include: how the larvae are being carried in the water currents, how fast the larvae are developing, and the swimming pattern of the larvae.

Larvae with large yolk sacs typically will sink to the bottom of the tank and initially cannot swim up into the water column but are able to wiggle along the bottom. This may result in larvae piling up in the corners of the tank and suffocation of some larvae. Semi-buoyant larvae need enough water current to keep the larvae up in the water column, otherwise they may settle on the bottom, get covered with sediment, or pile up in corners of the tank and suffocate. Placing air stones in the tank corners and providing enough aeration to create a water current to wash the larvae out of the corners can control this. Water currents created as part of the water exchange or through aeration can keep semi-buoyant larvae suspended in the water column. Semi-buoyant larvae will initially try to swim up in the water column then drift back toward the bottom. The deeper the tank the easier it is to keep semi-buoyant larvae suspended in the water column. The current pattern should wash any larvae off the bottom of the tank and up into the water column, keeping them from piling up in corners. The current pattern should not be too strong, causing the larvae to expend too much energy swimming against it. The more energy spent swimming, the less energy is available for larval development.

The larval holding tank should have a screen with enough screen area to minimize water pressure on the screen and reduce impingement. The mesh should be large enough to allow a good water flow and not clog easily but not allowing larvae to pass through. The best mesh size will vary with the fish being cultured. Cylindrical screens placed over a standpipe are effective especially when an air diffuser ring is attached to the base of the cylinder so air bubbles will rise up along the screen surface helping to reduce impingement of fish larvae and debris on the screen. In addition, screens should be cleaned at least twice daily to remove any debris accumulation. One of the most common problems during the larval holding phase is screens getting clogged resulting in tanks overflowing and larvae being lost.

Floating screen boxes or hapas can be suitable holding containers for fry when suspended in concrete or fiberglass tanks normally used for holding larger fish (Figure 6.1). The screen of the holding box and the mesh of the hapa will foul and limit water exchange. Screens need to be cleaned daily and the sides of the hapas scrubbed daily. Aeration should be provided inside

Figure 6.1. Hapa used for hatching of adhesive eggs or holding fry.

the container, but care needs to be given not to create currents that cause eddies and larvae to settle out or pile up on each other. Dissolved oxygen should be monitored frequently both inside and outside the larval holding container as the mesh may be clogged, resulting in little water exchange and a 2–3 mg/L lower dissolved oxygen inside the container than outside the container.

Larvae are more sensitive to ammonia than eggs and efforts should be made to keep ammonia concentrations as close to zero as possible. This begins with keeping the holding tank as clean as possible. Sides of the tank need to be cleaned daily to remove any organic film. The tank bottoms should be observed daily and any debris accumulation siphoned out. Water flow rates should be checked several times a day and adjustments made as needed.

6.2 Harvesting and Counting Larvae

During the period of yolk sac absorption the larvae are increasing in size and metabolic needs. Their density/L needs to be reduced if the fish are to be fed in the hatchery for any length of time. If the larvae are to be stocked directly into primary nursery ponds then the fish need to be stocked as soon as the fish are able to swim horizontally and mouthparts are functional. Soon after hatch, larval fish are often delicate and hard to handle safely and should be moved in water. The majority of fish have larva that are too small at the time of 1st feeding to be handled out of water. They are best harvested by dipping volumes of water from the larval holding tank or by siphoning them into a new container where the larvae can be temporally concentrated, enumerated, and then transferred to a new tank in the hatchery or to a primary nursery pond. The container used to concentrate the larvae in should be screened with a fine enough mesh material that will not allow larval escapement or impingement but will allow good water flow from the container. The container needs to be adequately aerated to keep oxygen levels near saturation, keeping the larvae suspended in the water column but not damaging the larvae.

Once the larvae are concentrated in the container they can be counted based on the number of larvae/unit volume of the container (Figure 6.2). The larvae need to be evenly distributed in the water column and 8 to 10 samples taken and counted. The volume of each sample should be large enough to contain 20 to 30 larvae. The samples should be taken at different depths and locations within the holding container. Often the larvae are not evenly distributed, contributing to variability among the samples, and to eliminate outliers the highest and lowest sample counts are discarded when calculating the mean number of larvae/unit volume. Small beakers of 100 to 250 mL are convenient sample containers. A sample of a known volume is poured into a petri plate in small aliquots, each containing less than 6 to 8 larvae/ aliquot, and the aliquot counted. Once the total volume of the sample has been counted then the sum of larvae from each aliquot is determined. The larvae are often very clear and best counted over a dark background. The total count of larvae in the holding container would be calculated based on the following calculation:

(number of larvae/ unit sample volume) × (holding container volume, expressed in the same units as sample volume).

When larvae are to be added to shipping containers or stocked into small nursery units (<100 m^2) then the appropriate volume containing the calculated quantity of larvae would be added to a transport container. When larger quantities of larvae are needed for stocking, then one or more holding containers would be counted and the appropriate volume added across several transport containers to obtain the quantity of larvae needed. For other counting techniques appropriate for eggs, larvae and larger fish see Chapter 17.5.

Figure 6.2. Volumetric determination of the number of Largemouth Bass *Micropterus salmoides* sac fry.

6.3 Management Alternatives

After hatching, larval fish initially depend on the yolk sac for growth and development. During that time mouthparts become functional and the larvae make the transition from endogenous to exogenous feeding. How long the larvae depend on endogenous nutrients depends on the species and the water temperature at which they are held. There is a transition period after mouthpart formation where the larvae are capable of exogenous feeding but are still dependent on endogenous nutrient reserves. This is a critical period in larval survival. Formulated feeds are not appropriate as a first feed for most fish larvae as their digestive tract needs more time to develop. In most cases, live natural foods of the proper size, at adequate densities and of sufficient nutrient quality must be available to ensure larval survival. These food organisms can be provided by stocking fish larvae soon after mouthparts are functional into properly prepared ponds (Chapter 7) or holding the larvae in the hatchery and feeding them selected types of natural food organisms.

There are trade-offs as to which approach to use. Using primary nursery ponds is a less intensive approach where the larvae are allowed to feed when they choose to feed, selecting the type of food organisms that they prefer. Once the pond is prepared and the larvae are stocked less labor is needed to manage the fish. The key to success is providing proper pond conditions. However, there can be considerable variation from pond to pond in the types and abundance of food organisms. Moreover, weather conditions from year to year can be variable and impact results. Mortalities due to predation are also a significant issue. The smaller and younger the larvae stocked into nursery ponds the lower the survival as well as more variability in survival from pond to pond can be expected.

Rearing the larvae in the hatchery does not necessarily improve the survival of young larvae while in the hatchery, but by rearing the larvae to a larger size in the hatchery before they are stocked into an outdoor setting, a higher and more consistent pond survival can be expected. In the hatchery the larvae may be reared in a clear water or green water setting. Most freshwater fish larvae are cultured in a clear water setting where live food or a formulated feed is added with a frequent water exchange. Many marine fish larvae are reared in a green water setting where single-cell algae are added to help make food organisms suspended in the water column more visible, help maintain water quality, and help maintain the food quality of the live food added to the tank. Often water exchange is held to a minimum in the first week or two of larval culture as the fish larvae may not be strong swimmers and to avoid washing away the live food organism.

Larvae reared in the hatchery are completely dependent on the hatchery staff to provide a suitable culture environment. This includes providing a suitable feed of adequate nutrition that is digestible, is of a size that can be consumed, is being offered at the right density, and is offered frequently enough to keep food in the digestive tract. In addition, the food must have an appropriate movement in the water column to attract the attention of the larval fish to induce a feeding response. Larvae reared in the hatchery are held at higher densities and food is offered in greater quantities than larvae stocked into nursery ponds soon after mouthparts are functional. This requires that the hatchery staff pay more attention to water quality issues that come with the intensive rearing of larvae in the hatchery.

6.4 Use of Live Food Organisms for Fish Larvae

6.4.1 Artemia

When the larvae of most freshwater fishes are ready to start exogenous feeding, their mouth opening is large enough to consume recently hatched *Artemia nauplii*. *Artemia*, often called brine shrimp, are a filter-feeding crustacean found in hyper-saline conditions. Under favorable conditions *Artemia* will reproduce asexually but under unfavorable conditions they reproduce releasing an encysted embryo. The cysts can be collected, stored, and hatched, and the larvae (*nauplii*) used to feed fish larvae. The desiccated cysts are commercially available and can be easily hatched in the hatchery (See Chapter 17.14 Toolbox/ Live feeds).

There are several considerations in purchasing and using *Artemia*. There are several species of *Artemia* and many sub-populations distributed throughout the world differing in the size of the 1st instar *nauplii*. The size of the 1st instar nauplii is important in that for some fish larvae a recently hatched 1st instar nauplii is slightly too large to be consumed as a first food for the larvae of that fish species. There is enough variation in the size of 1st instar *nauplii* among sources of *Artemia* that by selecting a source of *Artemia* cysts that give smaller *nauplii*, it may be possible to begin feeding *Artemia* sooner to such fish larvae. In other cases, a smaller food organism is needed as a first food for a particular fish species but the fish larvae grow rapidly and are able to consume *Artemia nauplii* within a week after the start of first feeding. The ease of use and availability make *Artemia* one of the most commonly used live foods in a fish hatchery.

The various sources of *Artemia* cysts differ in hatching rate and as such may differ in their commercial price. When selecting a source of *Artemia nauplii* the price is a consideration but hatch rate is a factor that also should be considered. Sources should be evaluated based on the price/quantity of *nauplii* obtained.

The various sources of *Artemia* also differ in their nutritional profile and that may be a factor in fish larval survival. *Artemia* are not a rich source of essential n-3 highly unsaturated fatty acids (HUFA) and should be enriched to make them a more suitable food (See Chapter 17.14). However, this enrichment process delays when the nauplii can be fed to the fish larvae resulting in a larger *nauplii* that is too large to be fed as a first food. As such, the innate nutrient quality of the 1st instar *nauplii* is a factor in fish larval survival. There is enough variation in nutrient quality among *Artemia* sources that it is possible to obtain a strain of a smaller size 1st instar *nauplii* with higher nutrient profiles. In many cases it is more practical to begin with un-enriched 1st instar *nauplii* and shift to an enriched *Artemia nauplii* as soon as the fish larvae grown enough to consume larger meta-*nauplii*.

When feeding a live food to larval fish the quantity of food given is typically not based on feeding a percent body weight but is expressed as the number of food organisms/unit volume of the container where the larvae are held. In the case of *Artemia nauplii* this might range from 1 to 10 *nauplii*/mL of tank water volume. The goal when feeding a live food is to maintain enough food organisms in the water column so the fish larvae can feed at any time that they desire while spending a minimum amount of energy to find the food. The natural environment for *Artemia* is hypersaline water and as such *Artemia* will not survive for an extended period in freshwater. When feeding *Artemia nauplii* to freshwater fish larvae the *nauplii* will die within about 3 h and settle to the bottom. The dead *nauplii* need to be removed before they begin to decompose and reduce water quality. New *nauplii* need to be added approximately every 3 h to the larval rearing tank to restore the desired *nauplii* density/mL. This requires that several sets of *Artemia* cysts be set up to hatch over the course of the day so that sufficient *nauplii* are available every three hours. Any delay in harvesting the *nauplii* soon after hatching will result in the nauplii going through additional larval stages each one larger than the last, resulting in *Artemia* that are too large to be consumed by first feeding fish larvae (Figure 6.3). One approach to having small *nauplii* available without having to hatch as many batches of *Artemia*/day is to slow down the development of the *nauplii* once they hatch. This can be done by holding the recently hatched *nauplii* in aerated saltwater in a refrigerator where the lower temperature will slow the development rate. Such *nauplii* can be harvested as needed, rinsed in freshwater and added to the larval rearing tank.

6.4.2 Rotifers

There are over 2000 species in the phylum Rotifera but only a few in the genus *Brachionus* are used in aquaculture. *Brachionus* (Figure 6.4) are smaller food organisms, (100 to 340 μm) than *Artemia nauplii* (1st instar, 400 to 500+ μm), that can be produced in a hatchery as a first food for many marine fish larvae as well as several freshwater fishes. The use of rotifers in the hatchery is not as straight forward as the use of *Artemia*. Rotifers are filter-feeding organisms typically cultured feeding single-cell algae or yeast. A production unit is seeded with adult rotifers that are allowed to reproduce until the desired density is reached then all or a portion are harvested for feeding larval fish. The life cycle is short, reaching adulthood in 0.5 to 1.5 days. From there females can lay eggs approximately every four hours. Under favorable environmental conditions *Brachionus* will reproduce parthenogenetically (asexually) leading to rapid population expansion but under stressful conditions

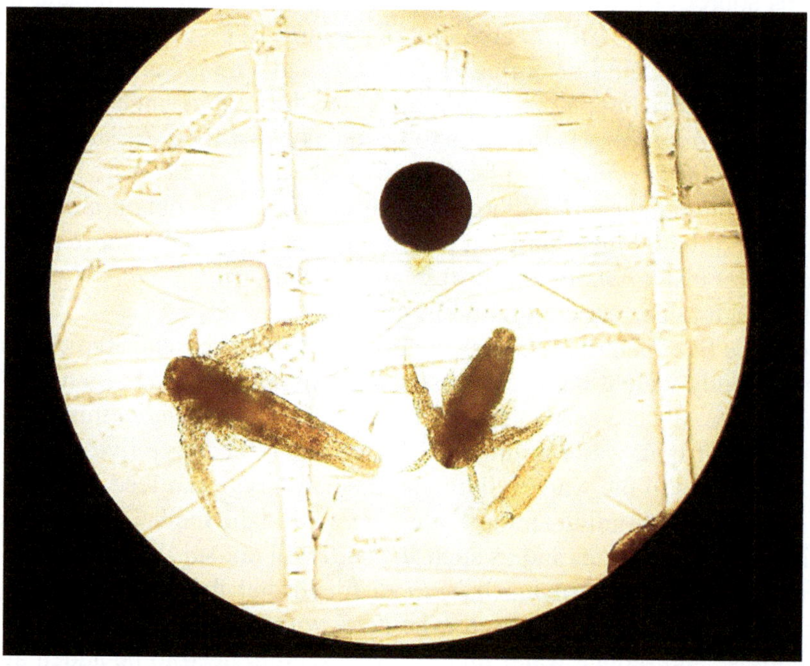

Figure 6.3. *Artemia* cyst, recently hatched *nauplii* and a 24-h *nauplii.*

Figure 6.4. *Brachionus* rotifers with eggs.

they reproduce sexually, producing an encysted resting egg. An indicator of population stress is when males start to become common and the number of eggs/female decreases. In the hatchery the goal is to keep the rotifer population expanding rapidly. To do so requires maintaining dissolved oxygen at 2 mg^{-L} or above and ammonia <1 mg^{-L}, while avoiding an excessive bacterial population that reduce water quality, as well as avoiding protozoans such as *Euplotes* that compete with the rotifers for food. Specifics on rotifer culture techniques are given in Chapter 17.14.

The nutrient quality of rotifers depends on what they most recently ate. The algae *Nannochloropsis* is commonly used to produce rotifers as it is easy to culture at high densities but it lacks in an essential fatty acid, docosahexaenoic acid (DHA), needed by rapidly developing fish larvae. One approach is to produce the rotifers on *Nannochloropsis* but once the rotifers are harvested their nutrient profile can be improved by allowing them to feed on a source rich in essential fatty acids before feeding the rotifers to the larval fish. This might be an alga such as *Isochrysis* that is more difficult to culture but richer in the essential fatty acids. There are also a variety of commercial products rich in HUFAs (Highly Unsaturated Fatty Acids) that can be used for rotifer enrichment (see Chapter 17.14).

There are several approaches to feeding rotifers. The most common method is to add enriched rotifers every 3 to 6 h to the larval rearing container at densities of up to 20/mL. Once the rotifers are added to the larval rearing tank, their nutritional value begins to decrease as the rotifers metabolize anything they have consumed. Rotifers periodically need to be removed from the larval rearing tank and new rotifers added. This can be done by slowly draining the tank halfway using a mesh size large enough for the rotifers to pass through while retaining the fish larvae, then refilling the tank and adding new rotifers. Another approach is to initially add the desired rotifer density and slowly drip in additional rotifers from a holding container having a high density of rotifers as well as *Isochrysis*. The addition of *Isochrysis* to the larval rearing tank provides food to help maintain the nutrient quality of the rotifers as well as helping to maintain a quality rearing environment for the fish larvae.

6.4.3 Other Live Food Alternatives

There are several other live foods that can be produced at the hatchery. Cladocerans, or water fleas, are small (0.2–6 mm) filter feeding aquatic crustaceans that naturally occur on most hatcheries. The most commonly cultured ones are in the genera *Daphnia* and *Moina*. They are larger than *Artemia nauplii* so are not used as a 1st feed for most larval fish but after a week or so the larval fish have grown to a size so they can consume *Daphnia* (1–5 mm) and *Moina* (0.7–1 mm). When feeding fish one of the objectives should be to offer the fish as large of a feed as they can swallow that way they maximize the nutrient intake for the amount of energy expended in finding the food. This holds true for live feeds as well as formulated feeds and is the contribution of *Daphnia* and *Moina*. They provide a larger feed size than *Artemia* when trying to produce a larger size fish in the hatchery building before they are stocked into nursery ponds.

Daphnia and *Moina* have similar life cycles. Under favorable environmental conditions they reproduce asexually but under unfavorable conditions reproduce sexually producing an encysted resting egg. *Daphnia* go through 4 to 5 molts after hatching and can reach maturity in two days at 25°C. *Moina* are able to reproduce within four to five days after hatch producing new broods every 1.5 to 2 days at 26°C. Overall, greater daily yields can be obtained with *Moina* (106–110 g/m^3) than *Daphnia* (25–40 g/m^3). Specifics on their culture are given in Chapter 17.14.

6.5 Environmental Settings for Indoor Larval Rearing

A significant consideration when feeding larval fish is light intensity and duration. Light intensity is commonly expressed in lux. Outdoor daylight values might range from 5,000 to 10,000 lux. Indoor lighting may range from 300 to 1000 lux. Appropriate light intensity for fish larva vary considerably by species and range from 10 to >1000 lux. The type of lighting provided varies from hatchery to hatchery and the preferences of the hatchery manager. Both incandescent and fluorescent lighting are common but LED lighting is becoming more popular as LED light intensity can easily be adjusted, a range of wavelengths can be provided, and are more energy efficient than incandescent and fluorescent lighting.

Larval fish are sight feeders and need light to help find food items, but too much light can wash out the color of the food items making the food difficult to find. Too high a light intensity can result in the fish larvae concentrating in the shadows of the rearing tank, increasing the competition for food in a limited area and the opportunity for cannibalism. Too little light makes food foraging more difficult. Lighting intensity is life-stage dependent. Some species do best at a low light intensity at early larval stages but better a higher light levels when they are older.

Tank color is another factor affecting appropriate light intensities. A darker colored tank will help provide more contrast to food organisms suspended in the water column. Adding single-cell algae to the rearing tank also helps to provide contrast making it easier for larval fish to see food organisms in the water column. Light colored tank walls and high light intensities make it harder for fish food organisms to be seen.

Photoperiod is also part of selecting an appropriate light intensity. Larval fish have short gut passage times and within a few hours the larvae are ready to consume more food. The longer the hours of light of an appropriate intensity, the more opportunity there is to feed. Extending the hours of light will require more food to be produced and more staff time to add the feed and keep the tanks clean. However, the more frequently the larvae feed, the more food is taken in, and greater growth is obtained often with improved survival. This better growth means the sooner the fish can be moved out of the hatchery.

6.6 Swim Bladder Inflation

Swim bladder inflation occurs soon after mouthparts are functional for most fish species. Most adult fish do not have a pneumatic duct connected to the swim bladder but do so for a short time as larvae. This duct attaches to the digestive tract and allows the larvae to gulp air at the water surface to inflate the swim bladder. Failing to inflate the swim bladder can be a common problem under hatchery conditions. This can result in more difficulty in swimming and capturing food, reduced growth and a reduced survival rate. Excessive surface tension can interfere with successful swim bladder inflation not allowing the fish to gulp air. Oil films that contribute to the surface tension in larval tanks can develop from decomposing egg shells and dead larvae, as well as oils associated with live feed enrichment. Oil films should be reduced several times a day using a surface skimmer or oil adsorbent pads. Placing an air stone just under the water surface will create a small spray of droplets that will break up the surface tension as the droplets fall back into the tank, resulting in an area where the larval fish can access the surface.

6.7 Considerations before Outdoor Stockings

How long the larvae are held in the hatchery depends on the hatchery goals and the facilities available. As discussed earlier, the fish can be moved as first feeding larvae, or held and grown to a larger size before being stocked elsewhere. Keep in mind that the larger the size of fish stocked into nursery ponds the greater the pond survival and the survival is more consistent from pond to pond but more effort is needed in the hatchery.

The longer the fish are held indoors in a nursery tank the greater the size variation within the population due in part to differences in feed consumption. There will be a portion of the fish that were not successfully weaned to accept a formulated feed while others quickly learned. In some species the slow learners can be graded out and pooled with other slow learners and retrained, otherwise such fish may be lost to cannibalism. Similarly, there may be some that outgrow their peers and should be graded out before the fish are to be stocked into outdoor ponds.

Before moving the fish to outdoor ponds, the ponds should be checked to see that the ponds are suitable to receive the fish. This includes checking the ponds for several days before the anticipated stocking to be sure that there are no chronic water quality problems. Dissolved oxygen values should not drop below 4 mg/L with no excessive supersaturations in the afternoons. The fish need to be acclimated before stocking to the new water quality conditions, particularly temperature. Ideally the hatchery building has two water supplies, the clean water source in which the larvae are reared and also a pond water source. The pond water source should be slowly added overnight to the larval rearing tanks before the fish are stocked into ponds the next morning. This will reduce the amount of acclimation time needed on the pond bank.

The zooplankton population in the receiving pond should be monitored for several days prior to fish stocking. For details on nursery pond preparation see Chapter 7. There should be expanding populations of zooplankton appropriate for the fish to be stocked. Minimum densities should be 500 to 750 rotifers or 100 to 200 cladocerans/L. Rotifers should reach these densities within 7 to 10 days of pond filling and cladocerans densities in 10 to 14 days depending on water temperature.

As part of the preparation to harvest nursery tanks in the hatchery the fish should be examined for parasites, particularly if they were reared in tanks supplied with surface water. Protozoan parasites can be common and should be treated with formalin before any fish are stocked into outdoor ponds where disease control can be more difficult.

6.8 Recommended Reading

See Chapter 17 and the following references for more information regarding larval management and feeding.

Allan, J. D. 1976. Life history patterns in zooplankton. American Naturalist 110:165–180.

Czesny, S. J., B. D. S. Graeb, and J. M. Dettmers. 2011. Ecological consequences of swim bladder noninflation for larval Yellow Perch. Transactions of the American Fisheries Society 134(4):1011–1020.

Davis, D. A., T. J. Derbes II, and M. E. Head. 2018. Culture of small zooplankton for the feeding of larval fish. SRAC Publication No. 0701. Southern Regional Aquaculture Center, Stoneville Mississippi. Available: https://www.ncrac.org/files/biblio/SRAC0701.pdf. (May 2020).

Dhert, P. 1996. Chapter 3. Rotifers, *in* P. Lavens and P. Sorgeloos, editors. Manual on the production and use of live feed for aquaculture. FAO Fisheries Technical Paper 361. Food and Agriculture Organization of the United Nations, Rome. Available: http://www.fao.org/3/w3732e/w3732e0c.htm#TopOfPage. (May 2020).

Doroshev, S. I., J. W. Cornacchia, and K. O. Hogan. 1981. Initial swim bladder inflation in the larvae of physoclistous fishes and its importance for larval culture. Pages 495–500 *in* R. Lasker and K. Sherman, editors. Second ICES Symposium, Woods Hole, 2–5 April, 1979, Copenhague, Danemark: Conseil international pour l'exploration.

Lavens, P. and P. Sorgeloos. 1996. Manual on the production and use of live feed for aquaculture. FAO Fisheries Technical Paper 361. Food and Agriculture Organization of the United Nations, Rome. Available:
http://www.fao.org/3/w3732e/w3732e01.htm#TopOfPage. (May 2020).

Rottmann, R. W., J. S. Graves, C. Watson, and R. P. E. Yanong. 2017. Culture techniques of *Moina*: the ideal *Daphnia* for feeding freshwater fish fry. UF/IFAS Extension Publication #CIR 1054. University of Florida, Gainesville. Available: http://edis.ifas.ufl.edu/fa024. (May 2020).

Sorgeloos, P., D. A. Bengtson, W. Decleir, and E. Jaspers, editors. 1987. *Artemia* research and its applications, 1987, Vol.3. Ecology, culture and use. Universal Press, Wetteren, Belgium.

Villamizar, N., H. Migaud, B. Blanco-Vivesa, and F. J. Sanchez Vazquez. 2011. Effects of light during early larval development of some aquacultured teleosts: a review. Aquaculture 315(1–2):86–94.

Watanabe, T., C. Kitajima, and S. Fujita. 1983. Nutritional values of live organisms used in Japan for mass propagation of fish: a review. Aquaculture 34(1–2):115–143.

Chapter 7

Extensive Pond Production: Nursery Pond Preparation

7.1 Introduction

Earthen ponds are used in the production of most warmwater fishes and several coolwater fishes cultured for food, bait or sport. Ponds are more appropriate than raceways for these species as ponds are more similar to natural habits of the fish, are a source of natural food, have lower construction costs and water requirements, allow simpler husbandry technology, and generally have lower production costs. Disadvantages include lack of direct control over both the fish and the environment and difficulty in monitoring the interactions of the fish with their environment. Ponds may be used as a natural spawning setting for species that will spawn in static water conditions. Ponds can be used for egg incubation, as well as larval and juvenile production. Larvae for stocking ponds can be obtained from induced or natural spawning in the hatchery, or from outdoor spawning ponds. The first 30 to 45 days in the life of the larvae are critical and is a point where significant mortality might be anticipated. During this period the fish make the transition from its larval food habits, feeding on zooplankton, to more adult food habits. Fish during this period can be managed in the hatchery, a combination of hatchery and pond production or ponds alone.

Ponds, if properly prepared can provide a very suitable culture environment for the first 30 to 45 days after the larvae have functional mouthparts and a complete digestive tract. Larval fish are zooplankton feeders, and ponds can provide rotifers, cladocerans and other suitable food items. However, there are limits in the quantities of zooplankton that can be produced and within 30 to 45 days post-stocking. The larval fish have grown, changed food preferences, and are consuming appropriate live food items faster than they can be replaced. Fish yields from such nursery ponds are often limited due to the amount of zooplankton that can be provided and are often less than 250 kg of fish/ha. The fish have developed from the delicate larvae at stocking to a much hardier juvenile fish that can be harvested safely from a pond. At that point the fish should be harvested and restocked at lower densities. This first pond culture period is called the primary nursery period. This first culture period is adequate to produce small fingerlings (26 to 60 mm) of many species suitable for distribution. However, larger size fingerlings are often needed and a second culture phase is needed (secondary nursery pond period).

The length of the secondary nursery phase depends on the size of fish needed for the market and the length of the growing season. In North America, many species of fish are spawned in the spring, nursed for a month or so in a primary nursery pond, then stocked into secondary nursery ponds and cultured into autumn when the water temperature falls and growth slows.

Densities stocked and yields obtained depend on how predacious the species is and whether it will accept a formulated feed. Some fish like hybrid Striped Bass are trained to accept a formulated feed in the primary nursery pond and thus are already feed trained when stocked into secondary nursery ponds. Others such as Largemouth Bass *Micropterus salmoides* are harvested from primary nursery ponds, feed trained in the hatchery, and then stocked into secondary nursery ponds for further growth (See Chapter 16.1.). Some species are produced in the same pond from fry to fingerling. This is the case with Channel Catfish *Ictalurus punctatus*, which are stocked as fry into ponds at low densities (<250,000/ha), are accepting formulated feed within 2 to 3 weeks of stocking and cultured in the same pond until the end of the growing season (See Chapter 13.1).

In some cases, a hatchery is called on to provide catchable size fish for sportfishing in locations that are being heavily fished or where there is limited natural production of catchable-size fish. To provide such fish, a third production cycle is needed. In temperate climates a second season of growth may be required where feed-trained fish are harvested from a secondary nursery pond and restocked at a lower density and fed a formulated feed to reach an acceptable size.

A successful hatchery production season depends on careful planning, including: what production technology will be used, what lead times are needed to prepare for each step, what quantity of fry will be needed for stocking into nursery ponds at specific points in time, and what growth and survivals should be expected. When using an extensive production approach a number of steps need to be taken before a nursery pond is ready to be stocked.

7.2 Pond Preparation

7.2.1 Drying

The first step in preparing a pond is drying it and making any necessary repairs. This allows any lateral seepage from adjacent ponds to be identified and corrected. Animals such as muskrats and crayfish can burrow in the pond levies weakening them to the point that the levies must be repaired before a pond can be safely used. All potholes and old nest sites in the pond bottom must be filled in and the slope of the pond bottom reset so that the pond drains smoothly into the catch basin. Depending on the soil type it may take three weeks or more to successfully dry the pond bottom. Clay soils should dry enough so that the soil cracks (Figure 7.1a) with no water left in former nest sites or the catch basin (Figure 7.1b). Drying for three months may be necessary if trying to sterilize the pond bottom. Some of the benefits of drying include:

1. Encystment of eggs and spores of food chain organisms so that hatching occurs when the pond is again flooded
2. Kills fish parasites
3. Oxygenates the bottom mud layer promoting growth of aerobic organisms and decomposition of organic matter in the sediments
4. Discourages regrowth of some aquatic weeds
5. Firms pond bottom making later harvesting easier
6. Eliminates wild fish living in potholes

Figure 7.1. Ponds should be dried thoroughly to the point where the clay soils cracks (a), with no water remaining in any nests or the catch basin (b).

Drying may be accelerated by tilling the pond bottom to a depth of 5 to 10 cm. This can be done with hand tools or by use of a mechanical spader, a disc harrow, or similar farm machinery. Sandy, rocky, or infertile soil beneath the surface layer of the pond bottom should not be disturbed by tilling as it might increase seepage rates and nutrient loss. Tilling should leave the bottom smooth and free of depressions. In deciding as to the length of time the pond remains dry, the benefits derived must be balanced against the unavailability of the pond for use during the fallow period. If the drying period is more than 3 to 4 weeks, terrestrial weeds may begin to grow and present problems. On occasions where portions of the pond, such as the catch basin or nest sites, are still holding water, fish larvae and predacious insects may survive in such pockets of water and must be eliminated prior to refilling the pond. This can be done by adding rotenone, quick lime, or chlorine at adequate dose rates to kill any organism in such pockets of water (Figure 7.2). When such wet areas are abundant it may be more practical to sterilize the entire pond as the pond is being refilled.

7.2.2 Sterilization

When chronic diseases or other related problems are associated with a pond a complete sterilization may be required. Drying the pond bottom may help sterilize the pond but the

Figure 7.2. Tilapia nests in a spawning pond, which were treated with quick lime to eliminate any fry or other potential pests before the pond is refilled.

use of quicklime, CaO, or hydrated lime, $Ca(OH)_2$, will be more effective than drying alone if enough lime is added to develop a pH of 11.5 or higher in the top layer of bottom mud. Liming for sterilization may be done either while the pond is drained or after refilling. Where these forms of lime are used, stocking should be delayed until the pH has dropped to a safe level (<9.5, species dependent) for fish food organisms and the fish themselves. Depending upon the characteristics of the bottom mud and the water supply this usually will take 10 to 25 days. The lower the alkalinity of the water the longer it takes for the pH to return to a safe level and up to 30 days or more may be required. The pH of the water should be checked in the afternoon for several days prior to the anticipated day of stocking fish to be sure that the pH is <9.5. The quantity of quicklime for sterilization may range from 400–605 kg/ha with the higher rates needed the higher the alkalinity of the water.

Sterilization can also be done with chlorine or potassium permanganate. These agents work best when the pond is filled with water. Calcium hypochlorite (bleach) is the best source of chlorine, adding enough to maintain a rate of 10 to 20 mg/L free chlorine for 24 to 48 hours. A similar rate of potassium permanganate is suggested. Both of these compounds are affected by the presence of organic matter. Where the level of organic matter is high additional sterilant is needed to establish the lethal level for unwanted organisms.

7.2.3 Liming

Alkalinity for productive pond water needs to be maintained above 25 mg/L to provide adequate buffering to reduce the magnitude of pH swings. Pond bottom soils, depending on nature of the water supply, soil type and cropping practices, should be analyzed every 1 to 3 years to determine their lime requirements. Samples should be collected from the top 15 cm of soil, approximately one sample/500–1000 m^2 of pond bottom. The samples should be dried, then the samples from a given pond combined to make a 500 mL pooled sample. The pooled samples should be sent to a laboratory capable of conducting a soil lime requirement analysis. If the soil test indicates a need for lime (pH lower than 6.5), lime can be applied to the dry pond bottom with an agricultural lime spreader. Supplying the recommended amount of lime will adjust the bottom soil pH to the optimum point (pH 6.5–8.5) for development of the fish food chain and physiological well-being of the fish crop. Usually liming for pH adjustment is done with ground limestone, $CaCO_3$, at 2,000 kg/ha depending on soil characteristics. Ground limestone is not toxic to plant or animal life and can be applied to the pond at any time with safety. It is quite insoluble and slow in its action. The full effect of applying a recommended dose does not develop until 6 to 12 months after application.

7.3 Fertilization

7.3.1 Role of Pond Fertilization

Nursery ponds must be fertilized to produce sufficient food organisms for the fish to be produced. The frequent draining and refilling of hatchery ponds does not allow for much nutrient accumulation in the ponds and a pond should be fertilized each time it is used. Proper fertilization can be expected to raise fish seed yields by 100% or more. Increased yields makes for better use of pond space and hatchery staff, and can improve the quality of seed produced. Harvesting conditions generally are better in a fertilized pond. The resultant phytoplankton bloom following pond fertilization will help shade out many aquatic weed problems that hinder seining, and movement of the fish to the catch basin when the pond is being drained. The natural food produced following pond fertilization is well suited for larval fish during the first weeks of development.

The first food of most fish larvae is some sort of zooplankton. Such zooplankton filter-feed on phytoplankton and bacteria. Pond fertilization provides the nutrients needed for increased phytoplankton, in turn resulting in increased zooplankton abundance. The process of primary production involves the action of green plants to convert water and carbon dioxide into plant material with the essential energy from the sun. Herbivores or saprophytic bacteria utilize the green plant material and are consumed in turn by organisms at a higher level in the food chain. Ideally the energy flow is from simple to complex with the target species being the final consumer in the chain. A measure of the effectiveness of the process is the difference between fish yields from unfertilized and fertilized ponds. Where little difference can be noted between the two yields, it may be that the fertilization protocol resulted in an inefficient pathway of energy flow and the types of fertilization material being used needs to be reviewed as well as the types of food organisms being consumed by the larval fish.

A variety of nutrient sources have been used in pond fertilization and are discussed in this chapter but there are many other substances that have possibilities and may perform as well or better in some situations. When evaluating materials to use as pond fertilizers they should be evaluated based on their nutrient composition, the cost/kg of nitrogen (N), phosphorus (P) and potassium (K) being provided, availability of the material, ease of transport and handling, any storage issues, and any post-application water quality issues.

7.3.2 Organic Fertilizers

Some of the earliest efforts to increase fish production in ponds involved the use of organic fertilizers. These materials included plant or animal wastes, green plants sometimes referred to as "fodder" or "green manure," and agricultural byproducts. Organic fertilizers are a source of carbon as well as N, P, and K. The release of these nutrients by the decomposition of the organic material helps to jump start the food chain relative to adding only inorganic fertilizers. Inorganic fertilizers provide N, P, and K for photosynthesis, and by increasing algae production contribute to organic carbon accumulation. A variety of aquatic organisms feed on the decomposing organic fertilizer. The associated bacteria are fed upon by zooplankton, and the zooplankton fed upon by larval fish.

Generally, organic products used for pond fertilization are low in the essential elements required for plant growth. Values for some of the materials are shown in Table 7.1. Agricultural byproducts such as oil seed meals (cottonseed meal, peanut meal) and animal waste meals (fish meal, meat and bone meal) are commonly used organic fertilizers. They are relatively high in nitrogen per unit weight, available as a meal or flour, low in moisture, and are easy to distribute and store. They are also important components of livestock and poultry rations. Consequently, the price and availability of a given type of material may vary from region to region and year to year.

Animal manures are organic fertilizers that can be used but are rather low in nitrogen and even lower in phosphorus and potassium. The quality of the manure depends on what the animal was feeding on and how much the manure has dried out and decomposed by the time it is added to the pond. In addition, foreign material such as straw, sawdust, and shavings may be mixed in with the manure lowering its quality as a fertilizer. Where the manure is exposed to rainfall during storage, leaching of soluble salts takes place. Storage for long periods may result in partial bacterial decomposition resulting in nitrogen being lost as a gas. Data given in Table 7.1 are general estimates of the nutrient content of those materials, and chemical analysis should be done to accurately know the nutrients available in a given lot of a source material.

In the southeastern USA, chicken litter from broiler rearing houses is generally available and has given good results as a pond fertilizer. Dairies and feedlot operations are locations where cattle manure is abundant. These and other animal manures can often be obtained at a fairly low cost. However, when the cost of labor to apply it is included, the cost/kg of applied nutrients can be quite high. In many cases when everything is considered, the cost/kg of applied nutrients from an oil seed meal may be cheaper to use when compared to animal manures. Use of bulk handling methods for manure handling can lower labor costs dramatically but there must be frequent usage of the required machinery to justify the high capital outlay required for such equipment.

Hay and other miscellaneous plant materials are often used as organic fertilizers. The decomposition of hay by bacteria gives a "hay infusion effect" resulting in a progression of protozoans, rotifers, cladocerans and copepods. Alfalfa and peanut hay have high nitrogen content and are often preferred when available (Table 7.1). Grass hay contains considerably less nitrogen. Where the carbon-nitrogen ratio in the material is wide (i.e. more than 17:1) the decomposition is slower and the inclusion of inorganic nitrogen along with the hay is desirable to give a C:N ratio in the range of 10:1 to 17:1. High C:N ratio fertilizers tend to favor crustacean zooplankton while those with lower ratios favor rotifer production.

Table 7.1. Amount of N-P-K Nutrients contained in various organic fertilizer materials (dry weight).

Source Material	Nitrogen (%)	Phosphorus (%)	Potash (%)
Alfalfa Hay	2.37	0.24	2.05
Grass Hay	1.12	0.21	1.20
Peanut Hay	1.62	0.13	1.25
Cottonseed meal *	5.54	0.83	1.22
Fish meal	10.22	2.67	0.40
Peanut meal	6.96	0.54	1.15
Rice Bran	2.37	1.70	1.49
Soybean meal	7.07	0.59	1.90
Cow Manure (fresh)	0.43	0.59	0.44
Chicken Manure (fresh)	1.31	0.40	0.54
Horse Manure (fresh)	0.49	0.26	0.48

* 36% protein

Growing a terrestrial plant in drained fish rearing ponds as a green manure has long been an accepted cultural practice in Europe and parts of Asia. Ponds are left fallow, gaining the benefits discussed earlier by pond drying. Once the pond is dry enough, it is planted with a crop such as vegetables, legumes, or cereal grains. The crop is then harvested and the unharvested vegetative portion left in the pond is flooded and allowed to decay releasing nutrients into the food chain. A better quality green manure can be obtained by planting a cereal grain, Italian ryegrass, or legumes such as vetch (*Vicia*) in the dry pond and allowed to grow several weeks and then flooded the entire biomass.

Growing a green manure crop in a fish pond removes the pond from fish production for a period of weeks or months. Any loss in fish production when a pond is taken out of production is offset by the improved condition of the pond bottom, recycling of adsorbed nutrients in the pond bottom soil, sterilization against disease organisms, and economical production of an excellent organic fertilizer. In temperate climates a green manure crop can be grown at a time when water temperatures are too cool for optimum fish growth. At the Marion Fish Hatchery, Marion Alabama, sowing Italian ryegrass seed a rate of 10 kg /0.4 ha in a dry pond in October or November would produce up to 2,500 to 3,000 kg (wet weight) of grass/ha after 4 to 5 months. At that time the pond is ready to be flooded for a spring crop of fish. Another approach appropriate for when working with egg scattering fishes is to lower the water level 20–30 cm of a pond holding brooders and plant a terrestrial grass along the exposed shoreline. Once the grass is established and the water temperature is appropriate for spawning of the fish held in the pond, the water level is raised flooding the grass. The grass provides a spawning substrate for the brooders. Once the brooders have spawned in the pond they can be removed and the pond will become a nursery pond taking advantage of the flooded grass as a fertilizer.

There are some disadvantages in using green manures. If the quantity of plant biomass produced is too great, as it decomposes it will create chronic low oxygen problems not acceptable for zooplankton or fish production, delaying when the pond can be used. If the vegetation has not decomposed sufficiently by the time of the fish harvest it will interfere with a successful harvest.

In tropical or sub-tropical climates where labor is often plentiful and cheap, the green manure crop can be grown on areas adjacent to the fish pond and transported to the pond as needed. This vegetation can be confined in one area of the pond to make a compose pile. For best results the plant material should be tender and succulent for rapid utilization by insects, bacteria, and zooplankton.

7.3.3 Inorganic Fertilizers

Although inorganic fertilizers may be cheaper/kg of nutrient and are easier to apply and store than organic fertilizers, they are rarely used as the only nutrient source for preparing primary nursery ponds. They are added to complement the nutrient profile of the organic fertilizer, facilitating its decomposition. However, in secondary nursery ponds inorganic fertilizers are often added to promote an initial phytoplankton growth. Nutrient release associated with feeding the fish in the secondary nursery pond will sustain the phytoplankton bloom. Relatively small amounts of inorganic fertilizer will produce an abundance of unicellular algae that in turn through photosynthesis provides oxygen and contributes organic carbon into the food chain. A major benefit of having an adequate abundance of phytoplankton is its shading effect, limiting light penetration to the pond bottom. Limiting light penetration to the pond bottom limits the growth of undesirable bottom dwelling algae and submerged weeds. However, keep in mind that excessive use of inorganic fertilizers can result in dense phytoplankton blooms, increased risk of plankton die-offs and low oxygen conditions.

An N-P-K ratio of 4-4-1 will provide the essential nutrients for many phytoplankton forms. In a pond setting this is equal to adding 9.1 kg nitrogen, 9.1 kg P_2O_5 and 2.27 kg K_2O/ha/application. These nutrients adjusted for percent active ingredient can be supplied/ha by 114 kg of a commercial grade of 8-8-2 granular fertilizer or 45.5 kg of a commercial 20-20-5 grade; or by on-farm mixing using 52.2 kg of ammonium nitrate (33.5% N), 45.5 kg superphosphate (20% P_2O_5) and 4.5 kg of muriate of potash (50% K_2O). Many inorganic source materials are available and have been used. Some of these are listed in Table 7.2 along with their nutrient levels.

A consideration when selecting an inorganic fertilizer is whether it is acid forming or not. It is better to use acid-forming materials in alkaline waters, and neutral and alkaline types of fertilizers in acid waters. Addition of organic fertilizers generally produces weakly acidic reactions. Pond soil should be testing for lime requirement every 3 to 5 years, and adding lime as needed will guard against acid conditions developing from use of an acid-forming fertilizer. Where the water supply has a high total hardness (> 150 ppm as $CaCO_3$) liming the pond bottom may not be necessary. Omission of potassium in the mixture is possible if the soil and water supply contain ample levels of this macronutrient.

Another consideration is the solubility of the source material providing the nutrient. Forms of phosphorus vary widely in solubility and some may be unsuitable for pond use because of low solubility. Two insoluble forms are rock phosphate (32% P_2O_5) and defluorinated rock phosphate (41% P_2O_5). Rhenania phosphate is not recommended for high calcium conditions and calcium metaphosphate also should only be used under acid and neutral conditions. In contrast, diammonium phosphate is completely water soluble, and a major portion of triple superphosphate (TSP) is water soluble. Inorganic nitrogen sources are highly soluble. Potassium sources also are quite soluble.

Table 7.2. Composition of some common inorganic fertilizer source materials.

Source material	% Available Nutrients N	P_2O_5	K_2O
Ammonium nitrate	33–35	---	---
Ammonium phosphate	11–16	20–48	---
Ammonium polyphosphate	10–13	25–34	---
Ammonium sulfate	20–21	---	---
Calcium metaphosphate	---	66–64	---
Calcium nitrate	15	---	---
Potassium muriate	---	---	50–62
Potassium nitrate	13	---	44
Potassium sulfate	---	---	50
Sodium nitrate	16	---	---
Superphosphate (ordinary)	---	18–20	---
Superphosphate (double or triple)	---	32–54	---
Urea	42–45	---	---

Liquid fertilizers in many cases have replaced the traditional granular forms of fertilizers for use in ponds. Liquid fertilizers give a quicker phytoplankton response as the nutrients are already in solution. They also come in more concentrated formulations than most granular fertilizers, in turn requiring less labor to apply the needed rate. Common grades of liquid ammonium polyphosphate fertilizer are 10-34-0 and 11-37-0. Liquid ammonium orthophosphate, typically sold as a 13-38-0 grade also can be used. Nutrients are found in a ratio of about 1:3 in liquid fertilizers. In some cases, addition nitrogen might be desired, and that can be accomplished using a commercially available solution of ammonium nitrate, urea, and water (Urea-ammonium nitrate). It is available as 28-0-0 (1.28 kg/L) and 32-0-0 (1.33 kg/L) concentrations.

7.3.4 Fertilization Application

When preparing a primary nursery pond there may be only a few days between filling the pond and stocking of the fish larvae. The succession of zooplankton is rapid and the nutrient supply should be at an optimum level as soon as possible if target densities of food organisms are to be achieved. This begins with scattering the first application of both organic and inorganic fertilizers over the dry pond bottom, making sure that the deep section of the pond receives a proportionate share. Once the pond is full, weekly or biweekly fertilizer applications should be made. The fertilizer should be widely scattered into the water along the shoreline of least one side of the pond. For a 30-day growing cycle, 3 to 5 applications should be made with the last application made 7 to 10 days before the pond is to be drained. Small but frequent applications of fertilizers are preferred as it helps to ensure a more steady concentration of nutrients is available when compared to applying the same total quantity of fertilizer in a lesser number of applications. A limited number of applications results in more a peak and valley pattern of nutrient availability, with the higher rate/application creating a greater oxygen demand at each application.

Organic meals and inorganic fertilizers are best spread over the bottom of a dry nursery pond just prior to filling. Once the pond is full, additional quantities of fertilizers can be scattered over the water surface using any wind to help carry it out over the pond. Hay can be applied in several ways. It can be chopped or ground for a rapid assimilation into the food chain and distributed like an organic meal. Loose hay is best scattered over the bottom in a thin layer over the bottom. Slower decomposition takes place when bales are only broken and the blocks loosened and scattered. Even slower breakdown takes place if the bales are left intact. Hay straw is often not completely decomposed when the fish production period is approximately 30 days. It may be necessary to rake the hay out of the pond prior to harvesting the fish. Animal manures are generally applied in 1 to 2 applications per crop either in piles or scattered over part of the pond bottom, adding 1/2 to 2/3 of the total quantity of manure to be used before the pond is flooded.

When the culture period is 8 weeks or more the best way to apply granular inorganic fertilizers is to use a fertilizer platform. This is a 1–2 m^2 platform set near the shore where the platform is approximately 10–20 cm below the water surface when the pond is full. Placing the platform just below the surface allows wave action to help dissolve the fertilizer over time, releasing nutrients directly into the photosynthetic zone. For shorter culture periods or smaller ponds, granular inorganic fertilizer can be broadcast along the shoreline. Broadcasting into deeper water results in a portion of the nutrients being tied up in the soil and less nutrients reach the photosynthetic zone.

Liquid fertilizers may stratify when stored for any length of time and the container should be stirred well before any is taken out for use. When applying liquid fertilizer to a pond it is best to dilute it 1:10 in water for better dispersal in the water column.

7.4 Factors Affecting Fertilization Success

The amount of fertilizer to use per application and per crop depends on the basic fertility of the soil of the pond bottom and the water supply. Occasionally a water supply will have near optimum amounts of one or more of the needed essential plant nutrients. It would be most unusual to have all essential nutrients present in optimum amounts, however. Analysis of the water supply and bottom soil from each pond are helpful in deciding upon a fertilizer program. A history of pond use and yields obtained may also help in selecting a fertilizer rate.

An effective fertilizer regime that has given good results in fish seed production at Marion Alabama includes a mixture of organic and inorganic fertilizers. Half the nutrients are from organic sources (114 kg/ha of cottonseed meal, soybean meal, meat scraps, etc.) and half from granular inorganic sources (9 kg N + 9 kg P_2O_5 + 2 kg K_2O) as the initial application. Half the rate is given again at 1 to 2-week intervals until 10 to 14 days of final harvest.

A mixture of liquid fertilizers can be combined to provide the desired quantities of inorganic nutrients along with organic fertilizers in the preparation of primary nursery ponds. They are used in secondary nursery ponds and other ponds where a phytoplankton bloom is desired at a rate of approximately 10 L/ha/application. The frequency of application might be up to 10 to 12 applications/growing season depending on the phytoplankton bloom obtained. In ponds where the fish are being fed, the ponds may receive liquid fertilizer applications initially to establish a phytoplankton bloom. But once the fish get larger and the feed rate is higher, there are enough nutrients resulting from feeding that no additional liquid fertilizer is needed to support a bloom.

Fertilization of ponds for fish seed production is not without problems. Probably the most common one in a temperate climate is failure of the desired food chain to develop. On one large fish hatchery where about 100 pond-cycles were done annually, 1 out of 3 pond cycles failed to respond to the fertilizer the way it was planned. Usually the failure was from aquatic weeds developing instead of the desired zooplankton or phytoplankton. Less frequently for unknown reasons, food organisms did not develop even though a weed problem did not exist.

When hay is used as an organic fertilizer in addition to establishing zooplankton blooms it also helps to remove clay turbidity from the water column and because hay decomposes slowly it is more difficult to establish phytoplankton blooms. The resulting clear water allows fast growing filamentous algae to better compete with phytoplankton for nutrients, making it more difficult to get a phytoplankton bloom. Limited phytoplankton blooms result in reduced abundance of desirable zooplankton. Filamentous algae also reduce the availability of suitable habitat for the larval fish, and make harvesting very difficult. When nursery ponds having an abundance of filamentous algae are drained for harvest the algae interferes with the flow of the fish into the catch basin resulting in many fish being trapped in the algae.

Copper sulfate is an effective herbicide for controlling filamentous algae but it is toxic to desirable zooplankton and should not be used in a nursery pond. When using hay as the organic fertilizer it may be necessary to increase the quantity of inorganic nitrogen added to help a rapid development of phytoplankton and in turn shading out filamentous algae on the pond bottom.

Successful oxygen management is part of establishing a desirable zooplankton bloom. Rotifers and cladocerans will shift to a sexual reproduction mode under chronic low oxygen conditions. This limits the population expansion rate and the abundance of food organisms/L. Low dissolved oxygen levels are also stressful to larval fish and may result in mortalities. A management goal should be to keep the dissolved oxygen level at $\geq$4 mg/L during the pond preparation phase as well as the larval rearing phase.

Several things will affect the impact of organic fertilization on the decomposition rate and the dissolved oxygen demand: the higher the nitrogen context of the organic manure the faster the decomposition; the smaller the particle size of the organic material the faster the decomposition; the thinner the layer of fertilizer applied to the pond bottom the faster the decomposition; and the higher the temperature the faster the decomposition. Over-fertilization can also contribute to excessive phytoplankton blooms resulting in wide diurnal changes in dissolved oxygen that can be harmful to the larval fish. The best way to avoid trouble from over-fertilization is to apply fertilizers in conservative amounts until first-hand experience with the water supply and pond system enables the manager to predict the effect of a given level of fertilization. Close watch over the rearing pond when trouble appears imminent is also helpful in avoiding losses due to low dissolved oxygen levels.

Highly acid or alkaline pond conditions may contribute to fertilizer problems on occasions. Water and pond soil pH levels in the range of pH 6.5 to 8.5 are desirable. Lime applications can correct acid conditions. Use of acid-forming fertilizers such as those containing ammonia or urea are helpful for alkaline soils and waters.

Ponds with excessive seepage have greater difficulty in establishing phytoplankton and zooplankton blooms, in particularly when relying on inorganic fertilizers to stimulate the bloom. The dissolved nutrients are lost in the seepage water as it filters out of the pond as well as the nutrients being absorbed into the bottom soils. A permissible amount of water loss for a hatchery pond has not been established. However, observation of many production units suggests that if a flow of more than 380 L/min/ha is required to maintain a stable water level in a pond after filling, trouble will be experienced in maintaining a phytoplankton bloom. Where seepage is excessive, use of organic fertilizer and supplemental feeding can to some degree offset the nutrient loss to seepage. However, for economical and efficient production of natural food organisms, construction of ponds that do not leak is highly desirable.

Excessive silt turbidity reduces productivity from fertilization because of insufficient light for photosynthesis. Also, under some conditions clay particles tend to adsorb phosphate ions before plants can use them. Organic fertilizer materials are helpful in settling suspended clay particles in addition to fertilizing the ponds. Light penetration of 15 cm or more is needed for adequate phytoplankton development. As unicellular algae begin to grow, they further stimulate settling of clay particles, giving still better light penetration. Vegetative cover on pond dikes and in the reservoir watershed offers the best long-term solution, reducing soil erosion and in turn turbid water coming into the ponds.

Another approach to dealing with problems of excessive filtration or turbidity is to learn the personalities of all the ponds at the hatchery. Ponds that present the least problems in developing plankton blooms should be considered for use as primary nursery ponds. Ponds that have chronic problems with seepage or turbidity can be used for secondary nursery ponds or brood holding if the fish are to be given a formulated feed. Similarly, ponds with chronic aquatic weed problems should not be used for primary nursery ponds but could be used for producing larger fish where weed control procedures might be tolerated without harming the fish production.

7.5 Recommended Reading

Boyd, C. E. 1979. Water quality in warmwater fish ponds. Auburn University, Agricultural Experiment Station, Auburn, Alabama.

Boyd, C. E., and C. S. Tucker. 2014. Handbook for aquaculture water auality. World Aquaculture Society, Baton Rouge, Lousiana.

Ludwig, G. M., N. M. Stone, and C. Collins. 1998. Fertilization of fish fry ponds. SRAC Publication No. 469. Southern Regional Aquaculture Center, Stoneville Mississippi. Available: http://agrilife.org/fisheries2/files/2013/10/SRAC-Publication-No.-0469-%E2%80%93-Fertilization-of-Fish-Fry-Ponds.pdf. (May 2020).

Mischke, C. C., editor. 2012. Aquaculture Pond Fertilization: Impacts of Nutrient Input on Production. John Wiley & Sons, Inc. Somerset, New Jersey.

Tave, D., and R. A. Anderson. 2018. Strategies and tactics for management of fertilized hatchery ponds, CRC Press, Boca Raton, Florida.

Wurts, W. A., and M. P. Masser. 2013. Liming ponds for aquaculture. SRAC Publication No. 4100. Southern Regional Aquaculture Center, Stoneville Mississippi. Available: https://agrilifecdn.tamu.edu/fisheries2/files/2013/09/SRAC-Publication-No.-4100-Liming-Ponds-for-Aquaculture.pdf. (May 2020).

Additional references are given in Chapter 17.15. Nursery Pond Preparation.

Chapter 8

Extensive Pond Production: Pest Control

8.1 Weed Control

Weeds occur in hatchery ponds because ponds provide suitable habitats for them to grow. Weeds make it difficult to harvest a pond, they provide shelter for predators, and make it more difficult to establish populations of desirable food organisms. The best way to avoid weeds is to eliminate suitable habitats. Aquatic weeds need light to grow and limiting access to light can help control them. Proper pond construction is the first step, deepening the pond edges to a minimum of 25 cm and having an adequate slope (≤2:1) so that a minimum depth of 50 cm is quickly reached. This combined with quickly filling the pond and establishing a phytoplankton bloom soon thereafter will help to limit light penetration to the pond bottom. By adequately shading the pond bottom it makes it difficult for many types of aquatic weeds to become established. Dyes can be added hatchery ponds in an attempt to limit light penetration in turn limiting emergent aquatic weed growth. Dyes are not appropriate in ponds where the fish are highly dependent on natural food organisms (i.e. primary nursery ponds). They are more appropriate for use in ponds where the fish are being fed a formulated feed. Aquashade® and other similar commercial products are effective in shading the bottom enough to limit the establishment of aquatic weeds in areas of a pond that are 50 cm deep or more.

8.1.1 Herbicides

Herbicides can be effective in controlling aquatic weeds, however, there are several things that need to be considered. Safely using the most appropriate herbicide is fundamental. Herbicides are designed to be toxic to aquatic plants but they are also toxic to a variety of other non-target organisms including humans. The applicator must read the label on the herbicide container and know the toxicity issues, safe handling procedures, modes of application, proper disposal, and other safety concerns. What herbicide to use depends on the type of aquatic weed to be controlled, its impact on non-target organisms, and any other associated risks.

Aquatic weeds can be divided into four groups: algae, floating plants, submersed plants, and emergent plants. Within each group there are specific herbicides that work best for a given type of weed. Knowing the specific type of weed to be treated and its characteristics is necessary for selecting the most appropriate herbicide. Aquatic herbicides are grouped by their mode of action. Contact herbicides kill the plant tissues that they come in direct contact with. They are not very selective as to the type of weed it kills and the response is generally seen within days. Such herbicides generally are not transported into other tissues of the weed. Spraying the leaves may not reach the roots and regrowth may occur. Examples of such aquatic herbicides would include: copper-based herbicides, diquat, and endothal. Another grouping, the systemic herbicides, are absorbed into the plant and transported to other tissues. They are generally more selective than contact herbicides, are slower acting, and are more effective in controlling perennials and woody plants. Examples of such herbicides would include: fluridone, glyphosate, and 2,4-D.

In thc USA, herbicides are regulated by the Environmental Protection Agency, and most states have specific information as to what herbicide is legal to use and how it can be used in that state. At any one time there may be some 200 EPA registered herbicides for use in the United States, but less than ten might be labeled for use in aquaculture settings. New herbicides may gain registration, others lose registration as new data becomes available or the manufacturer did not choose to re-register the product. Before selecting a herbicide, it is best to check with the local extension service to find out if a particular herbicide is legal to use and if it is effective under the local conditions. An applicator license may be required to purchase and use certain herbicides. The applicator is always responsible for the effects of herbicide residues on livestock and crops, as well as problems that could arise from drift or movement of herbicide onto other properties.

The herbicide selected must be labeled for use in ponds with fish. Most aquatic herbicides have post-treatment water-use restrictions as to if/or when the water may be used for other activities such as swimming, livestock watering, irrigation or off-farm discharge. In the USA a discharge permit may be required as part of the National Pollution Discharge Elimination System (NPDES). It is the responsibility of the herbicide applicator to stay up to date as to the restrictions associated with herbicide use.

8.1.2 Herbicide Use

There are several management considerations when using a herbicide in a hatchery setting. If a primary nursery pond is dried well before the pond is refilled most aquatic weeds will be slow enough to develop during a 30 d nursery cycle to not present much of a problem. However, filamentous algae can be a problem during the primary nursery period. Organic fertilizers can clear the water of a nursery pond allowing enough light to reach the bottom for filamentous algae to become established. In such settings they can outcompete phytoplankton for nutrients making it difficult to get the desired phytoplankton bloom and in turn the needed zooplankton bloom. Filamentous algae can be effectively controlled with copper-based herbicides, but effective dose rates for controlling filamentous algae will also kill desirable cladoceran zooplankton. In water of low alkalinity (less than 50 parts per million), the treatment rate needed to kill algae may be toxic to fish.

Other herbicides have been used to control filamentous algae applying them to the pond bottom prior to filling. Simazine (Aquazine®) and diuron (Karmex®) have been given as preflooding treatments to control the growth of filamentous blanket algae such as *Pithophora* sp. and *Hydrodictyon* sp. Both simazine and diuron have chronic post-treatment effects that may influence the productivity of the pond and may not be registered in your area for aquatic weed control. Penoxsulam is a selective, systemic herbicide that is currently the only aquatic herbicide that can be applied to exposed sediment after drawdown to inhibit the regrowth of susceptible weeds. It is a slow-acting herbicide that takes 60 to 120 days or longer to completely kill the target plants. Filling the pond will dilute the herbicide reducing its effectiveness.

Secondary nursery ponds and other ponds that remain full for >30 days can develop problems with aquatic weeds. If the proper herbicide is applied at the correct dose, to a known aquatic weed, the weed can be controlled. However, side effects may occur. Killing off large quantities of aquatic weeds at one time may create an increased demand for oxygen as the plant material starts to decay. This can cause a low oxygen condition in the pond that might stress or kill the fish being held in the pond. Treating only one-fourth to one-third of the total surface area of a pond at one time can reduce the risk of dissolved oxygen depletions associated herbicide applications. Keep in mind that the higher the temperature, the faster acting the herbicide might be, the faster the plant decay, and in turn, the faster the dissolved oxygen may drop. For details on aquatic weed identification, commonly used herbicides and how to apply them see Chapter 17.2. Aquatic weed control.

8.1.3 Use of Grass Carp for Weed Control

Grass Carp *Ctenopharyngodon idella* can be an effective aquatic weed control technique where the use of this exotic fish is allowed. Grass Carp are suitable for ponds holding fingerlings and larger fish if the ponds will remain filled for >60 days. Grass Carp are not suitable for use in primary nursery ponds as the culture period is too short for Grass Carp to be effective in eliminating aquatic weeds. Grass Carp also interfere with the safe harvest of the fish larvae that were stocked into the nursery pond. When Grass Carp are to be used, the density should be 50 to 125 fish/ha, stocking 10–15 cm fish. The higher stocking rates/ha are appropriate for smaller ponds because smaller ponds have a greater proportion of their pond area being shallow and more suitable for aquatic weed growth. If the ponds are used to hold predaceous fish then Grass Carp too big to be swallowed by such fish need to be stocked.

8.2 Animal Pest and Predator Control

Predation can be a major source of fish loss at a hatchery. This can begin with predacious insects feeding on fish larvae stocked into nursery ponds and continue on to birds, such as herons and ospreys, and mammals, such as otters, feeding on broodfish. A variety of pest and nuisance animals are found on a hatchery that may complete for food with the desirable fishes. In primary nursery ponds, fairy shrimp (Order Anostraca) and clam shrimp (Order Conchostraca) are too large to be early food items for the larval fish but they will compete for food with smaller zooplankton that would be an acceptable food for the larval fish. Wild fish left from a previous crop or introduced in with the water supply can be competitors or predators on the stocked fish. Muskrats are not predators on fish but will burrow through pond levies and cause them to collapse. This may result in fish being lost or fish from one pond becoming mixed with those in an adjacent pond. Predator and pest control is a year-around activity on a hatchery with a wide range of problems to solve. Specific techniques are discussed in Chapter 17.16.

8.2.1 Insects

There are several aquatic insects that as adults are capable of flying from pond to pond that are predators on or competitors with larval fish. They include giant water bugs (Belostomatidae), water boatmen (Corixidae), water scorpions (Nephidae), back swimmers (Notonectidae), and several beetles (Gyrinidae), where both the adult and larvae may prey on fish larvae. Other insects have larval stages that are predacious on fish larvae including phantom midges *Chaoborus*, and dragonflies. Mosquito populations can build rapidly in a nursery pond, going from egg to adult from the time you start to add water to the time fish larvae are stocked. Adult mosquitos can be vectors of diseases to humans and domesticated animals. The insects that are predacious as adults also can start to reproduce as soon as they arrive in a pond and the population can build rapidly. Backswimmers and water boatmen densities can build to the point of being a significant problem when a primary nursery pond is harvested. During the harvest, as the fish are more concentrated in the catch basin, the insect density is also increased making insects more of a problem. Having insects and other organisms mixed in with the fish when primary nursery ponds are harvested also makes it more difficult to sort the fish and estimate the number of fish harvested.

Drying the pond thoroughly, filling it quickly and stocking the larval as soon as appropriate zooplankton are available at an acceptable concentration is one of the best ways to control predacious insects in primary nursery ponds. Adult insects that might fly into the pond may consume some fish larvae. The adult insects may reproduce in the pond but by the third week of nursery the fish larvae have grown large enough to escape the developing larvae of most of the predacious insects. When large ponds are used as nursery ponds it often takes longer to fill the pond giving more time for aquatic insects to become established and become an issue in pond management.

Many aquatic insect pests are air breathers during the aquatic phase in their life history either coming to the water surface to take a bubble of air or have breathing tubes that can extend out of the water to take in air. These insects can be controlled in a nursery pond by increasing the surface tension of the pond water. The traditional approach to control air-breathing aquatic insects has been to use diesel fuel or kerosene at approximately 50 L/ha to create a film across the water surface blocking the aquatic insect's ability to come to the surface to take in air. The treatment should be applied several days before fish are to be stocked into the pond, doing so when there is little to no wind. After a few days the treatment will have disappeared on to the pond edges and the pond is safe to receive fish. Another group of chemicals Polydimethylsiloxanes (PDMS) also known as silicone fluids have a similar mode of action as diesel fuel in controlling mosquito larvae. One variant Aquatain AMF Liquid Mosquito Film has a low solubility in water and with a specific gravity of <1 that will float on water forming a film. It is effective in controlling mosquitos at <2 mL/m^2. Monomolecular films (MMFs) used for mosquito control are ethoxylated alcohol surfactants made from biodegradable plant oils. They act by significantly reducing the surface tension of the water and wetting mosquito structures, which leads to drowning. These materials have less environmental impacts than commercial insecticides that have been used to control predacious aquatic insects.

Dragonfly larvae (naiads) are voracious predators on a variety of potential food items for larvae fish as well as the fish themselves. Dragonfly larvae have gills and do not need to come to the water surface to breath. They cannot be controlled by putting a film on the water surface like is done with aquatic air-breathing insects. Female dragonflies, depending on the species, lay their eggs in the water, on vegetation or at the water's edge. Eggs may take 2 to 5 weeks to hatch during the summer. Some species lay eggs that can overwinter and hatch the following spring. Dragonfly naiads (larvae) go through 5 to 10+ molts over the course of a few months to a year or more, depending the species and latitude, before becoming adults. The naiads can over-winter in a moist pond bottom and undergo their final molting the next spring. Because of the extended period in which naiads can be present they can be a challenge to control. Thoroughly drying a nursery pond and filling it quickly can be a practical method of control. Thoroughly drying the pond can help eliminate any eggs or naiads that may have overwintered in the pond. Quickly filling the pond and stocking it once appropriate densities of zooplankton are reached will allow the fish larvae to outgrow any recently hatched dragonfly naiads reducing their susceptibility to predation. Keeping any vegetation growing at the water's edge or any emergent aquatic plants under control will reduce the availability of sites for egg laying by dragonflies.

If pond preparation can be carefully planned and an adequate water source is available there should be little need to use pesticides for insect control in primary nursery ponds. Predacious insects are not much of a problem in secondary nursery or production ponds as the fish are large enough to consume such insects, however, the later naiad stages of dragonflies are capable of feeding on 2 cm fish.

8.2.2 Control of Other Invertebrates

There are occasions where you have started a nursery pond to receive fish larvae on a certain date but the brood fish may have not spawned on schedule or there was a problem with the fish eggs or larvae. A delay in using the pond will result in inappropriate types and sizes of zooplankton developing in the pond by the time new fish larvae are available for stocking. Cladocerans may have become the dominant zooplankton but may be too large as a first food for the larval fish being stocked. Adult cyclopoid copepods may have become established that are capable of preying on larval fish and are too large for the fish larvae to consume initially. Organophosphate insecticides such as Trichlorfon at 1 mg/L have been used to restart a nursery pond, killing off most of the invertebrates except rotifers. Rotifers may become abundant but the trichlorfon treatment will delay the re-establishment of desirable zooplankton such as cladocerans, copepods and chironomid midge larvae that the fish would eat as they grow larger. In most cases the best alternative is to drain the pond and start over.

Another group of invertebrates that can be a problem in nursery ponds include fairy shrimps (Anostraca), clam shrimps (Conchostraca), and tadpole shrimps (Notostraca). The three are similar in that they are dependent on temporary pools of water (vernal pools) to complete their life cycles. They also are similar in their reproduction in that they can produce an encysted egg that can resist desiccation often for years. The cysts can be spread from location to location by animals and in a hatchery setting when moving equipment from pond to pond. Their life cycles are short, clam shrimp go from egg to reproducing adult in 4 to 7 days, fairy shrimp in 20 to 30 days and tadpole shrimp in 20 to 40 days. Fairy shrimp are filter feeders on suspended detritus, algae, protozoans, and bacteria. Clam shrimp consume similar items by grazing and filter feeding. Tadpole shrimp are benthic feeders consuming plant material, detritus, various zooplankton, insect larvae and possibly fish larvae. Tadpole shrimp densities can be high enough so that their feeding activities can turn the water turbid.

These groups of crustaceans present several problems at a hatchery. They are easily moved from pond to pond by normal hatchery activity. The resting eggs can remain on the pond bottom between fish crops and hatch once the pond is refilled. If the pond is being prepared as a primary nursery pond, fairy shrimp can become quickly established after the pond is flooded. In the time between when the pond was flooded and when fish should be stocked the fairy shrimp have quickly grown to be too large to be consumed by the fish larvae being stocked. In the meantime, the fairy shrimp, which are efficient filter feeders, outcompete smaller zooplankton, resulting in inadequate densities of acceptable size food organisms for the fish larvae you are trying to culture. These groups of crustaceans are generally not a problem in secondary nursery or production ponds in that the fish are large enough to feed on these crustaceans.

If hatchery ponds have problems with these crustaceans there are several approaches to their control. Dry ponds containing resting eggs can be filled for a week to allow the resting eggs to hatch then drain the pond soon thereafter before the crustaceans become sexually mature. It may be necessary to do this several times to ensure that all the resting eggs have hatched. Another alternative is to rotate the pond use. For example, if a pond has a history of producing fairy shrimp do not use it for a primary nursery pond. Instead, use the pond for a secondary nursery or production pond for one or more crops, noting each time it is filled whether fairy shrimp reappear. The fish will consume the fairy shrimp but some may escape and reproduce. It may take several production cycles to ensure that the fairy shrimp are controlled. Organophosphate insecticides will kill clam, fairy and tadpole shrimp but also will kill desirable food organisms that you are trying to establish in a primary nursery pond. Keep in mind that resting eggs can be moved from pond to pond on a muddy seine or other equipment and adults may be introduced if water from one pond is moved into another.

8.3 Lower Vertebrates as Pests and Predators

8.3.1 Fishes

Unwanted fish can come from several sources. When using a surface water source, it should be assumed that a range of sizes of undesirable fish including eggs and larvae can be in the water used to fill ponds. If surface water is used for nursery ponds, it should be filtered using a 250–500 µm mesh filter when filling the pond. Well water is typically free of unwanted fish and in most cases does not need to be filtered. Vigilance is needed when filling ponds through a filter sock. When the pond is filled the filter sock will become clogged with sediment and the sock must be cleaned periodically. If the sock is not cleaned, as it becomes clogged the increased water pressure on the sock may cause the sock to become loose from the water supply line and come off or cause the filter sock to rip. In either case, undesirable fish could enter the pond. After the pond is filled, it is best to remove the filter sock, take it back to the hatchery, clean it, and store it until the next time the pond needs additional water. If the sock is left at the pond once the pond is full, any fish trapped in the filter sock may result in animals such as rats and mice eating through the sock trying to get to any dead fish inside the sock. The resultant hole can allow undesirable fish to enter the pond through the water supply the next time the pond is filled using that sock (Figure 8.1)

A goal in preparing a hatchery pond is to dry it completely and fill it as fast as possible. However, sometimes there are areas of the pond that will not dry due to rainfall or seepage from other ponds. Such wet areas can harbor small fish and a variety of predaceous insects. Holes dug into the pond bank or bottom that retain water can also retain these animals. The larger the pond the greater the time needed to fill it and the more time is available for predators to become established. Watershed ponds depend on rainfall to fill the pond and it may take months before the pond is full. Adult backswimmers and water boatmen can fly in and begin reproduce resulting in high densities of these predators by the time the pond is full. Any ponds higher in the watershed that can overflow can be a source of wild fish that may move downstream into the pond you are preparing.

Figure 8.1. Fish trapped inside a chewed-into filter sock.

8.3.2 Frogs

Frogs in most cases are pests more than predators with the exception of some bullfrogs like the American bullfrog *Rana catesbeiana* that is an opportunistic predator occasionally eating a fish. The bigger problem is the tadpoles. Most frogs lay their eggs in water, often scattered over vegetation where they will hatch in a few days. The larvae (tadpoles) remain as larvae depending on the species anywhere from about a month for tree frogs to several years for bullfrogs. Tadpoles generally are not predacious but feed on many of the same food items as the fish being cultured. As such tadpoles are competing for food and can reduce fish growth and survival.

Perhaps the biggest issue with tadpoles is related to nursery pond harvesting. At harvest when tadpoles are mixed in with the fish the tadpoles increase the total weight to be removed from the pond and take up space in the transport tank and holding tanks, reducing the overall efficiency of the harvest. Having tadpoles mixed in with the fish at harvest makes it difficult to get an accurate estimate of the number of fish harvested and their average weight. The process of sorting out the tadpoles is time consuming as well as being stressful on the fish.

The presence of adult frogs can be minimized by keeping the shorelines of ponds and drainage ditches free of any vegetation that could provide shelter to a frog. This will make the frogs more susceptible to predation. Tadpoles can be kept under control by keeping the pond dry until it is needed. A common mistake is not drying or poisoning out the catch basin. A little bit of water in the catch basin can be enough for frogs to lay their eggs resulting in a large number of tadpoles ready to spread out once the pond is filled. Eliminating marginal aquatic vegetation along the shoreline of ponds will reduce appropriate habitats for frogs to lay their eggs. Tree frogs have a relative fast life cycle and this should be considered as part of your management plan for primary nursery ponds. If the water temperature is ≥20°C and the tree frogs lay their eggs as soon as the pond is filled, they may become juveniles that start hopping out of the pond before it is necessary to harvest the fish. By monitoring the pond to see if juveniles are appearing along the shoreline it may be possible to delay the pond harvest a few days to allow the remaining tadpoles to make the transformation, resulting in much less tadpoles to interfere with the fish harvest.

When the above techniques do not work a low fence around the pond may be necessary to limit access of adult frogs to the pond. This can be combined with a funnel trap set parallel alongside the fence or burying an 18.9 L bucket along the fence line so that the open end of the bucket is at ground level. Frog will drop into the bucket and cannot escape. Such traps will also collect snakes and turtles as they move along the fence line.

8.4 Chemical Control of Invertebrates and Lower Vertebrates

8.4.1 Use of Chlorine or Caustic Lime

For small areas that will not dry before it is time to refill the pond such as the catch basin or old nest sites, any residual animals can be killed using chlorine, quick lime or hydrated lime, or rotenone. Calcium hypochlorite, $Ca(OCl)_2$, is the best source of chlorine. It has excellent stability when kept in dry storage, maintaining its potency over time. Commercial grades of calcium hypochlorite generally contain about 70% available chlorine making it appropriate for both drinking water and wastewater applications. The goal is to apply enough chlorine to maintain 10 to 20 mg/L free chlorine for 24 to 48 hours. If quicklime, CaO, or hydrated lime, $Ca(OH)_2$, is used enough should be added in enough quantity to raise the pH to 11.5 or higher.

To treat a few areas in a pond it can be done dissolving the material in a bucket of water and add it to each wet area with a ladle. Chlorine, quicklime, and hydrated lime can cause skin and eye irritation and should be handled with care using gloves and goggles. When there are numerous wet areas in the pond bottom it is easier to use quicklime or hydrated lime spreading it over the pond bottom at approximately 400–600 kg/ha, then fill the pond until all the bottom is covered with water. If the pond is to be used as a primary nursery pond then the lime and organic fertilizer should be added about the same time and then flooded. After the pond has been flooded for at least 6 h, check to be sure that the pH of the water has risen to 11.5 or more. If the pH was not high enough additional quicklime or hydrated lime should be added. Completely filling the pond should dilute the effects of quicklime or hydrated lime and bring back the pH into a safe range. The pH of the water should be monitored daily. The pH should be consistently below 9.5 before any fish are stocked into the pond. In weakly buffered waters (<30 mg/L) it may take several days for the pH to drop to a safe level.

8.4.2 Use of Rotenone and Antimycin A, (Fintrol)

Undesirable fish are a common contaminate of hatchery ponds. Fish may have been left in the pond from a previous crop or entered the pond when it was being filled. Fish may have swum into the pond up through the drain overflow after a heavy rain or when one pond overflows into another. Watershed ponds often take several months to fill and run the risk of being contaminated with fish from the surrounding watershed after a strong storm. Any pond that has been filled for several weeks should be seined with a fine mesh seine to make sure there are no undesirable fish present.

Rotenone has been a traditional material used to control undesirable fish in hatchery ponds. It is available as Derris or cube powder that contains 5% rotenone. It is also available in an emulsified form that is convenient to use but does not store well in tropical settings. Regulations regarding rotenone use in the USA have been changing and currently it is listed as a restricted pesticide and should be applied by a certified pesticide applicator. A rate of 0.5 to 3.5 mg/L rotenone is recommended depending on the tolerance of the target species. Rotenone degrades fairly rapidly at temperatures above 15°C and can be neutralized with potassium permanganate when added at a rate twice the level of rotenone added. However, to ensure that all targeted organisms were killed the rotenone should be allowed 24 h to act before using potassium permanganate.

Another restricted pesticide available as a fish toxicant is antimycin A, (Brand name Fintrol). Its efficacy is less at lower water temperatures (<16°C) and higher pHs (>8). It can be used as a selective pesticide in Channel Catfish *Ictalurus punctatus* culture at rates of 7 to 10 µg/L to eliminate shad and sunfish contamination as catfish are tolerant to much higher concentrations of antimycin than the scaled fish. To renovate a pond including eliminating the catfish a dose up to 25 µg/L is necessary.

8.4.3 Other Toxicants

A variety of organic materials have been used elsewhere around the world to eliminate undesirable fish, predacious insects and other pests. A number of plants in the legume family have rotenone-like alkaloid toxins (flavonoids) that act by impairing oxygen consumption in cold-blooded animals. Another group of toxins that occur in 75 or more families of plants are water-soluble glycosides (Saponins) that when added to the water are taken in across the fish's gills destroying red blood cells. Among the better-known sources of saponins used as a fish toxicant are tea-seed cake at 10 to 25 mg/L, mohua oil cake at about 250 mg/L (composition 4–6% saponin), and sugar beet extract at 2 mg/L (as purified dry powder).

A number of more classical insecticides, particularly organophosphates, have been used as fish toxicants. They also have been used as insecticides as part of preparing nursery ponds. They are often lethal at low dose rates and decompose rapidly. However, many are also highly toxic to humans and many non-target animals. In most cases they are not approved for use in fish ponds.

8.5 Higher Vertebrates as Pests and Predators

8.5.1 Snakes and Turtles

Water snakes will eat fish when given the opportunity if the fish are small enough to be swallowed. Snakes can learn when it is fish feeding time. When fish are concentrated in a small area of a pond during feeding, snakes have been observed to swim up and pick off individual fish. Non-venomous water snakes can be beneficial in that they also consume frogs and tadpoles. Other non-venomous snakes feed on mice and rats common around feed storage areas and can help control these pests. The beneficial effects of non-venomous snake may outweigh their role as pests.

Venomous snakes are not uncommon around fish hatcheries and are human health risks. In the southeastern USA, cottonmouth water moccasins *Agkistrodon piscivorus* may be found around hatchery ponds and drainage ditches. Areas such as feed storage buildings that might attract mice and rats also may attract rattlesnakes, *Crotalus* sp., and copperhead snakes, *Agkistrodon contortrix*, that feed on mice and rats. The safest thing on a hatchery is to reduce the number of venomous snakes in turn reducing the likelihood of a snake bite. The difficulty in a work setting is being able to recognize which snake is venomous and which is not.

Turtles are common on a fish hatchery but in most cases are not a major pest or predator. In the USA there are three groups of turtles often found on a hatchery. Pond turtles/Sliders (Emydidae) are hard shell turtles often seen out of the water basking in the sunshine. They are omnivores, consuming plant material, a variety of invertebrates, dead and dying fish, as well as fish feed. Snapping turtles (Chelydridae) are more predaceous, feeding on a range of aquatic and terrestrial animals. Softshell turtles (Trionychidae) are predaceous, as well as scavengers. All will eat fish if given the opportunity. They can be pests as they try to reach a dead fish inside a cage, damaging the cage, allowing any fish being held in the cage to escape.

Turtle densities can be reduced as part of normal hatchery operations as they are often captured when seining or draining ponds. When turtles occur in ponds that are not drained routinely they can be trapped or fished out. Pond turtles/sliders can be trapped based on their characteristic of climbing out on a rock or tree trunk to sun then sliding back into the water. Several trap designs are given in Chapter 17.16. Snapping turtles and softshell turtles can be captured using cut fish as bait by hook and line, or in baited traps.

8.5.2 Birds as Pests and Predators

8.5.2.1 Types of Birds Needing Control

Birds present a variety of problems as pests and predators and can consume significant quantities of fish. They can also spread diseases from pond to pond by dropping sick fish from one pond into a new pond. Moreover, many birds are carriers of fish diseases that can be spread in the bird's feces. The presence of birds can intimidate fish at feeding time, reducing fish feeding activity, and in turn result in wasted feed and slower fish growth.

For hatchery management purposes, birds might be grouped as: fliers (eagles, gulls, kingfishers, ospreys, brown pelicans, terns); swimmers (anhingas, cormorants, ducks, grebes); and wader/stalkers (egrets, grackles, herons, ibises, storks). The fliers include mainly divers that fly over a pond, spot a fish, and dive in after it. Eagles, ospreys and pelicans are large enough to consume broodfish. Kingfishers and terns feed more on smaller fish. Gulls are hoverers, picking up food items near the surface (small fish or floating pellets). By hovering they can pick up and consume floating feed at the pond's surface and scare the fish from coming to the surface to feed. Swimmers may be a single bird such a grebe feeding on any aquatic animal it might find including fish, or a flock of hundreds of cormorants feeding as a group consuming 300 g of fish/bird/day. Wader/stalkers feed along the shoreline of a pond, as well as along the walls of raceways and holding tanks. Egrets, herons, and storks will wade into a pond to a depth just above their knee, feeding on any fish found along the shore. Great blue herons and storks are able to swallow >500 g fish. Grackles and similar birds will walk along the walls of tanks holding small fish, leaning over to pick off any fish that are near the surface. They will also feed on any feed floating in the tank, or any feed accessible in an open feed container.

8.5.2.2 Bird Control Methods

There are a number of approaches to reducing bird problems at a hatchery. In the USA, most of the problem birds are protected species as part of the Migratory Bird Treaty Act and cannot be harmed. However, there are exceptions that can be made when such birds are causing severe problems at a hatchery. Rules change and new techniques become available. It is always best to keep in contact with the appropriate Federal and State agencies as to how specific problems can be addressed.

Hatchery Design and Procedures.—There are several approaches that can be combined to help reduce bird problems. A starting point begins with the hatchery construction, design and maintenance. Ponds should be constructed to have as little of their depths <25 cm as possible. This will limit areas where wading birds can fish. Shorelines and ditches should be kept clean of tall vegetation where wading birds could hide. Overhead perches near the ponds should be avoided. This would include power lines, power or light poles, trees or any other structures that would allow a kingfisher or osprey to sit and look down over the ponds or anywhere else fish are held, then fly down to dive on the fish. Birds like cormorants and herons are colonial roosters that prefer to roost near the feeding area; i.e. the hatchery ponds. By frequently harassing the birds at night when they are on their roost, the birds can be driven to find other roosting areas farther from your facility and hopefully to new feeding areas. However, the birds will return to your hatchery if you allow them to re-establish roosting areas near you.

Another approach is to modify hatchery practices to make them less attractive to birds. This might include putting valuable fish that are more susceptible to predation close to where there is more human activity to scare off the birds. Daily patrols of the ponds, removing any dead or dying fish found, will make less food available for the birds, attracting less birds, and in turn reducing the risk that a bird may pick up a diseased fish and drop it into another pond. Floating feed might be preferred from a fish culture perspective but fish feeding at the surface makes them more susceptible to bird predation. In such cases it may be necessary to use a sinking feed. Gulls will rest a portion of the day and their habits should be learned. Floating feed can be fed to the fish during the gull resting period.

Limiting Access.—Limiting bird access can be practical when the ponds are relatively small (<1000 m^2) or the fish are in tanks or raceways by covering such units with netting of an appropriate mesh size to keep out the bird pest. When herons and other wading birds are a problem they can be fenced out of the area along the pond edge where they would normally wade. A net fence should be set on an angle from the pond bank outward covering the depths where that species of bird would normally wade.

When the problem is diving birds, they can be discouraged from diving by running a series of lines across the pond about 0.5–1.0 m above the pond surface. The lines should be pulled taut across the width of the pond, spaced to be less than the wingspan of the problem bird when it dives. The intent is to have the lines spaced close enough so that as the birds dive to get fish they will hit a line on the way down. After a few tries the birds learn that the fish in that pond are not accessible and the birds will move on to places where the fishing is easier. Motion-sensing water sprayers can be effective ways to keep herons and grackles from walking the tank walls of raceways and tanks to find fish.

Auditory, Visual, and Other Harassment Techniques.—There are a variety of auditory or visual harassment techniques that can be used to try to scare birds away from an area (see reference section of Chapter 17.16. Predator Control). Auditory devices might include propane-powered gas cannons set on timers to give a loud boom periodically (Figure 8.2), pyrotechnic devices that can be shot in the direction of the birds to explode in the air, sirens, and electronic noisemakers.

Visual harassment could include humans or dogs chasing the birds off, scarecrows, reflectors or laser lights, as well as aerial drones. However, birds are smart and quickly learn that if a specific harassment does not pose a true threat they will ignore it. A variety of harassment techniques need to be used and alternated frequently to avoid the birds becoming accustomed to them.

When bird depredation is causing major losses of fish at the hatchery it may be possible in the USA to get permission to kill problem birds as part of an overall control program. The starting point would be to contact your nearest USDA APHIS Wildlife Services office for help.

8.5.3 Mammals as Pests and Predators

8.5.3.1 Predators

Mammals can be both pests and predators at U.S. hatcheries. Otters *Lutra canadensis*, can be significant predators at many hatcheries. They are mainly nocturnal, primarily feeding on fish, but are opportunistic also feeding on shellfish, crayfish, amphibians, and reptiles. The presence of otters on a hatchery is often evident as worn pathways across the pond bank (slides), made by the otters as they repeatedly go from pond to pond or to a nearby stream. Evidence of their presence may include parts of their recent meal such as fish bones, parts of

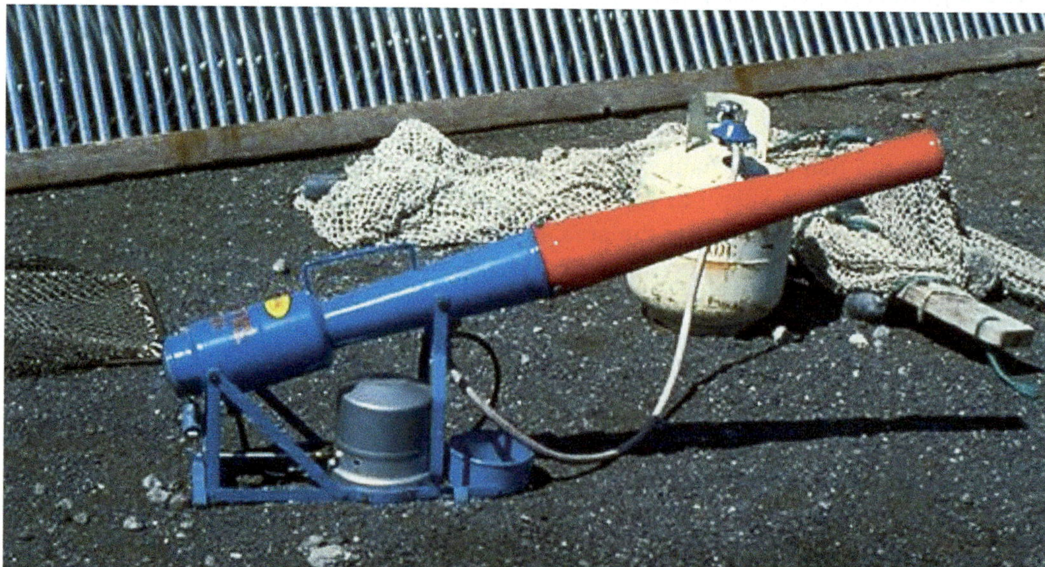

Figure 8.2. Propane powered gas cannon used to scare birds.

crayfish or similar hard items left on the pond bank or at the edge of the water. Feces on the pond bank may contain fish scales and bones.

Otters can be a problem particularly when fish are confined at high densities in a cage or in a raceway. Limiting access is the best way to control otters under such conditions. Fencing off fish holding areas using a wire mesh of 7.6 × 7.6-cm or smaller can help. Tight covers on cages to prevent access can be effective but otters are smart and will find a way in if the cover is not tightly in place or if there is a hole in the cover. Trapping to control otters is another alternative where it is legal (see Chapter 17.16). Otters are creatures of habit, frequenting the same pathways and feeding areas. Identifying these areas are the key to a successful trapping program.

8.5.3.2 Pests

Beavers and muskrats are not fish predators but are mainly herbivores, Muskrats feed primarily on aquatic plants. The biggest problem they create on a hatchery is their burrowing habits. They make dens in the pond bank with underwater entrances. The dens are often close enough to the top of the pond dike that the den collapses when a vehicle is driven on the dike. Sometimes one entranceway may lead from one pond to a den having another entranceway in another pond. This can result in fish moving from one pond to another. When one pond is drained, water may flow through the passageway into the pond being drained, causing erosion and collapse of the pond dike.

Beavers are also herbivores, feeding on woody plants and emergent aquatic plants. Their dam building habit is the biggest problem in a hatchery setting. They may dam small streams that flow into the hatchery's water supply reservoir. They also build dams across drainage canals, backing up water in the canal, flooding areas that normally would be dry, and making it more difficult to drain the hatchery ponds. Such dams create new habitat for other pests and predators that need to be controlled on a hatchery.

Both beavers and muskrats can be effectively controlled by trapping (see Chapter 17.16) but both will return, so trapping needs to be a part of the overall hatchery maintenance plan. Keeping emergent aquatic plants under control will reduce the food availability for muskrats and beavers and make it easier to see signs of animal activity. Muskrats will feed during the day and, if approached carefully, can be shot; however, beavers are more active at night and more difficult to shoot.

Raccoons, rats, and mice are pests on a hatchery, consuming fish feed wherever they can find it. Raccoons can lift the lids on containers to get to any feed stored there. Rats and mice will chew through bags or plastic containers to get to the feed. Raccoons also will pick up floating feed inside a feeding ring or any feed floating near the edge of a tank or the shoreline of a pond. In doing so, they scare off any fish that might otherwise come to the surface for the floating feed. Raccoons are nocturnal and very rarely seen during the day. They can be carriers of rabies. Infected animals may be seen during the day and often do not run from humans. Such animals should be shot on sight and disposed of without any direct human contact. Rats and mice also are carriers of several human diseases and likewise disposed of with care. Raccoons, rats and mice can be trapped using size-appropriate traps (see Chapter 17.16). The best way to minimize problems with raccoons, rats, and mice is to not allow them access to feed or feed ingredients. This includes using well-sealed feed storage containers, storing feed in buildings where the animals cannot enter, and not leaving waste feed on the ground or in a vehicle used for feed distribution.

8.6 Recommended Reading

Burleigh, J. G., R. W. Katayama, and N. Elkassabany. 1993. Impact of predation by backswimmers in Golden Shiner, *Notemigonus crysoleucas*, production ponds. Journal of Applied Aquaculture 2(3–4):243–256.

Curtis, K. S., W. C. Pitt, and M. R. Conover. 1996. Overview of techniques for reducing bird predation at aquaculture facilities. Jack H. Berryman Institute, Department of Fisheries and Wildlife, College of Natural Resources, Utah State University, Logan. Available: https://agrilifecdn.tamu.edu/fisheries/files/2013/09/Overview-of-Techniques-for-Reducing-Bird-Predation-at-Aquaculture-Facilities.pdf. (May 2020).

Gettys, L. A., W. T. Haller, and M. Bellaud, editors. 2009. Biology and control of aquatic plants: a best management practices handbook. Aquatic Ecosystem Restoration Foundation, Marietta, Georgia. Available: http://plants.ifas.ufl.edu/wp-content/uploads/files/caip/pdfs/bmp3rdedition.pdf. (May 2020).

Gettys, L. A. 2014. Aquatic weed management control methods. SRAC Publication No. 360. Southern Regional Aquaculture Center, Stoneville, Mississippi. Available: https://www.ncrac.org/files/biblio/SRAC3601.pdf. (May 2020).

Kane, A. S., R. Reimschuessel, and M. M. Lipsky. 1992. Effect of tadpoles on warmwater fish pond production. Fisheries 17(2):36–39.

Littauer, G. A., J. F. Glahn, D. S. Reinhold and M. W. Brunson. 1997. Control of bird predation at aquaculture facilities: strategies and cost estimates. SRAC Publication No. 402. Southern Regional Aquaculture Center, Stoneville Mississippi. Available: http://agrilife.org/fisheries2/files/2013/09/SRAC-Publication-No.-402-Control-of-Bird-Predation-at-Aquaculture-Facilities-Strategies-and-Cost-Estimates.pdf. (May 2020).

United States Department of Agriculture. Animal and Plant Health Inspection Service. 1997. Bird Predation and Its Control at Aquaculture Facilities in the Northeastern United States. APHIS 11-55-009 Issued June 1997. Available: http://fisheries.tamu.edu/files/2013/09/Bird-Predation-and-Its-Control-at-Aquaculture-Facilities-in-the-Northeastern-United-States.pdf. (May 2020).

For additional references on controlling aquatic weeds see Chapter 17.2 Aquatic Weed Control. For details on controlling specific vertebrate pest problems see the references given in Chapter 17.16.

Chapter 9

Extensive Production: Pond Management

9.1 Primary Nursery Pond Stocking

Successful pond management begins with good pond preparation and obtaining good quality larvae for stocking. Pond preparation may begin before even eggs are available as it can take 10 to 20 days or more to have pond conditions suitable to receive fish larvae. In the meantime, broodfish must be spawned, eggs incubated and hatched, and the larvae held until at least mouthparts are formed before they are stocked into nursery ponds. Each hatchery should develop a general schedule as to when larvae of a given species should be available, and in anticipation of those dates pond preparation procedures should begin. Management schedules need to be adjusted according to the time required to obtain adequate zooplankton blooms. For fish larvae that rotifers are the desired first food, target densities of rotifers should be >500/L in an expanding population of predominately egg bearing females. Where cladocerans are the target food organisms, an initial target density should be >200/L in an expanding population with no females holding resting eggs.

Stocking rates for primary nursery ponds range from 100,000 to 5,000,000 or more fish/ha depending on the species, pond fertility, and the length of the culture period. The higher stocking rates would be for first feeding fish larvae of species that need small organisms such as rotifers as a first feed. The lower stocking rates would be used for fish larvae that accept cladocerans as a first food or larvae that had been held and fed in the hatchery until they were large enough to consume cladocerans or larger food items that are found in the nursery pond. Keep in mind that growth rate in a primary nursery pond is a function of the availability of an appropriate size zooplankton and the quantity of zooplankton is finite. Lower fish stocking rates mean more zooplankton/individual fish and in turn a faster growth rate/fish. A higher stocking rate means less zooplankton/fish, a slower growth rate, and if the stocking rate is too high, will result in reduced survival. In primary nursery ponds the goal is to produce as many individual fish as possible that are large enough to be safely harvested and are acceptable in the market. It is best to stock individual lots of fish into separate nursery ponds. This allows tracking specific genetic groups, their performance, and disease susceptibility. If there is not enough fish to stock a nursery pond at the desired density on a given day, it is best to stock what is available if you are close to the target density. As the fish will grow faster and reach the target size sooner you should be prepared to harvest the pond sooner than planned. Stocking fry into the same pond in separate lots on different days is not recommended as it results in greater size variation at harvest, possible cannibalism, and creates more management problems. Each lot of fry held in the hatchery should be monitored as to mortality rates and percent deformities. Marginal quality fry in the hatchery should be discarded rather than be stocked into a nursery pond where their chance of survival is low.

Nursery ponds should be monitored for dissolved oxygen in the early morning for 3 to 4 days in advance of the anticipated stocking to ensure the pond environment has stabilized following organic fertilization. The ponds should be checked again in the afternoon the day before stocking to be sure the zooplankton abundance is adequate and all predaceous insects have been eliminated. On occasions, rotifer populations can build to the point that they are consuming algae faster than the algae can reproducc. This will result in low dissolved oxygen levels and a crash in the rotifer densities. A clear estimate of the number of larvae on hand to be stocked on a given day needs to be known to allocate ponds. These numbers can be estimated following the procedures given in Chapter 6.

Ponds should be stocked starting a few hours after sunrise to ensure dissolved oxygen levels are adequate, but not too late in the morning as the difference in water temperature between the hatchery and the pond gets greater requiring more acclimation time. A general rule of thumb is when fish larvae are going from a warmer water to a cooler water is to provide 5 minutes of acclimation for every degree down in temperature. When stocking into warmer water acclimate over 10 minutes for every degree up that the fish must adjust to. Acclimation to any major differences in other water quality such as pH or total hardness also should be made to avoid shocking the larvae at stocking.

Fish can be transported to the pond in plastic bags with water and oxygen. The bags can be set in the pond to allow the water temperatures inside and outside the bag to reach the same temperature. Once the temperatures in the bag and that of the pond water are nearing the same, the bags should be opened and pond water added to the bag to allow the fish to adjust to any other differences in water quality. Fish larvae can also be transported to the pond site in a small transport tank or similar tank. Using transport tanks allows dissolved oxygen to be added to the tank while the larvae are in transit by using bottled oxygen and air stones, being careful not to create strong water currents that would stress the fish. Electrical agitators should not be used with larval transport as the vigorous agitation may damage the small fish. Pond water should be added slowly to the transport tanks to acclimate the fish using a fine mesh screen in the tank to ensure no fish go down the tank drain. Acclimation should take place on the lee side of the pond to avoid any wave action that could harm the fish. The site of release should be free of filamentous algae and other aquatic weeds.

9.2 Primary Pond Management

Post-stocking care includes visual observation to note the extent of handling mortality, sampling to determine survival and growth, and a close watch over the developing food supply. ‘Did the fish survive stocking’ is one of the first questions to answer. This can be answered the first night after stocking using a flashlight as the eyes of small fish will reflect light at night. By checking along the pond bank in the area where the fish were stocked you should see some fish, looking for the reflection in the eyes. It is hard to quantify the survival with this technique, but if no fish are seen a wider area needs to be checked and if no fish are seen then the stocking survival may have been very low. A large-mouth plankton net can be pulled on a diagonal across the pond during the day to try and answer the question of fish survival. Again, this will not quantify survival but if some fish are caught then there was survival. Nursery ponds are a valuable resource and often the spawning season is short for a given species. Waiting two weeks or more to determine if there was survival is an opportunity lost. Each time a nursery pond is stocked it should be checked within a day or two to determine if there was survival. If there was no apparent survival then consideration should be given to draining the ponds and starting over.

For ponds that were successfully stocked, the ponds need to be monitored daily, including weekends and holidays, checking water quality, changes in water clarity and the occurrence of predacious insects. Fertilization protocols should be continued and additional inorganic fertilizer added if the pond has lost its plankton bloom. No new water should be added to the nursery pond for the first 10 days post-stocking. This will minimize the risks of undesirable organisms entering in from the water supply. If aquatic weeds develop, keep in mind that any chemical treatment must be applied well before the crop can be harvested. It should be anticipated that up to 21 days would be needed to eliminate submerged rooted weeds and 7 to 10 days for filamentous algae.

Nursery ponds should be sampled using a fine mesh, 3 m long seine within two weeks after stocking to have a general indication of fry survival and pond condition. A sample containing only a few but large fry would suggest poor survival while an abundance of fry but smaller than anticipated could indicate good survival or a poor food supply. Similarly, a wide range in sizes could indicate intense competition for a limited supply of food or an overstocking error. The occurrence of 20 or more backswimmers or water boatman would suggest a low fish survival. When predator species are being cultured, ponds should be inspected beginning with weekly seine samples starting about 2 weeks after stocking. If sampling reveals size differences that permit cannibalism, the pond should be scheduled for an early harvest.

Where efforts are directed toward obtaining the maximum number of a 25 to 50 mm size fingerling, harvesting should be done promptly when that size is reached. Once the fish population develops to the point where the natural food supply is becoming limited, the remaining adult zooplankton population is being consumed faster than they can replace themselves. On occasions, natural food supplies in a pond can change in 24 to 48 hours from where there are enough to support fish growth to where there are too little natural food organisms to meet the maintenance needs of the fish. When the fish stocking rate is high, a close watch must be maintained over food conditions.

Supplemental feeding during the primary nursery period can be appropriate for some fish species and should be done if natural food is depleted or prompt harvesting is not feasible. Training fish to accept a formulated feed while still in a primary nursery pond helps make the transition to being fed in a secondary nursery pond more successful. Hybrid Striped Bass can be trained to accept formulated feeds starting in the second or third week in the primary nursery pond. Supplemental feeding can be used to hold fish in a nursery pond after natural food availability has dropped too low to support fish growth. Fingerlings of some species can be held for days, weeks, or months on a maintenance diet without adverse effects while keeping them within the desired target production size. Some species must have ample food and other conditions for growth or they perish. This is particularly true of most carnivorous species.

Pre-harvesting activities should include ceasing fertilizer applications 7 to 10 days before the scheduled harvesting date. The increase of natural food organisms attributable to that additional fertilizer application is often minimum during those few remaining days of culture. Any supplement feeding should be stopped a few days before the harvest date. This will help reduce water quality problems at harvest associated with the decay of uneaten feed and feed the fish might spit back up when handled. It will also help get a more accurate estimate of fish weight if there is no feed in the intestinal tract when the fish are sampled.

If the abundance of air breathing insect numbers builds up during the culture period, an application of diesel oil or kerosene can be applied on a calm day a few days before harvest to control them. The catch basin should be clean and if need be, pumped clean of sediment. Holding tanks should be prepared and their water supply lines flushed ahead of time to ensure good quality water is available. If the fish are to be transported off-station, transportation schedules and supporting equipment need to be prepared, and the individuals to receive the fish notified.

Extensive culture of small fingerlings requires less labor than controlled culture systems. One or two trained fish culturists can care for up to 4 ha of ponds on a modern fish hatchery including production and distribution. An efficient production complex should range in size from 10 to 20 ha and should have a staff of 4 to 6 persons. Such a facility, depending upon the species and the nature of the distribution requirements, can produce two crops per year and have an output of 3 to 8 million fingerlings annually.

9.3 Secondary Nursery Pond Management

For many applications the fingerlings harvested from a primary nursery pond are of adequate size for stocking new bodies of water where there are no other fish, such as stocking into newly constructed farm ponds, or locations where there is little to no natural recruitment. It is easier to produce large numbers of small fish than larger fish for release into natural setting if the loss from predation is not excessive. However, in many cases it is better to produce a smaller number but larger fish for stocking as their survival will be greater. This is particularly the case where natural waters are to be stocked and potential predators are present.

In aquaculture one of the goals is to maintain standing crops near the critical standing crop (the standing crop/production unit when the growth slows from the maximum growth rate) to optimize the use of the production facilities. To go from larvae to a food-size fish may involve four or more production cycles with several of those taking place at the hatchery. As discussed earlier, the primary nursery period covers the transition from larvae to a recently metamorphosed juvenile, typically over a 30–45 day period. The length of a secondary nursery in a temperate climate may extend from the harvest of the primary nursery pond to when the water temperatures drops in the fall to a point where growth rates are slowed. In a tropical climate the length of a secondary nursery period is a function of the growth rate, the critical standing crop for a particular set of management procedures, the average harvest weight desired, and the initial number of fish stocked and their average weight. Common secondary nursery cycles in a tropical setting may run 60–90 days with one or more additional production cycles to produce a food-size fish.

Using only inorganic fertilizers as a nutrient source in a secondary nursery pond will result in a phytoplankton-based food chain where the greater the quantity of algae that is produced the less light penetration and photosynthesis becomes light-limited. This light-limited condition will determine the pond's carrying capacity. Adding additional inorganic fertilizer to a light-limited production setting will do little to increase fish production. When using only inorganic fertilizers to culture benthic feeding fish such as Common Carp, the foraging habits of the fish stir up the pond sediments, increasing the turbidity of the water, reducing light penetration and photosynthesis, resulting in reduced natural food availability, and in turn the carrying capacity of the pond. Inorganic fertilizer based carrying capacities are limited to about 1,000 kg/ha.

The application of organic fertilizers adds another carbon source to the food chain and will increase the carrying capacity of a secondary nursery pond. However, adding an excess of organic fertilizers will result in light-limited photosynthesis, an increased oxygen demand and an ammonia buildup. Carrying capacities of organic fertilized secondary nursery ponds might range from 2,000 to 5,000 kg/ha depending on the food habits of the species being cultured and the length of the culture period.

Ideally, the fish are already feed-trained when they are stocked into secondary nursery ponds. Many species of fish can be trained to accept a formulated feed during the last weeks of culture in a primary nursery pond, others can be taken back to the hatchery after being harvested from the primary nursery pond and feed-trained there. After being feed-trained in the hatchery the fish can be released directly into a secondary nursery pond, or to re-enforce the feed training, the fish can be stocked into a cage set in the secondary nursery pond and fed for another week before being released into the pond.

If the fish are successfully feed trained and a quality feed is available, then the issue becomes how much feed can be fed before water quality issues develop. In static water ponds feeding rates of 25–35 kg of feed/ha/day typically can be given without creating water quality problems. The risk of low oxygen conditions may become more frequent as the feeding rate approaches 50 kg/ha/d, and chronic low oxygen conditions should be expected at feeding rates of ≥100 kg/ha/d. Depending on the species tolerance to low oxygen conditions, emergency aeration needs to be available when feed rates approach 50 kg/ha and aeration may be required every night at feed rates of ≥100 kg/ha/d.

In secondary nursery ponds and later culture cycles a polyculture may be justified, particularly where the primary fish is fed a formulated feed and herbivorous fish (Grass Carp *Ctenopharyngodon idella*) are added at a much lower density to control aquatic weeds. This approach also applies to brood ponds where two year-classes of fish of distinct sizes are used.

Ponds used for a secondary nursery are typically larger than those used for primary nursery ponds and larger quantities of fish will be harvested. Such ponds should be seinable as standing crops at harvest time may be more than 2,000 kg/ha. A secondary nursery pond may be partially harvested as early as halfway into the production cycle using a selective seine cropping off the larger fish. By lowering the density, the remaining fish in the pond can grow faster and reach an acceptable size by the end of the culture period. Keep in mind that when harvesting a secondary nursery pond, the harvest techniques used should try to ensure that the harvested fish are healthy and can be released into another body of water with the expectation of a high survival rate.

A third nursery cycle might be used in a temperate climate. Fingerlings have a higher metabolic rate than adults and may need to be fed during the winter to avoid weight loss. This might be where small fingerlings are stocked at a high rate that will not allow them to reach the target weight in one growing season. A nursery pond with such fish would be harvested in the fall, inventoried and restocked at a lower density and cultured into the late spring of the next year to the desired size and harvested. A variation on this approach is to harvest the secondary nursery pond in the fall of the first growing season, and grade out any fish that reached the target size. The sub-target sizes are restocked for an additional nursery period until the fish reach the target size. Attention needs to be given to any nursery period that extends through the winter. Temperatures are not constant during the winter and fish might not feed on the colder days but will if the temperature rises a few degrees. The hatchery staff needs to pay attention to these changes in water temperature and not feed on those colder days but be sure to offer feed on the warmer days. Overfeeding during the winter can result in a buildup of unconsumed feed in the pond that will decay rapidly in the spring, causing low oxygen and an increase in nitrites to toxic levels.

Another issue with an over-winter nursery is the predation risks. Often in the winter, otters are more active, predatory birds are more common, and hatchery staff are not out on the ponds as frequently. The ponds should be patrolled daily looking for signs of otter activity and flocks of cormorants, and appropriate measures taken to control the problem. Predator control measures are covered in Chapter 8.2.

9.4 Production of Large Fingerlings

Rearing large fingerlings in ponds on natural food is an extension of the process of rearing small fingerlings. Some modification is required to accommodate space and food requirements of the larger size. Longer production cycles require more attention to weed control. The need for weed control dictates use of a fertilizer plan to produce a phytoplankton bloom that will shade the bottom. A food chain supplying food larger than zooplankton also is needed for much of the growth period. The food habits of the fish may change during the culture period. The production of larger fingerlings is enhanced by using supplemental feeds, increasing per unit yield without requiring more pond space. As an example, when growing Largemouth Bass fingerlings from 5 to 15 cm, the yield per unit area can be increased 400% by feeding. The potential yield increase for Channel Catfish *Ictalurus punctatus* from artificial feed is even greater.

Space requirement per fish is much greater in large fingerling production. To determine stocking rate the desired size, yield and mortality are considered. For rearing Channel Catfish to a size of 15 cm TL (38 fish/kg) in a pond which will support 2,260 kg/ha with an estimated loss of 10% during a 90-day growth period, the pond should be stocked with about 95,000 small fingerlings 3 to 5 cm TL/ha. If the size desired is to be attained, growth and survival projection must be realistic as density of fish in the pond is a significant influence on the size reached and the rate of growth to reach it.

To illustrate this further, to produce 3 to 4 cm TL Largemouth Bass fingerlings, rearing ponds are stocked at a rate of 175,000/ha. Survival averages 75 to 80% giving a yield/ha of small fingerlings from 130,000 to 140,000 fish. To grow large fingerlings to a size of 10 cm on a diet of natural food (insects) requires that the stocking rate be reduced from 175,000/ha to 25,000/ha. A survival of 80% can be anticipated with a return of 20,000 bass/ha being obtained. Because of the longer production period involved, production (kg/ha) would be 50 to 100% more when the large fingerlings are produced.

If the species being reared to large fingerling size is carnivorous, care must be taken to minimize losses to cannibalism. Stocking fish of uniform size, refraining from overstocking, providing proper nutrition and controlling weeds are means of minimizing cannibalism. Periodic (e.g. 2-week intervals) seine samples to determine size variation enables the fish culturist to tell when losses to cannibalism are likely. When cannibalism becomes a problem, the fish should be harvested, graded into uniform size groups, and restocked in separate ponds or shipped.

When formulated feeds are used to produce larger fingerlings, dissolved oxygen levels should be watched closely. Large fingerlings are expensive to rear dictating that every effort be made to prevent losses once the larger sizes have been reached. Growing two or more species in the same pond or several sizes of the same species together will make for more efficient use of the food supply. Combining phytophagous and insectivorous feeding species is an example of this.

9.5 Recommended Reading

Additional information regarding nursery pond management is available in Chapter 17 and the following references.

Asian Development Bank, Manila, Philippines. Available: https://pdf.usaid.gov/pdf_docs/PNAAV716.pdf. (May 2020).

de Graaf, G., and H. Janssen. 1996. Chapter 5: Fry nursing in earthen ponds in Artificial reproduction and pond rearing of the African Catfish *Clarias gariepinus* in sub-Saharan Africa—a handbook. FAO Fisheries Technical Paper. No. 362. FAO, Rome. 1996. Available: http://www.fao.org/3/W3595E/W3595E00.htm. (May 2020).

Hodson, R. G., and M. Hayes. 1989. Hybrid Striped Bass pond production of fingerlings. SRAC Publication No. 302. Southern Regional Aquaculture Center, Stoneville, Mississippi. Available: http://agrilife.org/fisheries/files/2013/09/SRAC-Publication-No.-303-Hybrid-Striped-Bass-Pond-Production-of-Food-Fish.pdf. (May 2020).

Jhingran, V. G., and R. S. V. Pullin.1985. Chapter 5, pages 61–73, Postlarvae and fry rearing *in* A hatchery manual for the common, Chinese and Indian major carps.

Mischke, C. C. 2011. Summary of current research on catfish fry pond management. MAFES Bulletin 1196. Mississippi Agricultural and Forestry Experiment Station, Stoneville, Mississippi. Available: https://extension.msstate.edu/sites/default/files/pdf//b1196.pdf. (May 2020).

Chapter 10

Feeds and Feeding

10.1 Formulated Feeds in the Hatchery

Dry formulated feeds are not generally given as a first feed for larval fish as the intestinal tracts have not developed enough to digest these complex feeds. As the fish get older and the intestine is more developed they should be offered dry formulated feeds. Formulated feeds have a number of advantages for use in the hatchery. First, they are easier to use than live feeds. Little effort is needed to prepare the feed compared to the effort and facilities needed to produce live feeds. Second, the feed quality is more consistent than live foods and no on-site enrichment is needed before the feed can be used. Formulated feeds are available in sizes as small as a rotifer, and in a variety of progressively larger sizes that make it easier to ensure the fast-growing larvae have an appropriate size food particle to consume. If the fish can be trained to accept formulated feeds in nursery ponds, the production can be increased over ten times what is possible where the fish are dependent on only in-pond production of natural food organisms resulting from pond fertilization. .

10.1.1 Feed Training

Most larval fish feed on some type of zooplankton as a first food. However, as the larvae get older, the larvae of most fishes can be trained to accept formulated feeds. This training is referred to as the weaning period, the transition from one type of food to another or a smaller size food to a larger size food item. The new feed item is added initially at 25% or less of the total quantity of feed added/feeding, and each day thereafter the percentage of the new feed item is increased until only the new feed item is being given.

Larval fish are sight feeders and the swimming pattern of the live foods used in the hatchery help attract the attention of larval fish. Formulated feeds sink over time and do not show any of the erratic swimming patterns seen with live feeds that stimulates the fish larvae to strike at the food item. The larvae need to be fed small quantities of formulated feed frequently so they have a chance to strike at a food particle as it falls through the water column. Formulated feeds need to stay in the water column as long as possible to give the larval fish an opportunity to strike at the food particle. Feeds that sink slowly are preferred. The deeper the container where the larvae are held gives more opportunity for the larval fish to strike at the food particle as it drops through the water column, as most larval fish do not actively feed on formulated feed once it settles to the bottom of the tank.

Weaning from live food to a formulated feed can be challenging. Not all fish within the same set learn at the same speed to accept formulated feeds, and some fish never learn. This results in increased size variation in the fish population, increasing the likelihood of aggression and the potential for cannibalism. To minimize this, it may be necessary to grade the fish into less variable groups during the weaning process if the fish can be handled safely. Close attention needs to be paid to maintaining water quality. To train the fish to feed, an excess of feed is added with much of the initial offerings going uneaten, settling on the bottom, where it will decompose and reduce water quality. It is best to offer the formulated feed six to ten times a day in small quantities until a strong feeding response is seen. Any waste feed collecting on the bottom of the holding container should be removed by siphon at least once a day. Frequency of feedings can be reduced as the fish become larger.

10.1.2 Feed Management in the Hatchery

Formulated feeds should be fed frequently but not continuously. Feeding a small amount of feed continuously can lead to some fish dominating the feeding area, consuming most of the feed, and growing larger than their peers. This can result in poorer feed conversion and cannibalism. A better approach is to divide the total daily quantity of feed to be given into 6 to 10 feedings (photoperiod dependent) so there is more feed given at one time than the dominate fish can consume, resulting in more feed being available for the less dominate fish.

Waste feed will clog the drain screens and the screens should be cleaned several times a day. The key is to observe the fish as they feed as to the quantity of uneaten feed left over after a feeding, and then adjust the quantity of feed offered so all fish have an opportunity to feed but with a minimum quantity of waste feed. Young fish held in the hatchery are typically fed to satiation a high-quality feed that may be expensive/kg but the total weight of feed used is not that great and in most cases the cost of such feed is justified considering the short time it is used and the improvements possible in fish growth and survival.

The frequent feeding can be very labor intensive with feed often needing to be given outside normal office hours of the hatchery staff. Labor requirements can be reduced using some type of mechanical feeder, most commonly a belt-driven feeder or a vibratory feeder. With a belt feeder, a wide belt coiled around a shaft is manually pulled out horizontally and the feed placed on the top of the belt. A clock mechanism or a small electric motor slowly pulls the belt back so it coils in from the top side the shaft. Feed is placed across the belt in 6 to 10 rows. As the belt is pulled forward, any feed on the top of the belt will drop into the water when the belt begins to coil around the shaft. Placing the feed in rows across the top width of the belt allows for a pulse of feed to drop at a time. This pulse of feed gives more fish an opportunity to feed at a given point in time. Placing the feed in a single strip going the length of the belt will result in a more continuous but smaller quantity of feed dropping in at a time often resulting in a few fish dominating the feeding area not allowing smaller fish to feed.

Another style of feeder is a vibratory feeder. Such feeders have an adjustable plate on the bottom of a cylinder holding the feed. The distance between the plate and bottom of the cylinder controls how much feed is dispersed when an electric powered vibrator vibrates the feeder for a controlled period of time. A timer controls how many times/day feed is released.

An issue with both of these styles of feeders and other similar feeders is that a hatchery is a moist environment and when a finely ground formulated feed is used it will adsorb moisture, cake up and become fungused. Such fouling makes it hard to dispense the feed. Attention must be given to keeping the feeders clean to ensure that they are dispersing the proper quantity of feed and no spoiled feed is being consumed by the fish. When using covered belt feeders running an air stone air line into the feeder will bring warm dry air from the blower into the feeder and the warm air can help keep the feed dry.

10.2 Feed Use in a Pond Settling

10.2.1 Supplemental Feeds

There is a limit in the quantity of natural food organisms that can be produced in a pond based on the addition of fertilizers. To increase fish production, outside sources of nutrients need to be added. Natural food organisms are typically high in protein (>50%).

When fish consume only natural food organisms produced in the culture pond part of the consumed protein is metabolized into the building blocks for growth but part of that protein is used to meet the fish's energy requirements. Adding a supplemental feed to help meet the energy requirements can increase fish production over that obtained by fertilization alone. Materials like rice bran may be only 11–12% protein but can be an economical source of carbohydrates to increase fish production. Supplemental feeding in ponds has been used with Channel Catfish *Ictalurus punctatus*, bait minnows, carps and other fishes beginning as soon as the fry start feeding. Feed not consumed by the fish serves as high quality organic fertilizer and thus is not totally wasted. Such materials often are not costly, but in many cases if you are going to put in the labor needed to feed the fish, the use of a more complete feed that provides both protein and carbohydrates as well as lipids will increase fish production enough to justify the higher costs of a complete feed.

10.2.2 Formulated Feeds

Formulated feeds designed to meet the general nutritional requirements of fish are available for some species starting in the hatchery with larvae soon after their mouthparts are functioning. Formulated feeds are often given to fish in primary and secondary nursery ponds, fish in production settings, as well as for broodstock development and maintenance. The type of feed offered, its particle size, its nutrient quality, and the quantity offered will vary with the species and life stage of the fish.

A complete feed should meet all the nutrient requirements of the life stage of the fish being cultured and be of a particle size that can be swallowed completely by the fish being cultured. Larval fish grow rapidly and require high protein feeds. Feeds for these early life stages may contain 45 to 65% protein, much of which is from an animal protein source. Some fish such as channel catfish and tilapia can be given a formulated diet as a first feed and acceptable growth and survival obtained. Others require a natural food initially and as the digestive tract becomes more sophisticated can digest a formulated feed. As the fish gets larger the maintenance requirements increase and growth rates are reduced. Diets for larger fish will have more carbohydrates added to provide energy for maintenance and the protein level reduced as the need for building blocks for growth is reduced. Depending on the species, a complete diet for adult fish may contain 35% protein or less.

10.2.3 Feed Size

Fish spend energy trying to find feed. The smaller the feed particle the more energy is spent finding the feed than when the same total quantity of nutrients is consumed in fewer but larger particles. An optimum size feed particle is near the maximum that a fish can swallow, approximately 25% of the mouth gape of the fish being fed. As fish grow rapidly, the size of feed should be increased to keep up with the growth. Formulated feeds as small as 100–200 microns (rotifer size) and larger (artemia-size) are available for use in reducing the quantity of live feed used in the hatchery phase of larval rearing, and to train fish to feed on formulated feeds.

Once the fish are accepting a formulated feed, then a range of feed sizes are available. Commercial diets are typically categorized as: meals (sometimes called powders), crumbles, or pellets. Meals and crumbles are the result of crushing a pelleted feed that was made from finely ground ingredients, then sieving the crushed materials into several size categories. The crumbled feed is sold by crumble number: 00 (<0.8 mm); 0 (0.6–0.8 mm); 1 (0.8–1.0 mm); 2 (1.0–1.4 mm); and 3 (1.4–1.7 mm), 4 (1.7–2.8 mm). Pelleted feeds are available in sizes of: 1/16" (1.6 mm); 3/32"(2.4 mm); 1/8"(3.2 mm); 5/32" (4.0 mm), 3/16" (4.8 mm); 1/4" (6.5 mm); and 3/8" (9.5 mm). Pelleted feeds may be pressed through a die to form sinking pellets or steam-extruded to give a floating or a slow sinking feed. Although sinking feeds are often cheaper, floating feeds are widely used because of the ease in observing the feeding response.

10.3 Determination of How Much to Feed

10.3.1 Satiation Feeding

How much to feed can be approached in several ways. Feeding to satiation (all they will eat in one feeding) is effective in feed training small fish as well as when growing larger fish. For small fish this should be done in a clear water setting where fish feeding activity and accumulation of waste feed can be observed. Larger fish should be fed a floating feed to facilitate observing feeding activity. The fish should be fed over a several minute period, initially offering a portion of feed and observing the feeding response. If the feed is eaten rapidly then additional feed is offered. This might be repeated several times until only a weak feeding response is seen. A variation of satiation feeding is to feed to satiation for a day or two, recording the quantity of feed offered. The daily quantity consumed during those days is used as a baseline for later feedings. For the next feeding 80 to 90% of the feed consumed the day before is offered. Based on the feed response additional feed is offered until feeding activity is reduced, recording the total quantity of feed offered. The total quantity of feed offered that day will serve as the baseline for feeding the next day.

Satiation feeding is not the most efficient way to feed, as when fish are fed to satiation some of the feed offered is not consumed and what is consumed is not efficiently digested.

10.3.2 Feeding Based on Percent Body Weight

Feeding based on percent body weight is a common practice. The percent body weight offered/day is reduced as the fish grow. Using a low moisture formulated feed, larvae and young juveniles might be offered 15–25% body weight, juveniles 10–15%, sub-adults 5–10%, adults 3% and brooders 1.5–2%. The quantity of feed can be calculated based on percent body weight as follows:

$$\begin{aligned}\text{Quantity of feed} &= \text{Fish biomass} \times \text{feed rate (\%)}\\ &= (\text{Av. body weight} \times \text{no. fish}) \times \text{feed rate (\%)}\\ &= [\text{Av. Wt.} \times (\text{no. fish stocked} - \text{mortality})] \times \text{feed rate (\%)}\end{aligned}$$

The biggest challenges when feeding based on percentage body weight are knowing the average body weight and the number of fish on hand. At the time of stocking the number and size of fish are known, but every day thereafter these change as the fish grow and as mortality may occur. Mortality is difficult to determine in most production settings. Trying to remove and count all dead fish is good management, but many fish may be missed, particularly in pond settings where dead fish may be consumed by birds and other animals or are not accessible for collection. The best estimates of mortality can often be based on past records of mortality rates where similar production practices were used. Samples will also suggest magnitudes of mortality and when it occurred. The more difficult it is to collect a sample, and the greater the size of the fish collected compared to the expected size, suggest a greater mortality. Larger than expected fish of similar size suggests mortality occurred soon after stocking and there is abundance of food and little competition for the food. A larger than expected size of fish but of variable size suggests a high enough density initially to give the size variation with a later mortality in the culture period.

Growth rates are not constant but vary with fish size, feed quantity and quality, and environmental quality. Average weights can be estimated by sampling if a fair sample can be obtained. Often a sample size of 10% is recommended, but in many cases this is not practical when working with a large population. A practical approach to establish a proper sample size is to take three samples of a common size and compare the means. If they are different, increase the sample size and repeat the process until a sample size is found where the three samples have similar mean individual weights. Once an appropriate sample size is found for a given size fish in a specific production setting, use that sample size for future sampling. How you take the sample can result in a sample mean weight that is different than the population as a whole. An estimate of average weight is influenced by the equipment used to collect the sample and whether all fish sizes are evenly distributed within the production setting. When collecting the sample avoid picking out individual fish to make up a sample as this might bias the sample mean weight. It is best to collect sufficient fish in one scoop to make the sample, then counting the number of fish in the sample.

10.3.3 Feeding Based on Anticipated Growth

Ideally, we could avoid over-feeding associated with satiation feeding and more accurately feed as a percentage of the biomass on hand, but determination of biomass has its problems. Sampling is stressful to the fish and can lead to reduced growth rates and disease outbreaks. The frequency of sampling can be reduced by keeping good records of past performance and the environmental conditions under which the fish were reared and use that information to predict growth rates and in turn biomass. Growth rate can be anticipated based on actual growth rates recorded when fish were previously produced under conditions equal to the current conditions, however any variation in those conditions might result in a slower or faster growth rate. Some sampling will still be needed to determine how close the actual average weights and the predicted weights match. Adding the anticipated weight gain per fish since the last sample to the average weight at the last sample, a new average weight can be calculated and used to determine the new quantity of feed to give.

Quantity of feed =
= Today's Fish biomass × feed rate (%)
= [{(Initial Av. body weight + (daily wt. gain × no. days)} × no. fish}] × feed rate (%)
= [Estimated new Av. Wt. × (no. fish stocked – mortality)] × feed rate (%)

The issue with this approach is actual mortality is still not known and has to be estimated from previous production efforts.

10.3.4 Feeding Based on Feed Conversion

Using the feed conversion records is another approach to estimate biomass, and in turn the quantity of feed to offer. Where the quantity of feed needed to obtain a kilo of gain of fish biomass (Feed conversion ratio, FCR) is known it is possible to estimate the gain in biomass based on the total quantity of feed given in a specific time period assuming that all the feed offered is consumed. That gain in weight combined with the initial biomass should equal the current standing crop, and from that the quantity of feed to be offered can be calculated.

Quantity of feed =
= Today's Fish biomass × feed rate (%)
= (Initial biomass + biomass gain) × feed rate (%)
= (Initial biomass + (Total feed fed/feed conversion ratio) × feed rate (%)

The FCR approach has the benefit of not being as dependent on adjusting for any mortality that occurred during the production cycle. The key is not overfeeding and maintaining good water quality so the feed conversion will be similar to past production cycles.

This approach works particularly well where the fish are confined with little to no natural food available for consumption. Natural food contributes to the FCR making the conversion of formulated feed appear to be more efficient than it actually is. The availability of natural food can vary from production cycle to production cycle and impact the apparent FCR.

10.4 How to Feed in a Pond Setting

Even though young fish were accepting a formulated feed in the hatchery they may not initially show much of a feeding response once they are stocked into a pond. The fish often become dispersed foraging on natural food. When the formulated feed is first offered it should be scattered along the shallow areas of the pond so the fish can find it. Once a feeding response is consistently seen, the feed can be scattered over a smaller area of 2–3 m. As the fish get larger the area should be expanded to ensure that all fish have an opportunity to feed. The fish can be trained to come to a specific area to feed. Preferably, that area should be on the lee side of the pond where the prevailing winds would not quickly blow the feed up on to the shore. The feeding area should also be near the deeper area of the pond where the fish often are found. Aquatic weeds need to be kept under control as they can limit access to the feed.

Small fish (<5 cm) in ponds should be fed four times a day, 5–10 cm fish 3–4 times/d, 10 cm fish, 2 times/d, and larger fish once a day when temperatures are optimum for that species. Feed quantities and feeding frequencies should be adjusted as a function of water temperature. The lower the temperature is from the optimum growth temperature the lesser quantity of feed should be offered and at a reduced frequency, feeding during the warmest portion of the day. When the water temperature is above optimum the fish may be stressed and consume less feed. In that case it can be beneficial to feed earlier in the day if the water temperature is lower.

Mechanical feeders are available that can broadcast feed out over the pond multiple times a day. Such feeders can be set to offer feed at specific times in specific quantities. Where labor is limited or remote ponds need to be managed, mechanical feeders can be appropriate. It is important that such feeders are designed to be waterproof otherwise the feed may get wet and spoil. Another issue can be wild animals such as raccoons or rats that may try to enter the feeder to obtain food. A combination of manual feeding for the first feeding in the morning and the last feeding of the day along with a mechanical feeder used for any additional feedings during the day may be practical in some cases. Floating feed should be used so that any earlier uneaten feed can be observed before the last feeding of the day is given. Mechanical feeders should be checked daily to ensure that there has not been a loss of electricity to the feeder and that the timer is set correctly.

Dissolved oxygen concentrations should be considered when selecting a time of day to feed. Often, overnight dissolved oxygen values are below saturation and may remain so for an hour or more after sunrise. Fish appetite is reduced at low oxygen concentrations and feeding should be delayed until dissolved oxygen levels are at saturation or above. This also applies to late afternoon feedings. If dissolved oxygen levels are below saturation, fish will feed less actively and any waste feed may contribute to even lower dissolved oxygen levels overnight.

The increased nutrient load in a pond associated with feeding will result in increased phytoplankton, an increased oxygen demand in the pond, and an increased risk of a plankton die-off. Depending on soil and water quality characteristics, emergency aeration may need to be available when feeding rates reach about 50 kg/ha and nightly supplemental aeration be provided when feeding rates near 100 kg/ha.

Other combinations of water quality also may suppress fish appetite. In a highly fertile pond setting afternoon temperatures can be high and the pH may increase as a result of photosynthesis. This will shift the NH_4-NH_3 equilibrium toward more of the toxic NH3, stressing the fish and reducing the appetite. Water quality parameters should be considered when selecting a feeding time.

Feed is often the most expensive part of the variable costs in producing a crop of fish. Tables 10.1, 10.2, and 10.3 can be developed as guidelines for feeding, but a well-trained hatchery staff that can observe the daily activity of the fish, and monitor water quality, weather conditions and other influences that may affect feed consumption, are essential for a successful hatchery.

Table 10.1. Guide for feeding Oregon Moist Pellet to Largemouth Bass *Micropterus salmoides* at water temperatures in the range of 26– 31°C.

Fish size (cm)	*N*/kg	Feed Size (mm)	% Body weight/d	Feeding frequency/d
5.0	>1000	1.6	12	5–6
7.5	220–1000	2.4	10	5
10.0	66–220	3.8	8	4
12.7	44–66	4.8	5	3
15-25	6–44	6.4	3	2
>25	<4	12.7	1–2	1

Table 10.2. Feeding guide for Channel Catfish *Ictalurus punctatus* according to temperature and size.

Average fish size		Mean water temperature °C.								
		18.3	19.4	20.6	21.7	22.8	23.9	25	26.1	27.2
Length(cm)	Weight(g)	% body weight/day to be fed								
7.6	3.5	1.2	2.4	3.6	4.8	6	7.2	8.4	9.6	10.8
10.2	8.4	0.9	1.8	2.7	3.6	4.5	5.4	6.3	7.2	8.1
12.7	16.1	0.7	1.4	2.2	2.9	3.6	4.3	5	5.8	6.5
15.2	27.2	0.6	1.2	1.8	2.4	3	3.6	4.2	4.8	5.4
17.8	43.3	0.5	1	1.5	2.1	2.6	3.1	3.6	4.1	4.6
20.3	63.8	0.5	0.9	1.4	1.8	2.3	2.7	3.2	3.6	4.1
22.9	90.9	0.4	0.8	1.2	1.6	2	2.4	2.8	3.2	3.6
25.4	123.3	0.4	0.7	1.1	1.4	1.8	2.2	2.5	2.9	3.2
27.9	170.2	0.3	0.7	1	1.3	1.6	2	2.3	1.6	2.9
30.5	228.4	0.3	0.6	0.9	1.2	1.5	1.8	2.1	2.4	2.7
33	296.3	0.3	0.6	0.8	1.1	1.4	1.7	1.9	2.2	2.5

Table 10.3. Feeding schedule for Channel Catfish *Ictalurus punctatus* fingerlings.

Size of Fish (mm)	age (days)	feed size (per day)	feeding frequency	% body wt/day
9–10	5–8	dust	5–6	15
12–15	10–12	starter	5	15
18–22	21	#2 crumble	4	10
25–31	31	#3 crumble	3	5
40–50	45	crumble	3	4
70–80	60	crumble	2	3

10.5 Recommended Reading

Additional information regarding feeds and feeding is available in Chapter 17.9 and the following references.

Crag, S., L. Helfrich, D. D. Kuhn, and M. H. Schwarz. 2017. Understanding fish nutrition, feeds, and feeding. Virginia Tech Publication 420–256, College of Agriculture and Life Sciences, Virginia Tech, Virginia. Available: https://fisheries.tamu.edu/files/2019/01/FST-269.pdf. (May 2020).

Davis, D. A., editor. 2015. Feed and feeding practices in aquaculture. Woodhead Publishing, Technology & Engineering, Cambridge, UK.

Gatlin III, D. M. 2010. Principles of Fish Nutrition. SRAC Publication No. 5003 Southern Regional Aquaculture Center, Stoneville, Mississippi. Available: http://fisheries.tamu.edu/files/2013/09/SRAC-Publication-No.-5003-Principles-of-Fish-Nutrition.pdf. (May 2020).

Halver, J. E., and R. W. Hardy, editors. 2002. Fish nutrition, 3rd edition. Academic Press, San Diego, California.

Holt, G. J., editor. 2011. Larval Fish Nutrition. Wiley-Blackwell, Chichester, UK.

Rosentrater, K. A. Commercial aquaculture feed production (floating feeds). 2013. Available: https://www.youtube.com/watch?v=KoEVYwFPCx8. (May 2020).

Chapter 11

Harvesting and Handling

A major distinction between a hatchery and many other forms of aquaculture is that a hatchery's goal to produce a live healthy product. When fish are produced for consumption they need only to remain alive a limited time after harvest but in a hatchery the fish must be harvested and handled in a manner so they will remain alive and healthy for an extended period of time. A successful harvest requires advanced preparation and scheduling. The recipients of the harvest must be notified and be prepared to receive the fish, holding facilities at the hatchery must be available, and a distribution schedule prepared. The method of harvesting is an important management decision that must be made well in advance of the harvesting season when fish are to be produced in ponds. Methods of harvesting ordinarily used are seining (pond full or nearly so), trapping, draining, or a combination of these.

11.1 Harvesting Techniques

11.1.1 Seining

Seining is the method of choice when the number of fish needed is relatively small or when ponds cannot be drained, and the fish are large enough to be handled safely. Seining is used to reduce the standing crop in a pond when the standing crop is too great to be safely harvested by draining the entire quantity of fish into a catch basin for harvest. Seining of a full pond can be efficient for larger fish that tolerate handling and can be concentrated in a small area in response to being offered a formulated feed or are a species that moves in a compact school.

Primary nursery ponds are usually not harvested by seining until the pond water level is lowered and the fish concentrated in the catch basin. At harvest the standing crop in a primary nursery pond is not too high, and the fish may be too delicate to be seined. A seine can be used in a primary nursery pond to crowd the fish, allowing the fish to be scooped up in water for transfer to a transport tank.

Some fish species are more easily seined than others. Fast swimming or bottom dwelling species often cannot be harvested efficiently by seining. Larger ponds become increasingly difficult to harvest. Considerable manpower or a mechanical seine puller is required for large ponds. Ponds with notable amounts of filamentous algae and rooted weeds can be very difficult to seine. Appropriate vegetation control should be applied far enough in advance to eliminate such weeds and allow water quality to return to normal before the start of the harvest. Seining is stressful to the fish, particularly when water temperature is high. Seining is not sufficient to completely harvest most ponds, and at some point the pond mush be drained.

The type of seine used impacts the efficacy of seining and the health of the fish post-harvest. The mesh size should be the largest possible to facilitate pulling the seine through the water, but not too large to avoid the fish's head being caught in the mesh of the net as if it was a gill net. Too small a mesh size will make the net too hard to pull, and allow it to fill with filamentous algae and mud that can suffocate the fish and make harvesting very difficult. The texture of the seine material can also impact the health of the fish being harvested. Smooth woven netting is preferred for fingerlings of scaled fish over knotted netting, as knotted netting is more abrasive on the skin of the fish. Fish with sharp serrated dorsal and pectoral spines such as catfish are more easily harvested using a net which has been coated with a tar-like compound making the netting stiffer and the weave less likely to be penetrated by a spine.

The depth of the seine should be approximately one third more than the depth of the pond. As the seine is pulled through the water, the water pressure will push on the mesh causing it to bow. If the depth of the seine is not sufficient, the bowing of the net may cause it to be pulled off the bottom and below the surface, allowing fish to escape. The length of the seine should take into consideration the quantity that can be expediently harvested, and the size and shape of the pond. When seining you want to catch only as much as you can efficiently collect from the seine and transport without causing undo stress on the population. In some instances, a shorter seine catching only a portion of the population per seine haul and making multiple seine hauls is preferred over using one long seine catching all the fish in one seining. Where all the fish are desired in one seine haul the length of the net should be one third greater than the width of the pond to allow for bowing of the seine as it is pulled while making sure that there is enough of the seine to reach both shorelines. If the pond bottom is very soft a mud or rope line should be added to the bottom lead line of the seine to minimize it digging into the mud (Figure 11.1).

Figure 11.1. A mud line added to the bottom lead line of the seine to minimize the seine from digging into the mud.

11.1.2 Trapping

Harvesting fingerlings by trapping is a useful pond harvesting technique when harvesting delicate species or where only a small quantity of fish are needed at one time. If used properly, traps can harvest fish with a minimum stress to the fish, an important advantage when having to harvest in warm water. However, they are not an effective way to completely harvest a pond.

Effective trapping requires a good understanding of the fish habits and what stimuli they respond to. The more common stimuli include response to light (both positive and negative), response to food, attraction to an inflowing current of water, following an outflowing current, and schooling.

Water flow into or out of a pond can stimulate the fish to congregate in specific areas. Striped Bass *Morone saxatilis* will respond to flowing water and can be captured by placing a trap in front of the water inlet where a current can flow through the trap. Opening the drain will attract 5–6 cm Largemouth Bass *Micropterus salmoides* to gather near the drain. Sometimes it takes a combination of factors to make a fish easier to trap. A slow release of water from the pond when the air temperature is below freezing has been found to stimulate Bluegill *Lepomis macrochirus* to congregate near the drain, facilitating their trapping and seining. Schooling fish that cruise along the shoreline are particularly susceptible to trapping as they can be guided into a trap by setting a section of netting out into the pond to act as a wing to direct the fish into the trap. Fish can also be lured into a trap by placing feed within the trap. A combination of trapping and seining can be used by placing a seine several meters offshore parallel to the bank and the fish are trained to feed between the seine and the bank. Once the fish are accustomed to feeding in that area, a harvest can be made placing feed in the area, allowing the fish to feed, and then closing the seine around them.

A variety of traps that have been used successfully in production hatcheries to harvest substantial numbers of fingerling fishes. A widely used trap in the United States is the glass V-trap. This trap is effective in capturing several species under a variety of conditions, is quite portable (Figure 11.2) and can be operated by one person. Up to 80% or more of fingerling fish in a pond have been harvested with this type trap.

Advantages of trapping include minimal stress on the fish being harvested, efficient use of labor when the trap is working successfully, and their suitability in ponds where aquatic weeds are abundant or cannot be drained or seined efficiently. The primary disadvantages of harvesting by trapping include a lack of predictability as to supplying needs for a scheduled delivery, and cost of building and maintaining the traps needed for a large-volume operation. Traps may be used to harvest as many fish as possible from the rearing pond but the pond still has to be drained to make a complete harvest.

11.1.3 Draining

Draining is generally the most economical, rapid, and efficient means of harvesting a pond. Primary nursery ponds are best harvested by draining into a catch basin supplied with clean water because the fish being harvested are often small and delicate to handle. Secondary nursery ponds and ponds with high standing crops of large fish might be seined or trapped to reduce the standing crop but the pond has to be drained to complete the harvest. To effectively drain a pond, it should be properly sloped to drain into the catch basin, be free of weeds, and without potholes. The drainage system to receive the pond effluent should be functioning properly so that it does not impede pond draining.

Figure 11.2. Glass V-trap with a wing to guide fish to the trap.

Careful planning is also a part of a successful pond draining. It is critical to know the approximate time needed to drain a pond so the water level can be low enough to begin the harvest at precisely a given time. Drain screens need to be cleaned periodically to keep the pond draining on schedule. Lowering the water level too fast may create a water current, causing some species of fish to swim against the current and become stranded out in the pond. Draining a pond too slowly during the day can result higher water temperatures and more stress on the fish. When parts of the pond bottom become exposed the outflow rate may need to be reduced to avoid stranding fish in flat areas or in potholes. Ideally the pond should be harvested before mid-morning, before the water temperature starts to rise rapidly. Often it is desirable to lower the pond volume during the day, stopping during the night at a volume that can be drained into the catch basin within an hour starting soon after sunrise. Having a lower volume during the night will allow the water to cool down, in turn, reducing fish metabolic rates and oxygen demand. However, lowering the water makes the fish more susceptible to predation. If herons and other birds are abundant it may require hatchery staff to periodically be at the pond during the night to chase off predators. During the night the dissolved oxygen concentrations should be monitored. Low dissolved oxygen in the catch basin is common when draining ponds with highly organic sediments. Having an oxygen bottle available on the pond bank with an air stone and sufficient tubing to reach the catch basin is good insurance for having a safe harvest from the catch basin. Having a source of clean cool water available to dilute the poor quality pond water in the catch basin is important to minimize stress on the fish being harvested and to improve survival. Stress can be minimized by careful handling of the fish and providing the best water quality possible. Loud noise, or sudden or rapid movement can frighten the fish, adding to their stress causing them to jump out of the catch basin.

Silt and high water temperatures can stress fingerlings to the point where direct loss occurs or disease develops.

Small fish are best harvested from a catch basin, scooping them up in buckets of water and immediately moved to the transport tank. Fish that are more tolerant to handling can be harvested from ponds without catch basins, first lowering the water level of the pond, then using a large seine to crowd the fish together at the end of the pond nearest the transport tank. Within the area surrounding the large seine the fish can be harvested using a smaller seine to further crowd the fish together so they can be scooped up and quickly be moved to the transport tank.

11.2 Harvesting Considerations

A key to a successful harvest is harvesting what can be safely handled at one time. This includes not seining up more fish than what can be expediently handled, and using an appropriate size transport container/s for the quantity of fish being harvested. Pond banks may be slippery and muddy, and often a stairway adjacent to the catch basin going up the pond bank is desirable to avoid spilling a bucket of fish (Figure 2.1). A backhoe tractor with a boom may be used to lift a bucket of water containing fish or a net basket of fish to the transport tank. The fish should be handled in small enough quantities in the collecting net, carrying containers, and transport tank to hold stress to a minimum.

It is important in minimizing handling stress to unload the transport container frequently. Many a fish has been lost after being stressed while held in the transport tank waiting for the last few fish to be collected from the pond. Delays in unloading, which creates stress, can result in delayed mortality from diseases even though no immediate damage to the fish is evident. It is often desirable to have two or more transport tanks so once one is full it is taken to the holding tanks and another transport tank is available at the pond bank to receive more fish. With only one transport tank you have to wait while the transport tank is taken to the holding tanks and unloaded, then taken back to the pond. In the meantime, the fish held in the catch basin or in the seine are becoming stressed. You should always think, "what is the best for the health of the population?" When more are caught in the seine than can be transported properly, it is best to release the excess to be harvested later. This also can mean taking a full load of fish to the holding tanks versus taking a few more minutes to pick up two to three fish that might be stranded in a mud puddle while risking the lives of the rest of the population already loaded into the transport tank.

Careful planning is needed to assure a successful harvesting operation. A trained crew with suitable collecting gear, machinery, time, and working conditions is needed. Once the fish start accumulating in the catch basin, there is no time to go find something you forgot or find a replacement for a broken piece of equipment that you forgot to test beforehand. Seines, nets, hauling containers, and other equipment should be located and inspected well in advance to be sure everything is available and in working order. Weather conditions should be anticipated. Extremes of heat or cold can stress both the fish crop and the hatchery staff resulting in fish losses and lower efficiency, respectively. Holding facilities should be ready to receive the fish being harvested. Water in supply lines may go anaerobic if left stagnant for extended periods of time. Water supply lines to the holding tanks should be flushed, and dissolved oxygen concentrations in the holding tanks checked before any fish are placed in the holding tanks.

The amount of time required to harvest a rearing pond stocked with fingerlings of a particular species will vary depending upon the species, sizes involved, the total biomass on hand, pond conditions, time of year and other factors. Table 11.1 shows estimated time requirements for harvesting a 0.32 ha pond stocked with either small Bluegill sunfish or Largemouth Bass fingerlings. The total hours of labor needed to harvest the Bluegill pond is more than three times more than needed to harvest the same size pond of bass. However, when viewed as the number of fish harvested/hour of labor spent, the labor requirements are similar.

The size of the transport tank should be in proportion to the quantity of fish to be harvested. Ideally, the fish are held in the tank at the pond for less than 15 min before being transported to the holding tanks. A 2.4 m^3 tank mounted on a pickup truck is commonly used for transporting the fish (Figure 11.3). See Chapter 12 for more details on fish transport. The transport tank should be filled with clean clear water, ideally of similar quality as that of the holding tanks.

11.3 Holding and Cleaning the Harvest

When fish are harvested from ponds they need to be held and prepared for on-station distribution or to be transported off-station. A typical pond harvest will contain a range of sizes of the desired fish and may be contaminated with undesirable fish, tadpoles, aquatic weeds, and other undesirable material. The fish will have been stressed during harvest and often transported in less than optimum water quality. They will need a chance to recover from the harvest and be prepared for the next phase in their use.

11.3.1 Holding

Holding facilities as described in Chapter 2.3 are needed where the harvested fish can be held for several days, the results of the pond harvest evaluated, and the fish prepared for distribution. The number and size of tanks depends on the number of ponds to be harvested at one time, the number of species to be harvested, and the quantity of fish expected/pond. Where large quantities of one species are harvested and distributed at once then a few large tanks are

Table 11.1. Estimated time requirements for harvesting fingerlings of Largemouth Bass *Micropterus salmoides* or Bluegill *Lepomis macrochirus* from a 0.32 ha earthen pond by draining.

	Person-hours by species	
Operation	Largemouth Bass	Bluegill sunfish
Preparing for draining	0.5	0.5
Lowering water level	0.5	0.5
Cleaning the harvest basin	0.2	0.2
Capture and transfer of fish	6.0	10.0
Cleaning up work site	0.2	11.4
Total hours required for harvest	7.4	22.6
Average no. fish harvested, (thousands)	50.0	160.0
Number harvested/person-hr.	6,757.0	7,080.0

Figure 11.3. A small transport tank for on-farm use and local deliveries (Photo courtesy of Charles Weibel, Kentucky State University).

adequate. Where various species are produced or various sizes are to be held for distribution, then a number of smaller tanks are appropriate. A rectangular tank design is convenient as it allows an efficient use of floor space, ease of harvest, and fish grading. Ideally the vehicle with the transport tank can drive up to the holding tank with the drain of the transport tank being close to the holding tank and higher than the water level of the holding tank. Using a pipe or chute connected to the drain of the transport tank, opening the drain will allow the fish to flow by gravity into the holding tank, and minimize the handling of the fish. The loading rate in the holding tank depends on the species and its size, the water temperature and water quality. Table 11.2 gives examples of some holding rates. Typical rates range from 10 to 50 kg/m^3 in tanks with a twice per hour water exchange rate. Electric agitators or air blowers are used to ensure that dissolved oxygen concentrations in the holding tank water are near saturation.

After the harvested fish are unloaded into the holding tank they should be left undisturbed for several hours so that they can recover from the shock of being moved from the pond. Oxygen demand of the fish will be high at that point and extra effort is needed to ensure that an adequate oxygen supply is available. The length of time the fish are held depends upon local conditions. The greater the harvest stress the longer the resting time should be before starting to prepare the fish for re-stocking. The cooler the water temperature the longer it will take for the fish to empty their digestive tract and be ready for transport. Figure 11.4 shows the steps involved in a typical harvesting, processing and distribution operation.

11.3.2 Cleaning

A strong water flow in the holding tank may be used initially after the fish are unloaded into the holding tank to wash out mud and fine particulate matter. Once that is done the water

Table 11.2. A guide for loading holding tanks with fry or fingerlings[1]

Species	Size	No./m^3	Holding time (days)	Purpose
Channel Catfish	sac & swim-up fry	100,000	3–10	yolk sac absorption and first feeding
Common Carp	"	125,000	" "	" "
Largemouth Bass	"	100,000	" "	" "
Striped Bass	"	125,000	" "	" "
Tilapia	"	150,000	" "	" "
Channel Catfish	7.5 cm	50,000	10–14	temporary holding[1], light feeding
Common Carp	5.0 cm	75,000	10–14	" "
Largemouth Bass	5.0 cm	50,000	3	temporary holding
Striped Bass	5.0 cm	50,000	3–7	temporary holding, light feeding
Tilapia	5.0 cm	75,000	3–7	temporary holding, light feeding

[1]Well oxygenated water supply, 2 exchanges/h

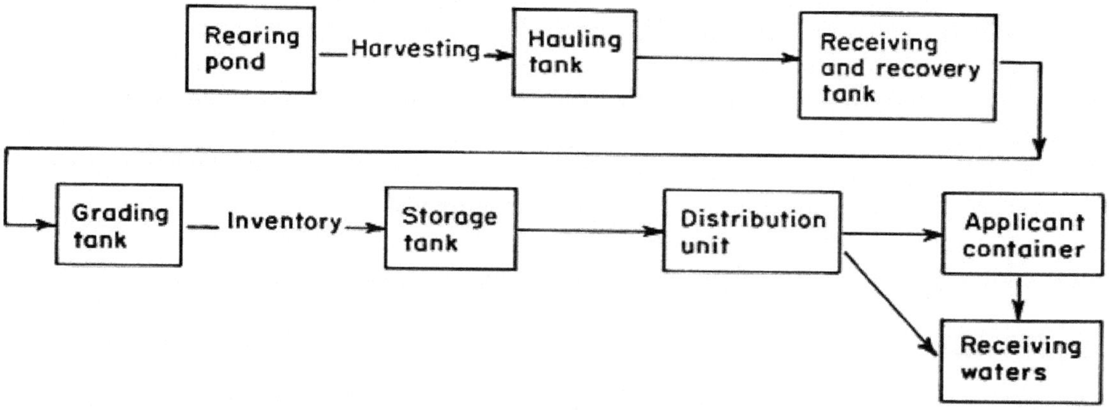

Figure 11.4. Pathway from harvest to distribution of fish.

flow rate can be reduced and additional cleaning begun. Extraneous plant and animal material interferes with grading, counting, and weighing and must be removed. Fish counts extrapolated from number-weight relationships are not accurate without first removing any extraneous material. Many undesirable weeds or animals have been spread from one location to another because adequate attention was not paid to cleaning the fish before their transport. A hatchery has an obligation to not introduce any exotic plant or animal when transporting fish to a new location. Extraneous material mixed in with the cultured fish is unsightly, may introduce an unwanted plant or animal, interferes with fish handling, and raises questions by the recipient as to the quality of the fish received.

The approach to cleaning a tank of fish is dependent on the nature of the contaminant. Some weeds and debris can be allowed to settle to the bottom of the tank and be removed with a strong siphon. Tadpoles can be exposed to a formalin treatment causing them to come to the surface in distress where a dip net can be used to remove them. Green Sunfish *Lepomis cyanellus* fingerlings (a common pond contaminate in the southeastern USA) can be separated from Bluegills by raising the water level within 2 cm of the lip of the holding tank and holding it at that point for several hours. The Green Sunfish will jump from the tank to the floor where they can be collected and disposed of. If the contaminant material cannot be removed from the fish, sometimes the fish can be removed from the contaminant by sorting them over a counting table. One quickly learns the value of doing a good job in pond preparation, preventing the development of weeds, insects, tadpoles or unwanted fishes, to avoid having to spend hours cleaning a load of fish before it can be shipped.

11.4 Grading

Grading or sorting of fish into various size classes can be done after or during cleaning. Grading reduces size differences, which makes numerical determination based upon number-weight estimates more precise. Reducing the size variation in a lot of fish will help to reduce cannibalism. Grading improves the appearance of a lot of fish as presence of large fish makes smaller ones less attractive. Grading also reduces the magnitude of size variation later in their growth. Large fish are more valuable than smaller ones if the fish are being sold, and should be sorted out. As an example, Channel Catfish *Ictalurus punctatus* fingerlings are sometimes sold on the basis of length. A 7.5 cm fingerling may be worth 50% more than a 5 cm one. If 20% of the pond yield were 7.5 cm fingerlings, grading the fish so the 7.5 cm fish could be sold at a higher price would improve the overall profitability of the pond harvest.

Grading can be done in several fashions depending on the body shape of the fish, whether that species tries to swim laterally or downward when confined, and how sensitive the fish are to handling. A bar grader is designed to fit the inside width of the holding tank, typically with vertical bars evenly spaced over a frame extending from the tank bottom to out of the water. The frame may have an outside rubber boarder to ensure that there are no gaps between the tank wall and the frame of the grader.

Several grader panels with different bar spacings are needed to separate the various size-groups of fingerlings. For small fingerlings, spacings of 4.8 mm, 6.0 mm, and 7.1 mm are used most often. Other desirable sizes are 4.0 mm, 8.3 mm, and 9.5 mm. To make a grader which will fit a tank 76.2 cm wide and 66 cm deep, a panel of 6, 25.4 x 40.6 cm grader grills (commercially available) are welded together and fitted with a handle of 9.5 mm diameter rolled steel rod (Figure 11.5a).

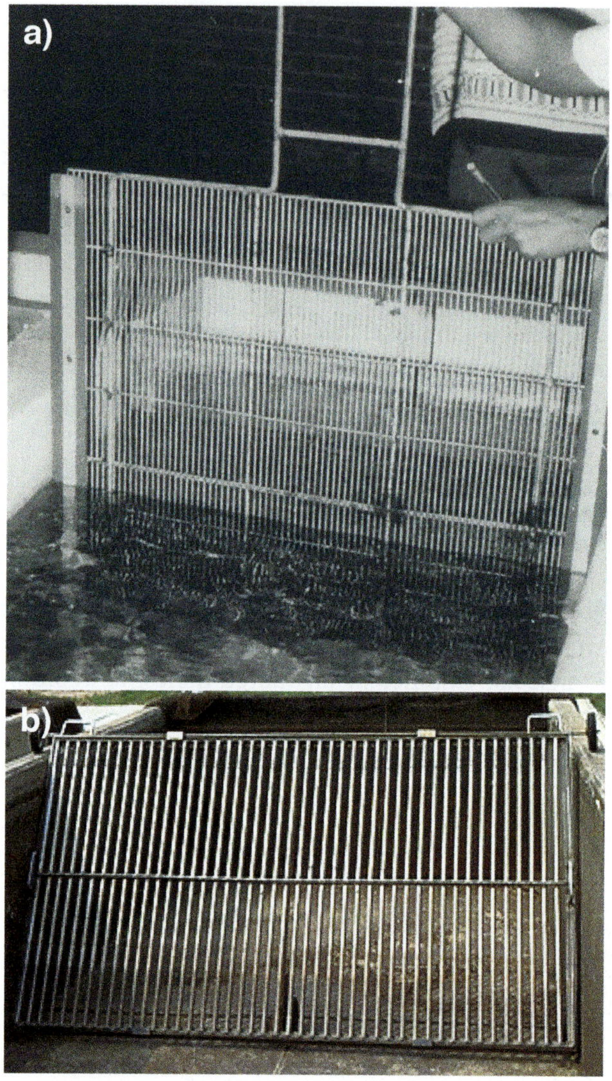

Figure 11.5. a) Small tank bar grader; b) large tank bar grader.

The grader is placed across the tank at the drain end of the holding tank and slowly pushed up the tank, allowing the smaller fish to swim through the bars while crowding the larger fish into the upper end of the tank. After forcing as many small fish as possible through the grader, a tank net is dropped between the grader and the end wall of the tank, capturing the larger fish that could not pass through the grader bars and transferring them to another tank. This can be repeated several times using graders of progressively smaller bar spacing to separate the fish into several size categories. This approach minimizes the amount of handling needed.

A simple and effective way to grade fingerling fishes is using a series of 2, 3, or 4 nested grading boxes. These boxes are enclosed on 5 sides with galvanized hardware cloth or plastic-coated wire cloth. The plastic-coated wire is the preferred material as it is more durable and less likely to injure fish. Generally, a series of 3 or 4 boxes will be built to nest one inside another with enough space between boxes to hold 5 to 15 kg of similar size fish. The body width and depth of fish determines its passage through a given mesh size. Commonly the inner box is 2.54 cm mesh followed in turn with boxes of 1.27 cm mesh, 0.84 cm mesh and 0.64 cm mesh. To use the grader, the nesting boxes are placed in the grading tank and ungraded fish poured into the inner box. The fish pass through the various meshes until they cannot pass through a mesh opening. The smallest size may be collected outside the outermost basket. When enough fish of a given size accumulate to make removal necessary, the grading boxes are separated and the collected fingerlings of each size are weighed into a holding tank.

A similar approach is using a series of floating grader boxes where an interchangeable set of grader bars are used. Having interchangeable grader bars allows you to grade a group of small fish and then using the same set of floating grader frames with a different set of grader bars a group of larger fish can be graded (Figure 11.6). Ungraded fish are transferred to a new tank or fine mesh hapa containing a floating grader with a bar spacing sufficient to allow the smaller fish to swim out but holds the larger fish. Fish are added to the grader and given a few minutes to sort themselves out. The grader may be gently lifted up to concentrate the fish and encourage the small fish to swim out. Fish remaining in the grader are passed to a new tank with a grader with a larger bar spacing where the fish are allowed to separate themselves. This process may be repeated several times resulting in 4 to 6 size categories being graded out. Bar spacing are measured in 64-ths of an inch. Common sizes available are 29, 31, 32, 33, 36, 38, 40, 42, 44, 46, 52, 58, and 62/64". However, using this type of grader requires more handling of the fish.

The most appropriate grid pattern and width depends on the cross-section appearance of the fish to be graded. Table 11.3 gives the approximate sizes of Bluegill fingerlings, a deep-bodied fish, that are separated by passing through the specified mesh size.

Figure 11.6. Floating bar grader.

Table 11.3. Sizes of Bluegill *Lepomis macrochirus* fingerlings that can be separated by grader mesh size.

Grader mesh size (cm)	Bluegill TL (cm)
0.64	2.00
0.84	3.50
1.27	5.00
2.54	11.00

11.5 Sampling, Disease Control, and Hardening

After grading, the number and average weight of the fish in each size category needs to be determined by establishing a number-weight relationship based on three or more random samples, each sample weighing 100 to 500 g or more depending on the fish size. Each sample should contain at least 20 fish to determine the average weight when working with larger fish and more fish if they are smaller. If the difference between samples is as great as 10 to 15%, an additional sample or two may be counted. Dividing the sample average weight into the total weight of fish in a grade category will give the total number of fish in that category. Summing the numbers and weights by grade category will give the quantity of fish by category available for distribution and restocking.

After being handled and graded, the fish have been stressed and can be susceptible to diseases. Prophylactic treatments to control bacteria and parasites should be administered at this point. Trying to prevent a disease outbreak at this point is much more practical and economical than to run the risk of a disease outbreak once the fish are stocked in a pond where control is more difficult and expensive. See Chapter 17.6 for details on disease treatments.

The fish should be held an additional 24 to 48 h to allow them to recover from handling and any associated mortality has ended. The holding period allows the fish to empty their digestive tracts before being distributed. Shipment of fish with empty stomachs reduces metabolic waste in the distribution container and possible water quality issues, and also reduces the possibility of transferring weed seeds and algal spores from the hatchery pond to waters receiving the fish.

Holding the harvested fish after processing, especially in cooler water, allows for what fish culturists refer to as "hardening." One aspect of stress is weakened control of the osmotic balance. Freshwater fish will absorb water when stressed causing the scales to not be held close to the body, resulting in scales being knocked off when the fish are handled. Holding the fish under good conditions allows the fish to recover from the handling stress, regaining control of their osmotic balance, and pulling the scales back against the body. Adding salt (NaCl) or calcium chloride at 1 to 5 ppt to the holding tanks will help reduce the osmotic gradient between the fish and the holding water, reduce the stress from handling and allow the fish to recover faster. Allowing the fish to "harden" will make the fish less subject to scale loss during transport.

The time required to harvest and process a given quantity of fingerlings is highly variable and is dependent on such divergent factors as the weather and the number of fish present per unit area of rearing pond. To ensure that distribution schedules are met, it is good to plan ahead, reviewing the approximate size and number of fish in each pond available for harvest. Experience as to how specific problems are likely to affect the harvesting and processing time schedule is helpful. Extra grading, cleaning, disease treatment or a delay to control weeds can change plans and in turn distribution schedules. Reducing the processing period from 48 to 24 hours gives some flexibility in an emergency. Another aid to help ensure a distribution schedule is kept is to have previously processed fish stored in a small, easily-drained pond where they can be obtained on short notice. Since the fish shipment often is scheduled days or weeks before the delivery is to be made, late developments can upset these plans if safeguards are not taken to avoid them. Cancellation of a distribution trip is expensive in terms of time of all concerned and should be avoided if possible. With good management and sound planning it is seldom necessary.

11.6 Recommended Reading

For additional information regarding harvesting, grading and transportation see the following references as well as Chapter 17.10.

Jensen, G. L. and M. W. Brunson. 1992. Harvesting warmwater fish. SRAC Publication No. 394. Southern Regional Aquaculture Center, Stoneville, Mississippi. Available: http://agrilife.org/fisheries2/files/2013/09/SRAC-Publication-No.-394-Harvesting-Warmwater-Fish.pdf. (May 2020).

Kelly, A. M., and D. Heikes. 2013. Sorting and grading warmwater fish. SRAC Publication No. 391, March 2013 Revision. Southern Regional Aquaculture Center, Stoneville, Mississippi. Available: http://agrilife.org/fisheries/files/2013/09/SRAC-Publication-No.-391-Sorting-and-Grading-Warmwater-Fish.pdf. (May 2020).

Killian, H. S., D. Heikes, P. Van Wyk, M. Masser, and C. R. Engle. 1998. Inventory assessment methods for aquaculture ponds. SRAC Publication No. 395. Southern Regional Aquaculture Center, Stoneville, Mississippi. Available: https://agrilifecdn.tamu.edu/fisheries/files/2013/09/SRAC-Publication-No.-395-Inventory-Assessment-Methods-for-Aquaculture-Ponds.pdf. (May 2020).

Ludwig, G. M., and N. Stone. 1997. Relation between bar grader spacing and Golden Shiner size. Progressive Fish-Culturist 59(4):312–316.

Zajicek, P. W., J. E. Hill, N. Stone, H. Thomforde, C. Ohs, D. Cooper, G. Flimlin, B. McLane, and W. D. Anderson. 2009. Preventing hitchhiking nonindigenous species in live shipments. SRAC Publication No. 3902. Southern Regional Aquaculture Center, Stoneville, Mississippi. Available: https://agrilifecdn.tamu.edu/fisheries/files/2013/09/SRAC-Publication-No.-3902-Preventing-Hitchhiking-Nonindigenous-Species-in-Live-Shipments.pdf. (May 2020).

Chapter 12

Distribution

Hatcheries not only produce fish, but in most cases also distribute them to the various user groups. What is distributed might range from eggs to broodstock. How far they are distributed might be only to another location on the hatchery or shipping the fish halfway around the world. How the fish are distributed depends on the life stage of the fish involved, the number of animals to be transported, the distance involved, and accessibility to where the fish are to be delivered.

Distribution schedules should be prepared in conjunction with those going to receive the fish. The hatchery manager should contact the individuals that are to receive the fish, confirming the type of fish, the number and size of fish to be delivered, the delivery site, and schedule a convenient time for both the hatchery staff and the recipients for the delivery. This needs to be done weeks ahead of the possible delivery date as the hatchery staff may need time to harvest the fish and prepare them for transportation as well as to ensure that the transportation equipment is available and in proper working order.

12.1 Modes of Transportation

Fish transportation systems can be grouped into two basic categories: open or closed systems, each with advantages and disadvantages. Open systems allow a continuous input of oxygen, and some gases such as CO_2 to be driven off. Such systems need supplemental oxygen to carry any significant weight of fish for an extended period of time. Closed systems do not allow gas exchange and can be supersaturated with oxygen, but CO_2 will build up, affecting water quality.

12.2 Aeration

12.2.1 Agitation

Agitators are used to mix the water of the fish transport tank, allowing oxygen to diffuse in and CO_2 to diffuse out. A common design is a 12-volt (DC) electric motor with an extended shaft where blades are attached (Figure 12.1) and is powered off the vehicle's electrical system. Such agitators are appropriate for small transport tanks and for emergency backup. Their power demand is such that the transport vehicle should have a heavy-duty alternator and the engine kept running whenever transporting fish. The agitator is mounted on the tank top with approximately half the blade extending down into the water. The spinning blade brings in atmospheric air into the tank and helps to prevent CO_2 buildup in the transport water. Most

Figure 12.1. 12-volt (DC) electric agitator. Minnow-Saver #3. used in fish transport (Photo courtesy of Brian Smith and Joshua Senkevech, Minosaver, MinoTech Inc.).

commercial brands of agitators have a mark on the blade to indicate how far the blade should extend into the water. This should be checked once the fish are loaded into the tank as their weight will displace water and cause the water level in the tank to rise. It is very important that the blade extends down to the correct depth in the water. If it extends down too far the agitator motor will be working too hard to make the blades revolve. If it does not extend down enough it will spin too fast. In either case it will damage the motor. Agitators are not good for mixing water in the entire tank but are effective in a localized area; however, the strong currents created may stress small fish. The blade must be screened to prevent fish coming in contact with the blade and being damaged. The mucous lost by the fish during transport can be whipped into a foam by an agitator. This foam prevents gas exchange in the transport tank; however, commercial anti-foaming products can be added to solve this problem.

12.2.2 Use of Pure Oxygen

Pure oxygen is often used as a stand-alone system or in combination with other aeration equipment. Higher oxygen concentrations can be more easily achieved in the transport tank when using pure oxygen as compared to pumping in atmospheric air. But because pure oxygen systems are efficient in providing oxygen they often do not agitate the tank water enough to drive off much CO_2. As the CO_2 concentration in the tank water rises during long transportation trips it interferes with the diffusion of CO_2 out of the fish's blood stream and oxygen uptake. A buildup of CO_2 also can lower the pH of the tank water and the lower pH may irritate the fish's gills leading to a mucous buildup that interferes with gas exchange. To minimize CO_2 issues, the water should be agitated to drive off the CO_2. Agitation can be provided using agitators as discussed earlier, or pumps to recirculate tank water, spraying it back into the tank, while pure oxygen is added to ensure safe oxygen concentrations are maintained. A water exchange during transport may also be needed to reduce CO_2 levels. Buffering the transport water not only helps the fish to maintain their osmotic balance, having an alkalinity of 100 mg/L or more will reduce free CO_2 in the water. Sodium bicarbonate (Na_2CO_3) at a rate of 264 g/m^3 of water is an effective way to increase the alkalinity of the transport water.

Bottles of compressed oxygen are widely available in much of the world for welding and can be a source of oxygen for use in fish transportation. The bottle size most frequently used is nearly 1.5 m tall (including cap) and weighs about 59 kg when full. A full bottle can contain 6.9 m^3 of oxygen at a pressure of 2200 pounds/square inch (psi) or 15,200 kilopascals (kPa) at 20°C. Flow rates used in fish transportation commonly range from 8–16 L/min of oxygen/m^3 of fish transport water. A compressed oxygen cylinder containing 6.9 m^3 of oxygen can provide up to 6 to 7 h of oxygen/m^3 of transport tank depending on the flow rate and type of air stone used. Multiple oxygen cylinders need to be available when using large transport tanks for an extended period.

An alternative when using large transport tanks or transporting fish for an extended period of time is a liquid oxygen cylinder (LOX). A typical liquid oxygen cylinder used in fish transport is about the same height but twice the diameter of a typical compressed oxygen cylinder. A LOX cylinder weighs about three times more than one compressed oxygen cylinder but can provide as much oxygen as 12 or more compressed oxygen cylinders.

A LOX cylinder is a more complex cylinder than a compressed gas cylinder. It has an inner container of stainless steel, and an outer shell of carbon steel, with insulating material held under a vacuum between the two. As the liquid oxygen warms up to –183°C (O_2's boiling point), it will begin boil into a gaseous state. A relief valve will vent the gaseous oxygen to the atmosphere, then the internal pressure rises above 235 psi (16,200 kPa). As such, a LOX cylinder will always be forming gaseous oxygen and a portion is vented to the atmosphere. As a safety precaution, keep the pressure-relief vent area dry as water can freeze and plug the vent causing a pressure rise in the cylinder. Although the loss rate is rather low (generally less than 2.8 m^3 per day), it is a continuous loss and LOX cylinders may not be appropriate on hatcheries where the LOX cylinder is not going to be used frequently. In a warm climate an unused bottle could empty itself in two or three weeks.

How effectively bottled oxygen is utilized depends on the type of regulator, flow meter, and air stone used. An oxygen regulator is used to reduce the pressure from an oxygen cylinder to a safe manageable level. A flow meter is needed to control flow rate. There are two basic types of gas pressure regulators. A single-stage pressure regulator reduces cylinder pressure to delivery or outlet pressure in one step. The regulator has two gauges: one displays the gas pressure in the cylinder, and the other shows the oxygen pressure exiting the regulator. As the pressure in the cylinder drops the level of pressure exiting the regulator changes, in turn changing the flow rate. Single-stage pressure regulators need to be monitored periodically and adjusted to ensure a consistent gas pressure is exiting the regulator. Two-stage regulators are actually two regulators housed in one body. The first regulator (first stage) is non-adjustable and reduces incoming pressure to an intermediate setting, typically 250 to 300 psig (psig = container pressure independent of atmospheric pressure). The second stage is adjustable and reduces intermediate pressure to the final desired delivery pressure. Because the second stage sees only relatively minor inlet pressure changes from the first stage, two-stage regulators have a stable exit pressure and do not require periodic adjustment. This is particularly beneficial when transporting fish on an extended trip.

Different types of bottled gases have specific thread patterns and require a regulator with matching threads. The threading pattern for a given gas may differ from one country to another as well. It is very important that the cylinder and regulator have matching threading and will seal tightly. Trying to force on a regulator with inappropriate threading is dangerous and will damage both the cylinder and regulator. Connections should be free of oil or grease because exposure to oxygen increases the flammability risk. Once the regulator has been connected to the oxygen cylinder, the valve of the cylinder should be opened slightly allowing the cylinder gauge to reach a steady pressure. At that point the valve should be opened completely. This allows the full pressure to the cylinder to help seal the regulator to the cylinder. Any time the cylinder is not being used the main valve should be closed, including when it is "empty."

Each cylinder actively being used should have its own regulator. Each transport tank should have its own flow meter; however, one regulator may supply several flow meters. A flowmeter directly measures the flow rate of a gas. Because gases differ in their densities, each type of gas has its own appropriate flowmeter. Oxygen flow rate should be measured using flowmeters calibrated for oxygen. To function accurately it should be mounted in a vertical position. The oxygen being released pushes up a metal sphere in the flowmeter making it easy to see how much pressure is in the line.

12.2.3 Air Diffusers ("Air stones")

The smaller the air bubble the greater its diffusion into the water. A variety of materials have been used as diffusers with the bubble size depending on the diffuser material and the in-line pressure. Oxygen supplied by compressed oxygen cylinders or LOX cylinders is under higher pressure than that provided by an air blower. Such pressure can force oxygen through a micropore (<0.3 μm) air stone producing very small bubbles. A high-quality ceramic diffuser, if proper care is given, will pay for itself in reduced oxygen usage when compared to carbon diffusers or soaker hoses that produce larger bubbles. The working gas pressure entering a micropore diffuser should be 25 to 35 psi. A pressure of >50 psi may break the ceramic material or cause the hose connection to become loose. An optimum flow rate is approximately 3–4 L/min not exceeding 6–8 L/min. The small pore size of a micropore diffuser can result in the diffuser becoming easily fouled. They periodically need to be soaked in a 50% bleach:water solution for 10–15 min while connected to a pressurized air line, then lightly scrubbed and rinsed well in freshwater. The bleach solution may seep into the diffuser and can be flushed out with pressurized freshwater until water appears on the surface of the diffuser. Other diffusers suitable for pure oxygen applications but with larger bubble sizes include: rubber poly (EPDM) diffusers, fused aluminum diffusers (producing a bubble size of 1–2 mm), and porous plastic diffuser tubes.

Air blowers do not produce enough pressure to force air through a micropore diffuser and a diffuser made of courser material is needed. A larger pore diffuser can also be used with compressed oxygen but the large air bubbles produced results in a less efficient oxygen transfer and a faster utilization of an oxygen cylinder to maintain a given oxygen concentration in the transport tank. Truck mounted air blowers, powered by enhancing the truck's electrical system, or a stand-alone gasoline powered motor, are often used for large transport tanks. Air is supplied through air stones or some type of diffuser hose such as low-pressure rubber/polyethylene diffuser hose. Diffuser hoses can be individually attached to an air supply line or connected in parallel to a manifold frame that will fit on the bottom of the transport tank. The air intake to the air blower should be filtered adequately to prevent dust and other fine particles from being blown into the diffusers and clogging the pores of the diffusers.

The warm air provided by an air blower may on a long trip slowly raise the water temperature of the transport water. An increased water temperature results in a decreased oxygen solubility and increased stress on the fish. When using a LOX cylinder and a well-insulated transport tank on a long trip the oxygen released by the LOX cylinder can lower the water temperature in the transport tank. These possible temperature changes should be monitored during transportation, controlled if need be, and taken into consideration when acclimating the fish to the water conditions at the receiving site.

Backup equipment should be carried when fish are being transported off-station. Extra compressed oxygen bottles should be available in case there is a delay in route or if aerators or air blowers fail. Replacement aerators, and fuses should be carried for the transport truck's electrical system. Other equipment that should be available includes a meter to measure oxygen and temperature, and for long-haul trips, a water quality kit for the measurement of ammonia, CO_2 and pH. A 12 V water pump and an adequate length of hose is useful for fish acclimation and any other water exchanges.

12.3 Open Container Systems

Open systems may range from a simple open container to sophisticated systems mounted on semi-trucks. For local shipments of small quantities of fish with transit times of ≤1 h, un-aerated containers such as 208 L oil drums, 95 to 132 L garbage cans or similar type containers that can be carried in a pickup truck are used. The containers are partially filled with water and the fish weighed for numerical determination and placed in the containers for immediate transportation to the receiving waters. The density of fish hauled in un-aerated containers varies depending upon the size and species of fish being transported. For a one-hour trip, 6 to 12 g of 5-cm size fish can be hauled per liter of water. Ice may be added in hot weather to lower the water temperature a few degrees, provided that the ice is free of chlorine.

A more typical approach used on most hatcheries consists of a rectangular tank having one or more compartments, each equipped with an electric agitator powered by the vehicle electrical system. Lids to the compartments permit access to the fish and for filling with water, and a drain is provided for emptying each individual compartment. The tank may be constructed of wood, aluminum or fiberglass and must be solidly built. The tanks should be well insulated as the tanks may be out in the sun for hours while the fish are being transported. The tank may have baffles extending down to within 20 cm of the bottom of the tank to limit wave action within the tank. Waves can be formed within the tank during transportation as the vehicle changes speed causing the fish to be damaged. The tank bottom should slope to the drain where the bottom of the drain is flush with the tank bottom. This will allow the tank to drain completely and all fish exit through the drain. When it is anticipated that only one large tank will be carried at a time, the drain on the tank should be located so the tank drains off the back end of the truck. When multiple species or multiple sizes of fish are to be transported at the same time a tank with several compartments, where each compartment has its own drain on the side of the tank, is desirable. An alternative is to have several rectangular tanks, each with a drain on one end, setting the tanks sideways on a flatbed truck or gooseneck trailer. A common tank size (1.2 × 2.1 × 0.7 m., with two 1.2 m^3 compartments) will fit a large pickup truck going the length of the bed (Figure 12.2), or be set cross-wise on a larger flatbed truck along with other tanks of that size. Each compartment should have one or two tight fitting lids large enough to allow a dip net have access to all of the compartment or a fish basket to be unloaded into the tank. As the fish are typically being transported in low salinity water any hardware used to fasten lids down, bolts, drain valves etc., to the tank should be made of non-corrosive metals.

The size of the tank selected is affected by the size of the truck to be used, the quantity of fish to be delivered to one location, the number of species or size groups to be carried at one time, and the accessibility of the site where the fish are to be unloaded. The truck may be as small as a 1/2-ton pickup or as large as a diesel-powered long-haul truck capable of pulling a 15 m trailer across the continent. Depending on how frequently fish are being transported at a hatchery a tank system may be set up that can be conveniently moved on and off the truck allowing the truck to be used for other purposes most of the time. If fish are being frequently transported off-station, a truck may be set up specifically for transport (Figure 12.3). An alternative is to set up a gooseneck trailer with tanks and aeration systems specifically for fish transport (Figure 12.4). This has the advantage in that the truck used for towing the trailer can be used for other things as well. A gooseneck trailer when transporting fish in saltwater helps to keep the saltwater off the tow truck, eliminating the associated corrosion and extends the life of the truck. For short hauls of less than 800 km, a truck size of 2.5 tons or less is best. Table 12.1 shows the general relationship between truck size and capacity of the tank it carries. Although rectangular tanks are the common configuration, circular tanks are useful for large fish sensitive to handling such as Striped Bass that transport best when allowed to swim continuously. Such circular tanks should be filled completely with no air space between the water surface and the top of the tank to avoid any wave action. A small open tower through the top of the tank should be added for gas exchange.

Figure 12.2. A transport tank with two 1.2 m^3 compartments set on a pickup truck along with associated equipment used with fish transport.

Figure 12.3. A dedicated truck with four side-draining tanks for off-station fish transport (Photo courtesy of Robby Mays, American Sport Fish).

Figure 12.4. A gooseneck trailer with a compartmentalized tank and aeration systems set up specifically for fish transport (Photo courtesy of Michael Matthews, Texas Parks and Wildlife).

Table 12.1. Relationship between truck size and quantity of fish hauled.

Truck Size*	Truck Capacity (L)	Weight of Fish hauled (kg)**
0.5 ton	946	135
0.75 ton	1,136	160
1.0 ton	1,514	225
1.5 ton	2,271–3,028	450
2.5 ton	4,542	1,360
Long-haul semi-trailer	19,000	4,500

*USA short ton.

**Approximate weight, exact weight depends on species, fish size, condition, distance being transported, weather and other factors.

A wide variety of fish sizes can be transported in an open system, with fingerlings (juveniles) being most often transported from a hatchery. The weight of fish transported might range from 60 g/L up to about 1,000 g/L. The quantity of fish transported/unit volume of water is influenced by: (1) fish species, (2) fish size, (3) fish condition, (4) weather, (5) distance hauled, (6) water quality, (7) personnel, (8) efficiency of aerating equipment, (9) stops and delays, and (10) design and performance of distribution unit.

The loading rate is influenced by the fish size. Whether it be an open or a closed container, an equal total weight of small or large fish differ in metabolic rate, with smaller fish having a higher metabolic rate. For example, on a 16 h trip an acceptable loading rate for 0.45 g Channel Catfish fingerlings held at 18°C would be 80 g/L., but using 1.1 g fish, their lower metabolic rate/g would allow a loading rate of 150 g/L of water. An arbitrary listing of loading rates would be impossible since so many variables are involved. Even when the species, size and fish condition are known, a variation of 200 to 300% may be needed because of the other factors. Since the effects of the various factors are so great, considerable experimentation is required for a hauling tank under specific operating conditions to establish the optimum loading rate.

The distance fish are hauled will have a bearing on the management practices and equipment used. In the Southwestern United States where trips of 24 to 36 hours are necessary, fish tanks are equipped with ice compartments and a water recirculation system, as well as aerating devices such as electric agitation or oxygen diffusers. When trips are 12 hours or less, the water circulation feature is often omitted and sometimes the ice compartment also. Aeration with oxygen, electric agitation or a combination of the two enables a load to be safely carried for a full workday of deliveries. With such equipment, up to 20% of the capacity load can be held overnight for distribution the following day.

For trips of 12 hours or less involving loads less than 275 kg of fish, a 1,500 L tank mounted on a 1-ton truck chassis has proven to be an efficient unit. The tank is insulated and made of either heavy gauge aluminum or stainless steel. Aeration is provided by a combination of agitators and oxygen diffusers. The tank is divided into 3 or 4 compartments. Up to 180 g of small fingerlings/L of tank capacity can be hauled for 6 to 10 hours. For large bass or Channel Catfish fingerlings, 240 g/L of tank capacity can be hauled for 10 or 12 hours without undue stress. Such equipment is ordinarily operated by one person, however, on more demanding trips a relief driver provides better assurance that the shipment will proceed without accident.

When larger quantities of fish are needed, a 1.5 ton truck carrying a 2,500 L, 5-compartment tank is recommended. This unit can carry 50% larger loads with the same operating personnel. For interstate deliveries involving large lots of fingerlings, a 2.5-ton truck carrying a 4,500 L tank is more economical. Such a unit can carry up to 1,360 kg of large size fishes or 680 to 970 kg of fingerlings. It can be equipped with pumps for water circulation, an ice compartment, electric agitators and oxygen diffusion system. All or part of this aerating equipment may be used depending upon the needs of a particular situation. For trips of 8 to 10 hours duration, a one-person crew is adequate, but on longer trips, a two-person crew is mandatory. A sleeper cab for the truck is recommended for frequent overnight trips.

Cross-country hauling of bait minnows, catfish fingerlings, or catchable size fish for fish-out operations is done in larger trucks individually designed for a specific purpose. Some are surprisingly simple adaptations, combining components of a smaller unit into a larger multiple unit operated by 1 or 2 persons. Other creations are ingeniously engineered systems that function to meet specialized needs such as the rapid delivery of bait minnows from production centers to sport fishing grounds 800 to 1,600 kilometers away.

When using unfamiliar equipment, new staff, or driving on unfamiliar routes it might be best to reduce the loading rate by 30 to 50% of the standard to ensure the arrival of high quality, healthy fish and only then increasing the loading rate when confidence is gained.

12.4 Closed Container Distribution

A convenient way to transport large numbers of fish eggs and larvae is to use a closed container. This approach has the advantage in that large numbers of seed can be transported at a minimum weight. One hatchery may specialize in fish spawning and ship eggs or larvae to other hatcheries for nursing the fish to larger sizes. Such an approach has allowed a thriving commercial business to produce trout eggs and ship eyed eggs throughout the world using specially designed containers to hold chilled eggs in thin layers in a moist environment during the period of transport. Millions of Striped Bass larvae are shipped in plastic bags each spawning season, and a substantial number of Walleye, Northern Pike and other larvae are also distributed by this means. The bulk of the ornamental fish are shipped as airfreight using closed containers, often with transit times of 48 hours or more. Packing in small units such as plastic bags allows small numbers of fish to be transported to isolated rural areas using motor bike, bicycle, boat, pack animal or on foot. Larger fish can also be shipped in closed containers but the cost of shipping/fish is too high for most applications.

12.4.1 Equipment for Closed Shipment

The most common container is a 3-5 mil polyethylene plastic bag placed in a Styrofoam box, protected by an outer cardboard box (Figure 12.5). When 3-mil thickness bags are used, two bags placed one inside another should be used to guard against accidental puncture or a faulty bag. This approach provides enough protection for the fish so that they can be shipped by bus, plane, rail, or even bicycle. A number of Styrofoam box sizes are used with a 42.5 ×42.5 × 25 cm box being very common. The thickness of the box wall may be 1.30 cm or 2.0 cm, with the thicker boxes being used for temperate climates or during the winter where more insulation is needed. Square-bottomed bags, available for specific sizes of boxes, have the advantage of fitting the box well with little lost space.

Local deliveries can be made using bags made from 3 mil poly tubing. One end of the tubing is tied tightly, water, fish, and oxygen are then added, and the other end of the tube is tied to seal the bag. This approach allows bags of various length to be made and can be appropriate for transporting larger fish. Laying the tube on its side exposes more of the water surface to the gaseous oxygen that was added to the tube (Figure 12.6). However, as the length of the tube might be variable, finding an appropriate size box to provide protection and insulation can be an issue. Covering the cylinder with several layers of cloth will provide some protection. Periodically wetting the cloth will provide some evaporative cooling to help the water temperatures in the bags from rising too much.

For local deliveries either on the hatchery or off-station, other types of closed containers have been used. The basic design is a cylindrical or cubic container with a removable top that can be sealed tightly to the container. The top may have two valves. One valve connects an airline to an air stone that will extend down into the water. Once fish are added, the container is sealed and oxygen is added. The other valve is to vent out atmospheric air while oxygen is

Figure 12.5. A 3- to 5-mil polyethylene plastic bag placed in a Styrofoam box, protected by an outer cardboard box used as a closed container for fish transport.

Figure 12.6. Transport bags made by cutting off sections from a tubular roll of polyethylene plastic and tying off each end.

being added to the container. Once the atmospheric air has been driven out, both valves are closed and the fish are ready to be shipped.

Items that are needed for closed container shipping typically include: appropriate shipping containers, a full oxygen cylinder and regulator to inflate the bags, sufficient airline to reach from the oxygen cylinder to the shipping containers, rubber bands or other material to close the bags and shipping container, a supply of clean water at the desired temperature, as well as any chemicals to be added to the transport water.

12.4.2 Preparation for Shipping

The objective when shipping fish in closed containers is the same as when using open containers: ship the maximum quantity of fish per unit volume of water that results in strong healthy fish upon arrival at their destination. Goals should include: keeping transit time to a minimum, being organized so once you start the loading time is minimal, and handling the fish kept to the least amount possible. However, open and closed fish transportation approaches differ in water quality management issues, and often who is responsible for the fish while they are in transit. When using open transportation systems, hatchery staff are often responsible for preparing the shipment and its safe delivery. Closed containers may be prepared by the hatchery staff but the safe delivery may involve a number of individuals that have no experience in the safe handling of fish. When preparing to ship fish in closed containers, particularly for airfreight shipments, several questions need to be considered.

1. Is the recipient ready to receive the shipment?
2. What are the conditions at the site where the fish are to be delivered?
   a. How will differences in water quality conditions affect acclimation?
   b. What time of day are the fish expected to arrive there and how will that affect acclimation?
3. How long will the fish be in transit from the time the fish are sealed in the bag at your hatchery to the time the fish are unpacked at their final destination? This includes allowances for likely in-transit delays such as:
   a. Availability and frequency of plane flights.
   b. Road conditions
   c. Weather conditions that could delay delivery
4. If the fish have to be transferred from one plane to another or one mode of transportation to another:
   a. Who is responsible for the transfer and how can they be contacted?
   b. While in transit, could the fish shipment be delayed because of arriving after closing hours of the freight company, the customs, or health inspection office? If so how long?
   c. If there are to be in-route transfers, will the fish be held at a climate-controlled settling while waiting for transfer?
   d. Are there oxygen and suitable water, and qualified people available to use it if there were any losses of oxygen from the shipping bags while en-route?
5. When shipping as airfreight, are live fish considered as priority cargo?
   a. Are there other items that seasonally may have higher priority that can delay the fish shipment?

b. Are there days of the week when a shipment of fish will have a faster transit time?

6. How can you track the progress of your shipment?
   a. Can it be tracked in-transit over the Internet?
   b. Who can you call to get updates on the shipment?
   c. Does the freight company notify the recipient of fish that the shipment has arrived?

Knowing the above information will allow you to adjust the start time when preparing a shipment, adjust shipping densities, modify shipping procedures (type and amount of insulation provided), and help ensure that the fish are appropriately acclimated upon arrival at the final destination.

Preparation should begin with making sure that the fish (whether eggs, larvae or larger sizes) are ready to be shipped. They should be free of known diseases and recovered from any handling stress. Often a formal examination and health certification documents may be required before a shipment can be made. The fish should be acclimated to the water in which they are going to be shipped at least 24 hours before being shipped so they can adjust to any changes in water quality. During that time the fish should not be fed, allowing the digestive tract to be cleared of feces.

There are time windows in which eggs and sac fry can best be shipped. Eggs can be best shipped after neurulation but must arrive at their destination before hatching begins. Younger eggs may not be suitable for shipping as they are often easier to damage, resulting in increased deformities and mortalities. If hatching begins while the eggs are in transit, the shell debris will cause the water quality to deteriorate rapidly. Sac fry are ideal for shipment as the fish do not need food and are usually tolerant of handling shock associated with packaging and transport. Individually they weigh very little so large numbers of sac fry can be shipped at a time. The window for shipping sac fry would be from shortly after hatching to just before mouthparts are functional. Older fish do not have such development window issues and can be in transit 72 hours or more as long as the water temperature and water quality remain within acceptable ranges, and nutritional reserves are adequate.

Prior to packing, dead eggs/larvae should be removed as such organic material becomes substrate for bacterial and fungal growth and its decay consumes oxygen, and releases ammonia and carbon dioxide. Only healthy fish should be shipped, as any fish that die in route jeopardize the survival of the remaining fish. A number/weight or number/volume relationship should be determined prior to starting packing so that the bags can be loaded quickly with the desired number of organisms.

The water in which the fish are transported should have an alkalinity of > 50 mg/L to help buffer the pH changes associated with CO_2 buildup during shipping (except fish accustomed to low pH and alkalinity conditions). A common stress response in fish is a weakening of osmoregulatory control. Salt can be added to the transportation water of freshwater fish at 1 to 9 g/L to reduce the osmotic gradient between the fish and the transport water.

While the fish are being held for shipment they should be acclimated to the shipping temperature. The desired shipping temperature needs to take several things into consideration. The lower the temperature at which the fish are held will reduce the metabolic rate and in turn reduce oxygen consumption and the release of CO_2 and ammonia. A lower water temperature will allow a higher fish loading rate. For eggs and larvae, it also slows the development rate in turn extending the time that they can be in transit. However, the water temperature can be too low causing stress and death. Appropriate temperatures for shipping tropical fish are 22–23°C while temperate climate fish might be shipped at 18–22°C or cooler depending on ambient temperatures and life stage. For example, cyprinid larvae should not be transported below 16°–18°C while juveniles will can be transported at <16°C.

Temperatures should be viewed not only as the temperature at which the fish were packed at for shipping, but the temperatures they are at during transit (on the plane or truck, during transit, in a warehouse or other location while waiting transfer, etc.). The water temperature at the final destination is also an issue. The more similar the water temperatures are for the fish in transit and the final destination, the less time is needed in acclimation of the fish to their new environment. A quantity of ice equal to 10–20% of the quantity of water in the transport bag is often placed in a separate sealed bag and placed in the transport box to help keep the water temperature from rising too much during transit. Similarly, hot packs are added to wintertime shipments to keep the temperature of a shipment from dropping too much.

12.4.3 Other Additives to Transport Water

A variety of chemicals have been added to the water of both open and closed transport containers as anesthetics, antibiotics or water quality regulators. Use of chemical additives has several issues: 1) do they have government approval for that application and 2) do they really work? There are a limited number of chemicals approved for use in aquaculture and some that are approved in one country may not be approved in another. Rules and regulations change. It is the responsibility of those shipping the fish that any chemicals that they add to the transport water are approved for that use, for that species and life stage of fish, particularly for any fish likely to enter the human food chain. A given chemical might be effective with one species under one set of conditions but toxic to another species under similar conditions. All chemicals used for shipping fish should be tested beforehand under local conditions to ensure that there are no detrimental effects.

Anesthetics have been used to slow the metabolic rate of the fish being shipped and in turn reduce oxygen consumption. Calming the fish down also reduces the risk of injury. Tricane methane sulfonate (MS-222) at 20–25 mg/L and quinaldine or quinaldine sulfate at 5–10 mg/L have been used. The dose needed for a light sedation varies with species and life stage and acceptable dose rates determined for the local conditions. These anesthetics can cause the pH to drop significantly in poorly buffered water and sodium bicarbonate is used to bring the pH back to about 7.

Antibiotics such as terramycin and nitrofurazone, and bacteriostatic dyes have been added to transport water to reduce the chance of bacterial diseases. However, the dose rate of antibiotics needed to not be toxic to the fish during a prolonged shipment are often not very effective in controlling bacteria. Use of low doses of antibiotics also increase the possibility of bacteria developing resistance to the antibiotic, and if a bacterial disease occurs later you may have to find another antibiotic to control it. Dyes such as acriflavine and methylene blue are commonly added to the transport water to control bacteria in the water column but generally are not very effective.

In both open and closed transport systems ammonia and CO_2 will build up in the transport water. Depending on how well an open system is agitated and vented, CO_2 levels can be controlled. In closed systems carbon dioxide cannot escape and can cause a major drop in pH. In either case the initial alkalinity of the transport water should be at least 50 mg/L. If need be the alkalinity can be increased by adding calcium carbonate or sodium bicarbonate. The transport water can also be buffered using tris buffer, tris [hydroxymethyl aminomethane], at 5–10 mg/L. The buffer keeps the pH above 7.5, allowing a shift of accumulating CO_2 to a non-toxic bicarbonate (HCO_3^-).

Ammonia is another metabolite that will build up during fish transport. Its accumulation can be reduced with the addition of clinoptilolite. Clinoptilolite is one of several microporous, aluminosilicate minerals called zeolites, that will bind with ammonium in freshwater. One gram of clinoptilolite can bind with about 2.2 mg ammonium-N. However, in very hard water conditions zeolites will also bind with calcium, magnesium and sodium and as such are less effective in binding with ammonia. When zeolite is used in seawater sodium is preferentially selected over ammonia and as such it is not effective in removing ammonia in high salinity waters.

For a given weight of clinoptilolite, the smaller particle size the greater the surface area and the greater the ammonia removal but smaller particle sizes can increase the turbidity of the transport water. Acceptable particle sizes are 2–5 mm but they should be rinsed before use. Clinoptilolite is best added to transport water in a net bag at 15–20 g/L of water or in a quantity equal or more than half the weight of fish being shipped. Over a period of hours clinoptilolite will become saturated with ammonia and cannot absorb more and fresh clinoptilolite needs to be added. Ammonia can be stripped off clinoptilolite by placing it in saltwater. The clinoptilolite can then be rinsed in freshwater and reused in freshwater settings.

Shipping fish in closed transport containers has some water quality issues often not seen with open transport containers. Oxygen is added to the container and the container sealed, typically resulting in an oxygen supersaturation in the water. If an appropriate quantity of fish is added to the container there is rarely a shortage of oxygen but because the container is sealed CO_2 and ammonia cannot escape. This presents several interesting interactions. Oxygen supersaturation reduces gill ventilation rate, resulting in CO_2 and ammonia concentrations in the blood to rise. As CO_2 levels build it reduces a fish's ability to adsorb oxygen. As CO_2 levels build the pH will drop depending on how well the water is buffered. A drop in pH to 6.5 is not uncommon in a closed shipment of fish by the end of a 24 to 48 hour period. As CO_2 increases, it will reduce the pH in the blood as well as the transport water. This drop in pH also shifts the ionic composition of the total ammonia concentration, reducing the percentage as the toxic unionized ammonia (NH_3) while increasing the percentage of the non-toxic ammonium ion (NH_4). Temperature also influences the NH_3/NH_4 ratio. The higher the water temperature, the greater the increase in NH_3. No one management technique will solve all these problems associated with fish transport, but by combining techniques they can be controlled. Transporting fish within the lower range of temperature that will not stress them will lower respiration rates, in turn reduce CO_2 and ammonia production. Using a well-insulated container and some ice will help keep the temperature down for a longer period of time. Respiration rates can be further controlled with an anesthetic. Buffering the water will help control major pH drops. However, by limiting a drop in pH it contributes to a greater ammonia fraction as NH_3, but by reducing the respiration rate and adding zeolite the total ammonia concentration can be limited.

12.4.4 Loading Procedures

Bags should be inflated before use to be sure they have no holes in them. Good bags are then placed in the shipping boxes and filled with known volumes of clean water of the same water quality as that used to hold the fish. Any chemicals to be used should be added to the water prior to adding the fish. The volume ratio of water to gaseous oxygen is typically between 20:80 and 35:65. All boxes to be shipped that day should be set up and ready to receive the fish before adding fish to the first box.

Before starting to add fish, a count relationship should be established. Larvae and fry are often too delicate to be handled in a net and are best moved in water. The number of larvae or fry/unit volume of the holding tank should be determined so the correct volume, i.e. the correct number, can be added to the shipping container. Larger fish can be safely handled in a net and stocked based on the number/unit weight of fish, stocking the proper weight into the shipping container.

Once the fish are added to the bag, an oxygen supply line is placed in the bag, and beginning at the water level the excess air is pressed out of the bags. At that point the oxygen regulator is slowly opened to add oxygen to the bag, keeping in mind the desired volume ratio of water to oxygen and that the sealed bag must fit inside the box when the lid is closed. When the desired volume of oxygen is reached the oxygen supply line is quickly withdrawn, and the upper end of the bag twisted several times and folded over to close it. The twisted, folded-over end is tightly secured using rubber bands, strips of rubber, or strong cord. A very effective way is to use thick rubber bands used for animal castration and a set of castration pliers. Using the pliers the rubber band can be easily stretched to fit over the twisted, folded end of the bag then released to tightly squeeze the bag closed. A small bag of ice (at approximately 10–20% of the volume of water in the bag); or other material to help stabilize the water temperature can be added to the shipping box, placing it under the bag of fish. A card should be prepared and added inside the box that gives contact information of the person shipping the fish and the person that will be receiving the fish. The card should say the type and number of fish in the box as well as any special care that they should receive when unpacking them. On the outside of each box there should be a large label saying that the box contains live fish (giving the species and number of animals), "handle gently," and other information such as “This Side Up,” and “Keep Warm,” or “Keep Cool” depending on the fish and where the boxes might be held in transit. Copies of health certificates might also have to accompany the shipment.

12.4.5 Loading Rates

It is difficult to give very specific loading rates that will cover all species as there are too many factors that can impact a successful shipment. However, information available for similar species, of similar life stages, under similar conditions can be used as guidelines (see Chapter 17.10.) and tested for their appropriateness for a specific hatchery and its need to transport fish.

Eggs.—Adhesive eggs attached to spawning mats can be transported on-station, collecting the spawning mats from the spawning pond and moving the mats to the hatchery. The mats can be held out of water as long as the mats are kept moist and are not in direct sunlight. The mats can be stacked five or more mats high. If the mats are kept moist and covered with a tarp or something similar, they can be transported off-station for a few hours.

Channel Catfish eggs held together as an adhesive mass can be shipped by vehicle or air-freight using plastic bags with oxygen and Styrofoam boxes. Two to three egg masses (500 g each) in 7.6 L of water can be shipped safely for 24 hours. The egg masses should be at least 24 hours old but still 36 h or more from hatching.

Catfish egg masses can also be carried at 24–27°C in transport tanks supplied with oxygen. Several egg masses can be held in a soft mesh bag that is suspended in the water column from a float. A total loading rate of 65 kg/m^3 of tank is possible using a number of bags and adequate aeration to create a water current in the tank so that the eggs are not held in static water.

Sac Fry.—Large numbers of sac fry can be shipped in closed containers by air or ground transportation. Striped Bass are shipped at 8,600 to 13,000/L for trips of 24 to 36 hours. A small amount of ice may be used in the insulated box if air temperatures above 20 to 21°C are expected. Largemouth Bass sac fry can be shipped at 4,000 to 6,600 fry/L for 48 to 72 hours without appreciable mortality if care is given to removing debris, egg shells, and dead fry before packing. Water temperature must be kept in mind when shipping. An appropriate density for shipping grass carp fry over a 24 hour period would be 1,500/L at 20°C but 750/L at 25°C. Length of the trip is also a major consideration. A shorter trip at 20°C of 4 hour would allow a Grass Carp fry stocking rate of 3,000/L.

Fingerlings.—Most ornamental are shipped in closed bags often for 48 hours or more. Loading rates are often conservative as the risks are high and delays unpredictable. For livebearers such as mollies or swordtails averaging 3.8 cm in length, rates of 43 fish/L might be used for a 48-hour shipment. Differences in physiology and environmental tolerances between species can change what would be an acceptable loading rate. For a 24-hour shipment of 2–3 cm Grass Carp, an acceptable rate might be 300 fish/L at 20°C and 200/L at 25°C. For a 24-hour shipment of 2–3 cm Common Carp, which is a more environmentally tolerant species, an acceptable rate might be 400 fish/L at 20°C and 300/L at 25°C.

12.5 Factors Affecting Transportation

12.5.1 Fish Health and Water Quality

There are several things that need to be considered when transporting fish. Only healthy fish should be transported. Transportation of fish for any distance is stressful and if a fish is already weak there is a high likelihood it will die during transportation or soon after being unloaded at the user's site. Fish dying during transportation can have severe effects on water quality, risking the health of the other fish. Weak fish can be vectors of diseases possibly infecting other fish in the transport tank. Releasing sick fish at the user's site can result in disease outbreaks at that site and have a very negative effect on the reputation of the hatchery.

The demand for oxygen during transportation varies by fish species and size. Smaller fish have a higher metabolic rate/g fish body weight, and in turn a higher oxygen demand/g body weight. In other words, for a given availability of oxygen/m^3 of transport container, a smaller weight of small fish (kg/m^3) should be transported versus the higher kg/m^3 possible with larger fish without causing stress.

12.5.2 Acclimation

Transporting fish is stressful and attention must be given to minimize that stress, particularly at the arrival site. Any rapid change in water quality on top of the stress associated with the act of transportation may be enough to kill the fish when they are released into their new environment. Differences in water temperature are often the most notable difference between the transport water and the receiving water. A slow transition from one temperature to another is needed to allow the fish to adjust their physiology to their new environment. A rough rule of thumb is to make the transition over 10 min. for every degree °C the higher the receiving water and allow 5 min. for each degree cooler or until the temperature of the bag is within two degrees of the receiving water. Ideally, by knowing the water temperature at the receiving site, acclimation can begin while the fish are still in transit using ice if the water in the transport tank is warmer than that of the receiving water.

In addition to temperature, acclimation should take in consideration differences in pH and alkalinity/hardness. Often the transport water in an open transport system may start with a pH near 7 and an alkalinity of >50 mg/L with little to no change in transit. However, in a closed system, total ammonia and CO_2 will build up and cannot escape. The carbon dioxide buildup will lower the pH and decrease the NH_3 fraction of the total ammonia. When a fish delivery is made in the late afternoon, the receiving water might be warmer than the transport water. If the receiving water is poorly buffered (alkalinity < 25 mg/L) the utilization of CO_2 during photosynthesis may result in afternoon pHs of 9 to 10. This can present some additional concerns when acclimating fish transported in closed systems, i.e. plastic bags.

To acclimate the fish to any temperature difference, the bag can be floated in the receiving water until the two temperatures are the same. However, the fish should not be released directly at that point due to possible differences in the pH within the bag and the receiving water. The bag should be opened and receiving water slowly added to correct for any differences in pH or other water quality parameter. Opening the bag will cause any excess oxygen and CO_2 to escape, shifting the pH upward. The increased pH and the acclimation to the higher temperature can cause a significant rise in the NH_3 fraction of the total ammonia, possibly into toxic levels. In that case it may be necessary to slowly dilute the water in the transport bag with clean water low in ammonia.

When using an open transport tank, one of the easiest ways to acclimate the fish to the receiving water is to use a small pump powered by the truck's electrical system to slowly pump water from where the fish are to be stocked into the transport tank allowing a slow exchange of water. This approach will gradually acclimate the fish to any difference in temperature or water chemistry.

Once the fish are acclimated, they can be released. Fish transported in bags can, after being acclimated, be released from the bag into the receiving waters, allowing the fish to flow out with the water. Ideally fish held in an open container can be released from the transport tank by gravity along with the tank water via a distribution pipe or chute without having to handle the fish directly. Keep in mind that if the fish are being stocked into ponds, a pond may be stratified, particularly in the deeper end where there will be cooler water. Using that water temperature as a target for acclimation is often beneficial in that it might shorten the acclimation time and it is often beneficial to stock the fish into such cooler waters as long as the oxygen levels are adequate.

12.6 Distribution Problems

Accidents can and do happen in fish distribution work. These can be minimized by careful planning and taking precautions against the more likely ones. Some of the more common accidents are listed below with suggested precautions.

1. *Mechanical failure*—This is the most common type of accident. Precautions include keeping the system simple in design with fail-safe features wherever possible. A systematic maintenance program should be followed to keep all systems in good operating condition. Monitor fish and equipment periodically in transit to ensure everything is functioning correctly. Operating personnel should be trained to handle commonly encountered emergencies. Backup parts and supplies for items most likely to fail should be carried along with appropriate tools for repairing such items. Backup aeration systems should be available and in-transit locations identified where oxygen cylinders can be obtained. Loading should not be so heavy as to create a crisis should a truck tire fail or the trip be extended because of detours or unexpected delays in transit. For shipment by commercial airlines, packing rates should take into consideration that cargo arrivals can be 4 to 24 hours later than the scheduled time.

2. *Overcrowding*—A modest loading rate should be used initially for a new species or new equipment. Loads may be increased as experience is gained. Where several compartments are used in a distribution unit, one compartment can be used to test heavier rates. Processing fish for shipment in a standard way is helpful. This is particularly true as it pertains to holding the fish and guarding against handling stress. Keeping careful records and using them as to what and how fish were transported will help ensure future shipments are made in a safe and effective way.

3. *Highway accidents*—Make sure personnel are skilled drivers familiar with the equipment being operated. The weight being carried by the transport truck when it is full makes it harder to brake and more roadway is required to stop the vehicle. Avoid setting up too tight of a delivery schedule. Build in allowances for driving conditions and other possible delays. Identify various water sources along the hauling route that can be used to replace or exchange hauling water. Have a way to communicate back to the hatchery and to the people expecting to receive the fish if there is a delay in transit due to an accident or other problem.

4. *Accidental poisoning*—Avoid using chlorinated ice. Carry sodium thiosulfate to neutralize chlorine if needed. Keep tanks tightly closed when driving through an area where aerial spraying of pesticides is being done.

5. *Feces contamination*—Not holding fish without food for enough time prior to being transported will result in the fish releasing feces into the transport tank. Such feces have an oxygen demand, contribute to ammonia buildup, and irritate the fish gills. Holding the fish long enough will reduce these problems. If water is being recirculated in the transport tank, a water filter can be added to remove feces and other particles.

12.7 Recommended Reading

For more details on fish transportation see the following references as well as those in Chapter 17.10.

Amend, D. A., T. R. Croy, B. A. Goven, K. A. Johnson, and D. H. McCarthy. 1982. Transportation of fish in closed systems: methods to control ammonia, carbon dioxide, pH, and bacterial growth. Transactions of the American Fisheries Society 111(5):601–611.

Berka, R. 1986. The transport of live fish. A review. EIFAC Tech.Pap., (48):52 p. Food and Agriculture Organization of the United Nations, Rome. Available: http://www.fao.org/3/af000e/AF000E00.htm#TOC. (May 2020).

Bower, C. E., and D. T. Turner. 1982. Ammonia removal by clinoptilolite in the transport of ornamental freshwater fishes. The Progressive Fish Culturist 44(1):19–23.

Harmon, T. S. 2009. Methods for reducing stressors and maintaining water quality associated with live fish transport in tanks: a review of the basics. Reviews in Aquaculture 1:58–66. Available: http://citeseerx.ist.psu.edu/viewdoc/download?doi=10.1.1.502.4939&rep=rep1&type=pdf. (May 2020).

Neiffer, D. L., and M. A. Stamper. 2009. Fish sedation, anesthesia, analgesia, and euthanasia: considerations, methods, and types of drugs. ILAR Journal 50(4):343–360. Available: https://vpresearch.louisiana.edu/sites/research/files/NeifferFishSedation.pdf. (May 2020).

Sampaio, F. D. F., and C. A. Freire. 2016. An overview of stress physiology of fish transport: changes in water quality as a function of transport duration. Fish and Fisheries 17(4):1055–1072.

Section II

Specific Hatchery Techniques by Species

Chapter 13

Catfishes (Siluriformes)

Catfishes are found on all the continents except Antarctica, with more than 3,000 described species. In this chapter, hatchery techniques for three groups of the most widely cultured catfishes representing different modes of reproduction and culture are discussed.

13.1 Channel Catfish *Ictalurus punctatus*, Rafinesque

13.1.1 Biology

Channel Catfish is an important food and sport species in the family Ictaluridae, native to North America. It was first propagated under hatchery conditions in 1916. Its basic cultural techniques were developed at public fish hatcheries in the mid-west from 1925 to 1950. By the mid-1950s hatchery techniques and food fish production procedures were established enough to support the growth of a commercial industry. Channel Catfish has since become the most widely produced warmwater food fish in the USA. It has been introduced into over 40 countries.

Channel Catfish are a relatively long-lived species that may attain a maximum size of more than 27 kg under natural conditions, and an age of >20 years. When cultured they can grow to 20 to 25 cm in the first 4 to 5 months after hatching. A market size fish of 340 to 680 g can be produced within two years after hatching. Channel Catfish can reach sexual maturity at an age of 2 years when growth conditions are favorable. This is especially true of males. However, 3-year old fish are preferred as first time spawners if dependable spawning is to be obtained. Females continue to produce eggs for a number of years with gravid 20 kg individuals being occasionally encountered in wild populations. Females at sexual maturity can be expected to weigh 1.3 kg while males of the same age will usually be slightly larger at 1.3 to 1.5 kg.

As the spawning season approaches, sexual dimorphism is apparent in mature fish. Males have a broader and more muscular head than the females (Figure 13.1.1). Pigmentation of males tends to be darker especially as the fish age. The genital orifice of the male is small in diameter and located at the tip of the genital papilla. The papilla is easily seen in mature males (Figure 13.1.2). The villiform nature of the testes does not present enough tissue mass to allow sperm to be released when the abdomen is squeezed. As such, mature males cannot be identified based on whether or not semen is released when the abdominal area of the fish is squeezed.

Females are typically lighter in color, with a narrower head than the male. Well-conditioned females have noticeable gonadal development in the abdominal region when ready

Figure 13.1.1. A male Channel Catfish *Ictalurus punctatus* in breeding season is typically dark in color and has enlarged head pads.

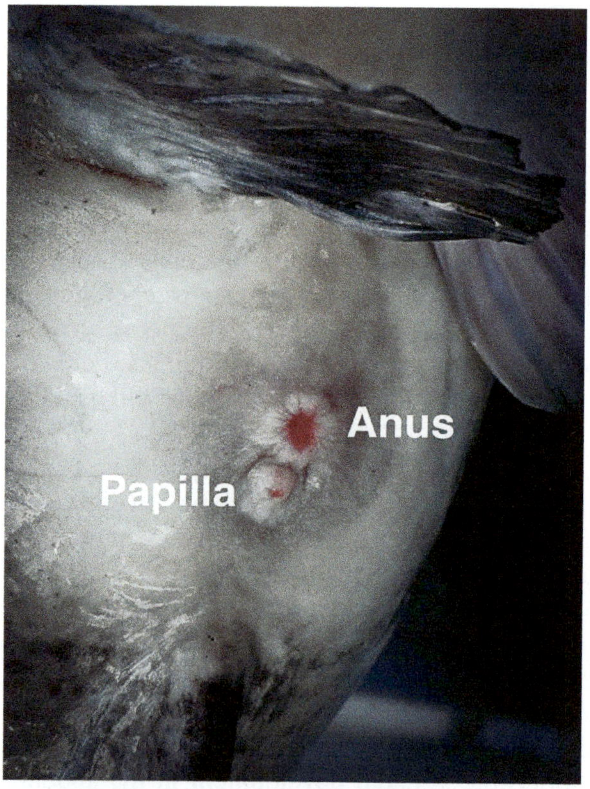

Figure 13.1.2. Papilla of a mature male Channel Catfish during breeding season.

for spawning. When placed on a table, the abdomen of a gravid female should bulge to the sides giving a triangular cross-section appearance (Figure 13.1.3). The genital opening (Figure 13.1.4) is a slit posterior to the anus. The opening becomes enlarged, reddish in color, and swollen as spawning nears. Sometimes the opening is not readily visible but can be distinguished by gently sliding a straw or pencil point down the length of the ventral side of the fish posterior of the anus, and the straw or pencil should enter the slit.

In north temperate climates, Channel Catfish spawn during late spring as water temperatures warm up to a stable level of about 24°C (range 21 to 29°C). Older and larger individuals tend to spawn earlier in the spawning season than smaller and younger brooders. There is more variability in spawning among poorly conditioned fish than in those that have been well fed prior to spawning. The spawning season can be manipulated by regulation of light and temperature under controlled conditions. In subtropical and tropical zones, often there are not the distinct seasonal changes needed to synchronize maturation and spawning of a Channel Catfish population. The fish will grow well and become mature, with a few fish being suitable for spawning over much of the year unless there are distinct seasonal changes to help synchronize brood maturation.

Channel Catfish prepare a nest site for spawning. The male sweeps out the spawning site and will guard the eggs and sac fry. Under natural conditions, protected sites such as rock crevices, sunken hollow logs, a hole in the bank, depressions under tree roots, or similar protected areas are chosen as nest sites. In a hatchery setting containers are added as nest sites. A male will attract a female to the nest site and the pair will begin a courtship that may continue off and on, for six hours or more. Courtship includes swimming side by side, curling the caudal fin around the head of its partner and quivering. As ovulation approaches the female will flare her pelvic fins and sweep the bottom of the nest site. Once she starts to ovulate she will

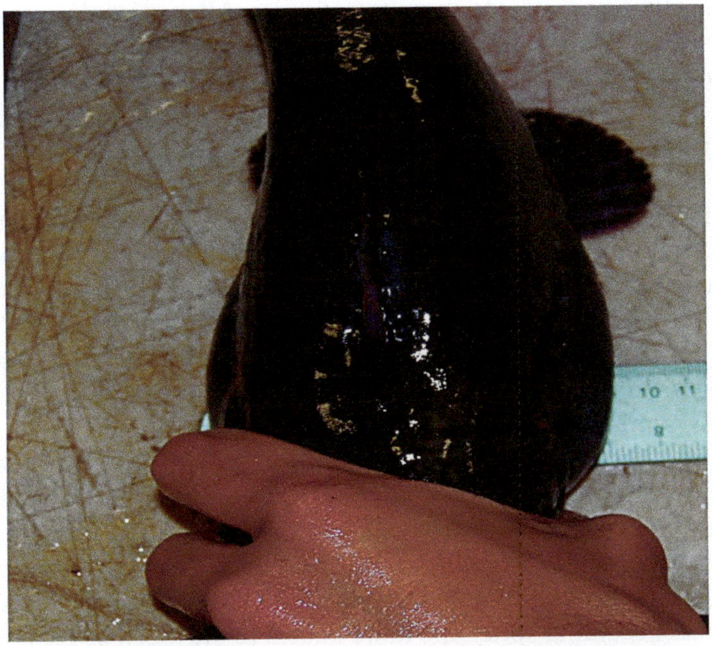

Figure 13.1.3. A mature female brood Channel Catfish is typically gray in color and should have an abdomen that spreads laterally when the female is placed on a flat surface.

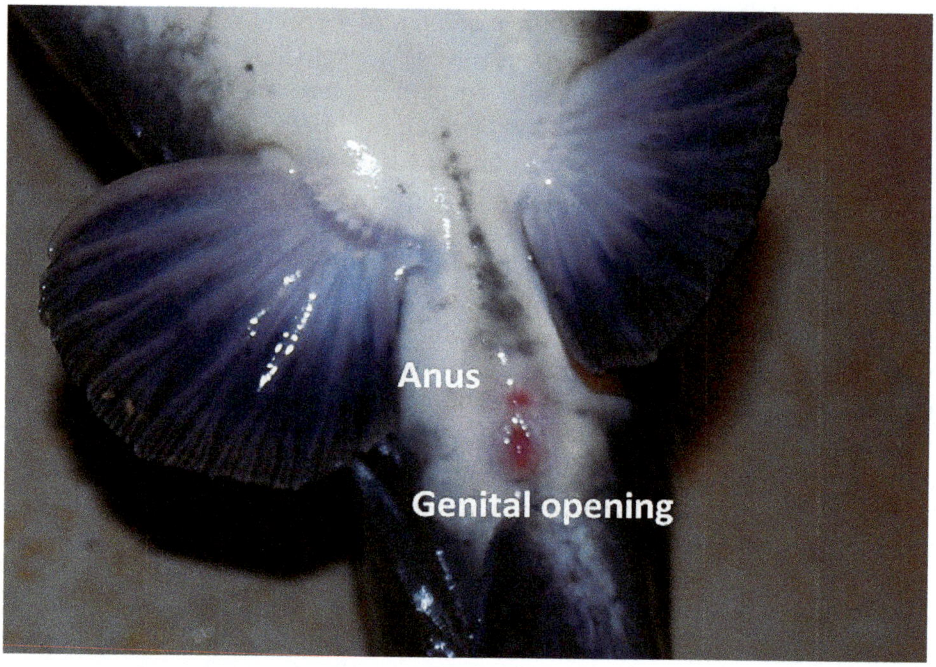

Figure 13.1.4. Female Channel Catfish genital opening becomes distinct during the spawning season, appearing swollen and reddish in color.

release a quantity of demersal, adhesive eggs and the male will pass over the eggs releasing a small volume of sperm to fertilize the eggs. The pair will repeat the process over a 6–12 hour period laying new eggs over old. The eggs become adhesive when in contact with water and adhere together in a jelly-like matrix to form an egg mass. After spawning, eggs are cared for by the male. His care includes driving away predators, circulating fresh water over the egg mass, and massaging the mass to aid in aeration.

Typical fertilization rates are ≥90% but on occasions the male may miss fertilizing an egg release or some eggs are scattered away from the main egg mass and do not get fertilized. Unfertilized eggs will absorb more water than fertilized eggs and within an hour or two will be twice as large as a fertilized egg and have a clear milky color. The mass may be up to 6-8 cm thick and up to 30 cm or more in diameter depending on the female size and the nature of the spawning site (Figure 13.1.5). Egg mass weights may range widely, from <300 g to >3,000 g, with the number of eggs released/kg female ranging from <700 to >12,000. The number of eggs/g of egg mass varies with the age of the egg mass. A recently laid egg mass might average 23 eggs/g of mass but the same egg mass could average 30 eggs/g of egg mass when ≥4 d old due to a loss of matrix over time.

The age of an egg mass can be estimated based on egg color and embryo development (Figure 13.1.6). At 26–27°C recently spawned eggs from well-fed females are bright yellow in color, with a neural ridge evident in the first 24 hours post-fertilization, by day 2–3 the eggs appear yellowish and the head of the embryo is evident. As embryo development proceeds, the egg mass changes color, taking on a rosy or reddish hue as the blood vessels of the embryo develop, with the mass becoming softer and more flaccid to the touch as the matrix is lost. By day 4 to 5 the eyes start to take on pigmentation and the eggs appear orange/red. As hatching

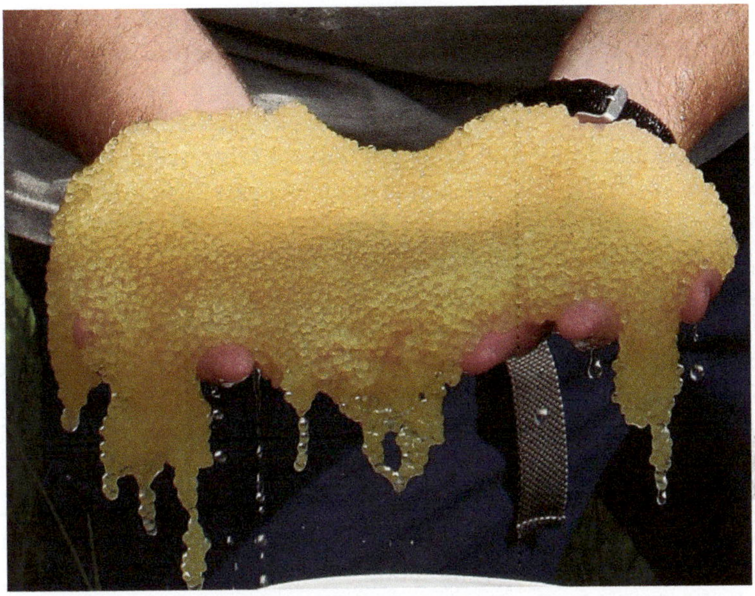

Figure 13.1.5. A large Channel Catfish egg mass where the eggs are held in a gelatinous matrix.

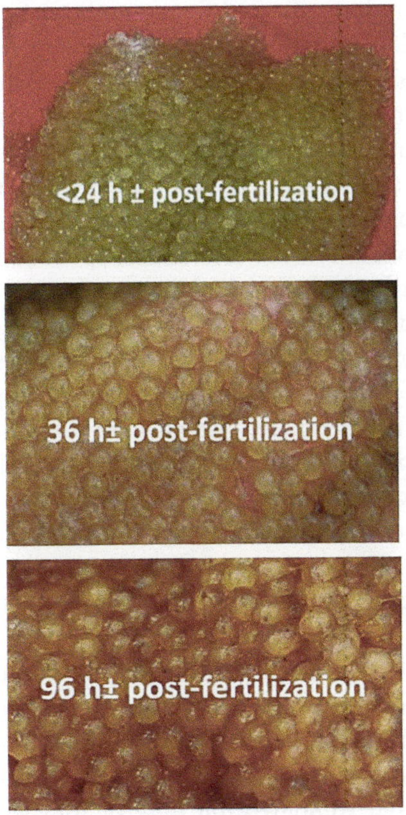

Figure 13.1.6. Change in color of Channel Catfish eggs over the incubation period.

nears the mass takes on a dark red/brown color, the eyes are dark and periodically the embryo is actively moving in the egg. Time to hatch is very temperature dependent, and the above development stages would occur sooner at higher temperatures and slower at lower temperatures. At 22°C it might take >12 days to hatch but at 28°C hatching would occur in <6 days.

The fry at hatching are 6 to 9 mm in length, have a large yolk sac and appear a creamy-yellow/red color. The size at hatching depends upon egg size. The number of fry/g of fry may range from 20 to 50/g soon after hatch. At hatching the large yolk sac weighs too much for the fry to swim up into the water column. After 4 to 6 days, depending on water temperature, the fry have grown, absorbed a portion of the yolk sac, and start to swim up into the water column looking for exogenous foods. During this transition the fry change color to a slate gray to almost black color. Depending on the hatchery, the sac fry are sometimes referred to as red fry and the swim-up fry as black fry.

In a natural setting the swim-up fry will feed on zooplankton as a first food transitioning to larger zooplankton and insect larvae in the first 10 to 15 days. As juveniles they are insectivorous, omnivores as adults, with larger adults also being piscivorous. Under hatchery conditions swim-up fry will accept formulated feeds as a 1st food, and although they may feed initially on zooplankton when stocked into ponds they will go back to feeding on formulated feed as the zooplankton population is reduced.

13.1.2 Broodstock Acquisition and Maintenance

Channel Catfish have been domesticated for a number of years. Many of the strains can trace their genetic history back to fish collected from the Denison Dam area of Lake Texoma, in 1949. Other stocks have also been collected from the wild over their natural distribution range, became domesticated, and were crossed with other stocks to form specific strains (strain = a breeding population having a similar history and possessing unique characteristics). Among the widely used strains are: Harvest Select, NWAC 103, and Kansas. Selection of a given strain might be for one or more desirable traits such as growth rate, feed conversion efficiency, water quality tolerance, disease tolerance, reproductive success, or harvestability. However, no one strain is best in all desirable traits.

Knowing the history of a population is important in selecting brood fish whether introducing them from another location or selecting fish from your own stock. Often stocks of fish are introduced from one hatchery to another with the intent of improving genetic diversity but in many cases the stocks are more closely related than the hatchery manager realizes. Disease history is another important consideration. Channel Catfish virus can be a very virulent disease often resulting in major mortalities. It is a disease that is only evident when there is an outbreak underway. Survivors will be carriers of the virus but the virus cannot be isolated from such fish for verification. However, survivors can have antibodies against the virus and potential brooders can be screened for such antibodies. Populations where individuals have such antibodies should not be considered as brood fish.

Wild fish are more likely to be difficult to spawn than domesticated strains. They can be vectors of diseases that hatchery stocks have not been exposed to. Wild fish may not have immunity to diseases prevalent on the hatchery and will have unknown growth and survival characteristics unless their performance has been measured. Third generation domesticated brood fish are likely to be easier to spawn but may or may not possess desirable survival, growth, and disease resistance characteristics.

Brooders are best developed from known stocks, from as many brooders as possible. For the first and second year the fingerlings and sub-adults can be cultured as a mixed sex population by year class. At the end of the 2nd year future brooders should be separated out, separated by sex and held separately for the next spawning season. Keep in mind that males often initially grow faster and selecting the largest fish in the population may skew the sex ratio highly to males. Ideally brooders are held separately by sex and year class in holding ponds until the start of the spawning season. During that time the fish are fed a 32%+ protein commercial ration with a portion of the protein provided by fishmeal. A diligent feeding program in the fall and early spring are the keys to good success. The feeding rate is temperature dependent as illustrated in the following guidelines:

<8°C do not feed
8–16°C feed 0.5% of body wt./d, 2 d/week
16–24°C feed 1.5% of body wt./d, 3 d/week
24°C feed 2.0% of body wt./d, 5–6 d/week

Feedings rate may be reduced to <1% in the heat of the summer and increased again in the fall. Some hatcheries will add forage fish such as Fathead Minnows, Golden Shiner, Goldfish, or Threadfin Shad to the brood holding ponds. Forage fish provide the advantage in that they are available in the pond when the brood fish are ready to feed such as a warm day in the winter when you might not otherwise be giving a formulated feed. Goals should include feeding conservatively but enough to have weight gain, but not too much to have excessive body fat or create chronic low oxygen problems in the brood holding ponds.

An appropriate brood holding pond would be a 0.5 to 1.0 ha earthen pond 1 to 2 m deep supplied with water of moderate hardness (75 to 100 ppm as $CaCO_3$) and free from submerged rooted aquatic weeds and filamentous algae. Brood stocking rates would be 800 to 1000 kg/ha with a maximum anticipated standing crop of <1600 kg/ha. Allowance should be made for a weight gain from one spawning season to the next of 20 to 50%. Attention should be given to dissolved oxygen management trying to keep it at least 4 mg/mL or more. If fish go off feed the cause should be determined. Feeding should be suspended until the problem is corrected. A proven catfish broodstock is a valuable asset and should be given a high priority as to pond space and management between spawnings as well as at spawning time.

Catfish are subject to several troublesome external parasites, particularly in the early spring when the fish's immune system is not very active. Most of these can be reduced to a minimum by routinely treating brood fish with formalin at a rate of 125 to 150 mg/L for an hour anytime the brood fish are moved. The protozoan *Ichthyophthirius multifiliis* (Ich) can be a major disease problem on both young and adults. Separation of young from adults before embryos hatch and rearing young in an environment free of this organism is the best means of control. To eliminate possible infection, the water supply should be free of fishes that can be carriers and the rearing facility should be free of the infective stage of the parasite before fry are stocked. One major reason for using the egg transfer method of spawning is to break the life cycle of Ich. Broodfish are particularly vulnerable to Ich in the month or so prior to the start of the spawning season when the water temperature is in the 17–21°C range. During this time period, ponds should be examined daily as broodfish having Ich will often swim sluggishly along the shoreline and the Ich infestation can be observed on the fish.

Bacterial diseases often develop following stress. Management of broodstock obviously should include precautions to minimize stress as much as possible. Overstocking, heavy feeding, low DO, and careless handling are some common causes of stress. Once bacterial diseases develop, treatment with antibiotics administered as a bath or added to the formulated feed can be the best therapy (see Chapter 17.6 Disease Management).

Gravid broodstock should be handled carefully. The water level of the holding pond should be lowered by about two-thirds before seining begins. During seining, captured fish should not be crowded in the net or handled out of water any more than necessary. Burlap feed bags or similar coarsely woven sacks facilitate moving fish from the seine to the spawning facility. No more than 10–12 kg/bag should be carried to minimize stress. For transportation on trips of up to 24 hours duration, brooders can be kept in the burlap bags in the transport tank but the weight of fish per bag is cut in half. Confinement in the bag prevents fighting between incompatible fish and quiets the fish, facilitating handling in and out of the hauling tank.

Adult Channel Catfish for food or sport use are hauled in truck tanks at rates as high as 0.5 kg/L but broodstock loading densities should be no more than 0.25 kg/L with good aeration and other measures to provide suitable water quality. When held in tanks, covers should be provided to prevent jumping, and the sexes should be separated if possible. During the spawning season, fighting and injuries often occur when males and females are held together in a confined space such as spawning pens or aquaria. Frequent observation and separation of fish involved in fighting can minimize such injuries.

13.1.3 Hatchery Production

13.1.3.1 Spawning

Channel Catfish are primarily reproduced in a hatchery setting using an egg transfer procedure, but can be hormone-induced to spawn if need be. Ponds used for open pond spawning with egg transfer can vary from 0.2 to 10.0 ha in area although the most manageable size is 0.4 to 1.0 ha. Water depth need not be more than 1.75 m with an average of about 0.75 m. Brooders would be added at the start of the spawning season at a 2 male: 3 female ratio.

A variety of spawning containers can be used including: ammunition cans, milk cans, plastic barrels, and wooden boxes. The size of container should be such that fish can enter and leave readily. An acceptable internal volume would be 75 L with a single entrance having a diameter of 15–23 cm. It may be necessary to add weights on one side of the container to be sure it does roll or float away. It is important to know the origin of the containers as the containers previously may have contained substances that might be toxic to the fish or the pond ecology.

Once the early morning water temperature is consistently 22°C, spawning containers can be added to the spawning pond, adding one container for every 2–3 male brooder stocked. The containers should be set about every 10 m along the pond bank at a water depth of 60–75 cm with the mouth of the container facing out into the pond. The containers should be marked with a stake or a float so they can be easily found. The dissolved oxygen and temperature at the depth where the containers are set should be monitored daily. If dissolved oxygen values consistently drops below 4 mg/L where the containers are set they need to be moved to other locations within the pond with better dissolved oxygen. In stratified ponds the water temperature above the thermocline may get too hot for successful spawning as the season progresses and the water below the thermocline may be too low in oxygen. Such conditions will limit the length of the spawning season.

After brooders and spawning containers are added to the pond, the spawning containers should be checked twice a week in the late morning for eggs. This is done by lifting the container off the bottom, gently pouring out the water, and looking inside. Any fish inside generally escape when the can is lifted. The egg mass, typically attached to the bottom of the container, is carefully removed by hand off the bottom of the container. Young egg masses are often more adhesive and need more care when freeing them from the container. On occasions the mass will be loose and float out of the container as the water is drained off. Once the egg mass is free it can be floated into to a tub or bucket by pushing down the edge of the tub/bucket into the pond water allowing water and the egg mass to flow into the tub/bucket. Figure 13.1.7 shows an egg mass still attached to the bottom of the spawning can that will be collected and transferred to a tub of water. Egg masses can also be placed into a fine mesh bag suspended in the pond water and the bag pulled from one spawning can to the next. A series of spawning cans can be checked at a time, placing 5 to 6 egg masses in the holding container if attention is given to not letting the water temperature in the container become too hot or the DO go too low. It is best to keep the container covered and avoid holding any eggs collected longer than 15 minutes without supplemental oxygen. When a large spawning pond is being examined it is best to have a truck with an insulated hauler and oxygen available, and frequently transferring any masses collected to the transport tank.

Channel Catfish eggs are hardy with no adverse effects when exposed to direct sunlight when being collected. No tempering is usually done with older egg masses when water temperatures are in the 24 to 30°C range. However, when eggs less than 24 hours old are moved they must be acclimated to prevent a high mortality rate. If the water temperature of the pond, transport container, and hatchery differ more than 2 to 3°C (5 to 7°F), eggs should be tempered to the hatchery water temperature allowing 15 to 20 minutes for each 2°C (5°F) of difference. Eggs can be tempered by using a hose to slowly run hatchery water into the transport container until the water temperatures are equal.

Pen spawning is an alternative at hatcheries where 25 egg masses will meet the production goals, or when there is a need to confine the catfish spawning period to a limited period, or where eggs of known parentage is needed. Spawning pens (A varity of containers are used for Channel Catfish spawning where the eggs can be deposited and guarded by the male) are generally located around the margin of the pond in water 60 to 90 cm deep. Pens may be as small as 2.5 m^2 although the usual size is about 5 m^2 (i.e. 2 m × 2.5 m) (Figure 13.1.18). Pens may be constructed from a variety of materials such as mesh wire, wooden slats, or chain-link fencing attached to posts in the four corners. The mesh spacing should be appropriate to retain the brood fish but be large enough to allow as much water circulation as possible. The pen must extend 40 to 60 cm above the water surface to discourage fish from jumping out. Each pen would have one spawning container facing out into the pond. A pair of brooders would be stocked into the pen and allowed to spawn naturally or the female may be hormone-injected to spawn then paired with the male to help synchronize when spawning takes place. An effective protocol is to collectively hold the females in a holding tank, give each one a preparatory dose of Common Carp pituitary at 2 mg/kg followed with a resolving injection at 8 mg/kg. Following the resolving injection the female is stocked into the pen with a male. The spawning container should be checked for eggs every 24 hours for hormone-induced females and every two days when using non-induced females.

If the water in the pen is clear, it may not be necessary to feel inside the container each time a check is made. Sometimes both fish can be seen outside the containers suggesting that

Figure 13.1.7. Catfish egg mass collection from a spawning pond.

Figure 13.1.8. Spawning pens for spawning specific pairs of catfish.

there is not a spawn or that the pair may have spawned outside the container and the egg mass is not in the container. Both fish inside indicates that spawning may be in progress or that both fish have sought shelter for some reason. When the male is inside and the female outside, it is likely that spawning has been completed and the male has driven the female out of the container. For pen spawning it is convenient to use a spawning container with a lid on the top that can be opened. Being able to lift the lid allows the container to be checked to see if there are eggs without having to enter the pond.

In open pond spawning the brooders can easily escape from the container when it is checked. In pen spawning, the brooders are more confined and may try to use the container as shelter. Males guarding eggs may not leave the container without a fight and will bite a hand when it is placed into the container. It may be necessary to use a stick or something similar to chase the male out of the spawning container before trying to collect the eggs and cover the mouth of the container with a dip net to prevent the male from coming back into the container while it is being checked. Once collected, the eggs are handled as those from open pond spawns.

One advantage with pen spawning is once a female spawns she can be removed and a new female added. Males often will spawn numerous times in a spawning season. Nine spawns from one male have been documented when pen spawned with female replacement. Egg masses are occasionally lost from the spawning pen. If a period of 10 to 14 days passes without spawning activity in a given pen, brooders should be examined to verify their sex and condition and replaced if need be. Once the desired number of spawns has been obtained the brooders should be returned to a holding pond where they can regain the weight lost in spawning and to form eggs for the next season.

Eggs collected from ponds have been exposed to a variety of bacteria and fungi and should be disinfected before being incubated in the hatchery. This can be done using a formalin bath at 200–400 mg/L for 5–10 min. or a 10 mg/L potassium permanganate bath at 30–60 sec.

There will occasions when all the incubation facilities in the hatchery are full but the catfish are still spawning in the ponds or pens. In that case it is possible to allow the male to incubate the eggs in the spawning container and hatchery staff come back and collect sac fry after hatching. When there is no room in the hatchery for more eggs the spawning containers still should be checked following the normal schedule. When eggs are found a small sample of eggs should be taken and the age of the egg mass estimated. The location of the spawning container, date when the egg mass was found, and its age should be recorded. Based on the age of the egg mass the male is allowed to continue to incubate the eggs. The spawning container should be examined again when it is estimated that the eggs have hatched. The sac fry have a large enough yolk so they cannot swim up for several days and they will lay on the floor of the container. If the spawning container can be lifted up carefully to the surface the sac fry can be poured into a bucket and transferred to the hatchery for further rearing.

13.1.3.2 Induced Spawning

On occasions spawning may need to be synchronized to produce a large number of fry in a short period of time or where eggs from specific fish must be obtained. This can be accomplished through induced spawning. Channel Catfish respond to hormone treatment of common carp pituitary, hCG and LH-RH. Individual females may be paired with a male and allowed to spawn naturally or the females may be held separately and hand stripped.

Suitable environments for induced spawning can be pens as previously described, aquaria or small tanks. Aquaria or small tanks are preferred. Average size brood fish can be handled in 75 to 115 L aquaria. For large brood fish a container of 175 to 200 L may be needed. It is best to fit the aquaria or tanks with substantial covers to prevent brood jumping. Jumping often occurs at night, and has been the cause of loss of excellent broodstock. Quality water at a rate of 1 to 2 changes/hour should be provided. When spawning takes place, eggs are removed and the brooders removed. Placing a sheet of tarpaper on the bottom of the tank facilitates removal of eggs. The adhesive eggs will adhere to the sheet and not the bottom of the tank, avoiding the need to scrape the eggs off the bottom of the tank.

When the fish are allowed to spawn naturally, hCG or LH-RH can be used. Daily injections of hCG at 600 to 1700 IU/kg have been given with spawning occurring generally after the second injection. LH-RH at 100 μg/kg given in a single injection is also effective with spawning occurring 4 to 7 days later. LH-RH can be a practical procedure for induce spawning of fish held in pens when the water temperature is 24°C to 28°C, brooders can be injected just once at the time of stocking and the spawning container checked daily thereafter. Recovery of females from pens for repeated injections of hCG or pituitary is a time-consuming chore. However, one approach is to hold the females in the hatchery where they will receive both injections and then they would be stocked into pens for spawning.

Common Carp pituitary (CCP) is the material of choice when eggs are to be artificially fertilized. This technique is used in association with producing hybrid (Channel × Blue) catfish. Female Channel Catfish are injected first with 2 mg CCP/kg and 8 mg/kg twelve hours later. Ovulation can be expected 24–30 h later following the second injection at a water temperature of 28°C. LH-RH is also effective when given at a total dose of 100 μg/kg giving 10% in the 1st injection and 90% in the 2nd injection 12 hours later. However, ovulation success can be more variable, particularly at temperatures of 22–26°C.

When held in aquaria, the females can be observed and the onset of ovulation noted. Following the second injection there is a very noticeable distention of the abdomen as the ovary hydrates. As ovulation nears the female will repeatedly flare the pelvic fins and press the vent on the bottom of the aquarium, and when ovulation starts small quantities of eggs will be released. After some 50 eggs or so are observed the female is adequately ovulated that she can be removed from the aquarium and manually stripped.

Another approach for induced spawning with manual stripping is to hold each female in a mesh bag suspended in a holding tank. Females are given two injections of either CCP (10 mg/kg total) or LH-RH (100 μg/kg total). Holding the females in mesh bags has the advantage in that it is easy to collect the brooder for injection or to check to see if she has ovulated. When large numbers of fish are held a tank and cannot be individually monitored, each fish should be examined at about 28 hours after the resolving dose, looking for a free flow of eggs. Before stripping, females should be deeply anesthetized using 100 mg/L MS-222. Special care should be given to avoid losing eggs when removing the fish from the aquarium or tank on the way to being anesthetized.

Artificially spawning Ictalurid catfishes involves another step not needed when artificially spawning most other fishes. Ictalurid males cannot be manually stripped to obtain sperm. The testes of this group of catfishes is villiform and mature testes at the start of the spawning season often represents less than 1% of the male body weight. To obtain sperm, the male must be sacrificed, the testes removed and the lobes of the testes crushed to release sperm. Approximately one mature male is needed for every 3 to 5 females to be spawned. A male can be sacrificed beforehand and the testes, if kept dry, can be held under refrigeration for several hours.

The artificial spawning process is as follows. An ovulated female under deep anesthesia is carefully dried off, particularly in the vent area. A bowl is prepared to receive the eggs by applying a thin coating of petroleum jelly or vegetable oil to the inside of the bowl. Any contact with water will cause the eggs to become sticky. Eggs are stripped into the bowl using moderate manual pressure on the abdomen to obtain a flow of eggs (Figure 13.1.9). When no other eggs are obtained or if they start to become bloody, then egg collection efforts should be stopped and the female is returned to fresh water to recover. The eggs are fertilized by crushing a portion of the testes over the eggs using a physiological saline solution to rinse the sperm on to the eggs. Testes can be prepared ahead of time, crushing the testes and storing them under refrigeration in a Hanks balanced salt solution using 5 mL of solution/g testes. A gram of testes should be adequate to fertilize 250 mL of eggs. Eggs and sperm are mixed, fresh water added, and mixing continued for two minutes. After two minutes the eggs are rinsed in freshwater and excess testicular tissue remove. The eggs are covered with 2–3 cm of water and allowed to stand 10 minutes during which time they will become sticky, adhering to each other, resulting in an egg mass.

Artificial spawning of catfish is the technique of choice when producing Channel × Blue hybrids. The two species do not naturally hybridize readily and the only way to produce them consistently is through artificial spawning using the techniques described above. Once an egg mass is obtained the eggs are managed as a naturally produced mass.

It should be stressed that hormone treatment is no substitute for proper conditioning of brood fish. Poorly conditioned female fish may not have eggs mature enough to respond positively to hormone induction. Such brooders should be returned to the holding pond and may be of acceptable condition several weeks later. Under the proper conditions, induced spawning can give excellent results. It can synchronize spawnings, shorten the spawning season, and allow the production of hybrids that are otherwise difficult to obtain. However, the

Figure 13.1.9. Manually stripping eggs following exogenous hormone induced ovulation.

method is labor-intensive and demands a high degree of knowledge and skill not involved in the less sophisticated approach of pen or open pond spawning.

13.1.4 Egg Incubation

13.1.4.1 Paddlewheel Troughs

In hatcheries where catfish are one of the primary species being produced the eggs are generally incubated in paddlewheel troughs (Figure 13.1.10). A typical trough might be 45–60 cm wide and 25–30 cm deep. The length might range from 7 to >14 m. A rod extends down the center length of the trough above the water surface. Metal or plastic paddles are attached to the rod approximately every 30–40 cm, extending down to within 2–3 cm of the bottom of the trough. The rod connected to a gear-reduction electric motor, rotates at approximately 30 rpm and the paddles create a side-to-side current in the trough. Egg masses are held in 400 to 600 cm^2 baskets made of 0.4–0.6 cm mesh metal or plastic screen, suspended between the paddles. The basket should extend down from the surface to within 5.0 cm of the trough bottom. The trough should receive 25–28°C water at a rate of 19–20 L/min. The water source should have a pH of 6.5–8.0, a total hardness of >25 mg/L and dissolved oxygen levels maintained at 6 mg/L or more. Supplemental aeration in the troughs may be required, particularly as the eggs near hatching and for several hours after hatching. Care should be taken to ensure that egg baskets do not come in contact with the paddles. Baskets can become entangled with a paddle preventing the shaft from rotating and stopping the water currents over the egg masses. If this persist for more than an hour then the microclimate around the egg mass will deteriorate, oxygen levels will drop, and eggs will die.

Figure 13.1.10. Channel Catfish eggs in a paddlewheel trough.

Eggs masses should be stocked into troughs putting similar age masses into the same trough so they hatch at similar times. The age of an egg mass can be approximated based on color. Recently laid eggs masses are typically pale yellow and a dark orange as they approach hatching. Differences in age and color are illustrated in Figure 13.1.6. This facilitates management during incubation and harvest. The egg mass should be inspected and any dead eggs removed. The egg mass loading rate is approximately 500 g/basket. Unusually large (1 kg+) or thick (>7.5 cm) egg masses (Figure 13.1.5) may need to be divided in half or quarters to assure all eggs receive sufficient oxygen. Figure 13.1.10. illustrates Channel Catfish eggs held in a basket for incubation using a paddlewheel trough.

Catfish eggs are prone to get fungus particularly when the incubation temperature falls below 25°C. Eggs are treated with a fungicide daily when necessary. The 10 min. formalin bath described earlier can be used, or a stronger 400 mg/L trough flush can be substituted. Hydrogen peroxide at 750–1000 mg/L active ingredient for 15 min in flowing water is also effective. No fungus treatments should be applied within 12 hours of hatching.

Occasionally, bacterial infections occur, particularly at temperatures above 30°C. Bacterial problems are likely to be worse in a deep trough with a low rate of water exchange. One of the iodine compounds such as Wescodyne® or Betadine® should be applied for bacterial infections at 100 mg/L for 10 min. Twice daily flushes with 2–3 g/L salt is also an effective disease prophylactic for catfish egg masses for both fungus and bacteria. Keeping the paddlewheel troughs clean is one of the best ways to keep diseases under control. The inside of the trough should be wiped clean 1–2 times a day and any organic matter accumulating on the bottom siphoned out daily.

Once hatching starts it may take about 12 hours for all the eggs in the egg mass to hatch. The sac fry drop to the bottom of the tank and the quality sac fry are strong enough to wiggle to the edges of the tank. The weak or deformed larvae are not strong enough to move far and typically remain on the tank bottom under the egg basket. The healthy sac fry that have gathered along the tank edges and corners should be siphoned out of the paddlewheel trough twice a day, stocking similar age sac fry together into fry holding tanks. The sac fry that remain underneath the egg basket should be siphoned out and discarded.

13.1.4.2 Jar Incubation

When a hatchery is producing multiple species where their eggs can be incubated in MacDonald jars and similar style incubators it may be desirable to use that same equipment for incubating Channel Catfish eggs and not devote hatchery floor space to paddlewheel troughs. However, the eggs must be separated from the matrix once the egg mass is more than 24 hours old following the procedure described in Chapter 17.4. When manually stripping and artificially fertilizing catfish eggs, the eggs can be treated with a Fuller's earth slurry for a half-hour shortly after fertilization to obtain individually free eggs suitable for incubation in MacDonald jars.

A MacDonald jar should be filled with eggs to approximately half its volume to ensure a good mixing of eggs. This may represent two to three egg masses, trying to load the jar with similar age eggs. A water flow of 2–4 L/min. should be provided so that there is an obvious rolling of eggs at the bottom of the jar and movement of eggs throughout the jar. Dead eggs may float out of the jar or form a layer above the live eggs that can be siphoned out. Disease problems during incubation are often less when using jar incubation compared to paddlewheel incubation and chemical treatments may not be required. Eggs incubated in this fashion will hatch 12 to 24 hours sooner than if in a paddlewheel trough. As the eggs begin to hatch the shell debris will be washed out of the jar and the sac fry will remain in the jar. Once all the eggs have hatched the sac fry can be transferred from the jar to a fry holding tank. At that time sac fry can be counted to determine the percent hatch and the quantity of fry available to add to the holding tank.

13.1.5 Fry Management

The fry are quite hardy at 1 to 2 days of age, but care should be taken to prevent suffocation when weighing or making volume measurements of large numbers in a small amount of water. Use of a number-weight or a number-volume relationship gives numerical estimates of acceptable accuracy as long as if they are of similar age and have a similar brood history. A 12–24 hour difference in time of hatch can result in the average weight of the two fry having different average weights. Similarly, average weight of same age fry may differ depending on the age and condition of the female brooder. At hatch, the average weight of fry from 3 year-old brooders may weigh less than that of those from 4 to 5 year-old brooders. Because of variations in fry size due to age and parental effects, periodic sampling should be done to assure a realistic number-weight or number-volume relationship. Fry of a common age and common brood history should be harvested and held together in a holding tank/trough.

Sac fry can be stocked at up to 250 to 300 fry/L of trough where there are 1 to 2 water changes/hour. Ideally the trough size is adequate to be stocked completely with one day's collection of sac fry so similar age fry will be available to completely stock a pond. If the fry are to be held longer than 7 days the density should be reduced by half to allow for increased feeding and growth.

After 2 to 3 days post-hatch the fry will have utilized much of the yolk sac and mouthparts are functional. At that point they begin to swim toward the surface looking for food and are called swim-up fry. They can be stocked out into a well-prepared nursery pond at that point, or be fed in the hatchery for several more days and then stocked. Swim-up fry held in the hatchery should be fed a high-quality feed (No. 0, powder, 45–55% protein) 5 to 6 times/day. The initial feed rate is about 15% of the body weight/day. At this stage, the quantity supplied is less important than making the feed available to as many individual fish as possible. Care should be taken to getting feed to all sections of the trough occupied by fry to give each fish a chance to find it. If the fish are not being fed enough a significant mortality can occur 10 or 11 days after hatching. Troughs should be cleaned twice a day and any uneaten feed siphoned out. If the fish are fed in the hatchery for more than 3 days a buildup of parasitic protozoans, such as *Trichodina*, may occur. Twice daily flush-through treatments using formalin at 100 mg/L should control it.

How long to hold catfish fry in the hatchery after they swim-up has its tradeoffs. Stocking fry soon after swim-up reduces the risks of mortality in the hatchery and reduces hatchery labor, while stocking recent swim-up fry into ponds often results in lower and more variable pond survival. Holding the fish longer in the hatchery and stocking larger fry into ponds typically will improve pond survival. However, holding the fry longer in the hatchery increases the rate of mortality in the hatchery, and requires more facilities and labor.

13.1.6 Pond Culture

Channel Catfish are not typically produced using a primary nursery followed by a secondary nursery but are produced in a single phase from swim-up fry to fingerling. Swim-up fry are stocked in May or June into fertilized ponds when the water temperature is >21°C, given formulated feeds and harvested 3–6 months later. They also can be produced in raceways using a primary/secondary nursery approach but that is not commonly done.

The rearing pond should be fertilized with 1 to 3 applications of a suitable organic/inorganic fertilizer mixture during the first 3 to 4 weeks of operation unless natural fertility is sufficient to produce an abundance of natural food. A suitable pond preparation protocol is adding 500 kg cottonseed meal/ha to a dry pond, along with agricultural lime at 1,000–2,000 kg/ha depending on the soil characteristics and water hardness. The dry pond is then filled and a liquid inorganic fertilizer (10-34-0) is added at 19 L/ha. Often additional organic fertilizer may or may not be needed once the pond is full as the feed being offered the fish is also a source of nutrients supporting the natural food chain. A mixed population of natural food organisms will develop rapidly. For best results swim-up fry are stocked on day 5 (post-hatch) into a pond that was filled in 5 days. In many cases it is not possible to fill a pond rapidly and a variety of predacious insects can become established. In that case, trichlorfon at l ppm has been used within 3 to 7 days of stocking to eliminate competitive and predaceous invertebrates. Methyl parathion at a rate of 0.25 ppm applied 4 days before stocking has been used for a similar purpose. However, neither of these insecticides is approved in the USA for preparation of fingerling catfish ponds. These insecticides are not very selective and while controlling predacious insects the abundance of desirable food organisms may be also reduced.

Stocking rate varies depending on the size of fish desired at harvest and the length of the culture period and might range from 25,000 to 500,000/ha (Table 13.1.1). Once stocked, the fish are fed a commercial diet beginning with No. 0, powder, 45–55% protein feed. They can be fed based on feeding a percent body weight, adjusting for anticipated growth and survival according to guidelines shown in Table 13.1.2. To feed based on a percent body weight the fish need to be sampled periodically to determine the average weight and have some idea what the survival has been. Careful record keeping during previous production cycles will provide data as to what growth rates and survivals that might be expected. Another approach is to feed a progressively greater quantity until water quality risks begin to occur (Table 13.1.3).

Feed should be given beginning the day of stocking in the same area as where the fry were released into the pond. In a well-prepared fingerling pond the catfish fry will initially feed on natural food organisms and will not actively feed on the formulated feed for several

Table 13.1.1. Stocking rates for Channel Catfish fry stocked into earthen ponds, fed a commercial diet and cultured for 90–120 d.

Fry stocked/ha	Mean size at harvest (cm)
25,000	18 – 25
75,000	15 – 20
130,000	13 – 18
300,000	8 – 13
500,000	5 – 8

Table 13.1.2. Suggested plan for feeding Channel Catfish fry to fingerling size in earthen ponds on formulated feeds when the water temperature is above 26°C.

Fish size		Percent body weight	Feeding frequency/day
cm	*n*/kg		
0.8	40,000	15	5
2.5	4,400	10	4
5.0	1,100	7	4
7.5	220	5	3
10.0	110	4	2

Table 13.1.3. A feeding schedule for Channel Catfish fry and fingerlings based on anticipated biomass and water quality.

Week	Feed type & size	daily input, kg/ha/d	feeding frequency /day	Aeration need
1	no. 0 meal	10	3–4	rarely
2	no. 1 crumble	20	3	rarely
3	no. 1 crumble	30	2	rarely
4	no. 2 crumble or 3.2 mm floating feed	40	2	occasionally
4–8	3.2 mm floating feed	40–60	1–2	frequently
≥ 9	4.0 mm floating feed followed by a 4.8 mm feed	40–60	1–2	frequently

weeks until the natural food organism density is reduced. However, it is best to continue offering the formulated feed in the area where the fry were initially released. Some fish will remain in that area feeding on the formulated feed and as natural food organisms become scarce other fish will join them and begin to feed actively on the formulated feed. In ponds with good plankton blooms it is difficult to see the fish as they feed on the crumbled feed that sinks to the bottom. However, when they feed actively they will stir up mud and release bubbles from the sediment. The extent of bubble release can be used as an indicator of the strength of the feeding response. The feed should be distributed for several meters along the shoreline where the fish were stocked. Once the fish are trained to feed, the feeding area can be expanded. Keep in mind that once the fish are being offered floating feed the fish should be offered the feed from the upwind side of the pond so that any wind will blow the feed across the pond allowing the fish a greater opportunity to feed and less feed washed up onto the shoreline.

As the culture period progresses, the risks of low dissolved oxygen increases and should be monitored nightly. Low dissolved oxygen problems are rare in the first few weeks of culture but on occasions plankton die-offs will occur or zooplankton can be so abundant that they reduce the phytoplankton to the point there is a significant reduction in oxygen production. Some dissolved oxygen problems can be anticipated by monitoring dissolved oxygen concentrations in the late afternoon. If dissolved oxygen values are below saturation in the late afternoon emergency aeration may be required over the course of the night. Identifying a potential problem in the late afternoon will help to ensure that necessary equipment and personnel are available that night in case a problem occurs. As the feeding rate approaches 40 kg/ha/day low dissolved oxygen problems will become more frequently and nightly aeration may be required. Nightly aeration should be available for all ponds receiving higher feeding rates.

Low dissolved oxygen is probably the most universal hazard in rearing fingerling as evident by: (1) direct mortality if dissolved oxygen levels drop much below 1.0 ppm, (2) reduced feed intake and increasing feed conversion when dissolved oxygen levels are much below 2.5 ppm, (3) increased incidence of disease after fish are stressed by low dissolved oxygen levels, and (4) increased vulnerability of small fish to predation as they come to the surface and along the edge of the pond during times when oxygen is deficient.

The risk of diseases also increases as the BOD and temperature increases. In most cases it is best not to sample during the culture period as the stress associated with sampling may result in disease outbreaks of columnaris and *Edwardsiella* (ESC). Feeding response should be recorded daily and any ponds showing a reduction in feeding response should be monitored more closely. This includes monitoring water quality more frequently, checking for any increased bird activity around the pond, and patrolling the shoreline looking for dead or distressed fish.

13.1.7 Harvesting and Processing Small Fingerlings

Small Channel Catfish fingerlings are ready for harvest when the average size is about 7.5 cm. Growth to this size can be achieved in 60 days when water temperature is favorable and the ponds are not overstocked. Generally, fish hatcheries use a combination of seining and draining to harvest fingerling catfish. Up to 75 to 80% of crop has been removed from 0.3 to 0.8 ha ponds by seining after attracting the fish to the feeding station. This tactic is most effective the first time it is attempted. After a delay of 10 to 14 days it can be repeated although the catch is usually less on the second attempt. A lift net can be substituted for the drag seine and can be employed with a shorter time interval between attempts. Obviously, aquatic weeds should be absent for partial harvesting to be successful. It is desirable to suspend feeding for 24 hours before a partial harvest. In this way the catfish are more readily attracted to the feeding station, and their digestive tract is likely to be almost empty, thus minimizing the time needed for holding the fish to empty their digestive tract before being shipped off-station.

After a significant part of the biomass has been removed by seining, a complete harvest should be made by draining the remaining fish into a catch basin, from where they will be harvested. This is best done when the water temperature has dropped to between 10–20°C.

After removal from the pond, grading may be desirable, especially if the fingerlings are to be sold. Spaced bar graders are used for this operation. Table 13.1.4 gives the bar spacings for separating some common fingerling sizes.

Table 13.1.4. Bar grader widths to separate Channel Catfish fingerlings.

Width between bars		Size fish held
27/64"	(1.1 mm)	3" (7.6 mm)
32/64"	(1.3 mm)	4" (10.2 mm)
40/64"	(1.6 mm)	5" (12.7 mm)
48/64"	(1.9 mm)	6" (15.2 mm)
56/64"	(2.2 mm)	7" (17.8 mm)
1"	(2.5 mm)	8" (20.3 mm)

Adjustable bar graders are also commercially available if spacings provided in Table 13.1.4 are not appropriate. Many fingerling producers prefer to build their own graders that are designed to work in conjunction with a specific processing routine and tank size. The essential features of a grading system are to permit the needed sizing, with a minimum of stress to the fish, while efficiently using the available labor.

Catfish fingerlings are hardy and withstand harvesting and processing stress well. They can be held in tanks for several days at densities as high as 64 kg/m^3 with 3 to 4 water changes/hour. They are about as tolerant of handling stress as any of the important food or game species except Goldfish, Tilapia, and the air breathing species. Hardiness should not lead to abuse and poor fish cultural techniques that can result in disease outbreaks or other adverse effects.

13.1.8 Rearing Large Fingerlings

In the commercial production of Channel Catfish, a pond may be in production for several years, having numerous partial harvests of food fish, as well as being restocked several times with large fingerlings before a final harvest is made. In sportfish ponds, large Channel Catfish fingerlings are stocked into established ponds to diversify the fishing opportunities. In both cases fingerlings of a large enough size are needed that will not be preyed on when stocked into setting where there are larger fish present.

Large fingerlings can be produced in a single season by stocking catfish fry at low densities of 25,000 and culturing them for 5–6 months. However, this is often not an efficient use of pond space. In many cases it is more efficient to produce a large fingerling over two seasons. The first season fry are stocked at a high density (300,000– 400,000/ha) in the spring and harvesting fingerlings averaging 7.5–10 cm in the fall. Those fish can be restocked that fall at 120,000–140,000/ha to produce 15 cm fish or at 50,000–60,000/ha to obtain a 20 cm fish by the end of the next summer. When the temperature is 24 to 30°C the feeding rate for large fingerling should be 2 to 4% of the body weight of the fish/day. A feeding guide for Channel Catfish is given in Chapter 17.9. Production should be from 23 to 33 kg/ha/day with a feed conversion of 1.2 to 1.5 during this stage of growth. The rate of feeding is a compromise between the amount that will give the most efficient gain per feed unit (moderate underfeeding) and a rate that will give the highest gain/ha/day (slightly overfeeding). During the warmest portion of the production season standing crops often will be in excess of 3,000 kg/ha and as such the risks of chronic dissolved oxygen conditions and disease outbreaks are high. Feeding a 3,000 kg/ha standing crop at 4% body weight/d would be 120 kg/ha/d, which will require a nighttime monitoring of dissolved oxygen and in most cases aeration every night. Both large and small fingerlings need to be fed through the winter, particularly anytime when the water temperature is 15°C or more, otherwise they will lose significant weight.

13.1.9 Distribution

Once the harvested fish have been cleaned and graded, any needed prophylaxis or disease treatment given, and have been held until their stomachs are empty, the fish are ready for distribution. Truck mounted tanks of several designs and sizes are used for catfish transportation. The quantity to be shipped, the length of trip, size of fish, climatic conditions and the number of ponds to be stocked from each load are factors influencing the kind of tank used. For small loads, a tank suitable for a 1/2-ton pickup truck is a popular choice (Chapter 11, Figure 11.3). Aeration is provided by 12-volt electric agitators powered by the truck generator. Fish can be removed by netting with a hand net or released with water through the tank drain. Larger loads are transported in a tank of suitable size equipped with electric agitators, oxygen diffusers and in some instances a water circulation system. For hauling larger fingerlings, a loading rate of 300 to 500 g/L can be used. This contrasts with small fingerling hauling densities that are in the 200 to 300 g/L range with the lower limit used for trips longer than 12 hours.

Optimum water temperature for the distribution of catfish fingerlings is around 12 to 13°C. For long trips, the use of ice or a refrigerated tank to bring the water temperature into this range can reduce the distribution cost by allowing the load rates to be increased up to 50% compared to what would be normal loads when the water temperature is above 20°C. Toward the end of the trip, temperature of the tank water should slowly be adjusted to that of the receiving water in order to minimize the amount of tempering required at the receiving site. Sudden changes of more than 5°C should be avoided, especially from warm to colder levels. When ice is used, care should be taken to avoid ice made with chlorinated water.

13.1.10 Recommended Reading

For additional information regarding channel catfish see the following references:

Avery, J. L., and J. A. Steeby. 2004. Chapter 7, Hatchery management. Pages 145–165 *in* C. S. Tucker and J. A. Hargreaves, editors. Biology and culture of Channel Catfish. Developments in Aquaculture and Fisheries Science 34. Elsevier Science,Amsterdam, The Netherlands.

Bosworth, B., C. Nagaraj, and J. Avery, 2011. Producing Hybrid Catfish Fry: Workshop Manual. USDA–ARS Catfish Genetics Research Unit and the Mississippi State University National Warmwater Aquaculture Center, April 6–7, 2011. Available: https://tcnwac.msstate.edu/2011%20Hybrid%20Catfish%20Manual.pdf. (May 2020).

Chatakondi, N. G. 2014. Suspending Mammalian LHRHa-injected Channel Catfish, *Ictalurus punctatus*, in individual soft mesh bags reduces stress and improves reproductive performance. Journal of the World Aquaculture Society 45(5):604–612.

Chatakondi, N. G. 2017. Effective dose of salmon GnRHa for induction of ovulation in Channel Catfish. North American Journal of Aquaculture 79(4):310–316.

Dunham, R. A., and R. O. Smitherman. 1984. Ancestry and breeding of catfish in the United States. Circular 273. Alabama Agricultural Experiment Station, Auburn University, Auburn.

Jensen, J., R. Dunham, and J. Flynn. 1983. Producing Channel Catfish Fingerlings. Circular ANR·327. Alabama Cooperative Extension Service, Auburn University, Auburn.

Kelly, A. M. 2004. Channel Catfish broodfish management. SRAC Publication 1802. Southern Regional Aquaculture Center, Stoneville Mississippi. Available: http://agrilife.org/fisheries2/files/2013/09/SRAC-Publication-No.-1802-Channel-Catfish-Broodfish-Management.pdf. (May 2020).

Robinson, E. H., M. H. Li, and B. B. Manning. 2001. A practical guide to nutrition, feeds, and feeding of catfish, Bulletin 1113. Mississippi Agricultural & Forestry Experiment Station, Office of Agricultural Communications, Mississippi State University, Starkville, Mississippi.

Small, B. C. 2006. Managing hatch rate and diseases in catfish eggs. SRAC Publication No. 1804. Southern Regional Aquaculture Center, Stoneville Mississippi. Available: http://fisheries.tamu.edu/files/2013/09/SRAC-Publication-No.-18.04-Managing-Hatch-Rate-and-Diseases-in-Catfish-Eggs.pdf. (May 2020).

Steeby, J., and J. Avery. 2005. Channel Catfish broodfish and hatchery management. SRAC Publication 1803. Southern Regional Aquaculture Center, Stoneville Mississippi. Available: http://agrilife.org/fisheries/files/2013/09/SRAC-Publication-No.-1803-Channel-Catfish-Broodfish-and-Hatchery-Management.pdf. (May 2020).

Steeby, J. A., and B. A. Wagner, 2005. Channel Catfish hatchery and fry production practices in the U.S. catfish industry. World Aquaculture 36(4):14–19.

Wynne, F. S., and W. A. Wurts. 2009. Live Hauling Channel Catfish. Kentucky State University, Division of Aquaculture, Cooperative Extension Program, Frankfurt.

USDA. 2003. Part I: Reference of Fingerling Catfish Health and Production Practices in the United States, 2003 USDA:APHIS:VS, CEAH, National Animal Health Monitoring System, Fort Collins, Colorado.

13.2 *Clarias* Catfish

The family Clariidae is widely distributed in the tropical and sub-tropical areas of the world with some 15 genera and 93 species. The most widely cultured species are in the genera *Clarias* and *Heterobranchus*. There are approximately 60 species of *Clarias* distributed from Africa to Southeast Asia. The genus *Heterobranchus* has four species with its natural distribution limited to Africa. The basic life histories are similar for both genera. They are commonly found in slow-moving streams, swamps and backwaters. The arborescent organ/suprabranchial organ (accessory respiratory organ) allows the fish to survive in low oxygen settlings and survive out of the water as it moves over land from one body of water to another earning the name "walking catfish."

Spawning takes place during the rainy season. The brooders move into flooded areas where they excavate a nest in the submerged bank of a river, irrigation canal, or rice paddy dike, or clear out an area in the submerged vegetation. Most are pair spawners where a nest is built and the male guards the eggs and fry. Philippine Catfish *Clarias batrachus*, North African Catfish (Sharptooth Catfish) *C. gariepinus*, Bighead Catfish *C. macrocephalus*, and Magur Catfish *C. magur* and the African Giant Catfish *Heterobranchus bidorsalis* are among the commonly cultured species.

13.2.1 Brood Management

Clarias will mature under hatchery conditions. Depending on the species and environmental setting most will mature when 6–12 months old at a size of ≥100 g. *Heterobranchus* may require 2 to 3 years to reach sexual maturity in the wild but only 10 to 12 months under hatchery conditions. In natural settings the gonadal development peaks as the rainy season begins. In captivity both *Clarias* and *Heterobranchus* normally do not spawn in static water ponds as long as there is no significant increase in the water level of the pond. Female *C. gariepinus* have well developed ovaries with mature eggs year-round when held in ponds at a water temperature of ≥25°C. A fecundity of about 40,000 eggs /kg can be expected from an induced spawned and manually stripped female. *Clarias gariepinus* males can be sexually mature at an age of 8–12 months at a weight of approximately 200 g. Males having well-developed testes can be expected to produce sperm year-round when held in ponds at around 25°C.

Brooders should be held separately by sex and size group in ponds where there is some water exchange. Densities for larger *C. gariepinus* might be 0.5–1 fish/m^2 and for smaller *C. batrachus* males at 3 males/m^2 and females at 4–5 fish/m^2. Many producers feed a moist mixture of animal and plant materials such as trash fish mixed with rice bran. Ideally, the diet would contain 35–40% digestible protein, and 3–4 kcal/g of digestible energy. A pelleted feed is better to use but may not be available or be too expensive in many locations. Small tilapia can be added to the brood pond as forage. The number of brooders and their sex ratio depends on the spawning method used. When brooders are pair spawned and stimulated to spawn naturally the brood population should be slightly skewed to male as the males may fight among themselves causing mortality. Males must be sacrificed to obtain sperm when induced spawning and artificially fertilizing the eggs. Typically, larger males are sacrificed as the testes of one male will provide enough sperm to fertilize the eggs of 5–10 females. However, not all males will have adequately developed testes but that is not known until after the male has been sacrificed. In that case maintaining a 1 male: 2 female brood ratio would be acceptable, but keep in mind that a significant quantity of the brood males may be sacrificed each year and have to be replaced.

At temperatures of ≥25°C, female brooders quickly develop new batches of mature eggs and can be hormone-induced to spawn every 4–6 weeks without affecting the quality or quantity of ovulated eggs. However, expect some females to be lost in each spawning cycle due to handling and disease outbreaks. One problem with controlled natural spawning is that the male may fight with the female before the female is ready to spawn even to the point that she may die from her injuries.

Brooders can also be developed in an intensive flow-through system using 1.0–1.5 m^3 rectangular tanks stocked at up to 100 to 150 kg/m^3 with a twice/h water exchange rate. The tanks should have a solid cover over 80–90% of the tank, while covering the area near the drain with a 3 cm mesh material. Enough lighting should be provided to the area near the drain to observe the fish, remove any dead fish and siphon out any waste. The goal should be to keep dissolved oxygen levels at 3 mg/L or more, being alert to oxygen drops that will occur when feeding such a high biomass of fish. Fish should be fed a high-quality pelleted feed twice a day at 1–2% body weight/d.

There is a distinct sexual dimorphism in mature *Clarias*. The male has a large prominent elongated genital papilla tapering to a point. Females have a rounded papilla, which in gravid females will be swollen and reddish colored. Females selected for spawning should have a well-rounded abdomen where the impression of the ovaries is evident. They should release a few eggs out the papilla when the abdomen is lightly pressed. The eggs should be uniform in size and the nucleus may appear as a small dark point in the center of the egg. *C. gariepinus* weighing 0.5–1.0 kg are preferable as they are easier to handle than larger fish. Acceptable size brooders of *C. batrachus* would be 100 to 250 g, while fish 200 to 400 g would be acceptable for *C. macrocephalus*.

13.2.2 Spawning

13.2.2.1 Controlled Natural Spawning

There are several approaches to spawning *Clarias* under hatchery conditions ranging from an extensive pond system to hormone-induced spawning with manual fertilization. A basic extensive approach would be to use an earthen pond where an island is built in the center of a pond that would be submerged about 25–30 cm when the pond is full. Grasses are allowed to grow on the island. Potential nest sites are dug into the island, each about 20–30 cm in diameter and 15–25 cm in depth for *C. batrachus* or 5–8 cm deep for *C. macrocephalus*. A portion of the pond needs to be 1–1.25 m deeper than the island and large enough to hold the quantity of brooders to be stocked. The brood holding area is filled and brooders added at equal sex ratios at 1–2 fish/m^2 (600–700 kg/ha). Brooders are fed a 9:1 trash fish: rice bran ration at 5–6% body weight/d until the start of the rainy season. Once the rainy season begins the pond is flooded raising the water level enough to flood the island. The fish should spawn within two days, the eggs hatch in 20–24 h with the larvae swimming up within 3–4 days. The male will guard the eggs and fry. Once the fry swim up they should be captured and transferred to the hatchery or to a nursery pond. A harvest ranging from 2,000 to 15,000 one-week-old fry per nest might be expected. The challenge with this spawning method is that it is often difficult to collect all the fry in a spawn, some are damaged while being collected, and some spawns disappear before you can collect them.

A variation on this approach is to use ponds with concrete walls through which artificial concrete nests are connected. A pond about 200 m^2 is connected through the wall with a 10 cm ID PVC pipe set below the full-pond water level to a 0.4 × 0.4 × 0.4 m nest. The nest is lined with fibrous material and covered with a roof that can be lifted to examine the nest. Brooders are stocked at a 1:1 sex ratio at 1 fish/m^2 of pond when the pond is at half volume. The water level is increased to flood the nests and the brooders enter the nests and spawn. The adhesive eggs will attach to the fibrous material. The nest should be examined twice a day for seven days. The fibrous material should be removed if eggs are present and transferred to the hatchery for incubation. New fibrous material should be added to the nest in case of additional spawns.

Another approach to natural spawning is to inject the females using the same hormones and dose rates used for induced spawning with manual stripping but pairing an injected female with a non-injected male. This is done in the evening, stocking a pair of fish into a 2–3 m^3 fine mesh hapa (0.5 mm mesh) that is placed in a tank of clear water and the fish are allowed to spawn naturally. It is important to make sure that the hapa has enough freeboard (≥40 cm) and is well covered so the brooders cannot escape. The fish should spawn naturally in less than 24 h and once the brooders spawn they are removed. The adhesive eggs stick to the hapa and are incubated on the hapa in the tank and after hatch the larvae are collected from the hapa when they start to swim. The fry prefer to hide in the dark and can be concentrated by adding a tile on the bottom of the hapa to give them a place to hide. The disadvantage of induced natural spawning in hapas is not all pairs spawn successfully, the male may kill the female, and some larvae are lost when harvested. The hapas can also be placed in earthen ponds but are harder to manage, and the risks of disease problems are greater and harder to control.

13.2.2.2 Hormone-Induced Spawning with Artificial Fertilization

Hormone-induced spawning with artificial fertilization allows much more control over the production cycle. It helps synchronize time of spawning within a group of females, can increase egg collection from a given species, and allows genetic combinations that may not happen naturally. This is helpful with *C. macrocephalus* that does not readily spawn under hatchery conditions and in the production of hybrids (*C. macrocephalus* × *Clarias gariepinus*). The hybrid provides the preferred taste of the *C. macrocephalus* with the faster growth rate and greater environmental tolerances of *C. gariepinus*. As hatchery-held *Clarias* females can be gravid most of the year if the water temperature is ≥25°C, the combination of careful selection of brooders and the application of appropriate dosages of exogenous hormones will result in ovulation. Males typically have adequate gonadal development at ≥25°C and may not need exogenous hormones as further stimuli.

The most basic approach is to induce ovulation using fresh pituitary of the same species. In the case of *Clarias* and *Heterobranchus* one equivalent weight of pituitary is given in one injection to the female (see Chapter 4 for collection and use of fresh pituitary). Other hormones and dose rates that have been used successfully are given in Table 13.2.1.

If the female releases eggs when a light pressure is applied on the abdomen then she should be removed from the holding tank and prepared for stripping. The fish should be wrapped in a towel and dried carefully, not letting any water come in contact with the eggs. One person should hold the head up and a second person hold the body over a dry bowl and lightly press the abdomen. The eggs should flow freely as a dense stream with no clumps or blood. Expected fecundities of hormone-induced females are given in Table 13.2.2.

Females are given an intramuscular injection and held separately from the males. Time to ovulation may range from 7 to 21 h post-injection and varies by: species (*C. gariepinus* slower than most); Brood quality (better quality respond sooner), water temperature (faster at higher temperature, 12 h at 24°C vs. 8 h at 28°C); and hormone administered (fresh pituitary quicker than synthetics). As with any other fish the proper timing of stripping after the hormonal injection is critical in obtaining high quality eggs. Females should be checked starting at 8 h post-injection to see if the abdomen of the female has become softer or if she is releasing eggs. If the female only releases a small quantity of eggs she should be returned to the holding tank and examined again 30–60 min later. Ovulation is not instantaneous but takes place over several hours for individual eggs to undergo final maturation and be released into the lumen of the ovary. However, any significant delay in stripping the eggs will result in poorer quality eggs, particularly the higher the temperature. In the case of *C. gariepinus*, acceptable egg quality can be obtained if the female is stripped within 120–240 min post-ovulation for brooders held at 20°C, but at 25°C the period occurs earlier and the window is narrower (60–90 min) and at 30°C even earlier (30 min).

Sperm should be collected and prepared before the anticipated time of ovulation. If the female had been injected with fresh *Clarias* pituitary from a male donor, the same donor can also provide the testes needed for egg fertilization. Such testes should be cleaned and stored either whole or in a filtered 0.9 % saline solution of macerated testes, and refrigerated at 5°C for up to 12 h before being used. When using females injected with other hormones, donor

Table 13.2.1. Hormones and dose rates used to induce spawn and manually strip female *Clarias* or *Heterobranchus*.

Hormone	Dose/kg Body weight	Injection frequency	Time to ovulation	Reference
Acetone dried carp pituitary	4 mg	single	7 h at 30°C	Hogendoorn and Vismans 1980
	6 mg	single	17 h at 25°C	Zonneveld et al. 1988
Human Chorionic Gonadotropin (HCG)	1,500 IU	single	11 h at 30°C	Legendre and Oteme 1995
	2,000 IU	single	14–16 h at 27°C	Adebayo and Fagbenro 2004
Luteinizing hormone -releasing hormone analogue (LHRH a)	20 µg	single	16–18 h at 27–31°C	Ngamvongchon et al. 1988
	50 µg + 5 mg pimozide	single	12–13 h at 25°C	De Leeuw et al. 1985
Ovaprim (20 µg/ml salmon GnRH-a +10 mg/ml Domperidone)	1.0–1.5 ml	single	14–17 h at 27–28.5°C	Sahoo et al. 2007
	2.0 ml	single	15 h at 28°C	Dhara and Saha 2013
Ovatide	0.6 ml	single	10–11 h at 29–30°C	Chand et al. 2011
	1.0 ml	single	17 h at 27–28.5°C	Sahoo et al. 2007

Table 13.2.2. Number of eggs*/kg of female body weight obtained from different Clariidae when hormone induced to spawn and manually stripped.

Clarias gariepinus	90,000–120,000
Clarias lazera	52,000–69,000
Clarias macrocephalus	20,000–90,000
Heterobranchus bidorsalis	22,000–23,000
Heterobranchus longifilis	46,000–103,000

*Rounded to the nearest thousand

males can be sacrificed and the testes collected and prepared an hour before the anticipated ovulation. The testes should be removed, cleaned of any blood or attached fat, be placed in a mortar, sliced into small pieces and the pieces ground with an equal volume of 0.9 % saline. The fluid should be filtered through a clean piece of cloth to remove any suspended tissue and the fluid saved. The fluid can be further diluted with 0.9 % saline to obtain a 10-2 or 10-3 solution without harming fertilization success. The solution should be kept refrigerated until being used.

The eggs collected should be clean without any feces or blood, with no drops of water in the bowl before sperm is added. The sperm solution is added at approximately 0.1 mL/100 mL of eggs and gently mixed in with the eggs. An alternative is to hold the testes in a piece of mosquito netting and crush the testes over the eggs, adding about 3–4 large drops/100 mL of eggs. Once the sperm is mixed in, a volume of fresh clean water equal to the egg volume is added and stirred in for 1–2 min, keeping in mind that the eggs will start to become adhesive. After two minutes the eggs should be rinsed in clean water and distributed over a substrate where they will adhere and can be incubated.

13.2.3 Incubation and Larval Management

Suitable incubators include 1–2 mm mesh hapas where the eggs are slowly spooned out along the side of the hapa and stick to the netting where they will remain until hatch. The hapa is ideally placed in an indoor hatchery tank but can also be placed in outdoor ponds or tanks. The adhesive eggs can also be incubated on a horizontally stretched window screen (1 mm mesh) on a frame placed 5–10 cm below the surface of a hatchery tank. The eggs should be distributed in a single layer over the netting trying to avoid eggs sticking to each other (Figure 13.2.1).

The tank should be aerated or have a water exchange. A steady temperature is desirable. Hatching time varies by species and water temperature (Table 13.2.3). An alternative mode of incubation is to overcome the adhesiveness by coating the eggs with talcum powder or full-fat powdered milk for 15–30 min. This should begin within minutes after fertilization, gently stirring the eggs while adding a slurry of talcum powder or full-fat powdered milk mixed in water. During the treatment period the slurry should be exchanged several times and at the

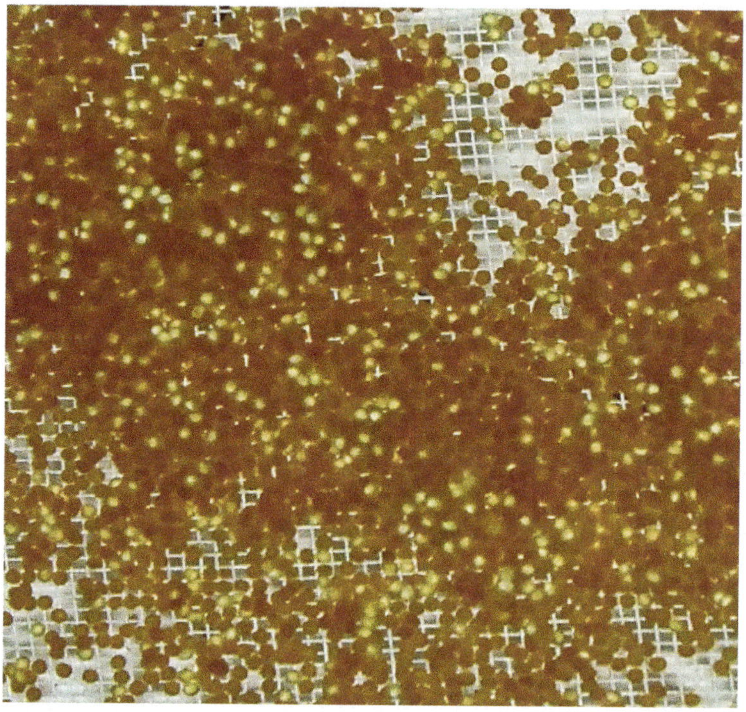

Figure 13.2.1. Manually fertilized, adhesive *Clarias gariepinus* eggs scattered over window screen and incubated indoors.

Table 13.2.3. Approximate hatching time, hours post-fertilization, by species and temperature.

Species	Temperature	Time to hatch	Reference
Clarias gariepinus	25°C 30°C	28–32 h 20 h	Janssen 1987
Clarias lazera	20°C 25°C 30°C	48 h 27 h 19 h	Hogendoorn 1979 Hogendoorn and Vismans 1980
Clarias batrachus	26–27°C 27–31°C	26–28 h 20–24 h	Sahoo et al 2004 Mahapatra 2004
Heterobranchus bidorsalis	27°C	24–28 h	Adebayo and Fagbenro 2004

end of the treatment period the eggs should be rinsed several times and any excess talcum powder or full-fat powdered milk washed out. After the eggs have been rinsed clean they can be incubated in MacDonald jars. A jar can be loaded up to about half volume with a water flow rate of up to 3 L/min. Hatch times and hatch rates (40–50%) are equal or better than that of adhesive eggs incubated on a screen. With good water quality and management hatch rates of ≥75% can be expected, but it is not uncommon that about 10% of the larvae are deformed.

When the eggs are incubated on a screen the larvae are able to drop through the screen to the tank bottom. The healthy larvae (approximately 4 mm total length) will move to dark areas of the tank bottom. Shell debris and dead eggs will remain on the top of the screen and can be removed by carefully lifting the screen out of the tank. The tank bottom should be examined carefully, siphoning out any shell debris, dead or deformed larvae. The healthy larvae can be transferred to a clean tank or left in the hatching tank until they begin to swim-up. When using a MacDonald jar the larvae can remain in the jar until hatching is complete and then the larvae can be poured from the jar into a larval rearing tank.

13.2.4 Fingerling Production

There are several approaches to rear the larvae to a 1 g fingerling. After hatching *Clarias* larvae take 2–3 days to develop functional mouthparts and begin to swim up into the water column searching for food. They can be stocked into carefully prepared nursery ponds soon after they are ready to feed or held in the hatchery and grown to a larger size before being stocked outdoors. Stocking the fry into earthen ponds can be practical if ponds are well prepared, a good zooplankton bloom develops, and predators and competitors are kept under control. Ponds are dried and fertilized with organic fertilizers (see Chapter 7) and filled 2–5 days before anticipated stocking. Stocking rates range from 100 to 250/m^2. Survivals depend on food availability, and how effectively predation and competition with other animals is controlled.

Amphibians, aquatic insects, birds and the *Clarias* larvae themselves are among the major predators and if they are not controlled survival rates of <10% are common. Control of amphibians and aquatic insects begins with thoroughly drying the pond and eliminating any aquatic plants. Drying the pond completely will eliminate any frog larvae (tadpoles) that might compete with the *Clarias* larvae for food and will kill most dragonfly larvae that may prey on the *Clarias* larvae. Keeping the vegetation cut along the shoreline of the pond will eliminate any hiding places for frogs, snakes, or birds. The pond should not be filled any sooner than necessary to get an adequate zooplankton population to provide food for the larval fish. Adult frogs should not have access to the pond. Nursery ponds should have a fine mesh plastic fence or other material that is buried into the pond bank at least 10 cm and extends upward 1 m to 1.5 m. The fence should be checked daily and any damage repaired. Any frog egg masses seen in the pond should be dipped out as soon as they are seen.

Once the pond is stocked, the fertilization protocol should be continued. Starting the second week the fish should be offered a formulated feed. This can be a diet made of local ingredients or a commercial feed. The feed size should be adjusted as the fish grow: 0.35–0.50 mm dia. for 50–100 mg fry; 0.50–0.75 mm for 100–250 mg fish; and 0.75 mm–1.25 mm for 250 mg–1 g fish. The feed should be offered at least 2–3 times/ day beginning at 50 kg/ha/d in week 2, 75 kg/ ha/d in week 3 and 100 kg/ ha/d in the 4th week. Dissolved oxygen should be monitored carefully, particularly in weeks 1 and 2 and the feed rate reduced those weeks if necessary. Chronic low oxygen conditions in those first few weeks will limit zooplankton production and can kill young *Clarias* as they have not developed the ability to use atmospheric oxygen until they are 10 to 14 days old.

Cannibalism is a major issue when producing young *Clarias*. It can begin within a few days after stocking if adequate food is not available and by 30 days post-stocking may result in a pond with a few small fish that have escaped being eaten and a few large fish that have grown rapidly, eating their siblings. There are several things that can be done to reduce cannibalism. Similar age larvae should be stocked. Stocking larvae that are a day or more different in age will result in two initial sizes and the opportunity for a high rate of cannibalism to develop as the size difference between the age groups becomes greater.

Reducing competition for food is a major part in minimized cannibalism. This includes keeping adequate zooplankton densities in the first two weeks, and providing enough formulated feed of the right size frequently enough so that all the fish have access to the feed. Monitoring growth and size variation will signal if the pond needs to be harvested sooner than anticipated. At harvest, the fish should be graded into two or more size categories, and restocked by size category into separate ponds. A size variation as shown in Figure 13.2.2 can occur within 30 days of nursery as a result of cannibalism. With good predator control and minimizing cannibalism a survival of 30–40% is possible by 30 days post-stocking. Pond survival rates can also be increased by stocking older, larger fry produced in the hatchery.

Indoor production requires better facilities and more attention, but survival over the first 30 days can be improved. *Clarias* respond best in low light setting. In outdoor ponds, the shading resulting from a good phytoplankton bloom provides a low light setting. In the hatchery building this can be accomplished by having low lighting in the building (<30 lux) and shading sections of the rearing tanks. Similar to brood holding tanks, most of a larval rearing tank may be covered. Healthy fish will concentrate in the shaded areas. The weaker fish are pushed by the water current toward the un-shaded area near the drain where they can be removed. The water depth should be set at about 15 cm. A flow rate of 3–5 L/min. of clean water at a temperature of 28–30°C should be provided with the goal of maintaining a dissolved oxygen concentration ≥5 mg/L. A filter screen should be set across the tank close to the drain pipe to prevent fish from escaping. This screen will also collect waste feed and other debris and needs to be cleaned several times a day.

Figure 13.2.2. Size variation among hatchery-reared *Clarias* contributes to cannibalism.

Recently hatched larvae are stocked into indoor troughs at up to 700/L. Once mouthparts are functional (day 2–3 post-hatch) the larvae should be given live zooplankton. *Clarias* have relatively large mouths as 1st feeding larvae and can accept *Moina* and *artemia nauplii* as a first food. *Moina* can be produced on-farm, and where artemia cysts are available, they can be hatched on-site to provide young *nauplii* (see Chapter 17.14). Live food should be provided every three hours with the goal of maintaining a zooplankton abundance of 1–2 organisms/mL of culture container. Water flow in the container might have to be stopped for an hour each time live feed is offered to avoid washing away food organisms before it can be consumed.

Clarias can be cultured on live foods for 30 days if progressively larger food items are given in adequate quantities. However, weaning the fish to accept a formulated feed is the best production practice. This should begin when the fish are 4 days old and continue until the larvae are 7 days old. Each day the quantity of live food should be reduced and the quantity of formulated feed increased. By day 8 the fish should be actively consuming only formulated feed.

The feeding behavior should be observed daily. Fish not getting enough food will be swimming up into the water column searching for feed while those that have fed to satiation will be schooling together on the tank bottom. Using a clear glass container, samples of fish should be scooped up in water at different points in the tank to see if all the fish are actively consuming feed. This would be apparent as the proportion of fish seen that have enlarged abdomens full of food. Ideally, all fish would have enlarged abdomens and feces visible in the intestinal tract. Keep in mind that the feed particle size and quantity offered should be increased as the fish grow. The larval rearing tanks need to be cleaned daily, siphoning out any waste feed or other debris, and cleaning the screen used to prevent fish going down the drain.

Figure 13.2.3. Manual stripping of *Clarias gariepinus* following induced spawning.

Disease problems are much more common in indoor culture of clarias than in outdoor ponds. Keeping the tanks clean is an essential part in keeping diseases under control.

Fish can be harvested after 10–14 days once the fish develop the brachial apparatus and can use atmospheric oxygen. At that point the fish may average 0.1–0.2 g or more with survival averaging near 80%. The fish should then be graded and restocked into ponds for an additional nursery period. A target size for stocking production ponds should be 5 g. This can be accomplished by stocking closely-graded 15 day old fish into outdoor secondary nursery ponds at 1,200/m^2 or tanks at 200-800/m^2 and harvesting them 15–20 days later before cannibalism becomes a major problem. The fish should be fed a formulated feed at 5–6% body weight/d given in 3 or more feedings/d. Survival in this second nursery period should average 70–80% of 1+ g fish. An additional nursery period may be necessary to produce the 5 g fish needed for stocking some production settings.

13.2.5 References

Adebayo, O. T., and O. A. Fagbenro. 2004. Induced ovulation and spawning of pond raised African Giant Catfish, *Heterobranchus bidorsalis* by exogenous hormones. Aquaculture 242:229–236.

De Leeuw, R., H. J. Th. Goos, C. J. J. Richter, and E. H. Eding, 1985. Pimozide-LHRHa induced breeding of the African Catfish, *Clarias gariepinus* (Burchell). Aquaculture 44:295–302.

Dhara K., and N. C. Saha. 2013. Controlled breeding of Asian Catfish *Clarias batrachus* using pituitary gland extracts and ovaprim at different temperatures, latency periods and their early development. Journal of Aquaculture Research and Development 4:1–9.

Eqwenomie, M., and A. Obi. 2012. Cost and effectiveness of Ovaprim, Ovatide and fish pituitary extract on induced spawning of African Catfish. Nigerian Journal of Fisheries 9(2):555–559.

Hogendoorn, H. 1979. Controlled propagation of the African Catfish, *Clarias lazera* (C. & V.). I. Reproductive biology and field experiments. Aquaculture 17:323–333.

Hogendoorn, H., and M. M. Vismans, 1980. Controlled propagations of the African Catfish, *Clarias lazera* (C. & V.): II. Artificial reproduction. Aquaculture 21:39–53.

Huisman E. A., and C. J. J. Richter. 1987. Reproduction, growth, health control and aquacultural potential of the African Catfish, *Clarias gariepinus* (Burchell 1822). Aquaculture 63(1–4):1–14.

Janssen, J. A. L. 1987. Chapter 3, Mass production of fry and fingerlings of the African Catfish *Clarias gariepinus*. *In* G. A. Delince, D. Campbell, J. A. L. Janssen, and M. N. Kutty, editors. Seed Production, FAO Working Paper Arac/87/Wp/13. Available: Http://Www.Fao.Org/Docrep/Field/003/Ac182e/Ac182e00.Htm#Toc. (May 2020).

Legendre, M., and Z. J. Otémé, 1995. Effect of varying latency period on the quantity and quality of ova after hCG induced ovulation in the African Catfish, *Heterobranchus longifilis* (Teleostei, Clariidae). Aquatic Living Resources 8:309–316.

Mahapatra, B. K., 2004. Conservation of the Asiatic Catfish, *Clarias batrachus* through artificial propagation and larval rearing technique in India. Aquaculture Asia. Oct.–Dec. 9(4):8–9.

Ngamvongchon, S., O. Pawaputanon, W. Leelapatra, and W. E. Johnson. 1988. Effectiveness of an LHRH analogue for the induced spawning of carp and catfish in Northeast Thailand. Aquaculture 74:35–40.

Olumuji O. K., and M. K. Mustapha. 2012. Induced breeding of African Mud Catfish, *Clarias gariepinus* (Burchell 1822), using different doses of normal saline diluted ovaprim. Journal of Aquaculture Research and Development 3:133 doi:10.4172/2155-9546.1000133. Available: https://www.omicsonline.org/induced-breeding-of-african-mud-catfish-clarias-gariepinus-burchell-1822-using-different-doses-of-normal-saline-diluted-ovaprim-2155-9546.1000133.pdf. (May 2020).

Sahoo, S. K., S. S. Giri, S. Chandra, and A. K. Sahu. 2007. Effect of ovaprim doses and latency periods on induced spawning of *Clarias batrachus*. Indian Journal of Experimental Biology 45:920–922.

Sahoo, S. K., S. S. Giri, and A. K. Sahu 2004. Effect of stocking density on growth and survival of *Clarias batrachus* (Linn.) larvae and fry during hatchery rearing. Journal of Applied Ichthyology 20:302–305.

Zonneveld, N., W. J. Rustidja, A. R. Viveen, and W. Maduna. 1988. Induced spawning and egg incubation of the Asian Catfish, *Clarias batrachus*. Aquaculture 74:41–47.

13.2.6 Recommended Reading

de Graaf, G. J., and J. A. L Janssen. 1996. Artificial reproduction and pond rearing of the African Catfish, *Clarias gariepinus* in sub-Saharan Africa–a handbook.

FAO Fisheries Technical Paper. No 362. FAO, Rome 1996. Available: http://www.fao.org/docrep/003/W3595E/W3595E00.HTM. (May 2020).

El Naggar, G. O., G. John, M. A. Rezk, W. Elwan, and M. Yehia. 2006. Effect of varying density and water level on the spawning response of African Catfish *Clarias gariepinus*: implications for seed production. Aquaculture 261(3):904–907.

Haylor, G. S. 1993. Controlled hatchery production of African Catfish, *Clarias gariepinus* (Burchell): an overview. Aquaculture and Fisheries Management 24(2):245–252.

Hecht, T. 2013. A review of on-farm feed management practices for North African Catfish (*Clarias gariepinus*) in sub-Saharan Africa. *In* M. R. Hasan and M. B. New, editors. On-farm feeding and feed management in aquaculture. FAO Fisheries and Aquaculture Technical Paper No. 583. FAO, Rome.

Isyagi, N. A., K. L. Veverica, R. Asiimwe, and W. H. Daniels. 2009. Manual for the commercial pond production of the African Catfish in Uganda. Auburn University, Auburn, Alabama. Available: https://www.yeanrwanda.org/images/E_library/Catfish%20ManualEntire.pdf. (May 2020).

Janssen, J. A. L. 1987. Chapter 3, Mass production of fry and fingerlings of the African Catfish *Clarias gariepinus*. *In* G. A. Delince, D. Campbell, J. A. L. Janssen, and M. N. Kutty editors. Seed Production, FAO Working Paper Arac/87/Wp/13. Available: http://Www.Fao.Org/Docrep/Field/003/Ac182e/Ac182e00.Htm#Toc. (May 2020).

Ogunsina, L. 2014. 19 Steps to efficient African Catfish Breeding—The Fish Site. Available: https://thefishsite.com/articles/19-steps-to-efficient-african-catfish-breeding. (May 2020).

Potongkam, K., and J. Miller. 2006. Manual on catfish hatchery and production. Available: http://www.fao.org/3/a-ak505e.pdf. (May 2020).

Sahoo, S. K., S. S. Giri, and A. K. Sahu. 2005. Induced spawning of Asian Catfish, *Clarias batrachus* (Linn.): effect of various latency periods and sGnRHa and domperidone doses on spawning performance and egg quality. Aquaculture Research 36:1273–1278.

Verreth J., E. H. Eding, G. R. M Rao, F. Huskers, and H. Segner, 1993. A review of feeding practices, growth, nutritional physiology in larvae of the catfishes *Clarias gariepinus* and *Clarias batrachus*. Journal of the World Aquaculture Society 24:135–144.

13.3 *Pangasius* Catfish

13.3.1 Introduction

The family of tropical catfishes Pangasiidae contains about 28 species in four genera with a natural distribution in southern Asia. The family has two important species that are produced as food fish *Pangasius bocourti* and *Pangasianodon hypophthalmus*. The nomenclature has changed several times particularly for *Pangasianodon hypophthalmus*. Previous names include *Pangasius hypophthalmus* and *Pangasius sutchi*. Common names for *Pangasianodon hypophthalmus* include: Striped Catfish, Iridescent Shark, Swai, and Tra. The common name for *Pangasius bocourti* is basa. In this chapter, the focus is on *Pangasianodon hypophthalmus* and will be referred to as pangasius unless otherwise specified.

Pangasius bocourti and *Pangasianodon hypophthalmus* are riverine species native to the Mekong and Chao Phraya river basins. In the Mekong system, they migrate up the Mekong river to spawn in the Khone falls area along the Cambodian/Lao boarder. Some fish may migrate as much as several hundred kilometers to reach the spawning grounds. They congregate in areas of deep rocky channels, sand bars, and rapids waiting for the start of the rainy season. During the rainy season the fish spawn over flooded tree roots and other vegetation releasing adhesive eggs that stick to such substrates. The eggs hatch in approximately 24 h and the larvae drift downstream where they are carried out into the floodplain. The flooded areas provide a nursery area where the larvae develop rapidly to juveniles. With the return of the dry season water levels drop and juveniles move out into deeper lakes and rivers.

A traditional fishery for fingerlings in the Tonle Sap area of Cambodia provided fingerlings for aquaculture in Cambodia and Vietnam as well as for the international ornamental fish trade. However, in 1994, the Cambodian government put a ban on the capture of pangasius fingerlings. The development of induced spawning and hatchery techniques overcame the wild fish ban and resulted in a rapid expansion of the pangasius aquaculture industry. Pangasius culture went from an annual production of <20,000 metric tonnes in 2000 to 140,000 metric tonnes in 2010 once hatchery-produced fingerlings became available.

Pangasianodon hypophthalmus have air-breathing organs and can tolerate dissolved oxygen conditions of <1 mg/L, making it tolerable of intensive pond culture conditions. Most of the commercial expansion has been in the production of this species. In addition to its culture in Vietnam it has been introduced for culture in Bangladesh, China, India, Indonesia, Malaysia, Myanmar and the Philippines, as well as out of the region into Nigeria and Puerto Rico. *Pangasius bocourti* has no accessory respiratory organ and requires an oxygen level higher than *P. hypophthalmus*, and typically is produced in cages in flowing water. Its hatchery production is less successful than *P. hypophthalmus*. Its production is more for in-country markets.

Pangasius are omnivores with younger fish feeding on algae, higher plants, zooplankton, and insects, while older fish consume crustaceans, fish and even fruit. Female pangasius reach maturity when 3–4 years old, weighing about 4 kg. Some males are mature when 2 years old. A four-year old female can have a fecundity of 112,000 to 138,000 eggs/kg. Mature eggs have a diameter of approximately 1.1 mm. Wild broodstock may spawn twice a year while gravid females may be found year-around when cultured in ponds. Under such culture conditions there may be some females that can be induced spawned every 3–4 months without any effect on egg quantities or quality.

13.3.2 Brood Development

Pangasius broodstock can be developed on-farm as they will become sexually mature when reared in ponds, however, they will not naturally reproduce in ponds. As it may takes 3 yrs for pangasius to reach maturity, a brood development plan is needed. Pangasius are very fecund and a limited number of brooders may be adequate to meet production goals but this may result in a limited gene pool and an accelerated rate of inbreeding. The history of the brooders needs to be documented so mating of closely related individuals can be avoided. This begins with knowing the parentage of the brooders to be used in a given season. The progeny should be tracked, beginning with where the progeny from a given set of brooders go, and how well the progeny grew and survived. Each set of fingerlings need to be tracked as to what production ponds they go into. Pangasius reach market size of about 1 kg at an age of ≤1 year, and still are not sexually mature. Fish of known genetic history need to be set aside from the production ponds for additional growth until they are three years old. Once sexually mature, a fish may be spawned several times a year. By the time the brooders are >5 y old they may be too large to be safely handled and should be replaced. Allowances also should be made to replace any yearly mortality of brooders. For example, if the goal is to have 500 4 y old brooders where there was an annual mortality of 20% then 900 1 y old fish need to be set aside. Careful record keeping is needed to follow the genetic history of a given set of fish. For young fish this means having enough ponds to keep the groups separate and tracking them by pond. When they are older, individual fish can be tagged as to what genetic group and year class it is, and then co-stocked with other identifiable genetic groups into common ponds.

Mature brooders should be held separately by sex and year class at standing crops up to 5,000 kg/ha. Fish being produced as food fish can be held at higher densities but brooders are held at lower densities to reduce the stress that can hamper successful reproduction. Brooders should be fed a ≥35 % protein feed once a day at 1–2% body weight/day. Overfeeding can result in excessive visceral fat that can hamper with successful reproduction. The feed can be balls of on-farm made feed composed of trash fish/fish meal and agricultural by-products, or a pelleted commercial feed. The pond should be flushed periodically with fresh water to maintain water quality and perhaps help stimulate further gonadal maturation.

When working with pond-reared brooders there will be some mature brooders available year-around but there is often a greater frequency of quality brooders in the early part of the rainy season, with egg quality also being about 20% higher at that time. In a survey of pangasius hatcheries in Vietnam, Hùng et al. (2008) documented seasonal differences in spawning success, fecundity, and hatching success with best results coming during February–March, with lower rates from March to September.

Where available, wild brooders can be collected and used, however, there is a risk of disease introduction, and the fish collected are genetically adapted to riverine life and not pond culture. Using wild fish makes hatchery planning more difficult as there is less control as to when brooders would be available. Genetic selection of domesticated brooders can improve the growth rate of pangasius when cultured in ponds. Vu et al. (2019) was able to improve harvest weight of a selected stock 27.9% after 3 generations of selection compared to the progeny of unselected wild caught brooders.

13.3.3 Brood selection

When brooders are selected for induced spawning they should be carefully seined from the brood pond. Gravid females should have large distended abdomens that are soft to touch with a reddish genital papilla. A catheter should be inserted up the oviduct into the ovary and an egg sample taken. The eggs should have a yellowish-ivory color and not appear whitish or translucent. The eggs should be similar in size and not stuck together. Minimum egg diameters vary by species and should be ≥1.0 mm for *Pangasianodon hypophthalmus*, ≥1.6 mm for

Pangasius bocourti and ≥1.7 mm for *Pangasius djambal*. Males should have a reddish genital opening and freely release sperm when a gentle pressure is applied to the abdomen.

13.3.4 Induced Spawning and Brood Handling

Pangasius brooders are selected at the pond then transported in water to the hatchery where they are best held in a flowing water circular tank at a temperature of 26 to 28°C. They can be induced to spawn with a number of hormones. In Vietnam the most widely used hormone used is hCG, given in a series of injections. Females receive four or five doses over a 56 or 66 h period. Males may receive a single 3000 IU/ kg injection at the time when the females are given their resolving injection. Injections are given intramuscular or intraperitoneal depending on the volume required. Further details on the use of hCG are given in Table 13.3.1 as well as other protocols that have been used successfully to induce ovulation in pangasius.

Careful monitoring of brood to know when ovulation has begun is key to successful pangasius spawning. The response time from injection to ovulation varies with the hormone and protocol used as well as water temperature. Brooders should be held at a stable temperature. The wider the daily water temperature fluctuation the more difficult it is to anticipate when ovulation will occur. The variation in response time also depends on the hormone used (Table 13.3.1). An 8-hour variation among females has been reported for Ovaprim, 3-hours with hCG, and a 1-hour variation when carp pituitary or Ovotide was used. Careful selection of similar quality females will also help reduce this variation. Holding females together of similar quality based on egg color and softness of the abdomen allows best quality females to be examined earlier post-resolving injection without having to disturb lesser quality fish each time.

Egg quality begins to deteriorate within an hour of ovulation with notable reductions in hatching rate and an increase in the frequency of deformed larvae by 2 h post-ovulation. Females should be checked starting about 3 h post-resolving injection to see if there is a free flow of eggs and every hour thereafter to see if a female has ovulated. Careful record keeping from previous attempts as to hormone and dose used, water temperature, brood history, brood weights, when females were stripped and the quality of eggs obtained will help guide the current spawning attempt as to when to anticipate ovulation.

Female brooders often are handled several times before they ovulate and need to be handled carefully. This can be done using a pillowcase slipped over the head or a small stretcher where the fish can be guided into and then closed around the fish. The abdomen should be massaged to see if it is hard and firm, or soft. If it is hard and firm, the female has not ovulated yet and can be released back into the holding tank. If the abdomen seems soft then the female should be rolled over upside down and the vent area uncovered, and a finger placed over the genital papilla. The female should be lifted up so that the vent area is out of the water while keeping the head in the water. When moving the finger off the papilla and putting a light pressure on the abdomen, a good quality female that has ovulated should release a free flow of yellowish-colored eggs. If some blood or clumps of eggs still attached to ovarian tissue are seen then the female may have not fully ovulated. These females can be returned to the holding tank and checked again about 30–45 min later to see if they have fully ovulated. If a mixture of yellowish and whitish eggs along with some clear fluid is seen then the female may have ovulated more than an hour before and the egg quality has begun to deteriorate.

Table 13.3.1. Hormone protocols used successfully to induce ovulation in *Pangasius*.

Hormone	Dose rate/kg	Injection time	Time to ovulation	Reference
HCG*	542 IU	0 h		Bui et al. 2010
	597 IU	23 h		
	893 IU	46 h		
	3,442 IU	56 h	**	
HCG*	500 IU	0 h		Bui et al. 2010
	500 IU	23 h		
	500 IU	46 h		
	1,500 IU	56 h		
	3,000 IU	66 h	**	
HCG	500 IU	0 h		Legendre et al. 2000
	2,000 IU	8 h	8–11 h	
Indian carp pituitary	2.5 mg	0 h		Chand et al. 2011
	10 mg	6 h	8– 9 h	
LH-RH*	20 µg	0 h		Saidin et al. 1988
	40 µg	8 h	**	
Ovaprim	0.3 mL	0 h		Legendre et al. 2000
	0.6 mL	8 h	3–11 h	
	0.5 mL	0 h	9– 10 h	Chand et al. 2011
Ovotide	0.6 mL	0 h	10– 11 h	Chand et al. 2011

*Data based on means obtained from a farm survey
** Data not available

13.3.5 Egg collection, Fertilization, and Incubation

Pangasius eggs are dry fertilized. Sperm can be collected an hour before anticipated ovulation following techniques described in the Chapter 17.13, or stripped directly from the male as the males produce relatively large volumes of semen. The vent area of the female should be dried very carefully, holding the tail up with a finger over the papilla to prevent egg release or water dropping into the collection pan. It is essential that no water drops into the pan where the eggs are to be collected. Drying includes drying the hands of the people holding the fish and wrapping a dry towel around the rest of the fish. Once everything is ready the tail is lowered, a dry pan is placed under the vent area of the female, and with a light pressure on the abdomen there should be a free flow of yellowish colored eggs (Figure 13.3.1). Pressure should be applied several times until there is only a slight flow of eggs left.

Pangasius are very fecund with a 10 kg female able to release over a million eggs/spawn. When working with large females it can be desirable to divide the collected eggs into four or

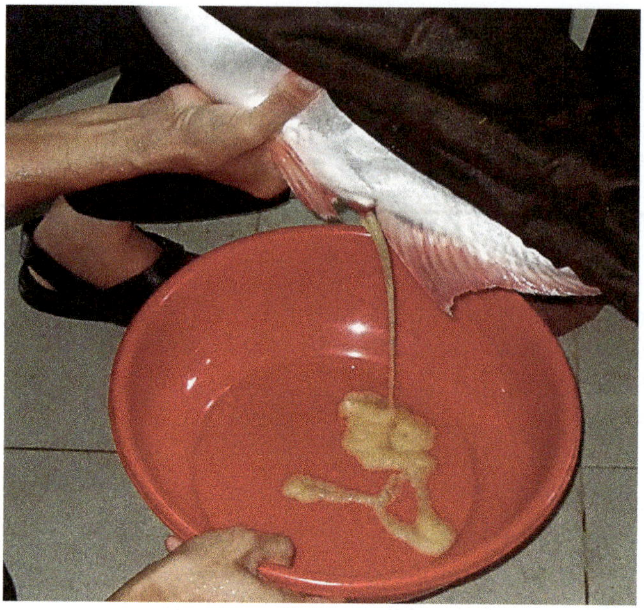

Figure 13.3.1. Manual stripping of HCG–induced female *Pangasianodon hypophthalmus*.

five subsets and fertilize each subset with distinct males to get as much genetic diversity as possible. The traditional approach often is to add the sperm from a number of males into one bowl holding all the eggs collected from a female. With that approach there is no guarantee that the sperm of each male fertilizes similar quantities of eggs. By dividing the eggs into several sets and fertilizing each set with its own male that will help ensure more genetic diversity.

With dry fertilization it is best to add the sperm to the eggs and mix it in with the eggs then add enough water to cover the eggs with 2 cm of water or more. One mL of sperm is adequate to fertilize a million eggs. Once the water is added the eggs should be gently stirred for two minutes to allow for fertilization. Pangasius eggs are mildly adhesive and can be treated with tannin, full-fat powdered milk or Fullers earth (see Chapter 17.4) to reduce the adhesiveness before loading the eggs into MacDonald jars, Zug jars, or other similar up-flow incubators for incubation (Figure 13.3.2). They are not highly buoyant and can be loaded up to half the incubator's volume. They can also be incubated without treating them for adhesiveness by gently distributing the eggs along the sides of the hapa as the eggs become adhesive and stick to the hapa.

Pangasius eggs are oval with little perivitelline space. The eggs will begin hatching within 24 to 26 h at 28 to 32°C and as late as 32 h at 20°C. All eggs from a female do not hatch at the same time. There may be 9 to 10 h from the start of hatching until 90% of the eggs have hatched. The larvae are approximately 3 mm in length at hatch and will be carried out in the water leaving the incubator. Larvae from eggs hatched in hapas will remain in the hapa until they are strong swimmers. When using upwelling incubators, the water carrying the larvae should flow into an aerated, circular tank. The tank needs a drain screen that has a fine enough mesh so larvae are not lost down the drain but not too fine a mesh to avoid it becoming clogged with egg shells and other debris. Larval densities up to 3,000–4,000 larvae/L of larval holding tank are possible during the first 24 h post-hatch as long as adequate water quality can be maintained and impingement of larvae on the drain screen can be kept to a minimum.

Figure 13.3.2. *Pangasianodon hypophthalmus* eggs incubated in an up-flowing incubator.

By 24 h post-hatch the larvae are 5.5 mm and mouthparts are forming (Figure 13.3.3). Near 48 h post-hatch the larvae may range in length from 6.2 to 6.8 mm and are active swimmers. At that point they have absorbed 90% of the yolk sac, have functional mouthparts, and the larvae need to be fed. Larvae can be held in indoor hatchery tanks at 10 to 50 fish/L and fed a live food starting at 36–48 h post-hatch. *Artemia* is the preferred live food for feeding pangasius larvae in the hatchery for 8–10 days post-hatch but in some cases its cost may be prohibitive. *Moina* is an acceptable live food that can be produced in hatchery ponds and harvested to feed pangasius larvae in the hatchery. Larval survival is similar to that obtained when *artemia nauplii* are fed but larval growth is slower than obtained using *artemia*.

Hatchery survival of *Pangasianodon hypophthalmus* larvae is more an issue than with *Pangasius bocourti* or *Pangasius djambal*. In the hatchery, the larval mortality of *P. hypophthalmus* can be >50% as the result of the larvae biting each other. Recently hatched larvae are not strong swimmers and depend on the water current to keep themselves up in the water column. They may strike at potential food items including other pangasius larvae that drift by even before the digestive tract is complete. They will grab their siblings and often are not able to release them and both fish die. A factor contributing this mortality is the presence of oral spines that by 36 h post-hatch hang out the mouth and interfere with the closure and reopening of the mouth. While the body of the larvae grows rapidly, there is less change in the length of the oral spines. When the larvae reach a length of 8.5 mm TL the oral spines do not extend out of mouth. By approximately 72 h post-hatch the larvae have grown enough that when they bite a sibling they are more likely to be able to able release their grasp of the larvae they bit.

Figure 13.3.3. *Pangasianodon hypophthalmus* larvae soon after hatch, dependent on water currents to stay up in the water column.

However, any larvae that were bitten are wounded and such wounds may become infected with bacteria resulting in another wave of mortality.

Two windows of larval mortality can be expected. The first occurs 48 to 72 h post-hatch, during which the larvae may have bitten each other and could not detach themselves, and the attached larvae suffocate. A second wave of mortality of lesser magnitude occurs around 120 to 168 h post-hatch, where wounded larvae succumb to bacterial infections. How wide these windows of mortality are depends on how fast the larvae are growing. This in turn is dependent on food availability, water temperature. water quality, and larval density. Holding the larvae at 29–30°C will speed up the larval development reducing the window when the larvae strike at each other and cannot free themselves. Bacterial infections in wounded larvae can be reduced with periodic baths of terramycin or formalin. Holding the larvae at light intensities of 1 Lux or less will help to improve larval survival. Adding *artemia nauplii* at 3–5 *nauplii*/mL of holding tank every 3 h while preventing water quality problems by removing any dead *nauplii*, will also improve larval survival. For more background regarding the mortalities of pangasius larvae see the references listed in this chapter.

Pangasius larvae can be reared by stocking the larvae into on-site primary nursery ponds or shipped to off-site nursery ponds as soon as they are ready to feed. The larvae are typically moved to nursery ponds in plastic bags filled with oxygen at densities of approximately 7,000–8,500/L depending on how long the larvae had been held in the hatchery and how much they had grown.

13.3.6 Nursery

Fingerlings are produced using a combination of primary and secondary nursery ponds. Primary nursery ponds are prepared as described in Chapter 7 using 1,000–5,000 m^2 ponds, stocking 3–4 day old larvae at 100–500/m^2. First feeding pangasius larvae have relatively large mouthparts and can consume *Moina* as a first food. To ensure an adequate quantity of food is available, ponds are often seeded with *Moina* at 20 kg/ha once the nursery pond is filled, with the goal of having 200 *Moina*/L of pond water when the larvae are stocked. No water should be added to the primary nursery pond for the first 10 days post-stocking to avoid the introduction of predaceous fish and other animals. Once the pangasius larvae have had a chance to grow for 10 days in the nursery pond they will be too big to be preyed on by anything that could be introduced in the water supply as long as the water was filtered through a 250–500 µm mesh filter screen.

Formulated feeds should be given starting week 1 using a slurry of boiled eggs and soybean meal or a 1:1 mixture of fish meal and rice bran. Feed should be given 4–6 times a day. In week 2 the fish should be offered a crumbled commercial feed, and by week 3 the fish should be large enough to accept a pelleted feed. Keep in mind that the change in feed type or size should be gradual over a few days so that the fish learn to accept the new feed. An aggressive feeding response should be seen (Figure 13.3.4). Another consideration with *P. hypophthalmus* is that by 12 d post-hatch the fish have developed a vascularized swim bladder and can use atmospheric oxygen, and thus can tolerate low dissolved oxygen levels. However, the reproductive rate of much of the zooplankton will be reduced under such low oxygen conditions. Supplemental feeding is essential during this period to maintain growth. Once the fish are large enough, clean fresh water can be added to the pond to improve the water quality. At two weeks post-stocking there might be enough size variation that in-pond grading may be appropriate.

Figure 13.3.4. Feeding response of pangasius fingerlings given a formulated feed.

After four weeks, the fish are large enough to be harvested by seining. The fish should not be fed the day before harvest so that the intestinal tract is free of feed when harvested. As such, the fish will handle better, and water quality in the transport tank and holding tanks will be better. The pond should be drained to 2/3 of its volume and seined to remove as much biomass as possible without significantly lowering water quality. The pond should be drained to harvest the remainder. If the pond is not needed for another primary nursery cycle, the pond can be partially harvested and the remaining fish cultured to advanced fingerlings. When harvesting a primary nursery pond the fish should average 0.3 to 1.0 g. Survival in primary nursery ponds should average 40–50 % but survivals of <30% are not uncommon. Yields of 100 ± fish/m^2 should be expected depending on the initial stocking rate and survival. The fish should be carefully handled and graded before being stocked into secondary nursery ponds.

Secondary nursery ponds are stocked at 150–200 fish/m^2. Fish should already be trained to accept formulated feeds when harvested from the primary nursery pond and feeding with a commercial pelleted feed should be continued in the secondary nursery ponds. The secondary nursery period may be for two months or more. Water exchange should be provided to help maintain water quality as standing crops before harvest may be as high as 21,000–28,000 kg/ha. After a two-month nursery the fish should have an average weight of 15–20 g. Survival rates should average 60–70%.

13.3.7 References

Bui, T. M, L. T. Phan, B. A. Ingram, T.T.T. Nguyen, G. J. Gooley, H. V. Nguyen, T. Nguyen, and S. S. De Silva. 2010. Seed production practices of Striped Catfish, *Pangasianodon hypophthalmus* in the Mekong Delta region, Vietnam. Aquaculture 306:92–100.

Chand, B. K., M. K. Singh, and M. Basudev. 2011. Studies on the breeding of *Pangasius sutchi* using different inducing agents. Journal of Applied Aquaculture 23(1):32–40.

Hùng, H. P., N. T. T. An, N. T. V. Triều, and T. T. T. Nguyen. 2008. Some technical and management aspects of catfish hatcheries in Hong Ngu district, Dong Thap province, Vietnam. Aquaculture Asia Magazine. October–December 2008.

Legendre, M., J. Slembrouck, J. Subagja, and A. H. Kristanto, 2000. Ovulation rate, latency period and ova viability after GnRH- or hCG-induced breeding in the Asian Catfish *Pangasius hypophthalmus* (Siluriformes, Pangasiidae). Aquatic Living Resources 13:145–151.

Saidin, T., A. A. Othman, and M. Z. Sulaiman, 1988. Induced spawning techniques practiced at Batu Berendam, Melaka, Malaysia. Aquaculture 74:23–33.

Vu, N. T., N. V. Sang, T. H. Phu, N. T. Vuong, and N. H. Nguyen. 2019. Genetic evaluation of a 15-year selection program for high growth in Striped Catfish *Pangasianodon hypophthalmus*. Aquaculture 509:221–226.

13.3.8 Recommended Reading

Aquaculture Stewardship Council. 2010. Better management practices for pangasius aquaculture: a tool to assist with compliance to the Aquaculture Stewardship Council (ASC) pangasius standards, Version 1.01. Published by the World Wildlife Fund Inc. Available: https://www.asc-aqua.org/wp-content/uploads/2017/07/ASC-Pangasius-Better-Management-Practices_v1.01.pdf. (May 2020).

Baidya, A. P., and S. Senoo. 2002. Observations of oocyte final maturation and eggs on *Pangasius hypophthalmus* under artificial rearing conditions. Suisanzoshoku 50(4):423–432.

Baras E., and M. Jobling, 2002. Dynamics of intracohort cannibalism in cultured fish. Aquaculture Research 33:461–479.

Baras, E., J. Slembrouck, C. Cochet, D. Caruso, and M. Legendre, 2010. Morphological factors behind the early mortality of cultured larvae in the catfish, *Pangasianodon hypophthalmus*. Aquaculture 298:211–219.

Baras, E., R. Hafsaridewi, J. Slembrouck, A. Priyadi, Y. Moreau, L. Pouyaud, and M. Legendre. 2010. Why is cannibalism so rare among cultured larvae and juveniles of *Pangasius djambal*? Morphological, behavioral and energetic answers. Aquaculture 305:42–51.

Baras, E., T. Raynaud, J. Slembrouck, D. Caruso, C. Cochet, and M. Legendre. 2011. Interactions between temperature and size on the growth, size heterogeneity, mortality and cannibalism in cultured larvae and juveniles of the Asian Catfish, *Pangasianodon hypophthalmus* (Sauvage). Aquaculture Research 42:260–276.

Bui, T. M., N. T. Phuong, G. H. Nguyen, and S. S. De Silva, 2013. Fry and fingerling transportation in the Striped Catfish, *Pangasianodon hypophthalmus*, farming sector, Mekong Delta, Vietnam: a pivotal link in the production chain. Aquaculture 388–391(1):70–75.

Bui, T. M., T. L. Phan, A. I. Ingram, T. T. T. Nyen, J. G. Gooley, V. H. Nguyen, T. P. Nguyen, and S. S. De Silva. 2010. Seed production practices of Striped Catfish, *Pangasianodon hypophthalmus* in the Mekong Delta region, Vietnam. Aquaculture 306(1–4):92–100.

FAO 2010–2015. Cultured Aquatic Species Information Programme. *Pangasius hypophthalmus*. Cultured Aquatic Species Information Programme. Text by D. Griffiths, P. Van Khanh, and T. Q. Trong *in* FAO Fisheries and Aquaculture Department [online]. FAO, Rome. Updated 14 January 2010. Available: http://www.fao.org/fishery/culturedspecies/Pangasius_hypophthalmus/en. (May 2020).

Hung, L. T., B. M. Tam, P. Cacot, and J. Lazard. 1999. Larval rearing of the Mekong Catfish, *Pangasius bocourti* (Pangasiidae, Siluroidei): Substitution of *Artemia nauplii* with live and artificial feed. Aquaculture Living Resources 12(3):229–232.

Islam, A. 2005. Embryonic and larval development of Thai Pangas (*Pangasius sutch* Fowler, 1937). Development Growth and Differentiation 47:1–6.

Legendre, M., J. Slembrouck, J. Subagja, and A. Hari Kristanto. 2000. Ovulation rate, latency period and ova viability after GnRH-or hCG-induced breeding in the Asian Catfish *Pangasius hypophthalmus* (Siluriformes, Pangasiidae). Aquatic Living Resources 13(3):145–151.

Morioka, S., K. Sano, P. Phommachan, and B. Vongvichith. 2010. Growth and morphological development of laboratory-reared larval and juvenile *Pangasianodon hypophthalmus*. Ichthyological Research 57(2):139–147.

Mukai, Y., A. D. Tuzan, L. S., Lim, and S. Yahaya. 2010. Feeding behavior under dark conditions in larvae of Sutchi Catfish *Pangasianodon hypophthalmus*. Fisheries Science 76(3):457–461.

Mukai, Y., A. D. Tuzan, S. R. M. Shaleh, B. Mabel, and M. Matsumoto. 2010. Development of sensory organs and changes of behavior in larvae of the sutchi catfish, Pangasianodon hypophthalmus. Fisheries Science 76:921– 930.

Nguyen, T. P., and T. H. O. Dang. 2009. Striped Catfish (*Pangasianodon hypophthalmus*) aquaculture in Viet Nam: an unprecedented development within a decade. Pages 133–149 *in* S. S. De Silva and F. B. Davy, editors. Success Stories in Asian Aquaculture. Springer Nature, Switzerland AG.

.Nguyen, T. P., V. H. Nguyen, T. M. Bui, T. L. Phan, M. S. Vo, N. Nguyen, N. L. Duong, T. T. Thuy-Nguyen, G. J. Gooley, B. A. Ingram, and S. S. De Silvia. 2011. Better Management Practices for Striped Catfish (tra) Farming in the Mekong Delta, Viet Nam, Version 3.0. Vietnam Ministry of Agriculture and Rural Development, Collaboration for Agriculture and Rural Development Program. Available: http://library.enaca.org/inland/catfishbmps/catfish_bmp_version3_final.pdf. (May 2020).

Nguyen, P. T., T. M. Bui, T. A. Nguyen, and S. De Silva. 2013. Developments in hatchery technology for Striped Catfish (*Pangasianodon hypophthalmus*), Chapter 16. Pages 498–518 *in* G. Allan and

G. Burnell, editors. Advances in Aquaculture Hatchery Technology, Woodhead Publishing Limited, Cambridge, UK.

Phan, L. T., T. M. Bui, T. T. T. Nguyen, G. J. Gooley, B. A. Ingram, H. V. Nguyen, P. T. Nguyen, and S. S. De Silva, 2009. Current status of farming practices of Striped Catfish, *Pangasianodon hypophthalmus* in the Mekong Delta, Vietnam. Aquaculture, 296:227–236.

Phu, T. M., N. T. Phuong, T. T. Dung, D. M. Hai, V. N. Son, A. Rico, J. H. Clausen, H. Madsen, F. Murray, and A. Dalsgaard. 2016. An evaluation of fish health-management practices and occupational health hazards associated with Pangasius Catfish (*Pangasianodon hypophthalmus*) aquaculture in the Mekong Delta, Vietnam. Aquaculture Research 2016 (47):2778–2794.

Slembrouck, J., O. Komarudin, Maskur, and M. Legendre, editors. 2003. Technical Manual for Artificial Propagation of the Indonesian Catfish, *Pangasius djambal*. 143 pp. IRD-DKP. Jakarta. Available: http://horizon.documentation.ird.fr/exl-doc/pleins_textes/divers14-02/010055573.pdf. (May 2020).

Slembrouck, J., E. Baras, J. Subagja, L. T. Hung, and M. Legendre, 2009. Survival, growth and food conversion of cultured larvae of *Pangasianodon hypophthalmus*, depending on feeding level, prey density and fish density. Aquaculture 294:52–59.

Subagja, J., J. Slembrouck, L. T. Hung, and M. Legendre, 1999. Larval rearing of an Asian Catfish *Pangasius hypophthalmus* (Siluroidei, Pangasiidae): analysis of precocious mortality and proposition of appropriate treatments. Aquatic Living Resources 12:37–44.

Van Zalinge, N., S. Lieng, N. P. Bun, H. Kong, and J. V. Jørgensen. 2002. Status of the Mekong *Pangasianodon hypophthalmus* resources, with special reference to the stock shared between Cambodia and Viet Nam. MRC Technical Paper No. 1. Mekong River Commission, Phnom Penh, Combodia.

Chapter 14

Carps—Cyprinidae

The family Cyprinidae, native to North America, Eurasia and Africa, contains of more than 2,400 species, which is approximately 25% of all the freshwater fish species in the world. They are cultured as food fish, sportfish, and ornamentals, and as such have been introduced into many other countries. Cyprinids are found in most freshwater habitats with some found in brackish water as well, and show a wide range in modes of reproduction. Depending on the species, mature adults range from 10 mm for *Paedocypris progenetica* to 300 kg for Giant Barb *Catlocarpio siamensis*. Hatchery techniques for several commonly cultured cyprinids are covered in sections:14.1 Common Carp, *Cyprinus carpio*; 14.2 Chinese Carps; and 14.3 Bait Minnows.

14.1 Common Carp *Cyprinus carpio*

14.1.1 Introduction

Common Carp is one of the most widely cultured fish in the world. It has been cultured for more than a thousand years resulting in many varieties of Common Carp differing in body shape, scale pattern, fin shape, and color. It is cultured as a food fish, as bait, and as an ornamental (Koi). Common Carp respond well to intensive culture systems, accept a wide range of natural and artificial feeds, grow rapidly to acceptable sizes, and is tolerant of a wide range of growing conditions. It can survive in both low and high temperature setting. In tropical counties with elevations of 1000–2000 m, it fits in a water temperature niche of 18–22°C where it can be too warm to culture trout but too cool for optimum Tilapia growth. Its long history of domestication, its flexibility in culture, and worldwide distribution has resulted in a number of reproduction and culture options.

Common Carp can have a long life span of >25 years and reach a weight of >40 kg. In a cool temperate climate, 3 years or longer may be required to reach sexual maturity but in a warm climate only 1 year or less is required. Under favorable growth conditions fish may not be sexually mature until reaching a weight of 1.2 to 1.4 kg. Under unfavorable growth conditions fish may reach sexual maturity at a weight of 120 g but such fish may be several years old. Common Carp are quite fecund, with the number of eggs/kg of body weight ranging from 80,000 to more than 120,000 eggs/kg.

Sexual dimorphism includes pearl organs on mature males, males being smaller than same-age females, and the papilla shape of males and females differing. The urogenital opening of males is in a slit while that of gravid females is swollen and protruding. Before and during the spawning season, males flow milt when stripped. Gravid females show noticeable ovarian distention and softening of the abdominal area just before spawning.

In temperate climates the spawning season may begin at temperatures as low as 16°C but more typically starting at 18–20°C and continues for 4 to 6 weeks. In temperate climates a female may spawn only once a year in the spring but in a tropical climate a female may spawn 3 to 4 or more times a year. The more frequent the spawning within a year, the smaller the number of eggs that are given per spawn.

Common Carp are egg scatterers. A female accompanied by several males, releases sinking eggs in shallow water while the males release sperm into the water to fertilize the eggs (Figure 14.1.1). The eggs become adhesive within 2 minutes of contact with water. Under natural conditions they stick to vegetation, debris, the pond bottom or any solid surface they contact. Egg diameter ranges from 1.5 to 2.0 mm when water-hardened. No parental care is given to eggs or fry. The eggs hatch in 3 to 4 days at a water temperature of 20°C. About 60 to 70 degree-days are required for incubation.

Size of fry at hatching is 5 to 6 mm. The sac fry have an adhesive disc on the head for attachment to stable objects during the yolk absorption period. Within 2 to 3 days after hatching (water temperature-dependent), the fry begin to swim horizontally and are searching for food.

Food habits of young Common Carp fry are similar to that of many freshwater fish larvae, consuming rotifers, cladocerans and small copepods. Some algae also may be ingested, but the primary food is some form of zooplankton. Use of artificial feed for early fry nutrition has largely been unsuccessful unless supplemented with natural food. However, improved quality diets are now available and should be considered.

There is no well-defined size for transition from young to adult food habits. When fingerlings are 5 to 10 cm they are able consume most of the organisms consumed by adults. Common Carp are omnivorous, with immature insects, particularly midge larvae, composing

Figure 14.1.1. Natural spawning of Common Carp *Cyprinus carpio* in a pond setting, scattering eggs over vegetation.

the major part of the diet of juveniles and adults when reared in ponds on natural food. Common carp are mainly bottom feeders and have mouthparts adapted for separating bottom fauna from the mud layer. This feeding action stirs up the bottom mud, dislodges rooted plants and creates a turbid water condition when fish density is high. Where 250 to 300 kg/ha or more of carp are present, most rooted aquatic plants and benthic algae will likely be eliminated if no supplemental feed is given. When standing crops approach 1,000 kg/ha, depending on the soil type, the pond may become very turbid, limiting photosynthesis, the production of natural foods, and oxygen. Artificial feeds will allow higher standing crops, but supplemental aeration may be necessary to provide sufficient dissolved oxygen.

Hatchery production of common carp can be accomplished in several ways. These include use of a spawning-rearing pond, a spawning-rearing pond with adult transfer, and egg or fry transfer. They can also be induced to spawn, and the eggs stripped and manually fertilized. Such eggs can be incubated in hatching jars, funnels, or trays if treated to remove the adhesiveness. They can also be scattered over a substrate, allowed to become adhesive and incubated in the hatchery without removing the adhesiveness.

14.1.2 Broodstock

There are several considerations when developing common carp broodstock. The goals for introducing Common Carp to a hatchery and their anticipated use needs to be clearly identified. There are many varieties available for selection, differing in body conformation, scale pattern, color, growth rate, and temperature tolerance. Often there are strong market preferences for specific body shapes, scale patterns or color. Improved growth or temperature tolerance has been obtained by crossing one variety of Common Carp with another, but this requires maintaining two separate strains and the F-1 generation may not breed true for the desired traits. Several desirable traits are single gene recessive and others are multi-gene linked. Selection of what variety of Common Carp to introduce to a hatchery needs to be given careful consideration as to the amount of time and effort needed to perpetuate the desired characteristics.

Disease history is a major consideration when introducing Common Carp to a hatchery. There are several viruses that can cause high rates of mortality in Common Carp and are difficult to control. Spring viremia of carp caused by a rhabdovirus (*Rhabdovirus carpio*) infects Common Carp as well as other cyprinids, including Chinese carps and Goldfish, causing mortalities of up to 70% in young fish. Outbreaks typically occur at water temperatures of 5–18°C. Survivors of an outbreak can carry the virus and later infect non-infected fish. Another virus, Koi herpesvirus, infects various ages of Common Carp. Disease outbreaks occur at water temperatures of 16–25°C with mortality levels as high as 100%. Any survivors may be carriers of the virus and can infect non-infected fish under favorable environmental conditions.

Survivors of outbreaks of either virus will have developed antibodies against that virus even though detection of the actual virus may be difficult. Laboratory techniques are available to determine if a particular fish has been exposed to either virus. Any potential introduction of Common Carp or Chinese carps to a hatchery should be reviewed as to whether the fish supplier's hatchery has had outbreaks of either virus. Where available, laboratory tests should be done to determine if the fish to be introduced to your hatchery have been exposed to either virus. In the USA, interstate shipping of cyprinids is becoming more regulated. Shipments need to be certified that a given lot of fish has no evidence of exposure to either virus. Likewise, international shipments may require similar certification. In the USA, a USDA APHIS Veterinary Services office should be contacted as to the latest updates on required certifications and as to where fish can be examined.

Upon arrival at the hatchery the fish should be quarantined for up to two months at temperatures favorable for outbreaks of both viruses. Naïve species of fish known to be susceptible to these viruses should be added to the quarantine unit and subsequently monitored for any onset of these diseases. Care should be given to keeping the quarantine unit isolated, which includes sterilizing any effluent, not transferring equipment or supplies out of the quarantine unit for use elsewhere on the hatchery, routine disinfection of the unit when not holding fish, and disposing of any dead fish found in the unit in a sanitary fashion. (e.g. incineration).

Broodstock can be developed from fry obtained from a reputable hatchery or by spawning in-house fish. Fry stocked at 3,500 to 5,000 fry/ha into a well-fertilized rearing pond will grow to a suitable size broodstock in 1 to 3 years. It will be necessary to provide supplemental feeding or divide the stock after the fish reach an average size of about 300 g. Brooders can also be selected out of production settings as long as the age and variety is known. Broodstock should be a minimum of 1 to 2 kg for good egg production even though some egg production is obtained from fish as small as 0.2 kg. The fish can grow to much larger sizes but they become more difficult to handle.

Proper broodstock maintenance between spawns is necessary to obtain good egg production. To make the most efficient use of available pond space, some form of supplemental feeding should be provided to condition the brood fish for the next spawn. From one spawning period to the next, females should regain any weight lost as ovulated eggs during spawning as well as gain additional somatic weight. An annual weight gain of about 50% is recommended. Standing crops of brooders should be kept at less than 75% of maximum carrying capacity of the holding pond. As an example, where the holding pond has a maximum carrying capacity of 3,000 kg/ha, not more than 2,250 kg (75% of max) should be present when the fish are ready for spawning at the end of the holding period. If a weight gain of 50% had been anticipated, the pond would have been stocked with 1,500 kg initially:

Final standing crop = 2,250 kg/ha = initial standing crop + a 50% gain
= 2,250 kg/ha = 100% + 50% gain

Weight gain = 33 % (50%/150%) × final standing crop = 750 kg/ha

Initial standing crop = Final standing crop – weight gain
= 2,250 kg/ha – 750 kg/ha weight gain
= 1,500 kg/ha

The hardiness of Common Carp makes frequent replacement of broodstock unnecessary. However, for ease in handling, fish less than 5 kg size are preferred. If broodstock are replaced after they reach this size, up to 5 years of service could be obtained from a female that spawned at 1 kg and gained 50% in weight each year. In such cases replacement of the broodstock after the fifth year or earlier would be made. By repeated use of a specific year class of brooders over several years the rate of inbreeding can be reduced. As inbreeding is measured per generation, use of the same year class of brooders over several years results in fish produced by the same generation of brooders even though they were produced in different years. To help maintain genetic diversity, each year different combinations of males and females of the same year class should be bred. This requires keeping good records and being able to identify individual fish.

Replacement brooders can be grown by year-class in separate ponds, or selected fingerlings can be tagged and stocked into a holding pond with tagged fish of other year-classes. They can also be held in a polyculture with Chinese carps. Where large numbers of brooders are replaced each year, it may be easier to hold the fish in separate ponds by sex and year class. If only a few brooders are replaced each year then it may be more efficient to tag each fish and hold several year classes in a common pond but keeping males and females in separate ponds. Keeping males and females in separate ponds is particularly important in temperate climates. In the spring the temperature may rise into a suitable range for spawning for a few days and then drop back to unsuitable temperatures. If males and females are held in the same pond they may spawn during those few warm days. However, when cool temperatures return, the embryo development is slow and fungus problems often occur. Such unplanned spawns create management problems and often few fry are obtained. Any female that spawned during that period would not be available for use later that spring when a stable temperature appropriate for spawning is reached.

Another issue in maintaining Common Carp brooders is the repressive effects of pheromones and other factors when Common Carp brooders are held at high densities in ponds or cages for an extended period of time. Under such conditions, even though mature males and females are held in the same environment and temperatures are in the right range they do not spawn. This can be used as a tool to have brooders available for a longer period within the spawning season as long as the fish are fed properly and do not become stressed due to low oxygen or high ammonia. Transferring brooders from such crowded holding conditions to new water where Common Carp have not been held often is enough to stimulate a natural spawn. Similarly, if mixed sex brooders are being held under the conditions that repress spawning, any major water exchange in the holding pond may dilute any pheromone effect and unwanted spawning may occur.

Brood selection should take into consideration the desired body shape, scale pattern and color pattern, as well as their degree of maturity and disease history. A mature male should release milt when gentle pressure in the vent area is given. A gravid female has a soft, swollen abdominal region (Figure 14.1.2). The urogenital opening of a female should be reddish and may become swollen when spawning is imminent.

Fish with anchor worms (*Lernaea* sp.) or hemorrhaged areas on the skin should not be used as such fish may be in poor health and may not spawn successfully.

14.1.3 Propagation: Spawning-Rearing

The simplest method of spawning Common Carp is the use of a spawning-rearing pond. Brooders in spawning condition are stocked into 0.01 to 1.0 ha ponds at 25–50 kg/ha using a 2 male: 1 female ratio once water temperatures remain at 18°C or above. If the brooders are mature and in good condition, their transfer to a newly filled pond is often enough to stimulate a spawn. However, injection of the females with either fresh or preserved fish pituitary at a rate of 1 to 2 mg dry pituitary/kg body weight is helpful in inducing overnight spawning. Good quality males will not need a pituitary injection.

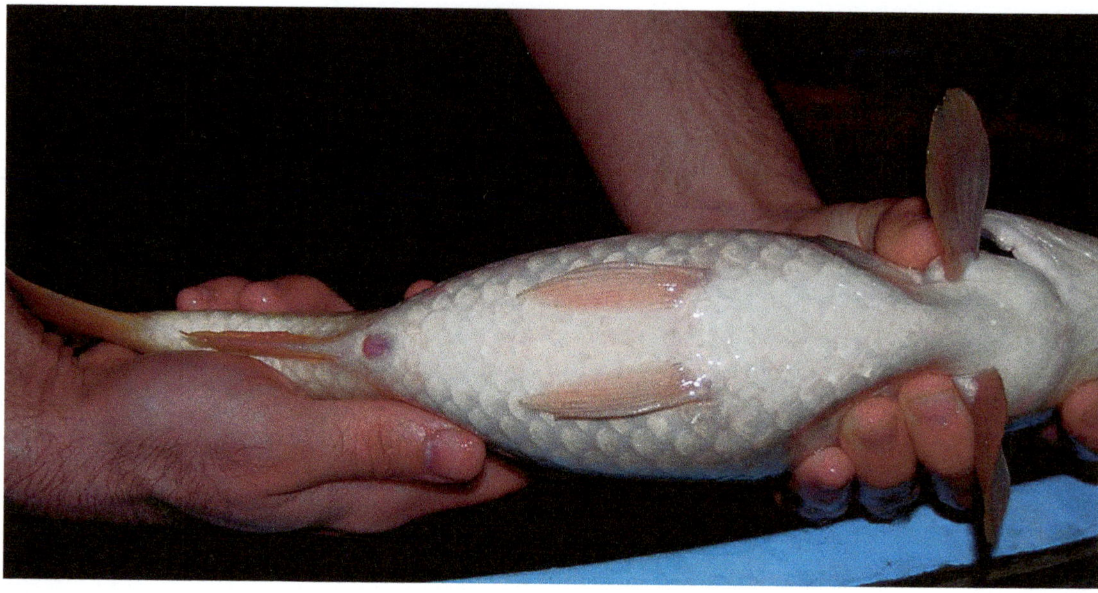

Figure 14.1.2. Acceptable Common Carp brood female with soft abdomen and enlarged genital papilla.

Another variation of spawning-rearing in a temperate climate is to draw down the brood holding pond in the late winter and allow vegetation to grow along the original shoreline. In the spring when the water temperature is 18°C, the water level of the pond is raised so that the grass along the shoreline becomes flooded. The rise in water dilutes any pheromone repression and the flooded shoreline grass provides a substrate for spawning.

With either approach the eggs are allowed to hatch in the pond and once the fry reach an acceptable size they are harvested along with the brooders. This may be after 1–2 months of culture. Partial harvests can be made to reduce fry density in the spawning-rearing pond and increase the growth rate of the remaining fry. Supplemental feed can be given to maintain acceptable growth rates.

The disadvantages of the spawning-rearing pond method include: no direct control over the number of fry produced or the size harvested; predation by adults on eggs and young; invertebrate predation or competition; and transmission of diseases from parents to young. These disadvantages are offset by the simplicity of the method and a lack of intensive management being required.

Using the spawning-rearing pond method, a well-fertilized pond should produce from 250,000 to 500,000 fingerlings/ha yielding 110 to 220 kg/ha provided that adequate fry recruitment occurs. The most economical pond fertilization approach probably will be the use of inorganic fertilizers or a combination of inorganic and organic fertilizers. Fertilization is continued at regular intervals until about 7 to 10 days prior to harvesting the crop. If conditions are favorable at the time the production cycle is started, 60 days should be adequate to produce a crop of 5 cm size fingerlings.

With the spawning-rearing approach the main activity after brood stocking is to apply the scheduled fertilizer applications. Additionally, some weed control may be necessary. While large carp fingerlings inhibit growth of most aquatic weeds, this is not the case with fry and small fingerlings. If the young are to be harvested by draining the pond, weeds should be controlled by the measures outlined in the Aquatic Weed Control section of Chapter 17.2.

14.1.4 Propagation: Broodstock Transfer

Yields using the spawning-rearing pond method are often unpredictable. Several other approaches can help improve yields and reduce the variability associated with the spawning-rearing pond method. They include: egg transfer, fry transfer, and brood fish transfer.

With broodstock transfer, spawning is often done in a small specially designed spawning pond (Dubisch pond and similar designs) that can be filled and drained rapidly. The pond has a shallow vegetated area that when flooded will be where spawning takes place with the adhesive eggs sticking to the vegetation. The pond also has a deep area not suitable for spawning but where the brooders can be held and later seined out after spawning.

Brooders are stocked in a freshly filled pond where the shallow area has been flooded. The fish will naturally move into the shallow area and spawn. Soon after spawning has occurred, the water level of the pond is lowered 30 to 60 cm to concentrate the brood fish in a collecting ditch where they are captured and removed from the pond. The water level in the pond is immediately raised after the broodstock removal to re-flood the eggs. The pond then becomes a nursery pond where the eggs hatch and the larvae develop. Once the fry are large enough to be safely harvested (12–15 mm, about 10 days post- hatch) they are transferred to other nursery ponds.

The drawdown period is not harmful to the eggs if the pond can be refilled as soon as the brooders are harvested, particularly if the weather is cool and drawdown is done in the late afternoon and early evening. Removing the brooders prevents eggs from being eaten and helps to prevent disease transmission from the adults to their progeny. Harvesting the brooders confines the spawning period within the pond resulting in more similar age progeny. It also allows brooders to be recycled, particularly the males who may be used again the same spawning season. Draining the pond just long enough to recover the brooders allows the spawning area to be observed and the magnitude of spawning to be estimated.

14.1.5 Propagation: Egg and Fry Transfer

Transferring a known number of eggs or fry to a nursery pond can result in a more predictable harvest. Eggs can be obtained from fish that have spawned on a substrate placed in ponds, tanks or hapas. Spawning ponds of <1000 m^2 are preferred for Common Carp spawning when using the egg transfer approach. Spawning ponds should be dried thoroughly and be free (including the shoreline) of any form of vegetation or fibrous material when the pond is completely full. When the water temperature remains at 18°C or above, the pond is filled and mature broodstock added at up to 500 kg/ha using a 2 male: 1 female ratio. Spawning substrates are added when or just before brooders are stocked.

Spawning substrates (Spawning mats) used in the USA are typically wire mesh rectangular frames filled with Spanish moss (*Tillandsia usneoides*) fiber. The size of the frame is approximately 30 × 75 cm. Other commercially available fibrous material such as used with air conditioner filters can be also used to fill in the frame. The same type of spawning mat is also used for other egg scattering cyprinids that have adhesive eggs such as Goldfish and Golden

Shiner. A series of mats are placed in shallow water on the lee side of the pond so the mats can be easily observed and there would be less sediment to settle out on the spawning mats. The mats are placed in the water at approximately 10–20 cm below the surface depending on the size of the brooders being used. The mat should be placed so that the brooders are able to swim over the top of the mat with the dorsal fin out of the water, as illustrated by Goldfish brooders in Figure 14.3.3. A frame may be needed to keep the mat at the right depth.

The mats should be monitored daily as to fish activity and the quantity of eggs on the mats. Ideally a mat would be covered in eggs within two days after placement and, if so, the mat should be removed and replaced with a new mat. If the mat is heavily covered in eggs, with eggs laying over each other, additional mats should be added to that pond so the fish have more substrate to spawn on. Allowing the eggs to be more spread out on the mat reduces the chance of egg suffocation and fungus problems. If the mat has only a few eggs, conditions in the pond might not be suitable for spawning. In such cases the mats should be removed from the pond leaving only one or two mats so any spawning activity can be monitored. Water quality should be monitored and temperature checked to be sure they are in acceptable ranges. When the test mats are covered with an abundance of eggs then additional mats should be added.

Mats having adequate egg densities can be removed and transported out of the water for a short period as long as the mat remains moist. The quantities of eggs/mat can be estimated by knowing the area of the mat, counting the eggs in five or more samples of known area, and then extrapolating the mean number of eggs/cm^2 of sample to the total area of the mat. When counting to get an estimate of the number of eggs/mat an estimate of the potential hatch rate can also be determined by counting the number of dead eggs of the total number of eggs in each sample. Subtracting the number of dead eggs from the total will give an estimate of the number of viable eggs. The count of viable eggs is best done when the eggs reach the eyed stage or later. The number of viable eggs may vary from mat to mat and after counting five or more mats, a mean number of viable eggs/mat can be estimated. The number of viable eggs/mat can be estimated by multiplying the mean number of viable eggs/cm^2 of mat × the total mat area. Based on the mean number of viable eggs/mat, the number of mats to transfer to a nursery pond can be calculated.

For example, if the goal is to stock 1,000,000 fry/ha into a primary nursery pond the number of mats needed based on the following:

Mat size = 30 × 75 cm = 2,250 cm^2
Sample size area = 10 cm^2
Mean total number viable eggs/sample = 41.6 eggs/sample = 4.16 eggs/cm^2
Mean total viable eggs/mat = Mean no. eggs/cm^2 × total mat area (cm^2)
= 4.16 eggs/cm^2 × 2,250 cm^2 mat
= 9,360 total viable eggs/mat
Number fry / mat = total viable eggs/mat = 9,360 ± fry/mat
No. mats/ha to stock = (fry stocking rate/ha) /(number fry/mat)
= (1,000,000 fry/ha)/(9,360 fry/mat) = 107 mats ± /ha

The above is only a rough estimate of the number of mats needed as there are a number of factors that can affect the estimate of fry/mat obtained. A portion of the eggs drop into the fibrous mat and are often overlooked when sample counts are made. Sampling can be a challenge, as often the eggs on a mat are not uniformly distributed and there also may be considerable variation from mat to mat in the number of eggs/mat. Related problems are found in estimating the number of fry obtained/mat due to sampling issues and differences in number of viable eggs at the time when the samples were taken and the actual number live eggs at the time of hatch.

The above approach of using mats, other than the depth of mat placement, is also used for pond spawning of Goldfish and Golden shiner. Details on these species are given in sections 14.3.2 and 14.3.3 of this chapter.

In Southeast Asia, an egg collector is often referred to as a "kakaban." It comes in two basic styles: 1) a fiber mat set horizontally on a frame, similar to the mats used in the USA, and 2) bundles of the fiber are tied to a pole so that the bundles of fiber will hang down into the water about 25 cm or more. Several plants provide suitable fiber for egg collectors including some palms (*Arenga* sp.), and sisal (*Agave* sp.). They also can be made from strands of plastic. Floating aquatic plants such as water lettuce/Nile cabbage (*Pistia* sp.), and water hyacinth (*Eichhornia crassipes*) that have roots extending into the water can also be used as egg collectors. When the fiber mats are set horizontally below the surface the brood fish will swim up on the mat and spawn. When the material hangs down in the water, the brooders will swim in among the fiber bundles and spawn. The adhesive eggs stick to the fibrous material and the egg-bearing kakabans are then transferred to a nursery pond. Mats placed horizontally are more two dimensional, allowing eggs to be counted as described earlier. The three-dimensional nature of kakabans where the fibers are hanging down in the water column and eggs are scattered within the bundle make it difficult to estimate the number of eggs on hand.

Common Carp can also be allowed to spawn naturally under more confined conditions both outdoors and indoors providing substrates as described earlier for egg collection. Hapas made of fine mesh netting ($\leq$1.5 mm) with dimensions of 90 × 90 × 125 cm or larger can be suspended on stakes in a pond at a depth of 70 cm using ponds where there is not a high density of Common Carp already present. A well-secured net cover is needed to prevent brooders from jumping out of the hapa.

Common Carp will spawn naturally in holding tanks, particularly when very mature fish are transferred into clean water where no other Common Carp have been held and a spawning substrate is available. Tanks with volumes of 2 m^3 or more are suitable, preferably with water exchange. Keeping the water level in the tank at 25–30 cm and the tank covered is necessary to prevent brooders from jumping out of the tank.

Brooders should be stocked into the hapas or tanks at up to 5 kg/m^3 at a 2 male : 1 female sex ratio. Spawning success can be increased giving the females a single injection of 2 mg dried Common Carp pituitary/kg female at the time when they are stocked into the hapas or tanks. Kakabans or mats should be added to the hapa or tank at the time when brooders are stocked providing approximately 1–1.5 m^2 of surface area of egg collection substrate/female. Stocking should be done in the late afternoon with spawning to take place overnight. The kakabans or mats are removed to a hatching pond as soon as they are loaded with eggs. A delay of several hours in moving eggs could result in eggs being eaten by the brooders.

Once spawning has taken place, an alternative, particularly when using kakabans composed of fiber bundles that hang down in the water, is to move the kakabans holding eggs to another tank or hapa for egg incubation. Hatching eggs on mats or kakabans in the hatchery has the advantage in that the eggs can be treated with formalin or other chemicals in an attempt to control fungus and other disease problems.

Where spawning has taken place in a hapa, eggs are often adhering to the sides of the hapa. In such cases the brooders should be removed and the eggs allowed to hatch in the hapa and the fry collected for transfer to nursery ponds. Having the eggs hatch under the more controlled conditions of a hapa or tank allows the larvae to develop to a swim-up stage (Figure 14.1.3) where they can be counted. This allows more accurate stocking of nursery ponds compared to stocking of egg-bearing mats or kakabans into a nursery pond. After the fry become free swimming the egg collectors should be removed, rinsed in clean water, allowed to sun-dry, and reused.

Nursery ponds should be prepared ahead of time to receive mats or fry. When the egg collectors are moved to the hatching pond they should be supported off the bottom in shallow areas where good water circulation is likely to occur but not along a shoreline where wave action creates suspended sediments that can settle on the eggs. The nursery pond should be prepared as described in Chapter 7 with a target of having an abundance of rotifers and other small zooplankton when the larvae are ready to start exogenous feeding.

Stocking densities of nursery ponds can vary depending on the size of fish desired, and the space and labor available. A density of 2 to 5 million fry/ha can be maintained for 10 to 14 days in a properly prepared pond. Once the fish are 1.5 to 2.0 cm they are large enough to be handled and a portion of the fry can be harvested and transferred to another nursery pond a rate of about 400,000 fry/ha. They should reach a size of 3.0 to 5.0 cm in 15 to 30 days. If growth is slow, the fry can be fed a powdered commercial feed at 2.5 to 5 kg feed/ha/d divided into 2 to 4 feeding/d. An initial fry density of 1 million/ha of nursery pond should yield 2.5 cm fish in 30 days with a 50% survival. Larger fish can be produced in the same time period using a quality feed when stocking fry at 500,000/ha (Figure 14.1.4).

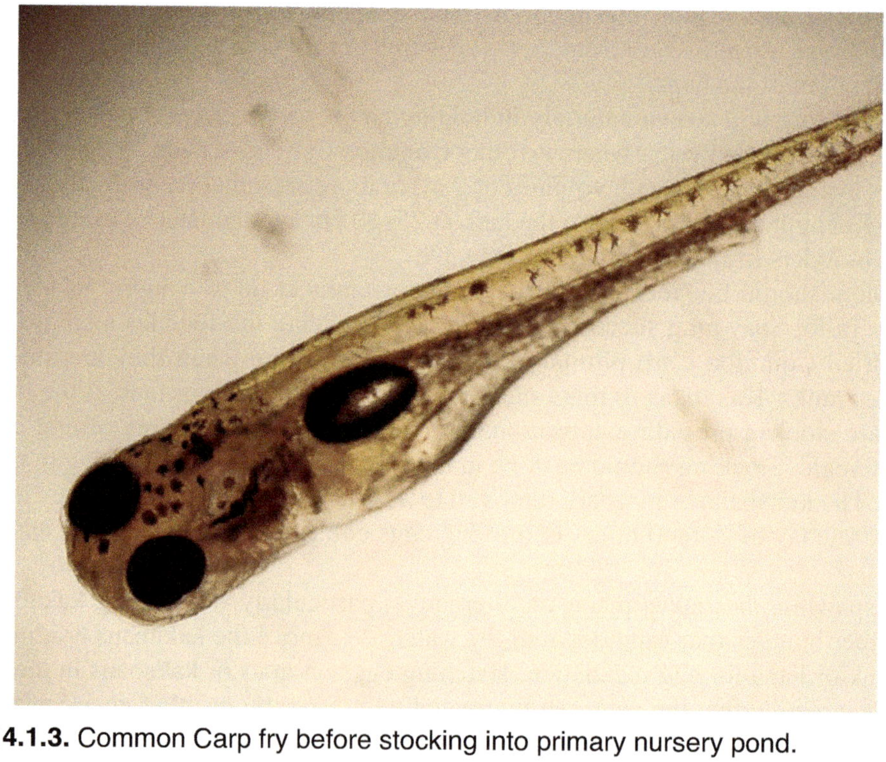

Figure 14.1.3. Common Carp fry before stocking into primary nursery pond.

14.1.6 Propagation: Artificial Spawning

Common Carp can be hormone-induced to spawn naturally or spawned under more controlled conditions where the eggs are stripped from the female and manually fertilized. Hormone-induced ovulation with natural fertilization helps to synchronize when the fish spawn and can increase the number of females that spawn. Hormone-induced ovulation with manual fertilization allows more genetic control by being able to spawn a specific female with a specific male.

Brooders are selected for induced spawning with manual fertilization based on females having well developed abdomens that are soft to the touch, and having a prominent genital papillae. Males should be running milt when a light pressure is applied to the abdomen near the genital papilla. Induced spawning with manual fertilization is stressful on the brooders and only healthy fish free deformities should be used. Males and females should be held separately in easily accessible covered tanks. Hapas are not preferred for brood holding as common carp brooders have hard serrated dorsal and anal spines that can get caught in the netting.

If eggs are to be stripped, both males and females are injected with carp pituitary. Females are given 4 mg carp pituitary/kg body weight giving 10% in the initial injection and the remainder 6 to 12 hours later. A 6-hour interval is acceptable for tropical settings while the 12-hour interval is more appropriate for a temperate climate. Male fish receive 1/4 to 1/2 the dose given the female. Ovulation may be anticipated as early as 6 to 8 hours after injection for fish held under tropical conditions. In temperate climates at water temperatures of 22–26°C the time to ovulation following the resolving injection of Common Carp pituitary can be anticipated within 240–260 degree (C°) hours. For example, at 24°C ovulation will occur between 10 h (=240 °h/24°C) and 10.8 h (=260 °h/24°C). Occasionally there will be a

Figure 14.1.4. Common Carp fingerling after a 30-day primary nursery.

female that will ovulate in response to the preparatory dose and the hatchery staff should be prepared to fertilize and incubate such eggs. However, egg quality may be an issue as the time of ovulation is unknown.

Several synthetic formulas of GnRH are effective when combined with a dopamine antagonist to induce spawning at success rates equal or greater than that obtained with Common Carp pituitary. However, the response time to ovulation may be slower with a wider variation among individual fish compared to Common Carp pituitary (Table 14.1.1). Common Carp do not respond well to synthetic formulas of GnRH without the addition of a dopamine antagonist, particularly early in the spawning season.

Milt production by common carp males can be improved using half the dose rate of the same hormone being used with the females. The hormone should be given at the time of the female's resolving injection or about 12 h before the anticipated ovulation. Sperm should be collected about an hour before the anticipated time of ovulation following procedures described in Chapter 4.

Females that are nearing ovulation can be identified by adding small mature males to the tank with the females. As a female nears ovulation she releases pheromones into the water, and the males will be attracted to such females. Once a female is identified she can be captured and examined. Females releasing eggs are removed, anesthetized, and stripped.

When females are held in clear water, the tank bottom can be checked periodically to see if any eggs have been released and have stuck to the tank bottom. Common Carp are very fecund and once ovulation begins the lumen of the ovary fills quickly and some eggs will escape out the oviduct into the tank. Egg quality will drop the longer the eggs are held in the lumen of the ovary and will have dropped 50% by 50–80 min. post-ovulation. By periodically examining the tank bottom you will know that a female had ovulated but if there are more than one fish in the tank you do not which fish had ovulated. Each female needs to be examined, and handled very carefully to prevent egg loss. The water level in the holding tank can be lowered and all the fish anesthetized to facilitate handling but this may slow down the time to ovulation of fish that had yet to ovulate. Another approach is, prior to injecting the female, sew the vent so that it is almost closed and only a few eggs escape when the fish is handled. Once a female releasing eggs is identified, she is removed from the tank, anesthetized, the threads closing the vent removed to allow a free flow of eggs, and then stripped. However, if the female is gently captured and handled, sewing the vent closed is not necessary. Instead, the vent of a female running eggs can be plugged with a finger, the female transferred to a tank containing MS-222 or other suitable tranquilizing agent, and once anesthetized the fish is stripped.

Dry fertilization is the preferred approach as the eggs will quickly become adhesive once they are in contact with water. The anesthetized ovulated female is dried with a towel and the vent region cleared by pulling pressure on the vent area with the forefinger and thumb to eliminate fecal material and urine. It is essential that no water comes in contact with the eggs until after eggs and milt are mixed together. If the fish is 2 kg or smaller, one person can probably do the stripping more efficiently than two. Larger fish may require the combined efforts of 2 or 3 persons working as a team.

An effective method to handle the fish being stripped is to hold the caudal peduncle up with the left hand and use a finger from the right hand to block the vent. The head is placed under the left elbow between the arm and body with the body of the fish supported between the forearm and side. The fish and the hands of the person holding the fish are dried very

Table 14.1.1. Hormones and dose rates used to induce spawn and manually strip female Common Carp.

Hormone	Dose/kg Body weight	Injection time & quantity	Time to ovulation	Reference
Acetone dried carp pituitary	4 mg	0 h, 10% 12 h, 90%	11–12 h at 22°C 9–10 h at 27°C	Phelps, Unpublished data
	3 mg	0 h, 10% 12 h, 90%	8–12 h at 21–22°C	Brzuska and Biaowas 2002
Ovopel [D-Ala^6,Pro^9 Net] -mGnRH + Metoclopramide	18–20 µg	0 h, 17% 12 h, 83% 8–10 mg	6–14 h at 21–22°C	Brzuska and Biaowas 2002
LH-RHa AZAGLY-NAFARELIN [D-Nal(2)^6, aza-Gly^{10}] -LH-releasing-hormone + 5 mg pimozide	20 µg	0 h, 10% 12 h, 90%	9–13 h	Mikolajczyk et al. 2004
Kobarelin des Gly^{10} (D-Ala^6) GnRH-ethynamide + Metoclopramide	20 µg 10 mg	single	16–19 h	Brzuska 2000
Lecirelin [D-Tle^6, ProNHEt^9] -GnRH-a + metoclopramide	15 µg 10 mg	single	13–15 h	Brzuska 2004

Table 14.1.1. Continued.

Hormone	Dose/kg Body weight	Injection time & quantity	Time to ovulation	Reference
LH-RHa [D-Ser(tBu)^6,Pro^9-NEt] + domperidone	20 µg 5 mg	single	28–30 h at 24°C	Arabaci et al. 2001
[D-Ser(tBu)^6,Pro^9-NEt] + haloperidol	20 µg 0.5 mg	single	14–16 h at 22°C	Arabaci et al. 2001
Salmon GnRH analogue D-Arg^6,Pro^9-NEt + domperidone	10 µg 5 mg	single	28 h at 24°C	Drori et al. 1994
Salmon GnRH analogue [D-Lys 6]-sGnRH-X + domperidone	10 µg 20 mg	single	8–10 h	Alok et al. 1997

carefully. Once the fish is dried the tail is lowered and the vent unblocked. At that point eggs should flow freely from the vent. The thumb and fingers of the right hand are used to gently force the eggs from the ovarian region, starting near the vent and progressing anteriorly beyond the pelvic area. The eggs can be allowed to flow down the anal fin or drop directly into the pan whichever is most convenient. Common Carp eggs normally appear greenish-brown to gray-green in color when stripped. Stripping progresses until egg flow ceases. Care should be taken to avoid introducing urine, feces, or blood into the pan of eggs. Large fish can be set on their side at the edge of a table with the vent hanging off the table and stripped, slowly lifting the head and applying pressure to the abdomen. After 1 to 2 liters of eggs have been collected, milt can be stripped from a male and added directly to the eggs but there is the risk that water, urine or feces may also be added when adding milt directly from a male. Collecting milt ahead of time is preferred, adding about 3 milliliters of milt per liter of eggs.

The milt is thoroughly but gently mixed with the eggs using a feather or a soft rubber spatula, until the entire quantity of eggs appears to have changed color. A volume of clean, clear freshwater approximately equal to the volume of eggs is added to the egg/sperm mixture and the stirring is continued for a minute. Sperm cells are mobile for up to 90 seconds after water has been added. A fertilizing solution (3 g urea, 4 g salt/L freshwater) can be used to extend the time that the sperm is motile.

Common Carp eggs become highly adhesive within 2 minutes of being in water. Several treatments can be used to reduce the adhesiveness or the eggs must be quickly distributed over a substrate for incubation as described for tank or pond spawning. Chapter 17.4 gives detailed procedures to follow using several agents to prevent adhesion. Full-fat powdered milk, rennet, talc, bentonite clay and a solution of salt and urea are among several materials that have been used successfully to control adhesiveness of Common Carp eggs.

If eggs are to be incubated on fiber mats, eggs should be distributed quickly after being fertilized by scattering them with a nylon bristle paintbrush over a mat that is immersed in water. It is essential that the eggs be carefully distributed to obtain a layer only one egg deep. When eggs are stacked on each other access to oxygen is restricted and potential for fungus disease is increased. After the eggs have been spread over the mats, the mats should be held in water and left undisturbed for several hours then transferred to a hatching tank or pond. During that time the eggs will water harden and stick to the mat fiber. The quantity of eggs fertilized and exposed to water should be limited to the amount that can be spread in the time period before adhesiveness develops. With practice, the technique of scattering the eggs can be mastered so an even distribution is obtained before the eggs begin to clump and stick to each other. Unfertilized eggs can be held at room temperature for several minutes. This allows small batches of eggs to be fertilized at a time and distributed over the mats, reducing the change of eggs clumping together before they all can be distributed.

If eggs are treated to eliminate stickiness and clumping, they can be incubated in a hatching jar or cone. A 7 L hatching jar may be filled with 1.5 to 2 L of eggs and a 30 L cone with 3–4 L. Common Carp eggs are sinking and a sufficient water flow is needed to stir the eggs without suspending them high in the water column. Any dead eggs will be lighter than live ones and with a proper water flow the dead eggs will form a layer above the live eggs and the dead eggs can be carefully siphoned off.

Fungus is a common problem when incubating common carp eggs and the eggs should be treated twice a day with 250 mg/L formalin for 30 min. Eggs on mats held in tanks can be treated by adding the formalin as a static treatment for 30 min then flushing the tank with clean water. Eggs incubated in jars are best treated using a recirculating system containing 250 mg/L formalin and pumping the formalin-containing water through the eggs for 30 minutes. A temporary stand-alone recirculating system can be used when the number of jars to be treated are 10 or less. The jars are moved to the system and treated, then the jars are flushed with formalin-free water and returned to the larger incubation system. When a larger number of jars are being incubated on a flow-through system and need to be treated, the overall incubation system should be plumbed so that the eggs can be incubated on a flow-through system for the majority of the incubation period. But with a turn of valves, the eggs receive formalin-containing water from a separate closed recirculating system (but not with a biofilter, as the formalin will kill the filter) for the length of the treatment period. At the end of the treatment period the non-formalin containing water supply is restored. This approach allows the eggs to be exposed a longer formalin treatment period and receive a better control of fungus that obtained with more classical flush treatments.

Hatching should take place within 60–70 degree-days. Common Carp have adhesive head pads at hatching and will initially attach to the inside of the incubator or larval holding tank. Within a few hours the larvae will begin swimming up and down in the water column. The larval holding tank should be aerated and have a water exchange, creating a water current pattern that prevents the larvae from piling up on the bottom of the tank and suffocating. In another 60–70 degree-days the larvae have adsorbed much of the yolk sac, have started to swim horizontally, and the mouthparts are functional. At that time the larvae need to be fed in the hatchery or transferred to primary nursery ponds. The number of fry to transfer can be counted by number/unit volume estimates or by visual comparison.

The nursery pond should be prepared as described in Chapter 7 and stocked with swim-up fry at rates as described earlier in this chapter. A typical routine is to use a primary nursery pond followed by a secondary nursery period. In temperate climates, mean springtime temperature may vary several degrees from year to year and this can greatly influence the growth of Common Carp fry during the nursery period and the availability of natural food organisms. By the end of the second week of nursery the fry will consume a fine particle (0 or 00) commercial meal. Depending on the stocking density, food availability, and water temperature, 2.5 cm fish can be expected with a 40–50% survival after 30 days of culture. Fish in primary nursery ponds should be sampled a week before the scheduled harvest to be sure they have been growing at a rate that should result in acceptable size fish when the pond is harvested. Fish harvested from the primary nursery pond can be restocked at 100,000/ha in secondary nursery ponds and fed a formulated feed. After three months of culture an average weight of >50g can be achieved with a 75% survival. Expect that the pond will become turbid as the standing crop increases and supplemental aeration may be required.

14.1.7 Harvesting and Distribution

The preferred method of harvesting Common Ccarp is by draining the pond. Small numbers of fry and fingerlings can be collected by seining or trapping, but draining must be done for a complete harvest. The small fingerlings are hardy and present no major problems at harvest. However, often there is considerable size variation among same age fish from the same pond, making accurate estimates of average weights and number of fish on hand more difficult. Grading the fish into groups that are more similar in size will help to improve accuracy of numerical determination based on a number-weight relationship.

If the fingerlings were produced using the spawning-rearing pond method, the fingerlings should be treated for external parasites as part of the harvest/distribution protocol. The broodstock often carry external parasites such as *Trichodina*, *Dactylogyrus*, *Gyrodactylus*, *Argulus*, or *Lernaea* that could be passed on to the fingerlings.

For distribution to receiving waters, small fingerlings can be transported by tank truck with supplemental aeration at quantities of 0.12 to 0.24 kg/L. At these densities, the fish can be transported for up to 12 hours without overly stressing the fish.

14.1.8 References

Alok, D., D. Pillai, and L. C. Garg. 1997. Effect of D-Lys 6 salmon sGnRH alone and in combination with domperidone on the spawning of Common Carp during the late spawning season. Aquaculture International 5:369–374.

Arabaci, M., H. Cagirgan, and M. Sari. 2001. Induction of ovulation in ornamental Common Carp (Koi, *Cyprinus carpio* L.) using LHRHa ([D-Ser(tBu)6, Pro9-NEt]-LHRH) combined with haloperidol and carp pituitary extract. Aquaculture Research 35:10–14.

Brzuska, E. 2000. Artificial spawning of carp *Cyprinus carpio* L.: differences between the effects on reproduction in females of Polish and Hungarian provenance treated with carp pituitary and (D-Ala6) GnRH ProNHEt (Kobarelin). Aquaculture Research 31:457–465.

Brzuska, E. 2004. Artificial spawning of carp (*Cyprinus carpio* L.): differences between the effects of reproduction in females of Hungarian, Polish and French origin treated with carp pituitary homogenate or [D-Tle6, ProNHEt9] GnRH (Lecirelin). Aquaculture Research 35:1318–1327.

Brzuska, E., and H. Biazowas. 2002. Artificial spawning of carp, *Cyprinus carpio* (L.). Aquaculture Research 33:753–765.

Drori, S., M. Ofir, B. L. Sivan, and Z. Yaron. 1994. Spawning induction in Common Carp (*Cyprinus carpio*) using pituitary extract or GnRH superactive analogue combined with metoclopramide: analysis of hormone profile, progress of oocyte maturation and dependence on temperature. Aquaculture 119:393–407.

Mikolajczyka, T., J. Chyba, P. Szczerbika, M. Sokolowska-Mikolajczyka, P. Eplera, W. J. Enrightb, M. Filipiakc, and B. Breton. 2004. Evaluation of the potency of azagly-nafarelin (GnRH analogue), administered in combination with different formulations of pimozide, on LH secretion, ovulation and egg quality in Common Carp (*Cyprinus carpio* L.) under laboratory, commercial hatchery and natural conditions. Aquaculture 234:447–460.

14.1.9 Recommended Reading

More details related to Common Carp are available in the following references.

Balon, E. K. 1995. Origin and domestication of the wild carp, *Cyprinus carpio*: from Roman gourmets to the swimming flowers. Aquaculture 129:3–48.

Hartman, K. H., R. P. E. Yanong, D. B. Pouder, B. D. Petty, R. Francis-Floyd, A. C. Riggs, and T. B. Waltzek. 2013. Koi Herpesvirus Disease (KHVD). IFAS Extension. Fact Sheet VM -149. University of Florida, Gainesville.

Horváth, L., G. Tamás, and G. Coche, 1985. Common Carp: Part 1: Mass Production of Eggs and Early Fry. FAO Training Series 8. FAO, Rome.

Horváth, L., G. Tamás, and G. Coche, 1985. Common Carp: Part 2: Mass Production of Advanced Fry and Fingerlings in Ponds. FAO Training Series 9. FAO, Rome.

Horváth, L., G. Tamás, and C. Seagrave, 1992. Carp and pond fish culture: including Chinese herbivorous species, Pike, Tench, Zander, Wels Catfish, and Goldfish. John Wiley & Sons Inc., New York.

Hu, X., Y. Ge, C. Li, Z. Jia, M. Shng, S. Wang, and L. Shi. 2018. Developments in Common Carp culture and selective breeding of new varieties. Pages 125–148 *in* J. Gui, Q. Tang, Z. Li, J. Liu, and S. S. De Silva, editors. 2018. Aquaculture in China: success stories and modern trends, Wiley Blackwell, New York.

Hulata, G. 1995. A review of genetic improvement of the Common Carp (*Cyprinus carpio* L.) and other cyprinids by crossbreeding, hybridization and selection. Aquaculture 129:143–155.

Jeney, Z., and G. Jeney. 1995. Recent achievements in studies on diseases of Common Carp (*Cyprinus carpio* L.). Aquaculture 129:397–420.

Jeney, Z., and V. Bekh, 2020. Technical Manual on Broodstock Management of Common Carp and Chinese Herbivorous Fish. Fisheries and Aquaculture Circular No.1188. FAO, Ankara. Available: http://www.fao.org/3/ca5827en/ca5827en.pdf. (May 2020).

Linhart, O., S. Kudob, R. Billard, V. Slechtad, and E. V. Mikodina. 1995. Morphology, composition and fertilization of carp eggs: a review. Aquaculture 129:75–93.

Rahman, M. M. 2015. Role of Common Carp (*Cyprinus carpio*) in aquaculture production systems. Frontiers in Life Science 8(4):399–410.

Rothbard, S. 1981. Induced reproduction in cultivated cyprinids. The common and the group of Chinese carps. I. The technique of induction, spawning and hatching. Bamidgeh 33:103–121.

Woynarovich, A. 2013. Species profile: Common Carp (*Cyprinus carpio* Linnaeus, 1758). In Aquaculture Feed and Fertilizer Resources Information System (www.fao.org/fishery/affris/en/org/fishery/affris/en/). FAO, Rome. Available: http://www.fao.org/fishery/affris/species-profiles/common-carp/common-carp-home/en/. (May 2020).

Woynarovich, A., P. B. Bueno, Ö. Altan, Zs. Jeney, M. Reantaso, Y. Xinhua, and R. Van Anrooy. 2011. Better Management Practices for Carp Production in Central and Eastern Europe, the Caucasus and Central Asia. FAO Fisheries and Aquaculture Technical Paper No. 566. FAO, Ankara. Available: http://www.fao.org/3/a-i2409e.pdf. (May 2020).

14.2 Chinese Carps

14.2.1 Introduction

The three most widely cultured Chinese carps are Grass Carp, sometimes called white amur, *Ctenopharyngodon idella*; Bighead Carp, *Hypophthalmichthys nobilis*; and Silver Carp, *Hypophthalmichthys molitrix*. They are native to large temperate rivers of eastern China and parts of Russia. In the spring they leave deep lower areas of a river and migrate upstream in response to rising water levels and increased river flow into spawning areas when water temperatures reach 15–17°C and spawning occurs at temperatures above 18°C. This also coincides with an increasing day length. A large female (10 kg) may release a million eggs/spawn. The eggs are semi-buoyant and require a turbulent water flow to keep the eggs and larvae up in the water column for several days as they develop until the larvae are able to swim horizontally.

For about a thousand years Chinese carps have been cultured based on capturing larvae as they drifted downstream in the river currents from where natural spawning occurred. These species will grow and mature in static water ponds but will not spawn naturally in such settings. In the 1950s and 1960s hormone induced spawning techniques were developed that allowed their reproduction and culture under hatchery conditions. This resulted in larvae being shipped throughout much of the world where they survived, reached maturity, and were reproduced under hatchery conditions. With the development of hormone-induced spawning techniques, Chinese carps are now cultured throughout much of the world.

The three Chinese carps differ in their food habits. Adult Grass Carp are herbivorous and are very effective in controlling several groups of aquatic weeds that are pests in fish ponds. They can consume up to 100% of their body weight/d, but they have a relatively short intestine for herbivores. Their pharyngeal teeth are effective in crushing the plant material consumed, but because of the short intestine much of the material in the feces is undigested and acts as a pond fertilizer. Adult Bighead Carp have well developed gill rakers that are able to filter out material ≥60 µm. They primarily consume zooplankton but are omnivorous, consuming phytoplankton, detritus, and also formulated feeds. There are no well-defined food transition stages but rather a progression as the fish grows to utilizing larger food items. The primary food of adult Silver Carp is phytoplankton. The larvae feed on zooplankton but once they are ≥25 mm phytoplankton becomes a more important part of the diet. As the fish grows, the gill rakers fuse into a sieve capable of filtering out algae and detritus as small as 20–25 microns.

Grass Carp have a relatively long lifespan (>10 yrs.) and grow rapidly at low densities at favorable water temperatures. A maximum size of 50 kg has been reported for the species in Russia. At Marion, Alabama, 7.5 cm fingerlings grew to a size of 1.4 kg in 120 days. Females tend to outgrow males of the same age-class. Bighead Carp are capable of fast growth, in a temperate zone reaching 40 to 50 cm in the first year. Silver Carp also have a fast growth rate and can reach a maximum size of up to 20 kg.

Grass Carp are sexually mature in about 3 years in the tropics and 4 to 9 years in colder climates. Minimum size at sexual maturity is about 1 kg and 2 kg for females and males, respectively, but is highly dependent on food availability and length of the growing seasons. Bighead Carp reach sexual maturity when 2 to 4 years old depending on the length of the growing seasons, with females averaging about 4 kg at first maturity. Silver Carp may reach sexual maturity in 2 to 3 years in the tropics but may not be mature until 6 to 7 years old when cultured in the northern climates. In all three species, males tend to be slightly smaller than females and become sexually mature a year earlier.

Mature Chinese carp males can be distinguished from mature females in the spawning season based on body configuration, vent appearance, milt flow, and roughness on the outer surface of the pectoral fins. Grass Carp females will be deeper bodied with a wider genital papilla than males. However, actively feeding males and females will have full digestive tracts and will appear deep bodied making it more difficult to separate the sexes. Immature fish or adults in poor physical condition are difficult to identify as to sex.

Sexual dimorphism in Bighead and Silver Carp includes slight differences in body shape, the appearance of the vent of gravid fish, a sandpaper roughness feel of the pectoral fins of males during the spawning season, with mature males flowing milt in season. In the spawning season, mature females show a reddening of the vent, softness and swelling in the ovarian region, and are generally thicker in the middle part of the body. Recognizing males and females can be difficult out of the spawning season.

Under hatchery conditions, temperature appears to be the key environmental stimulus to bringing brooders into final maturation. In temperate climates the seasonal increase in temperature helps to synchronize maturation of brooders held in static water ponds. Mature fish may be found at 18°C, and can be induced spawned but the low temperature extends the time to ovulation, the time to hatch, larval development, and in turn presents more problems in the hatchery. The preferred hatchery temperatures for spawning are 22–26°C, but Grass Carp can be spawned and produced successfully at 28°C. By the time water temperatures reach 30°C many of the female brooders will start to reabsorb their eggs. In tropical settings there is less of a seasonal temperature change to synchronize brood maturation to within a narrow portion of the year and within the brood population there may be some suitable brooders available over a much wider period of time. In a temperate climate the growing season is often too short for the maturation of more than one set of eggs/yr. Under such conditions an ovary has two sets of eggs, one that will mature that season, and a set of immature eggs that will be mature the next year. In a tropical setting the growing season is longer and multiple sets of eggs in varying degrees of maturity will be found in the ovary. As such, a female is capable of spawning more than once/yr. In either case, females will mature in static water ponds.

14.2.2 Brood Development and Selection

As it may take three years or more for a female Chinese carp to reach maturity and as there will be yearly mortality of brooders, a well-planned broodstock replacement plan is needed. Induced spawning is stressful, particularly to the females, and new brooders will be needed each year. In the case of Bighead Carp and Grass Carp, 50% of the females and 25% of the males used each year for induced spawning may not survive being spawned. In addition, brood replacement is needed to account for natural mortality and the culling of brooders when they become too large to handle safely. Silver Carp are the most delicate Chinese carp to handle and it is best to euthanize any brooder after it was used for induced spawning. New Silver Carp brooders will be needed each season. In all cases, some new brooders will be needed each year.

Brooders can be developed on the hatchery as part of a polyculture program. Grass Carp fingerlings can be produced and be used for several years as part of an aquatic weed control program in hatchery ponds where other fish species are being cultured. Bighead and Silver Carp fingerlings can be polycultured with fingerlings of other species, and when larger, can be polycultured in food fish production ponds. As filter feeders, they can be produced at standing crops up to about 1,000 kg/ha of either one species or the two combined when in a polyculture with a species that is being fed a formulated feed. When Bighead or Silver Carp are to be held from one spawning season to the next, the initial biomass stocked should be low enough to allow for a 50% increase in biomass so the final biomass at harvest is about 1,000 kg/ha. Once mature, potential brooders can be removed from such production ponds and held in smaller ponds a month or so prior to the start of the spawning season.

Grass Carp will take commercial feeds when reared in mono- or polyculture. Brooders set aside for use in the current spawning season can be maintained on a formulated feed at 2–3% body wt./d while keeping the standing crop at 1,000 kg/ha or less. Some hatcheries try to feed fresh plant material to Grass Carp brooders the last two months before the start of the spawning season. An appropriate feeding rate would be 20% body wt./d of a soft plant material. However, this often presents challenges in being able to obtain enough plant material for feeding at that rate and the quantity of organic material being added can create water quality problems (feeding a 1000 kg/ha standing crop at 20% body wt./d = 200 kg plant material/ha/d).

Adult Bighead Carp will consume floating feed but they are not aggressive feeders. In a polyculture setting they may wait until the more active species is through feeding and then feed on the leftovers. They also may feed along the outer edge of the school of active feeders on any feed washed out in that direction. Silver Carp generally do not take a pelleted feed but can be trained to accept a powdered feed as small fingerlings and given a meal on through adulthood. However, good quality Bighead and Silver Carp brooders can be produced without a specific feeding program as long as the pond has a good plankton bloom.

Before brooders are selected for induced spawning they should not be fed for 3–4 days to ensure that an apparent female with a large abdomen is not just a fish with an intestinal tract full of feed. This is particularly true with Grass Carp. Brooders should be collected carefully from the holding ponds by seining. Grass Carp and Silver Carp can jump a meter or more out of the water when excited. Bighead Carp will also jump but not to the same degree. A deep seine should be used that will allow the seine to be held at least 0.5 m above the water surface during the last portion of the seining but still keeping the seine in contact with the pond bottom. Often it is necessary to use a knotted mesh seine to collect the brooders but the knots can be abrasive to the snout and other portions of a Chinese carp's body. A soft mesh seine would be preferred but often are only available in small mesh sizes. A large seine with fine mesh can be difficult to pull. In either case extra effort is needed to handle the brooders carefully to minimize stress. Fish should be confined in a 3–6 m^2 area of the seine and individual fish gently handled to determine sex and whether it is a suitable brooder. Holding the fish inside a cloth bag or a spawning cradle will calm the fish as well as provide a better way to handle them. A bag or cradle should be used whenever handling brood fish for weighing, injection, or transport. Chinese carp are subject to hemorrhaging, particularly in the head area from poor handling and abrasions from seining and the use of dip nets. The resulting stress can impede successful induced spawning. Best results are obtained with brooders that were not fed for several days in the pond, then collected from the pond, taken directly to the hatchery, weighed, and injected. The longer the fish are held in the hatchery tanks the more stressed they will become, reducing the chance that they will respond to induced spawning.

Brood selection should begin with selecting proper size fish. Grass Carp are powerful fish capable knocking down the person handling the fish. Female brooders weighing between 2–4 kg and males between 1–3 kg are preferable. Potential brooders should be held in the water in a cloth bag, rolled over and the abdomen examined. Males and females differ in their appearance around the anal area. In both sexes there is an evident ridge anterior to the anus often more evident in the males (Figure 14.2.1). A female may have a ridge anterior to the anus but also should show a rounded ridge on either side of the anus (Figure 14.2.2). The ovary of a mature female Grass Carp will fill the length of the body cavity but will be constricted by the pelvic girdle. A good female should show a bulge anteriorly and posteriorly to the pelvic girdle, suggesting a figure-eight appearance when the underside of the fish is viewed. Overall, the abdomen should be distended and feel soft to touch. The vent should be enlarged, often with a slight reddish coloration. On occasions females nearing their peak stage of maturity will have a pink luster to their scales.

Bighead and Silver Carp females should have a distended abdomen that is soft to the touch with a prominent genital papilla. A very mature Bighead female will show an enlarged area between the pelvic fins and the genital papilla. A gravid Silver Carp will have an enlarged abdomen running from the pectoral fins to the genital papilla. Because of the enlarged abdomen the keel will appear more as a ridge than a distinct keel typical of an immature fish.

Chinese carp males are generally smaller than females of the same age. Mature males will have pearl organs (nuptial tubercles) on the top of the pectoral fins giving the fin a sandpaper feel (Figure 14.2.3).

Figure 14.2.1. Male Grass Carp *Ctenopharyngodon idella* anus and papillae.

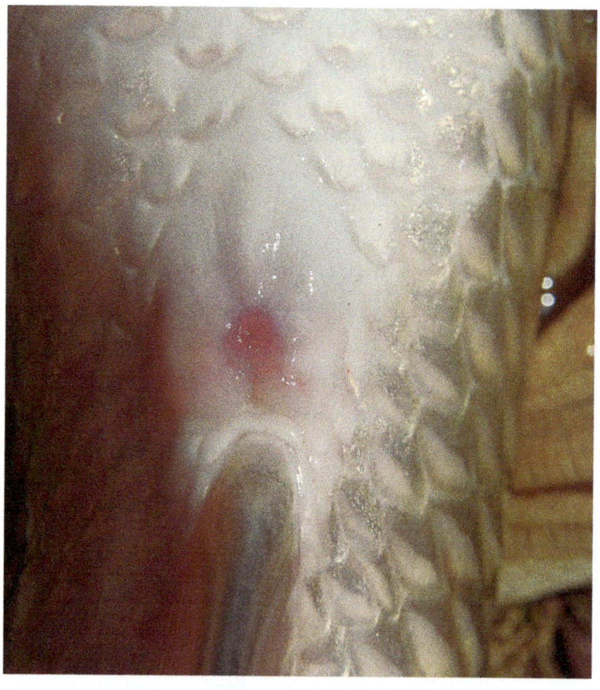

Figure 14.2.2. Female Grass Carp *Ctenopharyngodon idella* anus and papillae.

Figure 14.2.3. Male Grass Carp with pearl organs on pectoral fin evident by touch.

Males should release milt when the area around the vent is pressed. Adequate milt production often is a problem with male Grass Carp, and several males should be selected for each female.

Once brooders are selected, they should be tagged and weighed, and held in a circular tank having a depth of about 1 m and a freeboard of about 0.5 m, with a water exchange rate of 2 times/hour. The tank should have a tight-fitting net cover that will allow the fish to be observed. Males and females should be held in separate tanks, holding up to 4–5 fish in a 3 m^3 circular tank.

Each fish should be tagged so that the correct hormone dose can be given. Visible external tags are convenient so specific fish can be identified and collected with a minimum amount of stress. Pieces of bright-colored yarn as illustrated in Figure 4.3 are suitable for Chinese carp brooders. A 5–10 cm section of yarn is inserted around the first dorsal spine about halfway up the spine and tied tight.

14.2.3. Hormone-Induced Spawning Protocols

Grass Carp respond to several hormone spawning protocols (Table 14.2.1) and can be induced to spawn naturally, or be manually stripped and the eggs artificially fertilized. The range of hormones and protocols that have been effective in inducing ovulation in Grass Carp suggest that the hormone source and dose rate are not exacting. The selection of hormones, injection rates, and procedures is dependent on the costs and availability of hormones, and personal preferences.

Common Carp pituitary at 4 mg/kg of female is a commonly used procedure giving 10% in the first injection and the remaining 90% in the second 12 h later. Males are given 2 mg/

Table 14.2.1. Hormones and dose rates used to induce spawn and manually strip female Grass Carp.

Hormone	T. Dose/kg Body weight	Injection frequency	Time to ovulation	Reference
Acetone dried carp pituitary (CCP)	4 mg	0 h, 0.4 mg/kg 12 h, 3.6 mg/kg	220°h	Phelps, unpublished
	4 mg	0 h, 0.4 mg/kg 24 h, 3.6 mg/kg	≤ 14 h	Brzuska 1999
Human Chorionic Gonadotropin (HCG) + CCP	3,042 IU + 3.2 mg CCP (I.P.)*	0 h, 320 IU/kg 24 h, 2,722 IU/kg 48 h, 3.2 mg/kg	10–16 h	Bailey and Boyd 1970
	2,310 IU + 9.9 mg CCP	0 h, 440 IU/kg 24 h, 1,870 IU/kg 48 h, 9.9 mg/kg	---	Cassani and Caton 1985
Luteinizing hormone-releasing hormone analogue (LHRHa)	10 μg/kg	0 h, 10 μg/kg	13–23.5 h	Rottmann and Shireman 1985

Table 14.2.1. Continued.

Hormone	T. Dose/kg Body weight	Injection frequency	Time to ovulation	Reference
Luteinizing hormone-releasing hormone analogue (LHRHa) + dopamine antagonists	10 μg/kg +1 mg pimozide	0 h, 10 μg/kg +1 mg	---	Peter et al. 1988
	30 μg/kg +1 mg pimozide	0 h, 20 μg/kg 7 h, 10 μg/kg +1 mg	---	Peter et al. 1988
	10 μg/kg +5 mg reserpine	0 h, 10 μg/kg +5 mg 6 h, 10 μg/kg	4 h ±	Peter et al. 1988
	15 μg/kg +5 mg/kg pimozide (I.P.)	---	20–21 h	Brzuska 1999
Salmon GnRHa	20 μg/kg	0 h 2 μg 12 h 18 μg	---	Gasser et al. 2004
	20 μg/kg +5 mg 12 h 18 μg pimozide	0 h 2 μg +5 mg	---	Gasser et al. 2004
Ovaprim-C (20 μg/ml salmon GnRHa +10 mg/ml Domperidone)	0.6 mL	0 hr., 0.6 mL	---	Naeem et al. 2011

kg of male at the time of the second injection of the female. Our experience has been that 50–70% of the females will consistently ovulate near 220°h following the resolving injection within a temperature range of 21–27°C.

There are some constraints to consider when using LH-RHa. It is not very effective in inducing ovulation in Grass Carp held near 21–22°C. However, it is much more effective with brooders held at 25–26°C. Grass Carp are sensitive to handling and confinement, and in-turn, produce dopamines as a stress response. These dopamines are part of a negative feedback mechanism that can prevent ovulation. Dopamine effects can be minimized by careful handling of the fish, and the administration of anti-dopamine compounds such as pimozide (10 mg/kg of female) or domperidone (5 mg/kg of female) when using LH-RHa.

Bighead Carp and Silver Carp can be spawned successfully with the hormone protocols used for Grass Carp. In addition, they respond well to a combination of 1000 IU hCG and 2 mg dried Common Carp pituitary (CCP). Females are given 100 IU/kg as a preparatory dose and a resolving injection of 900 IU CG and 2 mg CCP/kg 12 hours later. Males are given 2 mg CCP/kg at the time of the resolving injection of the females. Silver Carp are the exception among the Chinese carps in that a fully matured female can be spawned with hCG alone when given at 250 IU in the first injection and 750 IU in the second, 12 h later. Male spawners are injected with about half the resolving dose at the time the resolving dose is given to the female.

Spawning success of Chinese carp can be increased by injecting brood fish as soon as they are collected from the holding pond, minimizing the time they are held in the hatchery before ovulation. Once the fish are weighed and tagged, and sufficient hormone is prepared at the proper concentrations, the females should be given their preparatory injection as soon as possible. The time interval between the preparatory and resolving injection often is 6–8 hours in a tropical setting and 12–24 hours in a temperate climate. Keep in mind that calm and gentle handling of the brooders is essential to good success in induced spawning of Chinese carps.

14.2.4 Ovulation, Fertilization, and Incubation

Ovulation of Chinese carps can be monitored in several ways. Using a degree-hour response time will help narrow down the time as to when the fish should be checked. The degree-hour (°h) response = (water temperature°C × hours to ovulation) will vary depending on the species, the temperature range, and the hormone protocol used. Chinese carp respond to 4 mg/kg of Common Carp pituitary, giving 10% in the 1st injection and the 90% in a 2nd injection 12 h later, at 200 to 220°h in a temperate climate. Silver Carp start to ovulate closer to 200 °h, Bighead around 210°h. and Grass Carp nearer 220°h when the temperature is in the 23– 27°C range. Keeping good records will allow a hatchery manager to develop a degree-hour response that is appropriate to that location's conditions and management practices. Degree-hour guidelines help to minimize stress on brooders as well as the staff. Use of degree hours helps to define when the fish need to be observed and eggs looked for, and helps hatchery staff to better plan their schedules.

One of the first things to look for starting at 200°h is a change in the swimming behavior. A female will release pheromones into the water as she gets closer to ovulation and if males are added to the female holding tank they will be attracted to that female. The males will swim close to the genital papilla of the female, occasionally nudging her. When such behavior can be seen, the female should be checked 25 to 30 minutes later for ovulation.

If females are in a tank without males another type of behavior can be monitored. In response to the hormone injections, the eggs of a mature female will start to hydrate. This makes it harder for the female to maintain equilibrium in the water column. The posterior portion of the fish may start to float up. She will swim back to a horizontal position, but if she stops, the tail may start floating up again. If the fish are held in clear water, this swimming behavior can be observed. This can help to identify which female in a tank should be examined first to see if she has ovulated. A change in body shape can also be observed. In Grass Carp, the abdomen will become enlarged, particularly the ventral area between the pelvic fins, and the vent will become noticeably enlarged.

The ovaries of mature Chinese carps are large and once a female begins to ovulate the lumen of the ovary fills and some of the eggs will leak out of the oviduct and into the water of the holding tank. As the anticipated time of ovulation approaches, the tank water can be sampled with a small fine mesh dip net to collect any eggs that might have been released. Recently released eggs will have just started to water harden and the perivitelline space has not increased very much. At that point, egg diameters are often 1.3 mm. Such eggs have a higher specific gravity than water and will settle to the bottom of the tank. Within an hour, egg diameters will have increased to 4.0 mm or more, and the specific gravity decreases as the eggs adsorb water. Such eggs are semi-buoyant and depending on the water flow rate and turbulence, eggs can be seen up in the water column. Using a small mesh dip net, the entire water column should be sampled and the results evaluated based on the quantity of eggs collected, the diameter of the eggs, and the variation in egg diameters. A few small diameter eggs suggest that a female may have just begun to ovulate. Two or more apparent sets of egg diameters suggest that more than one female has ovulated. Eggs having a large perivitelline space suggest that a female may have ovulated some time earlier and the eggs remaining in the lumen of the ovary of that female may be over-ripe and of poor quality. This sampling technique is appropriate for determining that one of the females in the tank has started to ovulate but does not identify which female has ovulated.

To know which female has ovulated, the water in the tank should be lowered and each fish physically handled to see which fish had ovulated. With careful handling a brooder can be slipped headfirst inside a soft cloth bag, covering at least from the head to the pelvic fin. This is a two-person operation, one to hold the bag open, and one to place their hands over the fish's eyes and holding the pectoral fins carefully guide the fish into the bag. Once in the bag the fish should be turned upside down, the bag rolled back to expose the vent and a light pressure applied to the area between the pelvic fins and the vent. Eggs should flow freely out the vent of an ovulated female. If that is the case, a finger should be pressed over the vent to stop any egg loss and the fish moved in the bag to a container of anesthetic. It is best to use one bag per fish, keeping any female that was found to not have ovulated in her bag until all the fish in the tank have been captured and examined. That will avoid capturing that female again and causing additional handling stress. Once all the females have been examined the non-ovulated females should be released from their bags back into the holding tank.

In the case of Grass Carp, even if no loose eggs are found in the holding tank at 220°h, the water level in the tank should be slowly lowered to a depth of about 15 cm so that the fish's back is only a few centimeters under the water. Keep in mind that Grass Carp are an excitable fish and should be approached and handled very calmly, gently sliding the fish into the bag headfirst. Grass Carp are different than Bighead and Silver Carp in that an injected Grass Carp that was not running eggs when pressure was applied to the vent area at 220°h may become excited and ovulate while being handled. The pelvic fins will become rigidly extended (Figure 14.2.4) and the body will begin to quiver, the abdomen swells rapidly, and as she ovulates, the eggs will be released out the vent as a pressurized stream. If the extended fins and quivering lasts for more than a few seconds, anticipate the ovulation and press a finger over the vent to prevent egg loss. If the quivering lasts for only a few seconds then set the female aside and examine her 30 min later.

Degree hours are a good guide to use when collecting sperm from males. It is best to collect sperm just before 200°h. This will allow you to have sperm on hand before the females start to ovulate, making things less hectic when females begin to release eggs. Grass Carp males can be inconsistent in providing adequate volumes of sperm even though the males were injected (typically a half dose of the total dose given the female). It is best to have 2–3 males for every female to be injected. Sperm can be collected and held as described in Chapter 4.

14.2.5 Fertilization and Incubation

Both males and females should be anesthetized before collecting sperm or eggs. MS-222 at 100 mg/L is an effective dose for Grass Carp. Quinaldine at 10–12 mg/L can also be used. To avoid repeated exposure it is better to not treat an entire tank of brooders all at one time. Not all females in a tank may have ovulated and anesthetizing a female that has not ovulated may delay or stop her from ovulating later. It is best to have a smaller separate tank to anesthetize the brooders. Individual fish can be fish slipped into a bag (as discussed earlier), examined in the holding tank and the selected fish transferred to the anesthetizing tank. This approach allows anesthetizing only the fish that need to be anesthetized. When working with large fish that might be difficult to transfer to the anesthetizing tank or when a quicker response to the anesthetic is needed, the gills of an individual fish can be sprayed with a MS 222 (1:1,000) or quinaldine (1:10,000) solution as long as care is given to prevent gill damage.

Figure 14.2.4. Female Grass Carp with extended pelvic fins, and starting to quiver, just prior to ovulation.

Eggs should flow freely from an ovulated female with little to no excess fluid (Figure 14.2.5). On occasions, a female has not fully ovulated when anesthetized and when stripping is attempted the ovary will become inverted and extend out of the oviduct limiting the flow of ovulated eggs (Figure 14.2.6). On such occasions it is best to discard such females and focus on the females yet to ovulate.

Manually stripped Chinese carp eggs are dry fertilized (see Chapter 3) adding approximately 3–5 mL of sperm/L eggs. Being highly fecund species, spawning 1–2 females may be adequate to meet production goals but can lead to a rapid rate of inbreeding. The rate of inbreeding can be reduced by dividing the total quantity of eggs collected into several sub-sets, and fertilize each sub-set with a different male. Enough sperm should be added to the eggs so that after the sperm has been gently mixed in there should be a noticeable change in the color of the mixture. Fresh water or a fertilizing solution (4 g salt + 3 g urea/L of water) is added at 100 ± mL/L of eggs and the eggs gently mixed for 1–2 min. The eggs should be rinsed in fresh water and at least 2 L of freshwater added/L of eggs. At that point the eggs are ready to be loaded into incubators (Figure 14.2.7).

Figure 14.2.5. Ovulated female Grass Carp with a free flow of eggs.

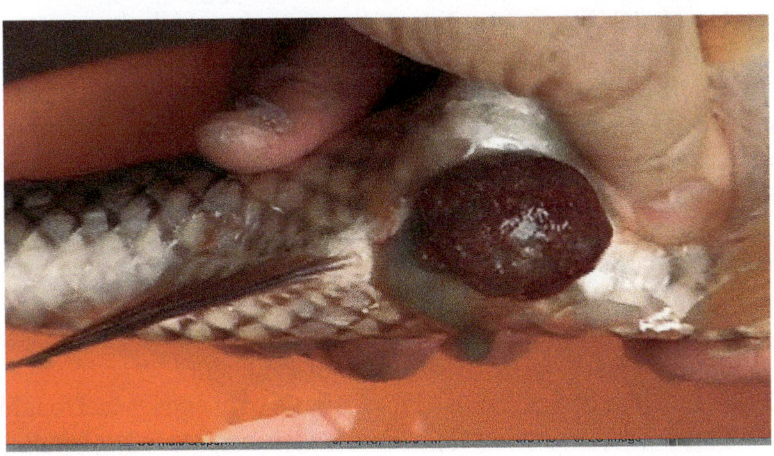

Figure 14.2.6. Female Grass Carp with an inverted ovary limiting egg flow.

Figure 14.2.7. Un-water-hardened Grass Carp eggs added to a fiberglass cone for incubation.

The eggs will begin to hydrate quickly and increase in diameter rapidly when in contact with water. Egg diameters before contact with water might be 0.9 to 1.2 mm but will have increased to 4–6 mm or more once fully water-hardened. The rapid increase in egg diameter must be taken into consideration when loading the incubator. An appropriate volume of recently fertilized eggs can be 25 mL of eggs/MacDonald jar. A large female may give such a quantity of eggs that it may take 10–15 minutes to load all the eggs into incubators. The eggs waiting to be loaded into an incubator continue to hydrate, increasing in diameter and an appropriate loading rate might be 50 mL of eggs/MacDonald jar by the time such eggs are loaded. When percent hatch estimates are desired the number of eggs/mL at the time of loading as well as the total volume loaded needs to be recorded as the number of eggs/mL will be changing over the course of the water hardening period. A goal should be to load a quantity of eggs so that when they are fully water hardened the volume of eggs would not exceed 2/3rd of the jar volume (Figure 14.2.8). A greater volume of water-hardened eggs/jar makes it difficult to provide adequate water exchange in the jar without the risk of washing eggs out of the incubator.

How much an egg hydrates, and how long it takes an egg to fully water harden, depends on the water hardness/total dissolved solids of the water in which the eggs are held. In hard water (≥100 mg/L) the eggs may become water hardened in about an hour with a final diameter of 4–5 mm. In soft water (around 20 mg/L) the osmotic gradient between the egg and the surrounding water is greater. It can take ≥2 h for an egg reach equilibrium and fully water-harden, and in the process, eggs adsorb more water and may reach a final diameter of up to 10 mm. Management of eggs in such soft water setting presents several problems. The larger the diameter of the egg the more the chorion is stretched and the more susceptible it is to being ruptured when handled. The more the eggs hydrate the more the specific gravity is reduced making the eggs lighter making it difficult to regulate the water flow rate in the incubator without damaging eggs or washing them out of the incubator. The larger the egg diameter also means a lower number of eggs can be incubated/L of incubator.

Several styles of incubators can be used for Chinese carp eggs. MacDonald jars (6–7 L in volume) are effective incubators but often 20 or more jars are needed to handle the quantity of eggs released by an average size female grass carp. Cones with volumes of 100–500 L, and an

Figure 14.2.8. a) Un-water-hardened Grass Carp eggs in a MacDonald jar, b) egg after water-hardening, with c) water hardened eggs filling approximately 2/3rds of jar volume.

up-flow of water from the point of the cone are also effective as long as the slope on the cone is steep enough. If the cone is too wide relative to its height, eggs may settle on the sides and die if they cannot be carried back into the water column by the water current.

Flow rates in MacDonald jars should initially be about 0.2 to 0.3 L/min. and gradually increased as the embryos develop and their oxygen demand increases. As hatching approaches the flow rate may be as high as 1 to 1.2 L/min. depending on the specific gravity of the eggs. Flow rates in cones should initially be just enough to push the eggs up 5–10 cm into the water column during the water hardening period. Once the eggs have water-hardened, the flow rate should be increased so that the current is strong enough to bring the eggs within about 10 cm of the water surface in the cone.

There are several management considerations during the incubation of Chinese carp eggs. Apparent fertilization rates, determined when the blastula stage is reached (4–6 h post-fertilization, temperature dependent) should be >70%. Egg quality is apparent several hours later when the gastrula stage is reached. The longer the eggs remain in the lumen of the ovary after ovulation the more egg quality will deteriorate. In the case of Grass Carp, egg quality will begin to drop in less than 30–40 min post-ovulation. Eggs left too long post-ovulation can be fertilized and cell division will take place but the embryo will not develop past the gastrula stage. This can result in a significant quantity of dead eggs. Dead eggs will have a greater diameter than live eggs. The embryonic cells will have ruptured, giving a whitest C-shape to the animal pole area.

Dead eggs are lighter than live eggs and in MacDonald jars will form a layer that can be siphoned off. In cones the dead eggs will be floating higher in the water column but the water current will also be carrying live eggs into the upper portion of the water column making it difficult to siphon dead eggs out. A common problem in cones is that the dead eggs clog the drain screen and can cause the incubator to overflow. Hatchery staff need to be available during this period of egg mortality in order to remove dead eggs and keep the drain screens clean. Oxygen demand increases as the embryos develop and needs to be monitored during the incubation period. It should be ≥4 mg/L both in the water supply and in the incubator.

14.2.6 Hatching and Larval Management

The incubation period can be short with eggs hatching in 18 h at 27°C but at 22°C it may take 30 h for a set of eggs to hatch. Egg shells can quickly clog the drain screens and the screens need to be cleaned frequently. The larvae are about 4.5 mm TL at hatch and rely on the water current to keep them suspended in the water column. The larvae will try to swim up into the water column for 10–15 cm and then drift back down. A water current pattern is needed that will help keep the larvae up in the water column and not pile up on the bottom where they will suffocate. A set of MacDonald jars can be arranged so their effluents carrying the larvae goes into a common channel that flows gently into a larval holding tank. Larvae in screened cones can be held there until mouthparts are functioning and the larvae are swimming horizontally or the screen can be removed and the larvae are allowed to flow out into a holding tank. Larval density in the holding tank during the yolk sac adsorption period can be as high as 200 fry/L if two water exchanges/hour are given. If fry are to be fed in the hatchery, the stocking rate should be reduced to about 100 fry/L.

The rate of larval development is temperature dependent, the larvae will be swimming horizontally with air bladder filled and mouthparts functioning at approximately 80 hours after hatching at 22°C and after 48 hours at 26 to 27°C. Supplemental aeration gives assurance that losses to low oxygen will not occur during the holding period. An ammonia free water supply is preferred for egg incubation and early larval development. It is essential that the holding container is free of organic matter and the resultant oxygen demand and ammonia buildup. Egg shells and dead fry in particular should be removed by siphoning or other suitable means. A slight buildup of unionized ammonia can quickly result in significant mortality. Recently hatched Bighead and Silver Carp larvae can tolerate un-ionized ammonia levels up to near 1 mg/L but 100% mortality can be expected as un-ionized concentrations approach 1.5 mg/L.

Counting Chinese carp fry for stocking primary nursery ponds is best done by concentrating the fry in the holding tank and determining the number of fry/unit volume. Based on the mean number of fry (8–10 sample counts)/unit volume, the appropriate volume of water holding larvae is removed from the holding tank for transport to the primary nursery pond.

14.2.7 Fingerling Production

Chinese carp larvae are typically stocked into primary nursery ponds soon after their mouthparts are functioning (Figure 14.2.9). The nursery ponds are prepared following standard methods to produce a zooplankton population dominated by rotifers at the time when the larvae are stocked (see Chapter 7). Larvae stocked at one million/ha should reach 2.5 to 3.0 cm in 30–45 days depending on the pond fertility and fish survival. The primary food for larvae in the nursery pond is zooplankton but they will accept a finely ground soybean meal, other oil seed meals, or rice bran as a supplemental feed. Survival should average around 50%.

Grass Carp are generally restocked into secondary nursery ponds at 125,000 to 185,000/ha and cultured to a size of 5 to 7.5 cm. Once the fingerlings reach 5 to 6 cm, the pharyngeal teeth are adequately developed to allow the fish to feed on aquatic plants. At that point, soft-tissued aquatic plants can be given ad libitum with the quantity depending on the type of plant, size of fish, water temperature and water quality. However, formulated fish feeds are readily accepted by all sizes of Grass Carp (except during the first 7 to 10 days during the primary nursery phase) and are more practical to use. With good culture practices the fingerlings should average 7.5 cm after 30 to 50 days at water temperatures of 25 to 30°C. Larger fish can be produced by restocking fingerlings from the secondary nursery pond at 8,000 to 12,000/ha. After 80 to 90 days of culture, the fish should average 25 cm in total length. Grass carp fingerlings after being harvested from the primary nursery pond can also be cultured in polyculture during the secondary nursery of non-piscivorous species. If stocked at 25,000/ha the grass carp generally outgrow the principal fish in the polyculture, and can be easily graded out at harvest.

Bighead Carp fry are harvested from the primary nursery and moved to a secondary rearing pond when 20 to 30 mm. Advanced fry and small fingerling bighead carp handle well and can be harvested from small (e.g. 0.04 to 0.1 ha) earthen rearing ponds by seining as the pond

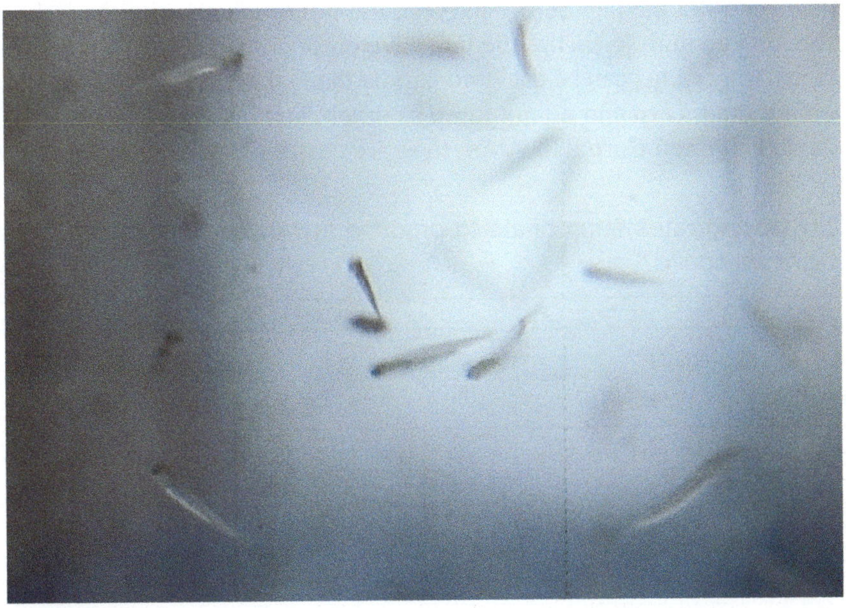

Figure 14.2.9. Grass Carp larvae ready to be stocked into nursery ponds.

is being drained. Care should be taken to move fish with a minimum amount of stress. The density for stocking the secondary rearing pond is governed by the size fingerlings desired and carrying capacity of the rearing unit. Assuming 80% survival of the fish stocked and an assumed carrying capacity, the stocking rates in Table 14.2.2 are suggested, depending on size of fingerling wanted. Silver Carp will give a similar growth performance as bighead carp in fertilized ponds but supplemental feeding of bighead and grass carp will increase their yields from secondary rearing ponds.

Polyculture with other species of fishes also is feasible to produce Bighead Carp fingerlings. They have been grown in association with Grass Carp, Silver Carp, Channel Catfish, and other non-predatory species. Obviously, the stocking density should be modified under such conditions. Bighead Carp fingerlings when stressed while being harvested or held in tanks waiting transport, often take on coloration similar to Silver Carp. It is difficult to separate the two species when produced in polyculture if stocked at similar stocking rates. They should be stocked at a rate that will result in a faster or slower growth rate than the other species in the polyculture making it easier to grade out that species from the other species.

14.2.7 Transport and Distribution

Chinese carp larvae are best distributed as soon after the swim bladder fills and the larvae begin to swim horizontally, which is approximately 48 hours after hatching at 24 to 25°C. Another 24 to 48 hours is then available for the larvae to arrive to their destination before the larvae begin to require food. The larvae are best transported in sealed plastic bags with oxygen at a density of 2,000 to 4,000 per liter of water. Lowering the water temperature to 20 to 21°C will extend the period they can be transported.

Harvesting and distribution of Grass Carp from primary and secondary ponds can be accomplished according to the general methods outlined previously. Grass Carp of 25 mm size and larger are relatively hardy and handle as well or better than the other species of carp. If fingerlings have been fed the day before harvest, the holding/hardening period prior to shipment should be long enough to ensure the fish have emptied their gut prior to shipping.

Bighead Carp fingerlings can be easier to handle than ether Grass or Silver Carp. They are less excitable than grass carp and do not jump as much, and do not escape under the leadline of the seine as do Common Carp. However, they are easier to stress than Grass Carp, but are

Table 14.2.2. Stocking rates appropriate to produce advanced fingerlings of Bighead Carp *Hypophthalmichthys nobilis.*

Average size fingerling wanted; length (cm)	weight (g)	Culture period (days)	Carrying capacity (kg/ha)	Suggested stocking (fry/ha)
5.0	2.1	60	1,200	715,000
7.5	8.0	70	1,400	220,000
12.5	34.0	90	1,800	66,000
15.0	63.0	100	2,000	40,000
17.5	103.0	100	2,000	24,000
20.0	134.0	100	2,000	18,000

more tolerant of handling than Silver Carp. At Auburn, post-harvesting mortality was 5–6 times higher for large Silver Carp fingerlings than for bighead of a similar size.

Advanced fry or small fingerling Silver Carp are sensitive to handling stress, particularly at water temperatures above 30°C. Harvesting early in the day, holding and transporting in cool water and avoiding crowding in transport vessels is recommended. Use of 0.2 to 0.5% salt (NaCl) in the holding-transport water can be helpful to reduce handling mortality. When handled carefully, survival of Silver Carp from secondary rearing ponds has ranged as high as 90% although the average is considerably less.

For transporting advanced fry and fingerling size Chinese carp, suggested packaging densities for closed containers are shown in Table 14.2.3. For transport in open fish hauling tanks a loading rate of 60 to 120 g/L of water is suggested for trips of more than 6 hours duration. For larger fingerlings, yearlings, and adults use of an anesthetic is helpful. MS-222 has been used at a rate of 1.6 mg/L.

Table 14.2.4 gives comparative data for transporting Grass, Bighead, Silver, and Common Carps in closed containers. The highest loading density is for Grass Carp with Common Carp being next. Bighead and Silver Carp are ranked slightly below Common Carp. Fry need to be carefully acclimated from their holding tank water quality to the water quality of the water they are to be transported in. If more than a 2°C temperature difference exists, it is recommended that 30 to 60 minutes be taken to equalize the temperature. Floating the container in the receiving water allows a gradual temperature acclimation. Other water quality differences can be equalized by opening the container, and gradually adding small amounts of receiving water every 3 to 5 minutes until conditions are similar. Advanced fry or small fingerling Silver Carp are sensitive to handling stress, particularly at water temperatures above 30°C. Harvesting early in the day, holding and transporting in cool water and avoiding crowding in transport vessels are recommended to avoid this.

Table 14.2.3. Number of Chinese carp fry that can be transported in a container of 18.5 L capacity containing 8.5 L of free oxygen (from Hora and Pillay 1962).

Size of fingerlings[1] or fry(mm)	Wt. Per 1000 fry or fingerlings (g)	*N* fry or fingerlings	Amount of free oxygen per g of fish (mL)
10–20	50–200	4000–5000	42.5
30	450	1500–2000	12.9
40	800	800–1200	10.6
50	1500	500–800	5.7
60	2500	400–500	3.4
70	3000	350–400	2.8
80	3500	250–300	2.4
90	4500	200	1.9
100	6000	150	1.4
110	8000	100	0.55
120	12,000	80	0.70

[1]By definition, young fish larger than 25 mm are termed fingerlings in this table.

Table 14.2.4. Guide for transporting small fingerling carps in a closed container supplied with oxygen (from Antalfi and Tolg 1968).

Species	Water Temp. °C	No./L/no. h of transport 2 h	4 h	6 h
Grass Carp				
	26	70	60	50
	24	100	90	80
	22	150	120	100
	20	180	150	120
	18	240	200	150
Bighead and Silver				
	26	40	30	20
	24	70	50	40
	22	100	80	70
	20	120	100	80
	18	160	130	100
Common Carp				
	26	50	40	36
	24	80	70	60
	22	120	100	100
	20	150	120	120
	18	200	160	120

In truck-mounted fish hauling units supplied with efficient aeration devices, a loading rate of 120 g/L of tank capacity is suggested at a water temperature of 18°C. At a temperature of 25°C the rate should be reduced to 60 to 70 g/L. Use of salt (NaCl) at a rate of 0.2–0.7% in hauling water should be used as an aid for increasing the density of fingerlings being transported. Special attention when transporting silver carp must be given to covering the container for all sizes larger than 2.5 cm. Some injuries occur when fish strike the walls of the container in their efforts to jump when excited. Use of a tranquilizing agent during transport can help in reducing the adverse effects of handling. Bacterial infections or fungal problems may occur from skin injuries during handling. An appropriate antibiotic treatment should be given for bacterial infections. Potassium permanganate is suggested as prophylaxis against fungus and general skin abrasions.

14.2.8 References

Antalfi, A., and I. Tolg. 1968. Herbivorous Fish. Mozogazdasagi Kiado, Budapest, Hungary.

Bailey, W. M., and R. L. Boyd. 1970. A preliminary report on spawning and rearing of Grass Carp (*Ctenopharyngodon idella*) in Arkansas. Pages 560–569 *in* Proceedings of the 24th Annual Conference of the Southeastern Association of Game and Fish Commissioners, Atlanta, Georgia. September, 27–30, 1970.

Brzuska, E. 1999. Artificial spawning of herbivorous fish: use of an LHRH-a to induce ovulation in Grass Carp *Ctenopharyngodon idella* (Valenciennes) and Silver Carp *Hypophthalmichthys molitrix* (Valenciennes). Aquaculture Research 30(11–12):849–856.

Cassani, J. R., and W. E. Caton. 1985. Induced triploidy in Grass Carp. Aquaculture 46:37–44.

Chen, Y., J. Li, H. Sun, and Z. Yang. 2013. Dose dependent effect of un-ionized ammonia on hatching success of fertilized eggs of Silver Carp *Hypophthalmichthys molitrix* and Bighead Carp *Hypophthalmichthys noblis*. Journal of Freshwater Ecology 28(1):39–46.

Glasser, F., C. Cauty, B. Mourot, and B. Breton. 2003. Disrupted sexual cycles in female Grass Carp (*Ctenopharyngodon idella*) raised in tropical conditions. Aquaculture 220:857–868.

Gonzal, A. C., E. V. Analar, and J. M. F. Pavico 1987. The effects of water hardness on the hatching and viability of Silver Carp (*Hypophthalmichthys molitrix*) eggs. Aquaculture 64:111–188.

Hora, S. L., and T. V. R. Pillay, 1962. Handbook of fish culture in the Indo-Pacific region. FAO Fisheries Biology Technical Paper 14. FAO, Rome.

Naeem, M., A. Zuberi, A. Salam, M. Ash. Ali, A. Ishtiaq, T. Malik, M. J. Khan, M. M. Ayaz, M. J. Iqbal, and B. Ahmad. 2011. Induced spawning, fecundity, fertilization rate and hatching rate of Grass Carp (*Ctenopharyngodon idella*) by using a single intramuscular injection of ovaprim-C at a fish hatchery Faisalbad, Pakistan. African Journal of Biotechnology 10(53):11048–11053.

Peter, R. E., H. R. Lin, and G. Van Der Kraak. 1988. Induced ovulation and spawning of cultured freshwater fish in China: advances in application of GNRH analogues and dopamine antagonists. Aquaculture 74:1–10.

Rashid M, M. H. Balkhi, G. A. Naiko, and T. Ahamad. 2014. Induced breeding of Grass Carp (*Ctenopharyngodon idella*) and Silver Carp (*Hypophthalmichthys molitrix*) using ovatide as synthetic hormone at national fish seed farm (Nfsf) Manasbal, Kashmir. J&K Fisheries and Aquaculture 5:110. doi:10.4172/2150-3508.10000110. Available: https://www.longdom.org/open-access/induced-breeding-of-grass-carp-ctenopharyngodon-idella-and-silver-carp-hypophthalmichthys-molitrix-using-ovatide-2150-3508.10000110.pdf. (May 2020).

Rothbard, S. 1981. Induced reproduction in cultivated cyprinids—the Common Carp and the group of Chinese carps. I. The technique of induction, spawning and hatching. Israeli Journal of Aquaculture 33:103–121.

Rottmann, R. W., and J. V. Shireman. 1985. The use of synthetic LH-RH analogue to spawn Chinese carps. Aquaculture and Fisheries Management 1:1–6.

14.2.9 Recommended Reading

Billard, R. editor. 1999. Carp: biology and culture: R. Springer–Praxis Series in Aquaculture and Fisheries. Praxis Publishers Ltd., Chichester, UK.

Coche, A. G., and D. J. Edwards, editors. 1989. Selected aspects of warmwater fish culture: a compilation based on lectures presented at a series of FAO-AGFUND international training courses in aquaculture hosted by Hungary in 1987 and 1988. GCP/INT/435/AGF. Fisheries Department. FAO, Rome. Available: http://www.fao.org/3/t8389e/T8389E00.htm. (May 2020).

FAO Fisheries & Aquaculture. 2005. Cultured Aquatic Species Information Programme—*Hypophthalmichthys molitrix* (Valenciennes, 1844). Text by N. Yang, N. *In* FAO Fisheries and Aquaculture Department [online]. FAO, Rome. Updated 26 July 2005. Available: http://www.fao.org/fishery/culturedspecies/Hypophthalmichthys_molitrix/en. (May 2020).

Horváth, L., G. Tamás, A. G. Coche, E. Kovács, T. Moth-Poulsen, and A. Woynarovich. 2015. Training manual on the advanced fry and fingerling production of carps in ponds. A handout for on-farm training workshop on fish seed production of Common Carp and Chinese major carps in Central and Eastern Europe, the Caucasus and Central Asia. Second revised edition. FAO-REU, Budapest. Available: http://www.fao.org/3/a-i4317e.pdf. (May 2020).

Jeney, Z., and V. Bekh, 2020. Technical manual on broodstock management of Common Carp and Chinese herbivorous fish. Fisheries and Aquaculture Circular No.1188. FAO, Ankara. Available: http://www.fao.org/3/ca5827en/ca5827en.pdf. (May 2020).

Jennings, D. P. 1988. Bighead Carp (*Hypophthalmichthys nobilis*): a biological synopsis. U.S. Fish and Wildlife Service, Biological Reports 88(29).

Jhingran, V. G., and I. S. V. Pullin. 1985. A hatchery manual for the common, Chinese and Indian major carps. ICLARM Studies and Reviews 11, Asian Development Bank, Manila, Philippines and International Center for Living Aquatic Resources Management, Manila, Philippines. Available: https://www.worldfishcenter.org/content/hatchery-manual-common-chinese-and-indian-major-carps-0. (May 2020).

Shireman, J. V., and C. R. Smith. 1983. Synopsis of biological data on the Grass Carp *Ctenopharyngodon idella* (Cuvier and Valenciennes, 1844). FAO Fisheries Synopsis No. 135. Food and Agriculture Organization of the United Nations (FAO), Rome. Available: http://www.fao.org/3/a-ap938e.pdf. (May 2020).

Szabóa, T., B. Urbányi, T. Müllera R. Szabób, and H. László. 2019. Assessment of induced breeding of major Chinese carps at a large-scale hatchery in Hungary. Aquaculture Reports 14, July 2019, 100193. Available: https://reader.elsevier.com/reader/sd/pii/S2352513418301753?token=21A2102A3379E4396F6294CDA4E0A8BA537D61D9369D523B5FBA176137CFA20B8929A7986546DA9A8A72EF682FACD199. (May 2020).

Woynarovich, E., and L. Horvath. 1980. Artificial propagation of warmwater fin fishes—a manual for extension. FAO Technical Paper 201. FAO, Rome. Available: http://www.fao.org/3/ac742e/AC742E00.htm. (May 2020).

14.3 Bait Minnows

14.3.1 Fathead Minnow *Pimephales promelas*, Rafinesque

14.3.1.1 Introduction

The Fathead is a small slender minnow used as fish bait in sport fishing, as forage for carnivorous species at a hatchery, and as a laboratory test animal. They are native to much of the mid- and upper- Mississippi river drainage in the USA, and because of its popularity as a baitfish it has been introduced to many other parts of the USA. They have also been introduced into Belgium, France, Germany, and the United Kingdom.

The life span of Fathead Minnows is often 3 years or less during which males seldom reach more than 8.5 cm and females 7.5 cm. They are sexually mature at 1 year of age or less depending on the length of growing season with males being approximately 5 cm in length and females 4 cm. The sex of sexually mature fish can be readily distinguished. In addition to size differences, male Fatheads have a distinctive color especially at spawning time (Figure 14.3.1). They are darker than females, being dark gray to black often with blotches of dark pigment around the head and upper part of the body. Males develop tubercles on the head along with a fleshy head pad that is used to massage the egg mass during incubation. These secondary sexual characteristics become evident about 30 days prior to the start of the spawning season. The female fathead lacks the distinctive male coloration and will be olive to dark gray or silvery in color depending upon the habitat. She will have a well-developed ovipositor at spawning time.

Figure 14.3.1. Adult male Fathead Minnow *Pimephales promelas* in breeding colors (Photo courtesy of Barry W. Smith, American Sport Fish).

Fathead Minnows are nest builders and intermittent spawners. Spawning commences in the spring when water temperature reaches 16 to 17°C and continues as long as conditions are favorable up to a temperature of 29°C. The spawning site is not a characteristic "nest," as the 1.4–1.6 mm diameter adhesive eggs are deposited on either a vertical or horizontal substrate such as a stone, log, post, stump, or similar object, often depositing eggs on the underside of the substrate (Figure 14.3.2). A male will clean the substrate, guard the nest site and a female will spawn on the substrate. After she spawns the male chases the female away from the nest and then guards the eggs until they hatch. Several females may deposit eggs in a nest, each contributing an average of 400¬–500 eggs/spawn. The number of eggs/nest can be highly variable. As much as 12,000 eggs have been found in a nest. A female may spawn every three to four days during the spawning season. A single female has been observed to spawn 12 times in 11 weeks producing a total of 4,144 fry. Average production per female over a 12-week period would be closer to 1,000 young depending on the food supply available and water temperature.

The egg incubation period is usually 4 to 7 days. Fry are about 3.5 to 4.5 mm in length at hatch. At 10 days post-hatch they should be 5 or 6 mm long. The fry do not school and no parental care is given after hatching. The yolk sac is absorbed in 2 or 3 days and first feeding begins at a size of about 5 mm.

Early food habits are similar to other larval fishes. The adults feed mainly on zooplankton and insects although they are somewhat omnivorous in their food habits. Growth rate of Fathead Minnows on natural food is given in Table 14.3.1. Fathead minnows readily take artificial diets and yields can be improved significantly by feeding.

14.3.1.2 Production

Spawning-Rearing Method.—The spawning-rearing method of propagation is well suited for fathead minnow culture. However, it has the challenge of trying to control the abundance of fry in a pond. One strategy as the spawning season approaches is to provide an abundance of natural food organisms through pond fertilization and provide supplemental feed for the brooders.

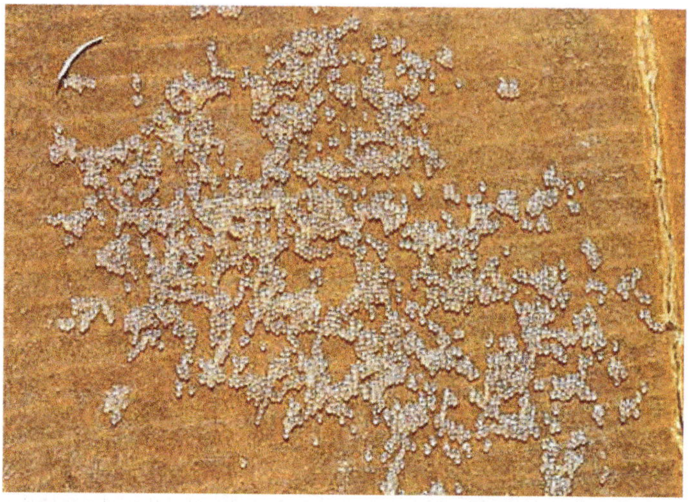

Figure 14.3.2. Fathead Minnow eggs deposited on a fibrous substrate. (Photo courtesy of Barry W. Smith, American Sport Fish).

Table 14.3.1. Growth of Fathead Minnows on a natural diet[1].

Number of days post-hatch	Length of fish, mm
10	5.0
20	10.2
30	14.7
40	19.3
50	24.9
60	29.2
70	34.3
80	38.9
90	44.5
100	49.5
110	54.6
120	58.9

[1]Dobie et al. 1956

This will help ensure that brooders are in very good condition and will actively spawn, creating a pulse of fry at the start of the spawning season. Stopping supplemental feeding of the brooders after that point will reduce the amount of spawning, and in turn reduce the quantity of new fry. This helps reduce the competition for food between the initial pulse of fry and any later fry production. As the season progresses and the water temperature falls below what is appropriate for spawning, supplemental feeding can be resumed. At that point the initial pulse of fry has grown large enough to accept a supplemental feed. This strategy helps to control fry density and allows the production of larger fingerlings. Fry abundance can be further controlled by removing the spawning substrates and brooders after the initial pulse of spawning has occurred.

When the spawning-rearing pond method is followed, ponds are prepared for stocking by eradicating wild fish, predators, competitors, and unwanted plant life. Shortly before or at the time when water temperature warms to 16 to 18°C, the spawning-rearing pond is filled and fertilized. If weeds are likely to be a problem, a pre-flooding treatment with an appropriate herbicide is good insurance to prevent weed development.

Pond fertilization should begin about 2 weeks before stocking brood fish. A mixture of inorganic fertilizers are used to provide 9.1 kg N, 9.1 kg P_2O_5, and 2.25 kg K_2O/ha/application. If a pre-flooding herbicide is used, an organic meal such as cottonseed meal should also be included in the first 2 or 3 fertilizer applications at a rate of 112 kg/ha. Fertilizer should be applied at 7 to 14-day intervals until plankton turbidity develops to the point where light penetration is less than 45 cm. Later applications are given when turbidity clears to more than 45 cm.

Supplemental feeding can begin when water temperature cools to the point where spawning is suppressed. A commercial feed can be used or a feed mixed according to the formula given in Chapter 17.9. A protein level of 30 to 35% should be adequate. An initial feeding rate of 10 kg/ha/day is given, divided into morning and afternoon portions. Feed should be given 6 days per week. The amount can be increased to 15 kg/ha after 2 weeks with later increases at biweekly intervals until the daily amount reaches 35 to 40 kg/ha, or the fish reach marketable size. When water temperatures drop to 10°C, feeding frequency should be reduced to one time per day and the quantity of feed offered reduced by 50 to 75%. Feeding can be suspended on days when sub-freezing temperatures occurs.

Use of the spawning-rearing pond method, that includes fertilization and cool season supplemental feeding, can be expected to support a gain of 5 to 10 kg/ha/day over a cool season feeding period of 110 to 190 days. A yield of 1,360 kg/ha after a 186 day feeding period was obtained by Prather et al. 1953.

Fry Transfer Method.—The foregoing method is applicable to ponds as small as 0.1 ha and as large as 10 to 15 ha. More control is gained by propagating Fatheads with the fry transfer method. Here the spawning pond should be suitable for seining or trapping. Preparation is the same as for the spawning-rearing pond but the stocking is higher, up to 62,500 brood fish/ha. The broodstock should include 40% immature and 60% mature fish, as mortality during spawning can be offset by the immature fish developing into spawners. Boards, concrete building blocks, tiles, natural stones or other course-textured material is placed in the shallow water around the perimeter of the spawning pond to provide nesting sites. Sections of polyethylene irrigation tubing (poly pipe) or wooden frames covered with shade cloth can be anchored in the shallow water and also be used. As a guide, 5 m of 2.5 × 10-cm board is needed for each 100 males. The spawning substrates should be located so that some areas of shoreline are left open to allow harvesting of fry by seining.

A sex ratio for intensive culture of 1 male: 5 females brooders is appropriate. Individual fish can be sexed by secondary sex characteristics described earlier. When large numbers are to be stocked, sexing is best done by grading. After the fish have been graded into size categories, several samples are counted, the male-female ratio determined and weighed for each grade category. Based on the sex ration of fish in each graded size category and their weight, an approximation of the desired number of males and females can be obtained by combining suitable quantities by weights of graded lots.

In addition to the higher broodstock stocking rate used with fry transfer, the brooders in the spawning pond should be fed. If a stocking rate of 37,500 to 62,000 fish/ha is used, the standing crop of brooders will be from 113 to 226 kg of brood fish/ha if brooders average 62 mm TL at stocking. A feeding rate of 2 or 3% of the body weight of the brood fish per day is desirable. If the fish are to be fed 6 days per week, the total ration for 7 days should be divided among the 6 days.

For example, at the lighter stocking rate and the lower percentage body weight rate, the daily ration (6 days per week) per hectare would be as follows:

(113 kg brooders x 2% feed rate x 7 days of feeding)/6 = 2.6 kg/day, 6 days/week.

If the higher stocking rate and the higher percent of body weight is fed, the daily ration (6 days per week) per hectare would be:

(226 kg brooders x 3% feed rate x 7 days of feeding)/6= 7.9 kg/day, 6 days/week.

After 2 or 3 weeks, the feeding rate can be increased so that young fish will grow rapidly. If a high density of fry is present, a feeding rate of 10 to 15 kg/ha/day can be given. After 6 weeks this rate can be doubled. Beyond this point, a daily ration of 20 to 30 kg/ha/day is given. The amount of fish present and the growth rate desired should dictate the amount of feed to supply.

For each spawning pond, 2 or 3 ponds of similar size to the spawning pond will be needed as rearing ponds. Preparation of the rearing ponds can begin when fry from the first wave of spawning are about 10 to 15 mm TL. The procedure outlined in Chapter 7 is followed to prepare the rearing pond for stocking. Transfer of fry to the rearing ponds can begin when the fry are 25 mm. It is desirable to use fry of uniform size if possible. Fry can be captured by seining along the edge of the pond, or by using lift nets or traps. Numbers can be estimated by visual comparison or the number-volume relationship. Care should be exercised in handling to avoid excessive mortality caused by stress. Collecting only in the cool hours of the day, moving the fry in a cushion of water, and avoiding over-crowding in containers for transport are precautions that can be helpful in reducing mortality.

A stocking rate in the rearing pond of 250,000 fry/ha is used in less productive ponds or when rapid growth is desired. The rate can be increased up to 750,000/ha when more time for growth is available and where higher feeding rates can be used. The maximum feeding rate should not exceed 34 kg/ha/day for more than a few days unless emergency aeration is provided. Under optimum growth conditions, the fish should grow from 25 to 50 mm in 50 to 60 days.

Fry are seined from the spawning pond as intensively as possible. As remnants of the first spawns accumulate in the spawning pond, it may be desirable to temporarily lower the water level of the pond to facilitate harvesting, returning the broodstock to the same pond or to another one for further spawning. Prior to draining a spawning pond consideration should be given to if the fish yield from the complete harvest will be greater than the loss of eggs and recently hatched fry plus the plankton present. Using the fry transfer method, Flickinger (1971) was able to harvest fry from small ponds at a rate of up to 7.5 million/ha.

Egg Transfer Method.—An egg transfer method for Fathead Minnows, based on a series of experimental trials, was described by Stone (2014). There, brooders are stocked at 247,000 to 371,000 fish/ha (560 to 780 kg/ha), which is about 5–6 times the stocking rate used with a spawning-rearing ponds with juvenile transfer approach. Brooders are fed daily a commercial 28–32% protein catfish feed during the spawning season.

For an egg transfer approach to be effective, the spawning pond must be free of other materials such as rocks, sticks or vegetation that could be used to deposit eggs on. Suitable spawning substrates should have a rough or textured surface for the eggs to attach to. The material should not degrade rapidly, be easy to clean, and be flexible so eggs can be removed. Used Sections of 15 cm wide conveyer belt, or 7.6 cm wide plastic GEOWEB ® material formed into series of cells have proven to be suitable.

Fathead Minnow eggs may hatch in 3–5 days. To avoid the eggs hatching in the spawning pond, the spawning substrates should be checked every other day when the water temperature is between 21–24°C. Whenever substrates with numerous clusters of eggs are found they should be collected and transported to the hatchery. The eggs are adhesive but can be loosened from the substrate with a 1.5 % sodium sulfite solution. The eggs should be held in the sodium sulfite solution for a few minutes while providing a gentle agitation. The eggs should drop to the bottom of the container and transferred to fresh water. The eggs can be incubated in MacDonald jars at a water flow rate of 2.3 to 2.6 L/min. Fungus on the eggs can be a problem during incubation, but it can be controlled with 15 min treatments with 500 mg/L formalin treatments daily. At hatch, the larvae should flow out of the MacDonald jar into a aerated holding tank having a fine mesh screen over the drain. Once the larvae have developed functional mouthparts, they are stocked into organically fertilized nursery ponds.

For an egg transfer approach to be effective, the spawning pond must be free of other materials such as rocks, sticks, or vegetation that could be used to deposit eggs on. Suitable spawning substrates should have a rough or textured surface for the eggs to attach to. The material should not degrade rapidly, be easy to clean, and be flexible so eggs can be removed. Used Sections of 15 cm wide conveyer belt, or 7.6 cm wide plastic GEOWEB® material formed into series of cells have proven to be suitable.

Fathead Minnow eggs may hatch in 3–5 days. To avoid the eggs hatching in the spawning pond, the spawning substrates should be checked every other day when the water temperature is between 21–24°C. Whenever substrates with numerous clusters of eggs are found they should be collected and transported to the hatchery. The eggs are adhesive but can be loosened from the substrate with a 1.5% sodium sulfite solution. The eggs should be held in the sodium sulfite solution for a few minutes while providing a gentle agitation. The eggs should drop to the bottom of the container and transferred to fresh water. The eggs can be incubated in MacDonald jars at a water flow rate of 2.3 to 2.6 L/min. Fungus on the eggs can be a problem during incubation, but it can be controlled with 15 min treatments with 500 mg/L formalin treatments daily. At hatch, the larvae should flow out of the MacDonald jar into an aerated holding tank having a fine mesh screen over the drain. Once the larvae have developed functional mouthparts, they are stocked into organically fertilized nursery ponds.

Choosing the method of culturing Fathead Minnows depends on the need for the product, the facilities available, availability of skilled manpower, and similar factors. The best return for the facilities comes from the successful application of the fry transfer method. However, some needs can be filled more economically using the spawning-rearing pond method. As of 2014, the egg transfer approach has yet to become widely adopted on a commercial scale. The costs associated with egg collection as well as total production costs need to be reduced for this approach to become more widely used (Stone 2014).

A detailed review of various aspects of fathead minnow life history and their commercial-scale production is given by Stone et al. 2019.

14.3.1.3 Harvesting

As soon as the fastest growing size group of Fatheads reaches a usable size they can be removed by partial harvesting. Fathead production is usually measured by weight, and whole-sale minnow sales may be measured by volume. Table 14.3.2 from Dobie et al. (1956) gives the expected length, weight, and volume relationships.

This species cannot be handled in large quantities in hot weather. When water temperature cools below 25°C, however, harvesting can safely begin. Seining and trapping are employed for partial harvesting. A net with a mesh size that will allow small fish to return to the open pond is one approach to sorting unusable sizes. A grading box with a bar spacing of about 5.0 mm will permit Fatheads smaller than 38 mm TL to pass while holding marketable fish. The market-size fish can be divided into small and large categories using a grader with 6.0 mm bar spacing. The 6.0 mm spacing will permit fish smaller than 50 mm to pass through the grader while holding the larger fish (50 to 85 mm).

It is well to dispose of brood fish as soon as possible after spawning is completed, as male mortality begins even before spawning stops. This is partly due to the stresses associated with preparing the spawning site by the male, caring for successive clutches of eggs, and defending the site against competitive males. Individual fish have been known to fight for extended periods attempting to guard a site they have selected.

Trapping with a wire trap baited with a ball of moistened fish feed is effective for partial harvesting. A lift net operated in the corner of a pond also can be used, attracting the fish over the net with feed. Fathead Minnows are also attracted to a stream of inflowing water. A glass-V trap works well when placed in a flow of water as a way to harvest Fathead Minnows.

Seining usually will harvest more fish than trapping but requires more labor and stresses the fish more. Seine selection is a consideration. Selection of a mesh size that will minimize fish being gilled by the net is desirable. A woven mesh net with a spacing of 4.8 mm will retain

Table 14.3.2. Fathead Minnow length and the number expected per kg or liter[1].

Length of fish (mm)	Number of fish /kg	/L
25.4	5,773	5,475
38.1	1,632	1,558
50.8	662	632
63.5	331	316
76.2	185	177
88.9	115	109
101.6	75	72

[1]Table modified from Dobie et al. (1956) to metric units.

some fish too small for sale, while the next mesh size, 6.35 mm, will permit fish as large as 50 mm to pass through. The 4.8 mm spacing or possibly one of 3.13 mm used with a grading box is the best combination. Seining a filled pond works best when feed is used to concentrate the fish. To harvest a pond completely the pond must be drained, but beforehand, as much of the crop as possible should be removed by seining or trapping. Ideally the pond should have a catch basin where fish can be concentrated while being held in good quality water. Seining is not a recommended harvest method for Fathead Minnows in hot weather.

Fathead Minnows are best moved in water, crowding the fish with a net, and using a bucket scoop up both water and fish. This approach instead of lifting them out using a hand dip net reduces handling stress. Fish harvested one day should rest until the following day before being graded. Holding without food for another day, preferably in cool water, is desirable before transporting.

14.3.1.4 Distribution

Fatheads are about average in regard to their ease of shipping. A loading rate of 80 to 120 g/L is suggested for trips of 8 to 12 hours duration in conventional truck-mounted fish hauling tanks. Use of salt as is done with some other species such as Golden Shiner minnows, Threadfin, Shad, and Striped Bass can help to increase the loading rate or to keep the fish in better condition.

14.3.1.5 Fathead Minnows as laboratory animals

Fathead Minnows are widely used as laboratory animals, particularly in toxicity studies. These types of studies require fish produced under very controlled conditions. A production system described by Denny (1988) is based on using glass aquaria as the culture units. The goal is to provide a continuous supply of embryos and older fish of known ages that are produced under known conditions. Specific ages include <24 h old embryos, 0–24 h old larvae, and 30 d old juveniles. Gordon et al. (2014) describes an improved laboratory production system designed to provide a minimum of 1,500 eggs/d. Stone et al. (2019) provides a good summary of the laboratory conditions and practices for laboratory production of Fathead Minnows. The most recent edition of Standard Methods for the Examination of Water and Wastewater should be consulted as to current standards for production of Fathead Minnows for use in toxicity studies.

14.3.2 Goldfish, *Carassius auratus* (Linnaeus)

14.3.2.1 Introduction

The Goldfish is a highly domesticated species cultured commercially as fish bait and for the pet market. Goldfish have been selected over 10 centuries for color, body shape, and fin patterns resulting in approximately 300 different ornamental varieties. Large Goldfish are also produced as a forage fish for brood fish of some predatory species produced on a hatchery. They have a longer life span than most of the cyprinids grown for fish bait, with fish older than 3 years not uncommon.

Sexual maturity should be reached at 1 year of age if fish have grown to 30 to 50 g Goldfish exhibit sexual dimorphism during the breeding season. Male fish develop pearl organs on the opercula, head, and pectoral fins. Female Goldfish tend to be larger than males as they grow older, and show more expansion in the abdominal and anal regions. When temperatures are favorable for spawning, milt can be stripped from mature males.

Goldfish scatter their eggs over weeds and other substrates located in shallow water around the margin of a pond. The eggs become adhesive within 2 minutes of being released into the water. No parental care is given to eggs or young. In fact, when adult fish are held at high densities they may eat their eggs or newly hatched fry. Goldfish eggs are 0.9 and 1.6 mm in diameter and number about 300/mL. The females are intermittent spawners, producing a clutch of eggs every 30 to 45 days as long as temperature and other conditions are appropriate. Data on number of eggs spawned/g female for Goldfish is scarce, but an average size female of 150 g would be expected to produce 10,000 eggs or more for the first spawn. The number of eggs in subsequent spawns will be considerably lower.

A water temperature of 16°C is generally considered to be warm enough for Goldfish to spawn. As long as the water temperature remains ≥16°C the fish will continue to spawn for at least 7 or 8 months. At temperatures of 16 to 18°C, incubation takes 6 to 7 days, and at 20°C, 5 days. At a mid-summer temperature of 29°C, incubation time may be 46 to 54 hours.

The fry at hatching can range in length from 3.5 to 5.0 mm TL. During yolk sac absorption they adhere to vegetation or other fixed objects by means of an adhesive gland on the head. In the spring, 3 or 4 days may be spent in this stage. Fry begin to feed at about 5.0 to 7.5 mm TL with suitable first foods being rotifers, copepod nauplii and other small zooplankton. If larvae are to be held in the hatchery a feeding protocol might begin with rotifers for the first 7 days, zooplankton and formulated feed for the next 3 days, and a formulated feed for the remainder of time in the hatchery. There does not appear to be a well-defined transition to adult food habits. Goldfish are omnivorous and readily accept formulated feed in later life stages. Cannibalism can be an issue. Advanced Goldfish fry will eat eggs as soon as they can swallow them. Small fingerlings will consume eggs and fry when accessible, and adults will consume eggs off of spawning mats.

14.3.2.2 Broodstock and Spawning

Because of the wide range of genetic characteristics that Goldfish can display, their genetic history and planned use should be considered in broodstock selection. Their disease history should also be considered. The goldfish herpesvirus (GHV), also known as herpes-viral hematopoietic necrosis virus or HVHN is common in many Goldfish populations around the world. Common signs of the disease include pale gills, lethargy, and anorexia. Outbreaks causing significant mortalities are rare, but in ponds 60% mortalities rates have been reported (Department of Fisheries, Government of Western Australia 2005).

Goldfish can also be a carrier for Koi herpesvirus, CyHV-3, resulting in serous mortalities of Common Carp or Koi carp when held with carrier Goldfish. Purchase of 1-year old sexually mature Goldfish fingerlings 12 to 15 cm or larger from a grower who has stock free from troublesome diseases is probably the quickest and best way to obtain a broodstock. If more time is available, young-of-the-year fingerlings 25 to 50 mm in size can be stocked into a rearing pond at a low density and then grown to spawning size.

Where possible, obtaining eggs from a nearby producer who has good stock is recommended. The eggs can be treated with iodine to eliminate organisms transmitted from the parent fish and then hatched in a clean environment. The resultant fry are then reared according to the procedure outlined below.

The most productive sex ratio for the broodstock will depend upon the spawning system. For the spawning-rearing pond method, a ratio of 1:1 is satisfactory. When using a spawning pond with egg transfer to a hatching/nursery pond, a ratio of 1 male: 3 females in the spawning pond is recommended. When tank spawning or stripping is used, 2 or more males:1 female is desirable.

It may be necessary to replace 30 to 50% of the Goldfish broodstock each year depending upon the spawning system followed. Major sources of loss include bird depredation and diseases. Goldfish are more susceptible to anchor worms (*Lernaea*) than Common Carp and brooders may be so infested that they are not in suitable condition for spawning. Between spawning seasons, the broodstock should be held in a broodstock holding pond at a rate not to exceed 1,700 kg/ha. Preferably, sexes are held in separate ponds to minimize wild spawning during late winter and early spring when the water temperature can rise above 15.5°C for 5 to 10 days. Replacement broodstock can be selected in the spring from large overwintered fingerlings that are approaching sexual maturity.

Good egg development is dependent on sufficient food availability for the brooders. If depending on just natural foods, well-fertilized ponds should be stocked with no more than 675 kg of goldfish/ha. Brooders can be stocked at 1,000–1,600 kg/ha when a formulated feed is given. A specific feed formulation for brood Goldfish has not been developed but a 36% protein ration has given satisfactory results under production conditions. A pelleted feed of a size appropriate for the size of broodstock being fed should be used. When the water temperature is 15–18°C the brood fish should be fed once daily, 5 to 7 days per week at 1.5 to 2% body weight/d. Feeding three times/week is adequate at water temperatures of 8–15°C. Feeding at <8°C should be limited to those days when the water temperature is near 8°C giving a maintenance level of 0.5%/body weight/d.

Goldfish are hardy and able to withstand handling stress and many common fish diseases. Sound fish culture and precautions against parasites such as anchor worms, Lernaea cyprinacea, and fish lice, *Argulus* sp., are the keys to successful Goldfish production. Losses to low oxygen levels are less common in goldfish culture as this species is tolerant of dissolved oxygen concentrations that can cause mortality of other cyprinids. However, this hardiness should not encourage the grower to overcrowd rearing ponds or become careless with feeding rates or other management practices.

Adult Goldfish are best harvested by draining the brood holding pond. Although seining is successful in removing up to 50 to 60% from a seinable pond, the fish not captured in the first seining quickly learn to avoid capture. Goldfish are particularly adept to going under the seine and instinctively go down to escape capture.

As with many other species of fishes, handling and movement of broodstock is facilitated by carrying them in wet burlap feed bags. Depending on the fish size, 50 to 100 fish can be carried in a 45 kg capacity bag.

In the temperate climates brood Goldfish form eggs in the fall and winter releasing the largest number of eggs in the first spawn in the spring. Egg laying may cease by mid-summer unless the spawning pond is kept free of small Goldfish and the broodstock remain adequately nourished.

14.3.2.3 Spawning Techniques

Several methods of producing Goldfish are effective, with the selection of a method depending upon the specific situation and the personal preference of the hatchery staff. Methods are: (1) spawning-rearing pond, (2) natural spawning in spawning ponds with egg transfer, (3) natural spawning in tanks or hapas with egg transfer, (4) induced ovulation with natural spawning in tanks or hapas with egg transfer, and (5) fry transfer after induced ovulation, stripping and jar incubation.

Spawning-Rearing Method.—The simplest spawning method is the spawning-rearing pond method. A suitable pond is stocked in the early spring at a rate of 740 fish/ha with equal numbers of male and female Goldfish weighing about 125 g each. The pond is well fertilized for a period of 4 to 6 weeks. When the temperature is in the right range spawning mats are added. Another approach is to allow vegetation to grow along the shoreline of the pond during the winter and flood the vegetation to provide substrate for spawning once the temperature is right. Spawning success is verified by seining around the perimeter of the pond with a fine-mesh seine. If the sampling shows large numbers of uniform size fry, supplemental feed is given to grow fry to 30–50 mm over a period of 60 to 80 days. The pond is then drained and the crop harvested. If the local climate permits another 60 to 90 days of growing season, the pond can be refilled and restocked with brood fish for a second crop. The second crop will be considerably less unless the gravid brood fish from several ponds are combined for the late spawn.

A variation of the above method is that once the young Goldfish reach a size of 25–30 mm, the spawning-rearing pond can be lowered to harvest the adults and/or small fish. The young are transferred into rearing ponds at a density of 350,000 to 500,000/ha for 40–50 days to grow to a length of 50 to 60 mm. This early harvest of broodstock allows them to be spawned again that season, and restocking the young fish at known densities will lead to a more predicable harvest and more efficient use of pond space.

Egg Transfer Method.—For natural spawning with egg transfer, the bottom of an earthen pond is thoroughly cleaned of grass and weeds that otherwise might provide a substrate for spawning. The shoreline should be bare soil with no vegetation for approximately 10 cm above the water level and the water level controlled so it does not rise into that zone. Spawning mats are placed around the margin of the pond in shallow water, placing up to 300 mats/ha of pond at a time. A convenient size spawning mat is a 45 x 90 cm wire frame, filled with Spanish moss, rubber-coated plant fiber, or similar fibrous material (Figure 14.3.3). The mats should set at a depth where the fish can swim up onto the mat to spawn. The depth depends on the size of the brooder but the mat should not be placed at more than 10 cm beneath the surface.

The spawning pond is stocked at a rate of 7,500 brood fish of average size/ha at a sex ratio of 1 male:3 females. The mats are examined daily and removed when eggs are scattered over most of the mat. The quantity of eggs on the mat is estimated by counting portions of the mat of known area and extrapolating the number of eggs/unit of known area to the total area of the mat as described earlier for Common Carp (Chapter 14.1). Mats with eggs are transferred to a hatching/nursery pond at a stocking rate ranging from 1 to 2.5 million eggs/ha. Preferably, a hatching or nursery pond is completely stocked with eggs spawned during

Figure 14.3.3. Goldfish *Carassius auratus* spawning on a spawning mat made of Spanish moss.

one 24-hour period to ensure uniformity of the fry. When this is not feasible, no more than 7 days difference should exist between the oldest and youngest eggs transferred. When the egg-bearing mats are removed from the spawning pond they are replaced with clean mats. Monitoring of the spawning pond is continued until egg laying stops or an adequate number of fry are produced. At that time, the mats are removed from the spawning pond and the brooders can be harvested.

If spawning and growing conditions are still favorable, the pond can be harvested, allowed to dry, then refilled with clean water and the brood fish restocked for further spawning. Brooders will need to be fed to maintain their condition and produce successive clutches of eggs.

The approach as described in the Golden Shiner section of mat transfer to the hatchery for hatching with fry transfer back to nursery ponds is also appropriate for Goldfish and Common Carp, stocking 1,000,000 swimup fry/ha.

Fingerling Production.—When using the mat transfer approach, the egg-bearing mats stocked into rearing ponds should be kept under observation to determine hatching success of the eggs. Fry typically remain on the spawning mats several days after hatching. When they are free-swimming, the mats can be removed, dried, and used again. The fry should be given supplemental feed within 10 days following hatching. A meal mixture containing 30 to 35% protein is suitable (see Chapter 17.9).

The feeding rate for a pond stocked with 1 to 2.5 million eggs with a 60% hatch should start at about 10 kg/ha/day. The daily ration should be divided into a morning and afternoon feeding and distributed along the windward side of the pond. The amount of feed supplied can be doubled every 10 to 14 days until the fry grow to about 5 cm. This size should be reached in 30 to 50 days. Their growth rate should be determined by seining at least bi-weekly. Unusually rapid growth is an indication of under-stocking or poor survival. Slow growth is a symptom of over-stocking, under-feeding, or low water temperature. Growth rates can be improved when temperature is favorable by reducing fry density or increasing the feeding rate.

At a suitable stocking density, a standing crop of 1,500 to 2,000 kg/ha should be present by the time the fish reach a size of 5 cm. With a 60% hatch and 85 to 90% survival from fry to fingerling size, 1 million eggs should yield about a half million fingerlings in 45 days. Allowing 10 days for harvesting and pond preparation, and another 10 days for egg collection and hatching, this would permit a crop of 5 cm fingerlings to be grown every 65 to 70 days as long as the brooders are still spawning.

Combination Method.—A more intensive and controlled method of obtaining eggs is to spawn brood fish in tanks or hapas. Where facilities are available, the recommended procedure is to hold the brooder separate by sex until you want them to spawn. When spawns are needed well developed males and females, averaging about 125 g each, are placed in the spawning container at a density of about 30 fish/m^2 at a 3 male:1 female sex ratio. Spawning mats or kakabans line the sides and bottom of the container as egg collectors. When gravid brood fish are placed in fresh water, spawning will usually take place during the early morning hours of the day after stocking. After 24 hours in the spawning container, the mats are removed. The spent females are removed to a holding pond and a new group of gravid females are added to the container if additional eggs are needed.

Males can be used 2 or 3 times if they are held in a temporary holding facility and given feed for 3 to 5 days before being used again. If the males are not fed for extended periods of time they may eat any eggs released in the spawning container.

Injecting the females will help to synchronize the timing of spawning and the number of females that spawn. Gravid females are given a single intraperitoneal injection of hCG at 660 IU/kg. Lesser quality females may need one or two preparatory injections 1/4 to 1/2 of the 660 IU/kg resolving dose. Spawning takes place 12 to 15 hours after injection at a water temperature of 18 to 20°C. Other hormones given as a single injection to successfully induce a natural spawn of Goldfish include: 2 mg/kg common carp pituitary, and LH-RHa at 0.1 μg/g body weight along with pimozide at 10 μg/g body weight.

In either case, after natural or hormone-induced spawns, the mats are examined and an estimate made as to the quantity of eggs collected. Egg-bearing mats can be transferred to nursery ponds or be held in the hatchery until after the eggs hatch and then stocking the larvae into nursery ponds once they are ready to feed.

Female Goldfish can be readily hand stripped after ovulation. In the spring it is possible to find brooders that are running eggs and such fish can be hand-stripped without hormone treatment. Another approach is to induce ovulation with an injection of hCG. Upon ovulation, the fish are hand stripped and the eggs fertilized in the conventional way. Keep in mind that, like Common Carp, the eggs become adhesive shortly after coming in contact with water. The eggs should be scattered over a spawning mat using a nylon bristle paintbrush before the eggs become adhesive. Holding the spawning mat at 10 to 15 cm under the water surface will help to facilitate that the eggs are evenly distributed over the mat. Care must be taken to avoid clumps or multi-layers of eggs as some eggs may suffocate. Dead eggs allow fungus to develop and spread to healthy eggs. Manual fertilization and scattering of eggs evenly over the mat so that no egg is in contact with another one requires a considerable amount of skill. The worker has about two minutes to activate the sperm, obtain fertilization, and spread the fertilized eggs evenly over the mat where there they will adhere.

Once the eggs have attached to the mat, the mat should be carefully placed in an aquarium, hapa, or tank and the embryos allowed to develop. After hatching when the fry are swimming up in the water column they can be transferred to nursery ponds. Fry of the more delicate varieties of Goldfish may need to be reared indoors and fed rotifers for several days then weaned onto artemia nauplii. Once they are actively accepting artemia they can be weaned to a powdered formulated feed.

When manually stripping and fertilizing the eggs, an alternative is to treat the eggs to remove their adhesiveness as described in the Chapter 17.4. Once the adhesiveness has been controlled, the eggs can be can be incubated in conventional incubation systems using hatching jars, cones or vertical incubators as done with Common Carp eggs.

The various options possible in goldfish spawning should enable the culturist to select a pocedure that is most applicable in a particular situation. For mass production at low costs, the egg transfer from spawning ponds to nursery-rearing ponds is probably the method of choice. For more extensive conditions, a spawning-rearing approach may be adequate as long as close control over the size and number of fish to be produced is not needed. Pond spawning with mat transfer to the hatchery for incubation can improve hatch rate and give a more predicable growth and survival in nursery ponds. Where facilities, skilled manpower and time are available, induced ovulation of brooders in a tank or hapa with the fish naturally spawning has the most advantages, particularly with high-value varieties of Goldfish.

14.3.2.3 Harvesting and Distribution

Goldfish fingerlings are hardy and withstand harvesting stress even during the hot months of the year. Partial harvests can be made by seining a full or partially drained pond especially when the fish can be concentrated in one area by feeding. To complete the operation, however, the pond should be drained. Grading may be necessary and can be done either with a wire cloth box grader or the spaced bar or grill type grader.

Loading rates of 5 cm size Goldfish fingerlings using conventional tank truck distribution equipment can range from 120 to 250 g/L of water. The exact rate depends upon the duration of the haul along with the condition of the fish, water quality, temperature control efficiency of the unit, and other factors. Goldfish fingerlings adapt well to high stocking rates and can be held for weeks in tanks or troughs where the rate of water exchange keeps DO up and the level of metabolites low. They accept feed under these conditions so that weight loss during holding is minimized. Food should be withheld for 24 to 48 hours prior to shipment to condition the fish to withstand handling and shipping stress.

14.3.3 Golden Shiner, *Notemigonus crysoleucas*, Mitchell

14.3.3.1 Introduction

The Golden Shiner is probably the most important bait minnow produced in the USA. It is native to most of the eastern USA and portions of Canada, and has been introduced in many other regions as bait. The young and juveniles are silver in color while the adults take on a more golden color (Figure 14.3.4). Males will show an orangish coloration on parts of the anal, caudal, dorsal and pelvic fins at the start of the breeding season. Mature females will be deeper bodied than the males. Females also grow faster and to a larger size than males. Adults may reach a maximum size of 25 cm, but are sexually mature at a size of 6.5 cm at one year of age in warm climates. In the northern part of its range, Golden Shiner may live up to 8 years.

Figure 14.3.4. Adult Golden Shiner *Notemigonus crysoleucas*. (Photo courtesy of Brian Zimmerman, Zimmermans Fish).

The Golden Shiner is an intermittent spawner when the water temperature is between 20 and 28°C. Spawning takes place over filamentous algae, submerged rooted vegetation, or other material that the adhesive eggs can stick to. No parental care given to the eggs once they are scattered over the substrate. Fecundity is about average for cyprinids, ranging from 88,000 to 110,000 eggs/kg of mature females. The eggs are approximately 1 mm in diameter. Eggs take 6–7 days to hatch at 20°C with fry developing to swim-up stage by the tenth day after spawning. At a temperature of 24 to 27°C, hatching should take place in about 4 days. Fry length at hatch may range from 3.5 to 4.5 mm and are 4.5 to 5.5 mm when ready for first food.

Golden Shiner are omnivorous, and feed on the variety of food organisms produced in a fertilized pond, and also readily accept slow-sink or floating formulated feeds. Suitable formulas for minnow feeds are given in the Feeds and Fertilizer section, Chapter 17.9.

14.3.3.2 Broodstock

Care should be taken in obtaining disease-free brood shiners as there are a number of diseases that affect their health and their marketing. Golden shiner can have a chronic problem with the microsporian parasite *Ovipleistophora ovariae*. This parasite damages the ovaries, reduces fecundity resulting in sterility in older females. Once a hatchery is infested with *O. ovariae*, the parasite is difficult to eliminate but can be managed by using only 1-year-old brooders as ovarian damage is less in young brooders. The exotic Asian tapeworm *Bothriocephalus* sp. has infected several species of cyprinid species in the USA including the Golden Shiner. Its presence may legally prevent the transport of infected fish into non-infected regions of the USA. Golden Shiner virus (GSV) disease is a cause of chronic mortality in hot weather with fish larger than 50 to 60 mm being more susceptible. Higher rates of mortality may occur when infected fish are handled. On occasions, common external parasites such as *Gyrodactylus* sp. and *Lernaea cyprinacea*, as well as bacteria such as *Aeromonas hydrophila* and *Flavobacterium columnare* can be a problem.

Any time new fish are being introduced to a hatchery, the disease history from where the fish were obtained should be known. If the fish species is a potential carrier of a disease not found at your hatchery then extra effort is needed to determine if the fish being introduced are carriers. This should include laboratory examination of the fish for Golden Shiner diseases not currently found at the hatchery. In addition, such fish should be held in isolation for at least 30 days under conditions favorable for such diseases to occur to help ensure that the fish are not potential carriers.

Although brood males and females can be separated based on appearance or body size, this is not practical because of the large number of brooders needed for pond spawning. Such handling also stresses the brooders and can lead to disease outbreaks.

14.3.3.3 Spawning Techniques

Spawning-Rearing Pond Method.—The spawning-rearing pond method works well where low or moderate yields are acceptable. Some of the disadvantages of this approach include: the adult fish expose the young to the diseases they carry, the adults eat eggs and small fry to some extent, and the quantity of fish being produced is unknown. The main advantages are its simplicity and minimal labor requirements.

If this extensive method of propagation is used, preparation of the pond begins about two weeks before water temperature is expected to reach 20°C. As with producing other baitfish, the pond should be free of all competition and predators of both adult and young fish. Generally, fertilization of the spawning-rearing pond using inorganic source materials is the most economical way to provide a food supply. If the food chain is slow in developing, including some organic nitrogen during the first 2 or 3 applications will be helpful. A phytoplankton-based food chain produced by fertilization alone will be adequate to produce up to 680 kg of shiners/ha.

Brood stocking rates with the spawning-rearing method are 2,000–2,500 adults/ha. Four or more spawns may be produced in the same pond over a 4 to 5 month period. If supplemental feeding is delayed until the water temperature is too cool for spawning it will help to ensure that the feed will contribute more to the growth of young fish rather than more spawning by the brooders.

The goal of the spawning-rearing pond is to grow the maximum number of usable size (7.5 cm TL or larger) Golden Shiner in the growing time available. If a yield of 680 kg/ha is assumed as a standing crop, and the average size of the fish is 7.5 cm, a yield of 175,000 fish/ha can be anticipated.

Egg Transfer Method.—Egg transfer is a more intensive method for propagating Golden Shiner. Brooders are stocked into the spawning pond by weight at a rate of 450 to 900 kg of brooder/ha. The spawning pond should be clean and free of debris and vegetation to a point above the water level. As the water temperature in the pond rises into acceptable range for spawning the brooders will start swimming along the shoreline. At that time spawning mats (same design as used with Goldfish) should be added to the pond, initially adding a limited number of mats to see how intensively the fish spawn. Enough mats need to be added so that in a 24 hour period the mat is covered in eggs (about 100,000–150,000 eggs/mat). If only a few eggs are seen on the mats, the mats should be removed and added back once the temperature has increased a few degrees or when water quality conditions have improved. If the mats are heavily coated with eggs to the point where eggs are stacked on top of each other then, more mats need to be added to the pond. The objective is to get a uniform distribution of eggs on the mat with enough spacing between eggs to keep fungus from spreading from dead eggs on to live healthy eggs.

The mats should be placed close to the shoreline at a depth of 3 to 6 cm so the brood Golden Shiner can swim up on top of the mat to spawn. Frames may be needed to hold the mats at the right depth. They should be placed on the sides of the pond where there is a minimum of wave action so the eggs do not get covered with sediment.

Depending on water temperature, egg-bearing mats are transferred to nursery-rearing ponds at least every 2 days, preferably once a day. To assure uniformity of the fry in a given pond, it is desirable to completely stock a rearing pond in less than a 7-day period. As soon as newly hatched fry swim off the mats, the mats can be removed, cleaned and returned to the spawning pond as needed to collect later spawns of eggs.

An estimate of the quantity of eggs being transferred can be made as done with Goldfish and Common Carp by counting the number of eggs/cm^2 of mat in several typical locations and multiplying the mean number of eggs/cm^2 times the total mat area. A hatch of 50 to 60% is considered about average. Of the number hatched, about 75% can be expected to survive to 7.5 cm.

As a guide to the number of eggs to stock, estimates should be made based on past performances under similar conditions. For example, past records might have shown that yields from rearing ponds dependent on in-pond production of natural food averaged 680 kg/ha with the harvested fish averaging 4.0 g. Based on that information and that similar production techniques are used, the anticipated number of fish in the next harvest can be calculated:

Anticipated number of fish at harvest = yield/average weight
= 680 kg/ha / 4.0 g
= 680,000 g/ha / 4.0 g
= 170,000 fish harvested

To produce and harvest 170,000 fish/ha, past records should be reviewed and back calculations made to determine how many eggs/ha will be needed and how many mats containing those eggs should be moved to the nursery pond. These estimates must take into consideration the mortality that will occur along the way.

Harvest results	= 170,000 fish with an average wt. of 4.0 g
Fry to harvest survival	= 75%
No. fry needed	= 170,000/75%
	= 226,667 fry
Hatch rate of eggs	= 50%
No. eggs needed	= 226,667/50%
	= 453,333 eggs
Av. egg density/mat	= mat size × no. eggs/cm^2
	= (53 × 76 cm) × 29 eggs/cm^2
	= 116,812 eggs/mat
No. mats needed	= No. eggs needed/Average egg density/mat
	= 453,333 eggs needed/116,812 eggs/mat
	= 4 ± mats (3.9 mats)

For higher returns from rearing ponds, supplemental feeding is necessary. Golden Shiner fry will consume finely ground dry feed soon after they have absorbed their yolk sac. For maximum consumption and minimal fouling of water, the feed should float until eaten. A starting rate of feeding for shiners is 2.25 kg/ha/day divided into 2 feedings. Feed should be distributed along the leeward side of the pond so that the film of feed drifts across the pond surface making it readily available to the feeding fry. The quantity of feed should be increased 50% each week until a rate of 25 to 30 kg/ha/day is reached. Supplemental feeding should enable a yield of about 1,000 kg/ha in a 70-day growing period.

Some of the disadvantages of allowing the eggs to hatch on the mats in a nursery pond includes the risk of fungal development during egg incubation, low hatching due to eggs becoming covered in sediment or exposed to low oxygen, having an unknown hatching rate in the pond, and a resultant unknown quantity of fry in the pond. An alternative is to collect the mats from the spawning ponds, incubate the eggs in clean water in the hatchery, then transfer the fry from the hatchery to outdoor nursery ponds.

Bringing mats with eggs for incubation in flowing water tanks in the hatchery gives an opportunity to treat the eggs with formalin to control fungus as well as providing a quality environment for egg incubation. After the fry hatch and swim-up they can be collected, counted, transported in oxygen filled plastic bags of water, and stocked into nursery ponds. Mats with eggs are collected from the spawning pond taking care not to crush the eggs or allow the eggs to dry out on the way to the hatchery, or temperature shock the eggs when loaded into the incubation tank.

Mats are added at approximately1 mat/50 L of tank volume, to tanks with volumes of 0.5 to 2 m^3. Mats should be examined before loading into tanks, trying to put mats with similar age eggs into the same tank. Occasionally some mats will arrive at the hatchery where the egg density is too high and the potential for egg suffocation and fungal problems become an issue. In such cases it is possible to rinse the mat in a 1.5% sodium sulfite solution for 1–2 min to reduce the adhesiveness of the eggs to the mat, then wash and shake the eggs off the mat for incubation in MacDonald jars or other similar incubators.

The improved hatching rate, the ease of fry handling and improved survival, while reducing on-pond labor, and overall efficiency of the technique has contributed to the majority of the commercial industry in Arkansas, USA adopting the egg transfer to the hatchery approach. For more detail on this approach along with an enterprise budget see Stone et al. 2008. This technique is also applicable to mat spawning of Common Carp and Goldfish. It is depending on the type of spawning substrate provided, it is also appropriate for fathead minnow production (Stone 2014).

Fry Transfer Method.—Fry transfer can be part of spawning–rearing or egg transfer to nursery ponds approaches, and is the justification for egg incubation in the hatchery. In spawning-rearing ponds the initial waves of fry are removed using a lift net or a small mesh seine once they are 2.0–2.5 cm. Harvesting this early wave of fry and transferring them to nursery ponds at known densities will allow the production of more uniform size fish suitable for specific markets. By reducing the standing crop of fry from the early spawns, it makes room for fry from later spawns in the spawning-rearing pond to have improved growth and survival. With egg transfer and partial harvests, mats are collected from spawning ponds and stocked into nursery ponds at higher egg densities than if the nursery pond was used just for fry production. Once the fry reach 2.0–2.5 cm as much fry as possible are harvested and stocked into other nursery ponds for additional growth. Those that remain may grow fast enough that season to provide a large shiner needed for a winter market.

When mats are incubated in the hatchery, fry are transferred to nursery ponds soon after they swim-up. It is important to carefully acclimate the fry to pond water conditions. This includes not only temperature but also dissolved oxygen and pH. Golden Shiner fry are particularly sensitive to pH changes. Well water often used in a hatchery may have a pH around 7 but the nursery pond pH may be closer to 9 or more. Ideally pond water can be introduced to the fry holding tank overnight and the fry are then transported to the pond in the same water. If not, then at the pond bank pond water needs to be slowly added to the container holding the fry to acclimate the fry to the pH differences.

The fry can be stocked at 2.500,000–7,500,000/ha with partial harvests made in the nursery pond starting after 30 d post-stocking. Nursery ponds are often large, 2–4 ha, and partial harvests will collect only a portion of the fish in the pond. Ponds need to be monitored as to fish growth rates to determine when additional partial harvests are needed and when fish are reaching market size.

The market size preference for Golden Shiner varies seasonally and also regionally. Smaller fish (about 70 mm) are often in more demand in the springtime, mediums (around 95 mm) to large fish (>105 mm) in the summer, and large fish in the winter months. To produce such a size range, several factors need to be controlled, including: stocking density, partial harvests, supplemental feeding, and timing of spawning. Fry are stocked into nursery ponds, often at 7,500,000/ha or more, and raised to juvenile size. After four to twelve weeks, the juveniles are transferred to secondary nursery ponds at varying densities with the intent of having a specific size of fish available over a wide period of time. Supplemental feeding will allow faster growth rates and the maintenance of high standing crops of fish. Some ponds may be stocked late in the spawning season at high densities with the intent of having small fish available in the early spring the next year.

Pond management to provide specific quantities of fish of a specific size on a specific date has often been called more of an art form than a science. Improved management comes with experience and careful record keeping. Often it is better to overstock a rearing pond than to under-stock. Seine samples taken around the margin of the pond within two weeks of stocking will be an indicator of how successful was the pond preparation and fry stocking. Each pond has its own personality and on-farm experience is needed in relating number and size of fry in the seine sample to the yield that might be anticipated at harvest. Sampling is helpful in deciding if the fish density is too high, too low, or about right. Slower or faster growth than normal is also a good indication of rate of survival. Where growth is slow, fish densities may be too high for the quantity of natural food available. Growth rates can be improved by increasing the food supply through supplemental feeding, or reducing the fish density by making partial harvests, or a combination of the two approaches. If the fish density is too low, often as a result of a low hatch rate following egg transfer, too few fish surviving the fry transfer, or water quality issues, it may be more profitable to drain the pond and prepare it again for a new set of fish.

14.3.3.4 Harvesting and Distribution

Golden Shiner are more delicate and easily stressed when handled (harvesting, grading, and transporting) than Fathead Minnows or Goldfish. They are more sensitive to low oxygen as well. Such stress results in reduced osmotic control and scale loss. Ideally the fish would be harvested during the cool months of the year but market demands and production practices require that fish be handled throughout the year. Harvesting is best done by draining the rearing pond, but that can result in too many fish to handle and market at one time. Partial harvesting by seining, trapping or use of a lift net is a key component in Golden Shiner production. Partial harvests allow adjustment of fish densities, provides fish for stocking other culture ponds as well as providing marketable fish.

Market demands often require that weekly harvests be made. Fish are baited into a corner of the pond for harvest using a sinking feed. The quantities of fish needed should be efficiently captured and moved in water to the transport tank in less than 30 min. A little extra care in capturing and moving the fish will payoff in reduced mortality and improved marketing quality of the fish. Transporting and holding harvested fish in water containing 0.5% salt (NaCl) has a beneficial effect in reducing the osmotic gradient between the fish's body fluids and the water the fish is held in. This helps to minimize mortality and allows the fish to better withstand the stress of harvesting, processing, and transportation.

Once at the hatchery building, the fish should be unloaded in water, ideally using a hose or pipe connected to the drain of the transport tank so the fish can flow with the water into the holding tank. The best combination is to fill the transport tank with pond water so the fish are not shocked by a water quality change when they are loaded on the truck. The holding tank in the hatchery that is going to receive the fish should be filled halfway with pond water. Once the fish are in the holding tank, water from the hatchery building water supply should be slowly added to acclimate the fish to the new conditions. Golden Shiners are excitable and will jump from containers or tanks when startled. Water levels should be maintained low enough in the tank to prevent fish from jumping out of the tank, being careful to not excite the fish when removing any aquatic weeds or debris that came into the tank as part of the fish harvest. The fish should be allowed to recover 24 h from harvesting before being handled again. During that period the fish will "harden," that is, the fish will regain control of its osmotic balance and pull their scales back in close to the body. Salt should be added several times a day to the holding tank to maintain a 0.5% salt concentration.

After being held a day, the fish can be graded into different size categories for distribution. Golden Shiner are graded into four basic size groups: 1) sub-market size, those that pass through a bar grader with a 5.16 mm bar spacing; 2) smalls, that pass through a 7.94 mm bar spacing but are held by a 5.16 mm spacing; 3) mediums, that pass through a 10.72 mm bar spacing but held are by a 7.94 mm spacing; and 4) large, that are held by a 10.72 mm spacing. Grading is stressful and the fish should be held an additional day after grading before being transported. That holding period is also an opportunity to apply a prophylactic treatment of formalin or other chemicals to help control diseases before the fish are to be distributed.

Using truck-mounted distribution tanks supplied with oxygen and electric agitation, a loading rate of 120 g/L can be used if the water temperature is below 18°C. As the temperature rises to 25°C, the loading rate should be reduced to about 60 g/L. For transport of small quantities over a period of 10 hours or less, a plastic bag with oxygen can be used. Loading rate should be 120–240 g/L.

14.3.4 References

Denny, J. S. 1987. Guidelines for the culture of Fathead Minnows *Pimephales promelas* for use in toxicity tests. EPA/600/S3-87/001. USEPA, Duluth, Minnesota.

Department of Fisheries, Government of Western Australia. 2005. NAAHTWG. Slide of the Quarter (April–June 2005): Cyprinid herpesvirus 2. Available: http://www.fish.wa.gov.au/Documents/aquatic_animal_health/goldfish_cyprinid_herpesvirus2_apr_jun_2005.pdf. (May 2020).

Dobie, J., O. L. Meehean, S. F. Snieszko, and G. N. Washburn. 1956. Raising bait fishes. U.S. Fish and Wildlife Service Circular 35, Washington, D.C.

Flickinger, S. A. 1971. Pond culture of bait fishes. Colorado State University Cooperative Extension Service Bulletin 478A, Fort Collins, Colorado.

Gordon, D. A., M. E. Smith, M. Wratschko, D. Agard, L. Holden, S. Wilcox, and J. M. Lazorchak. 2014. A new approach for the laboratory culture of the Fathead Minnow, *Pimephales promelas*. Environmental Toxicology and Chemistry 33(1):126–133.

Prather, E. E., J. R. Fielding, M. C. Johnson, and H. S. Swingle. 1953. Production of bait minnows in the Southeast. Circular 112, Alabama Polytechnical Institute, Agricultural Experiment Station, Auburn, Alabama.

Stone, N. 2014. An indoor hatching and intensive rearing method for Fathead Minnows. SRAC Publication No. 1203. Southern Regional Aquaculture Center, Stoneville, Mississippi.

Stone, N., C. R. Engle, and E. Park. 2008. Production enterprise budget for Golden Shiners. SRAC Publication No. 122. Southern Regional Aquaculture Center, Stoneville Mississippi. Available: https://fisheries.tamu.edu/files/2019/01/SRAC-0122.pdf. (May 2020).

Stone, N. M., L. A. Roy, and E. D. Park. 2019. Fathead Minnow reproduction: implications for commercial culture. Journal of the World Aquaculture Society 50(2):267–298.

14.3.5 Recommended Reading

Ankley, G. T., and D. L. Villeneuve. 2006. The Fathead Minnow in aquatic toxicology: Past, present and future. Aquatic Toxicology 78(1):91–102.

Burtle, G. J., L. W. Dorman, and D. L. Gray. 1994. Baitfish production in the United States, Aquaculture Technical Series, U.S. Department of Agriculture. Cooperative State Research Service and Extension Service, Washington, D.C.

Department of Fisheries, Government of Western Australia. 2005. NAAHTWG. Slide of the Quarter (April–June 2005): Cyprinid herpesvirus 2. Available: http://www.fish.wa.gov.au/Documents/aquatic_animal_health/goldfish_cyprinid_herpesvirus2_apr_jun_2005.pdf. (May 2020).

Engle, C. R., N. Stone, and E. Park. 2000. An analysis of production and financial performance of baitfish production. Journal of Applied Aquaculture 10(3):1–16.

Flickinger, S. A. 1973. Investigation of pond spawning methods for Fathead Minnows. Proceedings of the Annual Conference Southeastern Association of Game and Fish Commissioners 26(1972):376–391.

Giudice, J. J., D. L. Gray, and J. M. Martin. 1981. Manual for bait fish culture in the South. EC550, University of Arkansas Cooperative Extension Service and U. S. Fish and Wildlife Service, Little Rock, Arkansas.

Heil, N., editor. 2009. National Wild Fish Health Survey—Laboratory Procedures Manual. 5.0 Edition. U.S. Fish and Wildlife Service, Warm Springs, Georgia. Available: https://www.fws.gov/wildfish-survey/manual/NWFHS_Lab_Manual_5.0_Edition.pdf. (May 2020).

Huner, J. V., and H. K. Dupree. 1984. Production methods for baitfish: Golden Shiners and Fathead Minnows. Third Report to the Fish Farmers. U.S. Fish and Wildlife Service, Washington, D.C.

Lazur, A. M., and F. A. Chapman. 1996. Golden Shiner culture: a reference profile. University of Florida Cooperative Extension Service, Institute of Food and Agricultural Sciences FA-34. University of Floriday, Gainesville.

Lochmann, R., and H. Phillips, 2001. Nutrition and feeding of baitfish. ETB256 University of Arkansas at Pine Bluff School of Agriculture, Fisheries and Human Sciences, Pine Bluff, Arkansas. Available: http://fisheries.tamu.edu/files/2013/09/Nutrition-and-Feeding-of-Baitfish.pdf. (May 2020).

Ludwik, G. L., and N. Stone. 1997. Relation between bar grader spacing and Golden Shiner size. The Progressive Fish-Culturist 59:312–316.

Rice, E. W., R. B. Baird, A. D. Eaton, and L. S. Clesceri, editors. 2015. Standard methods for the examination of water and wastewater, 22nd edition. American Public Health Association, American Water Works Association, Water Environment Federation. Washington, D.C.

Smartt, J. 2001. Goldfish varieties and genetics: handbook for breeders. Blackwell Science Ltd, New York.

Stone, N., E. Park, L. Dorman, and H. Thomforde. 1997. Baitfish culture in Arkansas: Golden Shiners, Goldfish and Fathead Minnows. University of Arkansas, Cooperative Extension Program, Miscellaneous Publication 386, Pine Bluff. Arkansas.

Stone, N., A. Goodwin, R. Lochmann, H. Phillips, and C. Engle. 2005. Baitfish culture. Pages 607–643 *in* A. M. Kelly and J. Silverstein, editors. Aquaculture in the 21st Century. American Fisheries Society, Bethesda, Maryland.

Stone, N. 2014. An indoor hatching and intensive rearing method for Fathead Minnows. SRAC Publication No. 1203. Southern Regional Aquaculture Center, Stoneville, Mississippi.

Stone, N. M., A. M. Kelly, and L. A. Roy. 2016. A fish of weedy waters: Golden Shiner biology and culture. World Aquaculture Society 47(2):152–200.

Stone, N. M., L. A. Roy, and E. D. Park. 2019. Fathead Minnow reproduction: implications for commercial culture. Journal of the World Aquaculture Society 50(2):267–298.

Watson, C. A., J. E. Hill, and D. B. Pouder. 2004. Species profile: koi and Goldfish. SRAC Publication No. 7201. Southern Regional Aquaculture Center, Stoneville, Mississippi. Available: https://fisheries.tamu.edu/aquaculture/species-profiles/. (May 2020).

Chapter 15

Tilapias

15.1 Introduction

Tilapia are one of the most commonly cultured species in the tropics, adaptable to a variety of culture techniques. They are native to Africa and portions of the Middle East, but have been introduced into almost all remaining tropical countries as well as many temperate ones. They are true tropical species and will not survive at temperatures of 8 to 10°C and below. The name tilapia refers to over 100 species divided between the genera of *Oreochromis*, *Sarotherodon*, and *Tilapia*. Much of the earlier literature uses *Tilapia* as the genus name for the three groups now recognized as separate genera, but tilapia remains as the common name for all three.

Substrate spawners, such as *Tilapia rendalli* and *T. zilli* (Figure 15.1) have been cultured for aquatic weed control, or in production systems relying on the addition of macrophytes as a feed. Pond production of *T. rendalli* can be increased feeding a pelleted feed but yields are much lower that obtained with *Oreochromis* species. Few hatchery techniques have been developed specifically for *T. rendalli* or *T. zilli*. They are both prolific species capable of spawning numerous times a year in a static water pond, and if not harvested frequently will result in stunted populations. Stunted *T. zilli* females as small as 7 cm, weighing 12 g, have been found to be sexually mature, similarly males as small as 5 cm weighing <8 g were mature.

Pitted saucer-shaped nests are made by *T. rendalli* while *T. zilli* makes a cluster of 5 to 10 small holes in the pond levee or bottom that can damage the levee (Figure 15.2). Both parents will build the nest and guard the eggs. A female may spawn as frequently as every 26 days releasing on average over 3000 eggs/spawn. The eggs hatch in about 5 days. The male will aggressively defend the nest sites and the fry for several weeks. Most-appropriate hatchery techniques for producing *T. rendalli* or *T. zilli* are not well established. A spawning-rearing technique may be acceptable for the hatchery production of this group, but care must be given to not allow the pond to become overcrowded.

Among the mouth brooding tilapias *Oreochromis aureus*, *O. mossambicus*, and *O. niloticus* are the most commonly cultured tilapias, with *O. niloticus* the predominate species cultured. They spawn naturally in ponds with no manipulation of the environment needed to induce spawning when the temperature is 23°C and above. The males sweep out bowl-shaped nests in the shallow areas of a pond (Figure 15.3). A male and a female spawn in the nest over a period of up to two hours. A batch of eggs is deposited in the nest by the female, fertilized by the male, and then the eggs are picked up by the female and held in her mouth. This is repeated several times until spawning is complete. The female will then incubate the eggs in her mouth (Figure 15.4) until hatching and may shelter the fry in the mouth until through the swim-up stage. The reproductive cycle for a female is about one month, but can be shortened by collecting eggs or fry from the female's mouth.

Figure 15.1. Adult *Tilapia zilli*/*Coptodon zillii* (common name Redbelly Tilapia).

Figure 15.2. *Tilapia zilli* nests dug into the pond bank.

Figure 15.3. Nests prepared by male *Oreochromis niloticus*.

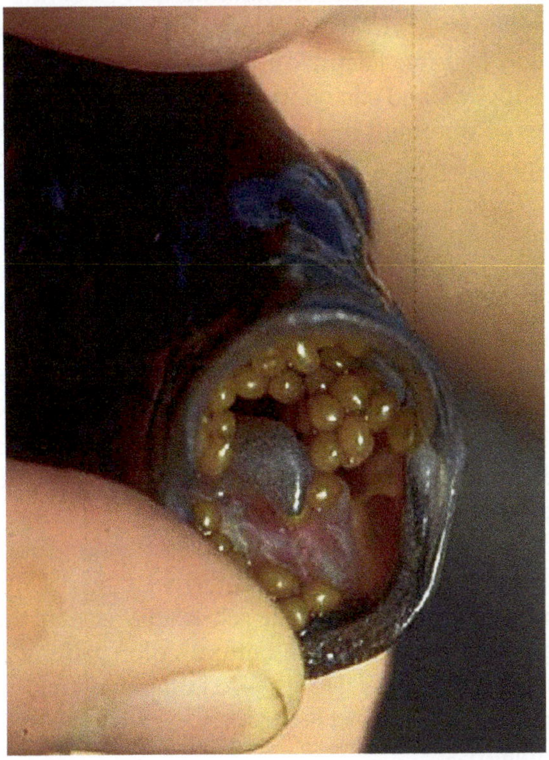

Figure 15.4. Female *Oreochromis niloticus* holding eggs in mouth.

15.2 Brood Management

Oreochromis tilapias are a paradox in reproduction. On a per spawn basis they are not highly fecund. *O. niloticus* produce between 6,000 and 13,000 eggs/kg of body weight/spawn. However, because they spawn frequently, a female's yearly quantity of eggs/kg can be equal to many group synchronous spawning species (Grass Carp 1,000,000±/kg). Fecundity/kg female varies by brood size. Larger females produce a larger egg but less eggs/kg body weight, with the percentage survival of fry typically higher. Smaller brooders produce more eggs/kg and though the fry survival may be less the total fry produced/kg female can be >25% greater than a kg of larger brooders. Females as small as 50 g may be mature depending on the growth rate, but working with such small fish requires much more fish to be handled. Large females may give more eggs/fish but their maintenance requirements are higher and are more difficult to handle. Females ranging from 100 to 200 g are the preferred size as broodstock.

Genetic history is a major consideration when establishing a brood population of tilapia at a hatchery. A species such as Nile Tilapia *O. niloticus* has a wide natural distribution in Africa but only a limited number of collections have been made from wild stocks to establish the founding populations used elsewhere in the world. The majority of the world production of Nile Tilapia is based on six initial sources: Ivory Coast, Ghana, Egypt-Manzala, Egypt-Ismailia, Sudan-Nile and Lake George, Uganda. Stocks originating from Aswan, Egypt, and Lake Turkana, Kenya are less common. These founding stocks collected in the late 1960s and early 1970s were often established from a limited number of wild-caught fish with their progeny being redistributed throughout the world. As a result, many hatchery stocks are highly inbred.

Genetic purity is also an issue when maintaining and selecting tilapia brooders. Many species in the genus *Oreochromis* will hybridize naturally with each other and special care is needed to maintain species purity. Tilapia often escape into streams supplying water to the hatchery and its drainage ditches, and can contaminate the spawning ponds. The fry are hardy and can survive in puddles of water left in a pond between crops. If two or more *Oreochromis* species are held at the same hatchery there will be a risk of hybridization and genetic contamination.

A "red" tilapia color pattern is popular variation in a number of countries. This mutation has occurred in both *O. mossambicus* and *O. niloticus* and efforts have been made to select for this color variation within the species as well as when hybridizing with other *Oreochromis* species. Most red varieties do not breed true to one color pattern, some may appear predominately white, pink, bronze, red-orange or red, others may be a mosaic of "red" colors with irregular black patches. Manual selection is needed to increase the frequency of red tilapia in a population. Natural colored fish will become more frequent if there is not an active selection for red colored ones. Depending on the genetic background of the "red" the growth rate can be as good as a normal colored *O. niloticus*, however, if the bloodline is predominately from *O. mossambicus* the growth rate may be slower. Active selection for red coloration can result in a very inbred population and reduced progeny survival.

A major consideration when working with red tilapia is that they are often more susceptible to bird predation than those with the natural color pattern. Brood ponds should be covered with netting to limit bird access. Holding red tilapia in tanks and ponds close to areas where there is frequent human activity will help to keep predacious birds chased away, in turn, helping to reduce red tilapia loss.

Because tilapia can be cultured in such a variety of conditions, a number of fingerling production techniques have been developed, but several constraints must be considered. With many species of fish produced in a hatchery, the difficulty is getting them to spawn successfully, but with tilapia, the challenge is to control excess reproduction. Tilapia will spawn in aquaria, indoor and outdoor tanks, earthen ponds, and any other containers where the eggs can be deposited on the bottom for a few seconds, as long as the fish are in good health and the water temperature is in the correct range. If eggs and fry are not harvested frequently and efficiently, fry production will decrease over time due to fry from earlier spawns out-competing and preying on fry from later spawns.

Another consideration with tilapia is that even though it is easy to produce large numbers of fingerlings, the fecundity of tilapia/kg female is low in comparison to other cultured fishes. As a result, larger numbers of broodfish are needed and more hatchery space must be devoted to broodstock and to spawning. Compared to many other cultured fishes, the larger number of tilapia brooders needed to meet production goals helps to keep the rate of inbreeding low. The bigger issue is often the restricted gene pool resulting from the low number of fish in the founding populations when they were first introduced from Africa, and further gene pool restrictions as small numbers of fish were often transferred from one hatchery to another to establish a new population.

Under suitable conditions, tilapia can in just one year grow from egg to being too large to serve as brooders. At least two age groups of brooders need to be developed each year to make up for brood mortality, and to replace fish that have become too large to be conveniently used. Future brooders can be selected from the harvest of secondary nursery ponds. At that point the fish will be large enough to be accurately sexed. The selected fish should be stocked separately by sex at a density of 5–10/m^2 in 500–2000 m^2 earthen ponds, holding 2–4 females for each male being developed. In a spawning setting, males will often beat up on some of the females, killing some, and reducing the condition of other to the point that they should be discarded as brooders. The fish should be fed twice a day a ≥30% protein feed until they reach an average weight of ≥100 g and then be fed once a day. Brooders can also be developed in cages, stocking 1,000 fish/m^3 and fed as described above.

As *Oreochromis* species are mouth brooders, female brooders feed less when held in a spawning setting and will lose weight if actively spawning. This is particularly true with the intensive spawning techniques described later in this chapter. In those setting, once a spawning cycle is over, the males and females should be removed and held separately and fed to satiation for 10–14 d or more before being restocked for spawning.

15.3 Seed Production Alternatives

15.3.1 Mixed Sex Production

Tilapia seed production can be directed to provide fingerlings for two basic applications: 1) mixed sex culture of food fish, and 2) monosex culture. In a mixed sex culture, 5–10 g fingerlings that are as young as possible should be used to stock production ponds or cages. Depending on the length of the culture period the fish may reproduce in the production setting. There should be market opportunities for the faster growing males, the slower growing females and any reproduction if a mixed sex culture is being considered. Mixed sex tilapia can also be used as forage when predacious fish such as snakehead *Channa* sp. are being produced. In markets where larger fish are preferred, male fingerlings can be obtained by manually separating males from females, hybridizing specific species that give predominately male offspring, or by sex reversal. A spawning-rearing procedure with some modification can be appropriate for these applications.

A basic pond spawning-rearing procedure for mixed sex seed production begins with good pond preparation, elimination of predators, and establishing a plankton bloom using organic manures. Broodfish are stocked at 1,000 to 5,000/ha using ≥100 g fish at a sex ratio of 3 to 4 females/male. The broodfish are fed at 1 to 2% body weight/d and additional quantities of fertilizers may be added to maintain the plankton bloom. Broodstock should begin to spawn within 5 days after stocking and the first spawn may reach 2.5 to 5.0 cm in 30 to 50 days. Once the first spawn grows above 1 g in average weight (about 4.0 cm), partial harvests can be made every 10 days using a 6 mm mesh seine. It is difficult to seine out all the ≥1 g fingerlings and those that escape will begin to prey on fry from later spawns. After 9 to 12 weeks, the yield of fry will begin to decrease and a complete harvest should be made. The overall fingerling production should average between 3 and 9 fish/m^2/month depending on the effectiveness of the partial harvests and the fertility of the pond. Where partial harvests are not possible the culture period should be shortened to 60 days and a lesser yield anticipated. At the final harvest, a variety of fingerlings sizes, from <1 g to >5 g, may be common. The harvested fish should be graded before restocking. An additional nursery phase is recommended for most of the production before the fingerlings are distributed.

A modification of the spawning-rearing procedure is to begin the partial harvest sooner using a 3 mm mesh seine. The pond is stocked at 5,000 to 10,000 broodfish/ha at 3 to 4 females per male. Less emphasis is given to ponds fertilization, but the feeding rate is increased to 2 to 3%/day. Fry are collected daily along the shore in the mornings starting about two weeks after the broodstock are placed in the pond. If the pond is partially harvested daily, it may be possible to keep the spawning pond in production for more than 3 months with a daily yield of 2–3 fry/m^2 of shoreline harvested. The fry collected may be 8 mm or larger with the smaller fry being much more sensitive to handling than fingerlings. A common mistake made when working with tilapia fry is not remembering that they are much more sensitive to handling than are juveniles or adults. The fry should be collected and moved in water whenever possible, and oxygen levels in the transport containers need to be kept near saturation.

The fry are graded into various size classes and restocked by size class into nursery ponds at rates dependent on desired goals and nutrient availability. Stocking similar size fish at the desired density on the same day is key to a successful nursery. The scale of fry production should take into consideration the size of nursery ponds so that a nursery pond can be stocked completely in one day. Multiple stockings over a several day period will result in increased competition during the nursery period, and increased size variation at harvest.

Nursery ponds are used for rearing small fry to 1 g, as well as rearing 1 g fingerlings to 10–25 g. Where fry are collected, they are stocked into organically fertilized ponds at 200 to 500 fry/m^2 and given a supplemental feed. After 30 days, the fish should average 1 g depending on water temperature, survival, and quality and quantity of feed and fertilizer used. Fertilization rates equivalent to initial applications of up to 2,000 kg/ha (dry wt.) of chicken manure are appropriate, with a maximum feeding rate of 50–75 kg/ha/d. With proper pond preparation survivals of 60–80% can be anticipated. Larger fingerlings are produced by restocking the 1 g fish at rates of 10–20 fish/m^2, and culturing the fish 60 to 90 d using organic fertilization and feeding.

The food supply, time available, size desired, and productivity of the facility should be considered in stocking the secondary rearing pond. As an example, assuming that: large 30 g fingerlings are needed; an adequate feed is available; ponds have a critical standing crop of 3,000 kg/ha; and a survival rate of 80% of the small fingerlings stocked can be anticipated, the stocking rate/ha would be based on the following:

Number to stock/ha = ((critical standing crop/ha)/ average wt. at harvest))/% survival

Number to stock/ha = (3000 kg critical standing crop/ 30 g average wt.)/80% survival

Number to stock/ha = 100,000 fish (30 g average wt.) harvested/80% survival

Number to stock/ha = 125,000 fingerlings to stock/ha (12.5/m^2)

The length of the culture period to reach the desired size fish is a function of growth rate. The growth rate is a function of life stage, stocking density, water temperature, other water quality parameters, as well as the quantity and quality of the food. Under favorable secondary nursery conditions 1 g tilapia fingerlings can reach 25–30 g in 60–90 days.

15.3.2 Monosex Production

The major problem with tilapia as a culture species is that they can mature and reproduce in a production pond before reaching market size. Energy is diverted into reproduction that more profitably should be spent in growth. In addition, the offspring remain in the pond increasing the competition for food, reducing the growth rate of the adults. Where reproduction is not controlled in a production pond, it is rare to have more than 20% of the individuals at harvest to weight 200 g or more.

Tilapia have a distinct sexual dimorphism in growth. As the fish start reaching maturity the males grow faster than the females. Thus there is an advantage to culture only males due to the faster growth and as a way to control reproduction. All or near all male populations can be created by three basic approaches: manual separation of males and females, hybridization, and sex reversal. In addition, as greater understanding of the genetic control of sex determination and the role of influencing environmental factors is gained, it may be possible through within-species selection to develop highly-selected lines of tilapia that consistently give near all-male progeny.

15.3.2.1 Manual Separation

Manual separation is based on external differences in appearance between males and females. The shape of the genital papilla is a key characteristic in identifying the sex of tilapia. As illustrated in Figure 15.5a, males have a common urogenital opening near the tip of the papilla, whereas females (Figure 15.5b) have a separate genital opening (oviduct) appearing as a transverse opening across the papilla and a urinary orifice near the end of the papilla. The papilla is adequately developed and of a large enough size that they can be sexed by visual observation when the fish reach 15–25 g. The papilla of *O. mossambicus* are often very distinct and easy to sex visually, the papilla of *O. niloticus* is less distinct, and larger fish are needed

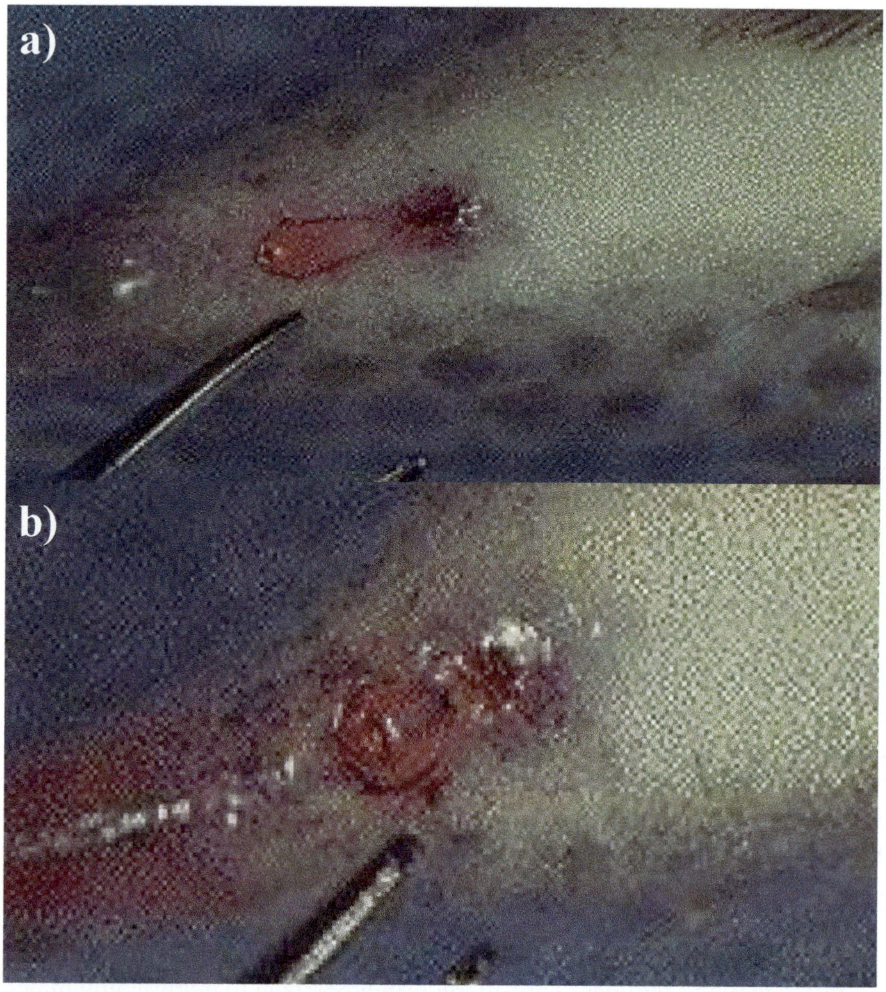

Figure 15.5. Genital papilla highlighted with carmine red of a male (a) and a female (b) tilapia used as a key for separating sexes. Note the slit of the oviduct across the papilla of a female.

when hand sexing. A dye such as methylene blue or carmine red can be applied with a cotton swab to the papilla to make the oviduct more distinct.

Manually separating the sexes is an inefficient way to obtain male tilapia because pond space must be devoted to spawning, rearing of mixed sex fingerlings up to 15–25 g, then discarding half of the fish produced. Fish used for sexing can be produced as described earlier, stocking mixed sex fingerlings into secondary nursery ponds. An experienced worker can sex between 2,000 and 3,000 fish/h, making large-scale production possible but it is labor intensive. In most cases it is not a commercially practical approach to obtain males for culture. However, hand sexing can be appropriate for very small-scale production operations where the farmer is limited in the availability of resources, and must be self-dependent, and any small female discards will be consumed. In either case, a 5% or more sexing error should be expected and that reproduction will occur in the rearing pond. A predatory fish may be needed to be added to control the reproduction in the production pond, or the pond should be harvested before the reproduction begins to limit the growth of the adults.

15.3.2.2 Hybrids

Some matings of one species of tilapia with another can give skewed sex ratios in the offspring and has proven to be, with some limitations, an effective approach to having near all-male populations. Male *O. hornorum* and *O. aureus* are thought to be homogametic for sex determination with female *O. niloticus* and *O. mossambicus* also being homogametic. When a homogametic male of one species is crossed with a homogametic female of another the sex ratio of the offspring is highly skewed toward males with up to 98–100% males reported from crosses of *O. hornorum* males with females of *O. niloticus* or *O. mossambicus*.

The spawning success in hybrid crosses is less than in pure line matings, and thus a lower number of fry will be produced. Spawning can take place in aquaria, hapas, tanks, and earthen ponds. In ponds, the reproduction cycle should be 90 days or less to avoid the offspring back-crossing with the parents. Hybrid tilapia are fertile and will reproduce. Care must be given to eliminate any fry remaining in a spawning pond from a previous cycle before beginning a new cycle in the same pond.

In earthen ponds, hybrid fingerlings are produced in 60 to 90-day cycles with a single complete harvest or along with a series of partial harvests. The average size of fingerlings will be larger with less frequent harvests, but weekly partial harvests beginning 30 days after stocking broodstock will result in much higher numbers of fry being produced. Broodstock are stocked at a rate up to 1 fish/m^2 using a sex ratio of 1 male per 1–4 females. A yield of 2–3 fry/m^2/d is possible with frequent harvesting. The fry and fingerlings can be managed in primary and secondary nursery ponds as described earlier.

The major difficulty for most hatcheries is to maintain pure lines of both parent species used to make the hybrid. This is particularly true where the fish are reared in open ponds and the water supply is a surface water source where tilapia occur. Brood stocks of both parents should be kept isolated in covered tanks or ponds supplied with a carefully filtered water supply, that are located as far away from other tilapia as possible. The hybrid is capable of spawning with the parent species but with a reduced rate of males in the progeny. It is not uncommon for a hatchery to initially produce hybrids that are 98+% male but with time the percentage of males becomes less as the parent stocks become genetically contaminated.

15.3.2.3 Hormone Sex Inversion

Male tilapia can be produced by controlling the gonadal development of tilapia at an early point in their life history. As tilapia fry begin to swim-up and feed, the gonadal tissue has not yet differentiated into ovary or testis and is sensitive to exogenous sex steroids that can be administered to direct the gonadal differentiation. Androgens are male steroids that when given correctly to fish with undifferentiated gonads can result in 100% male populations. The most commonly used androgens are methyltestosterone and ethynyltestosterone.

The key to successful all-male production using androgens is to begin hormone treatment with fish having undifferentiated gonads and continue the treatment until gonadal differentiation is complete. Tilapia as young and as small as possible should be used, with fry up to 12 mm total length being acceptable if their growth up to that point has been rapid. In general, best results are obtained beginning with fry that are 9–11 mm at the start of hormone treatment. Depending on the environmental conditions and the growth rate of the fry, gonadal differentiation takes place when fry are between 14–22 mm in length. Hormone treatment must continue until the differentiation is complete, which may take 2–4 weeks. A 4-week hormone treatment period is recommended, as it will help to ensure successful sex reversal and result in larger fry easier to handle.

15.4 Fry Production Procedures for Use in Sex Reversal

15.4.1 Pond Spawning

Fry used for treatment may be obtained by making frequent partial harvests of spawning ponds using a 1.6 mm mesh seine or scoop net following the procedures described earlier for daily fry harvest. Pond production of fry through daily fry collection has the advantage of being straightforward and requiring little extra equipment. The main disadvantage is that some fry from the early spawns will escape capture and become predators on fry produced in later spawns. As the abundance of these larger fish builds over time the yield of desirable size fry is reduced. This technique is labor intensive and the daily yield of fry is low (2–3 fry/m^2 of shoreline harvested/d). As such, it difficult to obtain enough fry of similar age and size on a given day to for large-scale sex reversal operations unless enough pond space is devoted to spawning and enough labor is available for such harvesting.

Complete harvests of spawning ponds on a short spawning cycle is another alternative. Large quantities of fry (50,000+) can be obtained at a time from 500 to 1000 m^2 earthen ponds that have a catch basin lined with 2 cm mesh netting to remove the broodstock at harvest. The pond is prepared by completely drying the pond or adding a fish toxicant to any remaining puddles to eliminate any fry from previous production cycles. Broodfish are stocked at 35 to 55 kg of females per pond at a sex ratio of 1 male per 1.5 to 2 females. Brooders are fed at 1–2% body weight for a 14 to 21-day period. At water temperatures of 28°C and above the broodfish will begin to spawn soon after stocking and the development period for the eggs and larvae is short. Under such conditions, a maximum number of suitable size fry can be obtained at approximately 16 days post-brood stocking. At lower temperatures the development rate is slower and at 22 to 24°C, a harvest after 21 days would be optimum.

The pond is drained slowly over a 5 to 15 h period using a 1.6 mm mesh screen with a surface area of 1 to 1.5 m^2 around the standpipe. A large surface area of screen helps to reduce the water pressure on the screen during draining, reducing the impingement of fry on the screen. In addition, the large surface area helps to prevent the screen from becoming clogged too fast. The timing of the pond draining and harvest should be scheduled so that the harvest begins within an hour after dawn while the water is cool, and be completed within an hour thereafter. Once the water level is down to the catch basin, the fry may come to the surface if the oxygen becomes too low. If so, the fry should be collected as they come to the surface using a 1.6 mm scoop net. The fry are delicate and should be collected quickly, skimming the fry from the water surface while avoiding dipping the net into the mud. The fry should be moved in water to an oxygenated transport tank.

Brooders can be removed before any fry are collected or after the majority of the fry were collected from the water surface. Lining the catch basin with coarse mesh netting as part of the pond preparation will facilitate the capture of brooders at harvest. A team of people should surround the catch basin, holding the edges of the coarse mesh netting, and lift the netting and brooders out of the catch basin. The broodstock can be transferred to another pond for spawning or be held in a cage or tank before restocking in the original pond. However, before the brooders are restocked for spawning, the mouths of the females should be checked to make sure they are not holding eggs or fry. A harvest equal 5 to 6 fry/m^2 per day of culture is common.

To optimize the harvest of appropriate size fry the timing of the harvest must take into consideration water temperature. A few degrees difference in water temperature can speed up or delay the rate of embryo and fry development. Harvesting too soon will result in many females still holding eggs or sac fry that will be lost during harvest. Harvesting too late will result in more fry that are too large for use in sex reversal. Green and Teichert-Coddington (1993) developed an equation to time the fry harvest in order to obtain the maximum number of fry of a sex reversible size. They found that a 195–220 degree-day period was optimum for best production of fry that are suitable for sex reversal when calculated as follows:

Degree-days to harvest = (Mean daily water temperature – a threshold temperature of 15°C) × number of days to harvest.

Restated from a hatchery perspective the number of days to harvest would range from:

Minimum days to harvest = Minimum degree days/(mean water temp – 15°C)
Maximum days to harvest = Maximum degree days/(mean water temp – 15°C)

For example, at 28°C the pond would be harvested between:

Minimum days = 195 degree-days/(28°C – 15°C) = 15 days;
Maximum days = 220 degree-days/(28°C – 15°C) = 17 days.

Complete harvests of spawning ponds if timed right will yield large numbers of similar size fry. However, some eggs and fry will be lost during the harvest, as well as in most cases the pond water. A spawning pond must be drained completely every 16–21 days, held dry for a few days and then refilled. Whether ponds are partially harvested daily or completely harvested, the harvested fry will need to be graded to obtain the proper size if they are to be sex reversed.

15.4.2 Hapa Spawning

In water-scarce regions, the frequent filling and draining of the spawning pond may be an issue and hapa spawning might be an alternative. Mouth-brooding tilapia need a substrate to spawn. It can be the bottom of a pond but they will spawn on netting. If the mesh is small enough, the eggs will not drop through the mesh and can be picked up by the female and incubated in her mouth. Fine mesh cages, often called hapas (Figure 15.6), are used as a spawning setting. They are suspended in ponds or tanks and, if properly set, allows easy access to the broodstock and their offspring, which facilitates the harvest. Hapas are made of 1.6 mm mesh netting, rectangular (2–5: 1 length: width ratio) in shape, and may range in volume from 2 m^3 to over 100 m^3 with an in-water depth of about 1 m. with a freeboard of >30 cm. Hapas are suitable for tilapia reproduction as the initial investment is low and they can be set in bodies of water otherwise not suitable for fry production. However, depending on the material that

Figure 15.6. Tilapia spawning hapas.

hapas are constructed of and how they are treated they will deteriorate and may need to be replaced every 6–12 months. A major issue with hapas is that the fine mesh fouls easily and water circulation can be blocked. This can result in low oxygen conditions inside the hapa that might kill eggs or fry, even though outside the hapa the oxygen level may be at saturation or higher.

Hapas are managed intensively. Broodfish are stocked at approximately 5/m^2 in a 1 male: 1–2 female ratio and fed a floating feed at 0.5–1% body weight/d. The frequency of harvest depends on whether the goal is to obtain eggs or a specific size of larvae at harvest. Hapas are harvested by starting at one end of the hapa with a pole at the water surface that goes across the width of the hapa. The pole is moved along the length of the hapa, passing the netting over the pole, to crowd the fish into one end of the hapa (Figure 15.7). Keep in mind that brood tilapia may jump out of the hapa when being crowded, particularly when confined in the last 3rd of the hapa. The last part of the crowding should be done calmly and slowly, trying to keep that portion of the hapa with as much depth as possible to provide a refuge for the brooders to gather in. Females holding eggs or fry often have a distinct buccal pouch, which makes them easier to recognize, and such females should be examined first. However, a buccal pouch is not distinct in all females and each female should be examined for eggs or fry. How well a species holds eggs or fry varies by species. When handled, female *O. mossambicus* will retain the eggs or fry in the mouth better than *O. niloticus*, which is more prone to release the eggs or fry when handled, making it more difficult to collect their eggs and sac fry. Females should be collected with a fine mesh dip net to prevent the loss of any eggs or fry from her mouth when handled. A two dip net system can be used to handle the brood females while collecting eggs and fry. A 6 to 12 mm mesh dip net is set inside a 3 mm mesh dip net. A female is collected from the hapa, held in the coarse mesh net with the fine mesh dip net underneath. The female's

Figure 15.7. Harvesting a tilapia spawning hapa.

mouth is held open with the index finger on the lower lip and the opercula held open with the thumb, and then any eggs or fry are rinsed into the fine mesh net. Each female, after a careful inspection of the mouth to ensure that she is still not holding eggs or fry, is transferred to holding facilities or returned to the other side of the spawning hapa for another cycle. Males, if in good condition, may be used in the same hapa for another cycle.

Harvests made every 14 to 21 days (depending on temperature) will produce predominately swim-up fry, which can be collected along the edges of the hapa with a dip net. The majority of such fry are typically of the correct size to begin sex reversal with but in some cases, it may be necessary to grade the fry to ensure they are of an appropriate size. Mouths of females need to be examined and any eggs or fry collected. Any eggs or sac fry found should be taken to the hatchery for further incubation. Once examined, the adults can be transferred back to the other side of the hapa or rested, and any remaining fry in the hapa collected. A yield of 50 to 60 fry/kg of female/day might be expected using ≥250 g females. Depending on female size and stocking density, approximately 40 fry/m^2 of hapa/d can be expected.

A more frequent harvest will give a greater total number of seed but requires that eggs and sac fry be incubated for up to 5 days before they can be used. Such an approach requires more facilities and labor. When hapas are to be managed for providing eggs, brooders are stocked at 5–6/m^2 at a 1:1 sex ratio and the hapas are harvested every 5–7 days. A range of 500–1000 eggs and sac fry/kg female/d can be expected. This would be equivalent to 75–125 eggs and sac fry/m^2 of hapa/d. If more than one spawning cycle is to be conducted then at the end of each spawning cycle, the brooders need to be replaced and new brooders added. The brooders that were removed are separated by sex and restocked into recovery tanks or hapas where they are allowed to feed actively and regain condition over a 10–15 d period. If the spawning hapas are to be harvested every 5 d then three sets of brooders are needed for each spawning hapa: one set that just finished spawning; one set that has been recovering for 10 days and are ready to be stocked into the spawning hapa; and another that need 5 more days of recovery before they can be stocked into spawning hapas. In addition, extra brooders need to be on hand to replace any escapement, mortality, or fish that are in too poor a condition to be reused.

Females should be examined as described earlier, checking one female at a time. Fresh water should be added periodically to the bowls holding the eggs, stirring the eggs occasionally while waiting to load the incubators. As the eggs of a given female are collected, the approximate age of the eggs should be estimated based on their color. Recently spawned eggs are light yellow in color and will transition to a dark orange color as the eggs approach hatching. Similar age (i.e. color) eggs should be incubated together, making three or more sets: light yellow, yellow/orange, and dark orange. Sac-fry should be held separately. This will facilitate egg management during incubation and help to ensure the similarity in age. The more similar the eggs are in age the more similar they will be in hatching time. This allows more efficient use of the incubators while reducing the time spend waiting for the last of the eggs to hatch. This also facilitates having larger numbers of similar age swim-up fry available at one time for use in sex reversal.

Eggs coming from the spawning hapas have been exposed to water-borne bacteria and fungus and may also be covered with mud and debris. They should be rinsed in clean water and treated with 200 mg/L formalin for 15 min. before loading into incubators.

Down-flowing incubators, like MacDonald jars, are preferred for incubating tilapia eggs. Loading rates of eggs range from 15–30 g of eggs/L of incubator. The flow rate in the incubator may initially be 5 L/min for recently fertilized eggs and increased to 10 L/min as hatching approaches. The down-flow of the water to the bottom center of the incubator should be strong enough to prevent the tilapia eggs from settling on the bottom and provide a vigorous roll of the eggs in the water column (Figure 15.8b). Optimum water temperature for egg incubation is 28–30°C. Recirculation systems can be used to help maintain a stable water temperature for egg incubation. However, most recirculating systems cannot manage the pulse in ammonia that comes from the decomposition of egg shell debris following hatching, Tilapia eggs tolerant 100 mg/L un-ionized ammonia during incubation, but young fry are much more sensitive, with negative effects on growth and survival occurring at values as low at 0.4 mg/L un-ionized ammonia. Additional clean water of the proper temperature is needed to dilute any ammonia buildup during this period.

The older the eggs when collected the better the hatch rate when incubated. Eyed eggs may have a hatch rate of ≥75% while un-eyed eggs average 50–60%. Hatching time for recently fertilized eggs may range from 3 to 6 days, depending on the water temperature. The sac-fry remain in the incubator for 1–2 d until they swim-up. Once they start to swim up, they are carried out of the incubator in the water current to a screened holding basket (Figure 15.9). Collection baskets should be placed only after hatching is complete to avoid egg shells and other debris from clogging the mesh of the basket. Just before the fry begin to swim-up, baskets are put in place to catch swim-up fry. One basket/incubator is typically used, but if several incubators have similar age eggs, their effluent can be diverted to flow into a common basket. Swim-up fry can be held at up to 5,000/L of basket when the water exchange rate is 2–3 times/min. Each basket should be harvested daily. A daily harvest of swim-up fry will yield fry similar in age and size, suitable for beginning sex reversal.

15.4.3 Grading

A key to successful sex reversal is using as young and as small a fry as available. Fry that are less than 15 days old and less than 11 mm are preferred. Fry that are 12 to 13 mm are acceptable if they meet the age requirement. Where eggs are collected and hatched, it is possible

Figure 15.8. Down flow incubator (a) for tilapia eggs providing (b) a vigorous roll of eggs.

Figure 15.9. Once tilapia larvae have developed enough they flow out of the incubator and are harvested daily from a mesh collecting box.

to have uniform age and size larvae for sex reversal. Where fry are collected from hapas or ponds, it is necessary to separate the fry into size categories. Such fry will be of different ages and size and must be graded to select fry of an appropriate size for sex reversal (Figure 15.10). A grader can be made of metal or plastic. A mesh size of 3.2 mm is generally appropriate for allowing fry ≤13 mm to swim through. If the mesh of the material is too large the mesh size can be reduced using a spray paint to lightly coat the mesh to make the grader more selective. Samples need to be measured to the nearest mm of fish that swam out of the grader to determine the largest fry to swim through the grader. Likewise, the smallest size fish retained in the grader needs to be determined. Graders are not absolute in their selectivity. An effective grader for tilapia fry should allow 85–90% or more of the harvested fry ≤13 mm to swim out of the grader. Tilapia fry swim more laterally than downward when held in a grader so a grader with more of its surface in a vertical position will be more effective. The grader should be set in a fine mesh hapa that is large enough so that there is ≥20 cm separation between the grader and the sides of the hapa. The grader should be gently lifted up periodically but still keeping the fry in the water. This will encourage the fry to swim out.

15.5 Sex Reversal Procedures

15.5.1 Feed Preparation

For sex reversal to be effective the feed must be consumed. Using the highest quality ration available will help ensure that the feed is palpable. Diets high in animal protein having a crude protein level of 35 to 50% are preferred. The diet may be supplemented with fish oil,

Figure 15.10. Screen grader for separating tilapia fry for sex reversal.

as well as vitamin and mineral supplements to improve its quality. The particle size should be equivalent to a no. 0 or 00 crumble or meal, using a 0.6 mm mesh screen to sieve the feed to obtain the proper particle size.

The hormone is administered in the feed by dissolving the hormone in alcohol and spreading the alcohol over the feed. The alcohol is allowed to evaporate off, leaving the hormone on the feed. A stock solution of methyltestosterone (MT) is first prepared by dissolving an equivalent of 6.0 g of the water-insoluble hormone in 1 L of 90–95% ethyl alcohol. The stock solution should be held under refrigeration in the dark, and if so, will have a shelf life of at least three months. The hormone is administered at a rate of 60 mg/kg of feed. Using the 6 g MT/L stock solution, 10 mL of the MT stock solution is added to 500 mL of either ethyl or isopropyl alcohol (>90% ethyl alcohol preferred). The dilute MT solution is mixed with 1 kg of finely ground feed until all the feed feels moist to touch. When making large batches of feed (≥25 kg), it can be mixed using a clean cement mixer, spraying the feed with 200 mL of an alcohol solution containing 60 mg methyltestosterone/kg of feed to be treated. Once the feed is coated with the MT-alcohol solution, the feed is spread out to a thickness of 3 to 5 cm and the alcohol allowed to evaporate off. This can be done (not in direct sunlight) at room temperature in the shade or in an oven at ≤60°C. The feed should be stirred 2 to 3 times during drying. Once dried, the feed should be stored in sealed plastic bags at room temperature, refrigerated, or frozen for later use.

15.5.2 Fry Stocking

Tilapia fry can be held in a variety of containers when being treated. In the hatchery they can be stocked into flowing water aquaria, troughs, or circular tanks at densities up to 10–20/L. Tilapia can be sex reversed outdoors in static tanks or earthen ponds. There the fry are confined in hapas at densities of 1000 to 5000/m^2 (Figure 15.11) or released into a tank or pond at densities of 150/m^2 or more. Fry are stocked into the container where they are going to be sex reversed by first preparing a standard of a known number then making visual estimates

Figure 15.11. Tilapia held in a hapa set up with a feeding ring and a belt feeder.

based on the standard (see Chapter 17.5, Counting Techniques). A subsample is often taken to determine initial length, and weight is estimated using a length-weight formula. The exact length-weight relationship varies slightly from site to site but a general formula acceptable for adjusting feeding rates is:

$W = 0.02 \times L^3$. W is the weight in grams of 1000 fry and L is the average total length/fish in mm.

15.5.3 Feeding

The environment in which the fry are held is not as important as the ability to train the fry to accept the hormone-treated feed. Feeding rings are needed to confine the feed to a limited area making it easier to train the fry to feed. Tilapia fry learn to feed very easily and should be feeding actively at the feeding station within three days after first receiving hormone-treated feed. Fry are fed a minimum of twice a day with four times or more per day preferred. The initial feeding rate is between 15 to 20% body weight/d. Once the fry are larger than 15 mm, the feeding rate is gradually reduced 1% per day until the fish are receiving 10% body weight/d. Daily adjustments of the amount of feed added are made accounting for growth. For fry ranging in total length between 8 and 12 mm a daily growth rate of 0.2 to 0.3 mm might be anticipated, 0.3 to 0.7 mm/d for those 13–17 mm, and 0.7 to 1.4 mm/d for those 18 to 25 mm. These daily changes in length then would be added to the initial length, new average weights calculated for each day, and feeding rates adjusted accordingly. The growth rate during treatment will vary depending on stocking rate, feed quality, temperature, and the health of the fry. The assumed growth rates should be verified weekly by measuring a minimum of 25 fry from each container and a new average length determined. Feeding rates are then recalculated using this new average length. Once experience has been gained and good records kept, feeding tables as in Table 15.1 can be developed as a feeding guide and reduce the need to sample as frequently at a given hatchery.

Table 15.1. Daily ration per 1,000 tilapia fry of known or assumed average length[1] for the first month of feeding.

Average Total Length (mm)	Daily ration per 1,000 fry (g)	Average length (mm)	Daily ration per 1,000 fry
9	2	18	17
10	3	19	21
11	4	20	24
12	5	21	28
13	7	22	32
14	8	23	37
15	10	24	41
16	12	>24	41
17	15	---	---

[1]The above table is based on a target feeding rate of 150 g per Kg fry biomass daily during treatment, which is an effective daily dose of 7.2 mg MT/Kg biomass, assuming fish consume 80% of proffered feed. The weight-length relationship employed is $W = 0.02 \times L^3$, where W = weight per 1,000 fry in grams and L = average total length of fry in mm.

At the end of hormone-treatment period, the fry should be at least 0.2 g and may be as large as 1.5 g depending on the production environment. These fish are then stocked into secondary nursery ponds as discussed earlier. At the end of the secondary nursery phase, a subsample should be taken and the percent males verified. A minimum acceptable rate would be 95% males with a higher percentage needed when stocked for growout cycles of 180 days or more.

15.6 Production of Sex Reversed Fingerlings to Stocker Size

Similar management as given to mixed sex fingerlings can be used with sex reversed fish. Generally, one to two nursery periods are used to produce advanced fingerlings. At the end of the sex reversal period, treated fish are commonly less than one gram and are nursed to 10 g or larger. Fish are stocked at 25 to 50/m^2 and cultured 45–60 days to reach an average weight of 10 g. Growth rate is highly dependent on temperature and food availability. If only organic fertilizers are being added, the lower stocking rate should be used. If fish are given a high-quality feed containing >32% protein and are fed three to four times a day at 5% body weight per day, then the higher stocking rate would be appropriate. Daily feed input rates of 100 kg/ha/day can be tolerated, but low oxygen values can be expected at dawn. Mortalities of tilapia resulting from low oxygen are rare, but forcing the fish to the surface in search of oxygen makes them very susceptible to bird predation. Chronic low oxygen will reduce growth rate and affect feed conversion. Temperature is a major factor controlling tilapia growth rate; at 30°C the growth rate can be twice that at 23°C. Production schedules must keep in mind the quantity and quality of nutrient input and the temperature at which the fish are being cultured.

Fingerlings larger than 10 g are used for stocking some growout setting but there are settings where larger fingerlings of about 50 g are needed to stock growout units. To obtain 50 g fish, an additional nursery period of 30–45 days is used. Ten gram fish are stocked at 10/m^2 and fed at 3% body weight until harvest. Fingerling survival from 1 g to 10 g may range for 60–90% and from 10 to 50 g is often >90%.

15.7 Within Species Selective Breeding

Genetic management is a key part of successful tilapia food fish production that begins at the hatchery. As discussed earlier, often tilapia populations used in aquaculture were established based on a limited number of fish and in some cases genetic bottlenecks have occurred further limiting the genetic diversity within a population. However, through careful selective breeding growth rates can be improved. The Genetically Improved Farmed Tilapia (GIFT) project formed in 1988 began with the goal of developing a faster-growing strain of Nile Tilapia (*Oreochromis niloticus*) suitable for both small-scale and commercial aquaculture. After six generations of selective breeding, an improved strain of tilapia was developed that grew up to 85% faster than the fish at the start of the breeding program. The GIFT strain is now being used in 16 or more countries. Details of this breeding program are available in Ponzoni et al. 2011, listed in the references of this chapter.

As discussed earlier, male tilapia are preferred for culture, and several alternatives for obtaining males have been reviewed earlier in this chapter. Another alternative is creation of brood populations through genetic selection and selective mating within the species that will give progeny with a very high proportion of males. The basic sex determination pattern in *O. niloticus* is thought to be females being homozygous (XX) and males heterozygous (XY) with the presence of a Y chromosome controlling the sex to be male. Based on that, if a fish had a YY genotype and was mated with a female of a XX genotype, the progeny would be all males (XY). To obtain a YY male, several steps would be needed. *Oreochromis niloticus* larvae would be fed an estrogen during the period of gonadal differentiation resulting in fish that are functionally female but genetically may be either XX or XY. To recognize the genotype of an estrogen-treated fish, it must be mated with a normal male (pure line XY). In such a population of estrogen-treated fish which developed as functional females, 50% would have a XX genotype and the other 50% would have the male XY genotype. However, the genotype of such fish is not obvious and each fish must be mated with a normal XY male and the sex ratio in the progeny determined based on the phenotypic sex to determine the genotype of the female parent. An estrogen-treated XX female spawning with a XY male should give 50% male and 50% female progeny as evident based on the phenotypic sex. A XY female mated with a XY male should give a population that is genetically 25% XX females, 50% XY males and 25% YY males but phenotypically would appear as 25% female and 75% male. To identify any YY genotypic males, each male should be pair spawned with a normal female and the sex ratio of the progeny determined. In theory, the progeny should have a 50:50 male:female sex ratio if the parent male had a XY genotype. But if the male had a YY genotype, the progeny should be "100%" male. In reality, sex determination is more complicated than a XX/XY sex determination, and many mating, and careful record keeping and brood selection is needed for a YY-male program to work consistently.

Sex in *O. niloticus* is not just a simple XX/XY determination, but there are several genes involved that influence sex determination, as well as environmental conditions influencing their expression. An *O. niloticus* population as a whole may collectively produce a sex ratio of approximately 50% male and 50% female in the progeny, but the sex ratio of individual pair spawns can be highly variable. Tuan et al. (1999) examined the sex ration of 95 pair spawns of Thai-Chitralada strain of *O. niloticus* and the average percent males for these spawns was 50.5%. However, the percent males in a given spawn ranged from 15.5 to 100% males. At Auburn University the sex ratios from individual spawns from three strains of *O. niloticus* held at 28–30°C were determined. The percent males in the progeny were determined for a total of 141 pair spawns with a grand mean of 51.4% male. The percent males by strain averaged 52.3% for 41 Egypt strain spawns, 54.1% for 41 Ghana strain spawns, and 49.0% for 59 Ivory Coast strain spawns. However, the percent males sired by one male could range widely from spawn to spawn while much less variable in the spawns of another male. In one case in the Auburn study, a male spawned four times and the percent males in the progeny ranged from 7 to 62% males, while in four spawns from another male gave a range of 43 to 53%. Where a given female brooder spawned four times with different males, the percent males in the progeny in a given spawn ranged from 16.3 to 55.0% and the percent males in the progeny from four spawns from another female ranged from 26 to 61.6% male.

To complicate the development of brood lines of tilapia that consistently breed true to a specific sex ratio, within a population of *O. niloticus* there may be individuals whose sexual differentiation can be altered in response to the water temperature in which they were held as egg or larvae. Through selective breeding, populations can be established that are highly skewed to male when larvae starting at 10 days post-fertilization are held at 36°C for 10 days. For more details on environmental and genetic interactions of sex determination see the reference section of this chapter. As genetic tools continue to advance it may be possible to develop lines of tilapia that have less modifier genes and will consistently produce progeny with sex ratios highly skewed to males under normal culture conditions. The challenge will be to use such fish on a commercial scale, keeping brood lines pure while having sufficient brooders to support commercial-scale production of food fish.

15.8 Harvest and Distribution

Tilapia are often considered to be extremely hardy, but this characteristic should not be abused. Young fry are sensitive to handing and should be moved with care, keeping them in clean well-oxygenated water as much as possible. Larger fish can be handled at high densities as long as the water temperature is between 23 and 30°C. Tilapia are vulnerable to handling stress, particularly at lower temperatures, and disease outbreaks can occur within a few days after being moved to the new location.

Fry can be easily transported in plastic bags with oxygen at densities of 50 to 100 g/L for 12 to 24 hours. Larger fingerlings are transported in open aerated systems at densities of 150 to 250 g/L. Salt at 0.25–0.50% can be used to help reduce stress during transport. Attention should be given during transport to prevent the temperature from dropping significantly, and should be of special concern when shipping fry in plastic bags with oxygen by airfreight.

15.9 References

Green. B. W., and D. R. Teichert-Coddington, 1993. Production of *Oreochromis niloticus* fry for hormonal sex reversal in relation to water temperature. Journal of Applied Ichthyology 9: 230–236.

Tuan, P. A., G. C. Mair, D. C. Little, and J. A. Beardmore. 1999. Sex determination and the feasibility of genetically male tilapia production in the Thai-Chitralada strain of *Oreochromis niloticus*. Aquaculture 173:257–269.

15.10 Recommended Reading

Ansah, Y. B., E. A. Frimpong, and E. M. Hallerman. 2014. Genetically-improved tilapia strains in Africa: potential benefits and negative impacts. Sustainability 2014(6): 3697–3721; doi:10.3390/su6063697.

Aquaculture Stewardship Council. 2011. Better management practices for tilapia aquaculture: a tool to assist with compliance to the International Standards for Responsible Tilapia Aquaculture Version 1.0., published by the World Wildlife Fund Inc. Available: https://www.asc-aqua.org/wp-content/uploads/2017/07/ASC-Tilapia-Better-Management-Practices_v1.0.pdf. (May 2020).

Bentsen, H. B., B. Gjerde, N. H. Nguyen, M. Rye, R. W. Ponzoni, M. S. Palada-de Vera, H. L. Bolivar, R. Velasco, J. C. Danting, E. E. Dionisio, F. M. Longalong, R. A. Reyes, T. A. Abella, M. M. Tayamen, and A. E. Eknath, 2012. Genetic improvement of farmed tilapias: genetic parameters for body weight at harvest in Nile Tilapia (*Oreochromis niloticus*) during five generations of testing in multiple environments. Aquaculture 338–341:56–65.

Brummett, R. E. 2008. Genetic quality of cultured tilapia stocks in Africa. World Aquaculture 39(2):46–49, 70–71. Available: https://www.worldfishcenter.org/content/genetic-quality-cultured-tilapia-stocks-africa-1. (May 2020).

Chen, J., Z. Fa, D. Tan, D. Jiang, and D. Wang. 2018. A review of genetic advances related to sex control and manipulation in tilapia. Journal of the World Aquaculture Society 49(2): 277–291.

Donga, H. T., G. A. Atagubac, P. Khunraea, T. Rattanarojponga, and S. Senapin. 2017. Evidence of TiLV infection in tilapia hatcheries from 2012 to 2017 reveals probable global spread of the disease. Aquaculture 479:579–583.

Dverdal Jansen, M., H. T. Dong, and C. V. Mohan. Tilapia lake virus: a threat to the global tilapia industry? Reviews in Aquaculture 11:725–739.

Lovshin, L. L. 1982. Tilapia hybridization, Pages 279–308 *in* R. S. V. Pullin and R, H. Lowe-McConnell, editors. The Biology and Culture of Tilapias. ICLARM Conference Proceedings 7. International Center for Living Aquatic Resources Management. Manila, Philippines.

Macintosh. D. J., and D. C. Little. 1995. Nile Tilapia (*Oreochromis niloticus*). Pages 277–320 *in* N. R. Bromage and R. J. Roberts, editors. Broodstock management, and egg and larval quality, Blackwell Science,, Oxford, UK.

Meyer, D. E., and S. T. Meyer. 2007. Practical manual: reproduction and rearing of tilapia fry Escuela Agrícola Panamericana Tegucigalpa, Honduras. Available: http://pdacrsp.oregonstate.edu/pubs/featured_titles/Fingerling_English.pdf. (May 2020).

Mohamed Din, M. S., and R. Subasinghe. 2017. Basic biosecurity manual for tilapia hatchery technicians in Bangladesh. Penang, Malaysia: CGIAR Research Program on Fish Agri-Food Systems. Manual: FISH-2017-10. Available: http://pubs.iclarm.net/resource_centre/FISH-2017-10.pdf. (May 2020).

Phelps, R. P., and T. J. Popma. 2000. Sex reversal of tilapia. Pages 34–59 *in* B. A. Costa-Pierce and J. E. Rakocy, editors. Tilapia aquaculture in the americas, Vol. 2. The World Aquaculture Society, Baton Rouge, Louisiana.

Ponzoni, R. W., N. H. Nguyen, H. L. Khaw, A. Hamzah, K. R. Abu Bakar, and H. Y. Yee. 2011. Genetic improvement of Nile Tilapia (*Oreochromis niloticus*) with special reference to the work conducted by the WorldFish Center with the GIFT strain. Reviews in Aquaculture 3(1):27–41.

Popma, T. J., F. E. Ross, B. L. Nerrie, and J. R. Bowman. 1984. The development of commercial farming of tilapia in Jamaica 1979–1983. Research and Development Series No. 31. International Center for Aquaculture, Alabama Agricultural Experiment Station, Auburn University, Auburn, Alabama.

Popma, T. J., and B. W. Green. 1990. Sex reversal of tilapia in earthen ponds. International Center for Aquaculture and Aquatic Environments Research and Development Series No. 35. Auburn University, Auburn, Alabama. .

Santiago, C. B., and O. S. Reyes. 1993. Effect of dietary lipid source on reproductive performance and tissue lipid levels of Nile Tilapia *Oreochromis niloticus* (Linnaeus) broodstock. Journal of Applied Ichthyology 9:33–40.

Siegfried, K. R. 2010. In search of determinants: gene expression during gonadal sex differentiation. Journal of Fish Biology 76:1879–1902.

Tsadik, G. G., and A. N. Bart. 2007. Effects of feeding, stocking density and water-flow rate on fecundity, spawning frequency and egg quality of Nile Tilapia, *Oreochromis niloticus* (L.). Aquaculture 272:380–388.

World Organization for Animal Health. Tilapia Lake Virus (TiLV) —a Novel Orthomyxo-like Virus. OIE Technical Disease Cards. Available: http://www.oie.int/fileadmin/Home/eng/Internationa_Standard_Setting/docs/pdf/A_TiLV_disease_card.pdf. (May 2020).

Yin, J., Q. Wang, Y. Wang, Y. Li, W. Zeng, J. Wu, Y. Ren, Y. Tang, C. Gao, Huzi Hu, and S. M. Bergmann. 2019. Development of a simple and rapid reverse transcription—loopmediated isothermal amplification (RT-LAMP) assay for sensitive detection of tilapia lake virus. Journal of Fish Diseases 42:817–824.

Chapter 16

Sportfish Species

Providing fish for sport fishing was a driving force behind the establishment of many fish hatcheries in the USA. Fish are produced and stocked into existing natural bodies of water, as well as man-made lakes and reservoirs. This tradition continues and a variety of species and life stages are produced to enhance fish populations and to provide sportfishing opportunities. Fishes being produced for sportfishing include Largemouth Bass, several other species of sunfishes, Striped Bass and White Bass. Several of these species are also being produced as food fish. These species are the focus of this chapter.

16.1 Largemouth Bass *Micropterus salmoides*, Lacepede

16.1.1 Introduction

The Largemouth Bass is considered the most important warmwater sport species in North America. It provides a range of sportfishing opportunities from small farm ponds to major reservoirs in 49 of the 50 states in the USA. It also has been introduced into many countries in Central and South America, Western Europe, Africa, and Southeast Asia, primarily as a sportfish. Largemouth Bass are effective predators of small forage fishes having phytophagous or insectivorous food habits, keeping their abundance in check. This can be in a farm pond setting to control Bluegill reproduction or in tilapia aquaculture setting to control any in-pond tilapia reproduction. Largemouth Bass can be trained to consume formulated feeds and be cultured intensively as a food fish.

Largemouth Bass are native to an area east of the Rockies from southeastern Canada to Mexico, but since about 1887, the species has been widely distributed elsewhere. There are two recognized sub-species of Largemouth Bass, the Florida Largemouth Bass that is native to peninsular Florida, and the Northern sub-species that is found over the rest of its natural range. The Florida sub-species is known for reaching a larger size, but is not as an aggressive feeder as the Northern sub-species. Because of the larger size obtained the Florida bass has been introduced in many areas outside its native range. However, the Florida bass does not have the cold tolerance as seen in the Northern sub-species. The two sub-species will cross where the native ranges overlap, or where Florida bass has been introduced into populations having Northern bass. The two sub-species are crossed under hatchery conditions and the progeny stocked into new farm ponds in the southern USA to obtain the growth characteristics of the Florida bass and the more aggressive feeding response of the Northern sub-species.

Largemouth Bass typically can be found in natural waters where the water current is slow or absent, bottoms are soft and the water temperature rises to more than 20°C for at least a few weeks during the year. Optimum temperature for growth is 25 to 29°C and where such warm temperatures occur throughout the year the growth is extremely rapid for a carnivorous species.

In the USA, Largemouth Bass have been stocked in more than two million recreational fishing ponds in the country, with more ponds being built each year. There is an annual need for Largemouth Bass to stock new ponds and older bodies of water that are being renovated to restore the sport fish population. In addition, Largemouth Bass are sometimes stocked into commercial catfish production ponds to control unwanted wild fishes and tadpoles.

The waters to be stocked with Largemouth Bass are usually of two types: 1) ponds or lakes where the previous population has been eradicated or never existed; and 2) ponds, lakes, and reservoirs where there is an existing fish population, including other carnivorous species. In the first situation, small fingerlings 2.5–5.0 cm TL can be used as they are being stocked into new ponds where no other predacious fish have been added. In the second case, larger fingerlings need to be stocked that are large enough to escape other predacious fishes. Hatcheries need to be able to provide Largemouth Bass of suitable sizes for both applications.

16.1.2 Biology

The life span of Largemouth Bass can be more than 15 years, but in the wild most fail to reach 1 year of age. In managed fish ponds, it can be expected that 30% of stocked fingerlings will die the first year and 15% will be lost the second. The growth of Largemouth Bass is rapid when optimum conditions of habitat, food and temperature exist. When fingerlings are stocked into managed farm ponds in the southern part of this species' range, they should grow to 0.45 kg by the time they are one year old. Under hatchery rearing pond conditions, swim-up size bass fry can grow to 5 cm in a 30 to 50 day period. Young bass begin feeding on zooplankton and feed on insects and small fish as they become larger. Adults eat mainly fish but also consume insects, crayfish, tadpoles, adult frogs and any other animal that will fit inside their mouth. Juveniles can be trained to accept formulated feeds and grown to adults on such feeds.

In the southern part of their natural range, Largemouth Bass will be mature when one year old. Secondary sex characteristics are poorly defined in immature Largemouth Bass. Mature fish can be sexed with reasonable accuracy just prior to spawning based on the males releasing small amounts of sperm when pressure is applied to the vent area. Females with well-developed ovaries are recognizable, particularly when held ventral size up. The two lobes of the ovary should be distinct giving the abdomen an appearance of having two round ridges running the length of the abdomen with a slight depression between. When looking at the ventral side of a male, the abdomen will have a more rounded appearance along the centerline of the abdomen. The shape of the papilla can also be used as a guide with females having a pear-shaped papilla while that of the male is more rounded.

A more accurate approach for sexing Largemouth Bass is holding the fish ventral-side up, and using a straw or a plastic catheter to probe the urogenital opening. In females, the probe should enter on a downward angle toward the head passing down the oviduct and contact the ovary. With males, the probe would enter the urogenital opening, but go on a flatter angle and after going a short distance enter the urinary bladder. This technique allows a more accurate sexing of brooders particularly when sexing fish out of the spawning season.

Largemouth Bass spawn naturally in static water and flowing water settings. They are nest builders. The male selects and prepares the site and guards it against all intruders. Preferred sites are protected, defensible locations in shallow water with a firm bottom or a fibrous substrate such as tree roots. They will also clear out nest sites among stands of aquatic plants or flooded grassy areas. Eggs are demersal and adhesive. Fertilized eggs range in diameter from 1.2 to 1.8 mm with smaller and younger fish producing the smaller eggs. This is illustrated by the data in Table 16.1.1 that also suggests that diet affects both egg size and number produced by individual fish. Number of fertilized eggs/mL of eggs averages 571 for 1-year old fish as compared to 374 for 3-year old fish. Table 16.1.1 also gives the number of eggs/spawn when fish are allowed to spawn naturally in a hatchery pond setting. However, the number of eggs/spawn is not necessarily the yield from a single female, as on occasion, more than one female may spawn in a nest, and a given female may spawn several times in different nests. Egg counts of samples from ovaries of gravid un-spawned fish indicate a potential to yield 65,000 to 100,000 eggs/kg of body weight.

Spawning activity of Largemouth Bass is closely related to spring temperature fluctuations. The temperature units (degree days) required for maturation and release of eggs has not been reported. In temperate climates the date of spawning can vary. Successful reproduction has occurred at the Marion Hatchery, Marion, Alabama, as early as January 18th during an unusually mild winter. February spawning occasionally occurred but the peak of spawning activity occurred in mid-March. A water temperature of 18.3°C is generally recognized as the minimum spawning temperature for this species. However, bass do not always begin to spawn when this temperature is reached. Often if the winter has been cold, spawning will not take place until the second or third time the temperature warms to this level. Characteristically, this species spawns as water temperatures rise. During the spring, shallow ponds show a fluctuation of temperature related to weather changes. The majority of the brooders may spawn as the water temperature rises to 18.3°C, but later spawning may continue for weeks after the "main spawn" has occurred as long as the water temperature remains above 18°C. Younger females tend to spawn later in the season than the older, larger ones.

Table 16.1.1. Characteristics of spawns from Largemouth Bass *Micropterus salmoides* given artificial or natural diets.

Year Class Diet	I OMP[1]	II Live Fish	II OMP	III OMP
No. Spawns	10	6	10	11
Av. No. eggs/ml	571	486	392	374
Range, eggs/ml	444–674	470–504	320–505	319–422
Av. No. eggs/spawn	9,551	19,410	21,744	15,223
Range, eggs/spawn	6085–14669	13115–29793	11,116–42,903	8,588–33,003
Av. % viable eggs	79.8	63.5[2]	91.0	92.1
Range, % viable	35.0–97.0	2.3–92.1	73.2–98.1	51.9–99.1

[1]Oregon Moist Pellet (37% protein, 30% moisture).
[2]78.8% excluding a spawn where survival was 2.3%

Incubation time varies with temperature and can be as long as 317 h at 10°C to as short as 49 h at 28°C. At hatching, fry may range in TL from 3.0 to 5.5 mm and continue to grow and develop through the yolk sac absorption period that may last 120 h or longer. By the time of first feeding, the fry are 5.5–6.5 mm and the body color has changed from amber or cream to some shade of brown.

Largemouth Bass larvae characteristically attack and try to swallow the largest food items possible when a choice exists. First feeding larvae prefer copepod nauplii over smaller zooplankton such as rotifers. As the fry grow, there is a rapid progression to larger zooplankton and small insect larvae, with larger insects and small fish being taken by the time the bass reach a size of 30 to 40 mm. It is believed that a TL of about 40 mm marks the point where the carnivorous food habits of the species become dominant. Lacking small fish for consumption, young bass can grow to a size of 150 to 180 g on a diet of insects and tadpoles. Food of the adults under natural conditions is primarily fish, but other animals are sometimes consumed somewhat incidentally.

Table 16.1.2 gives the size bass fingerling needed to consume common fingerling fishes of several typical species that might be present in a given environment. Fingerling bass up to a size of 7 to 10 cm can be produced on a diet of insects. For Largemouth Bass above a size of 10 cm TL, a live fish diet or formulated feed should be employed to achieve the desired growth.

16.1.3 Broodstock Management

In the culture of any species, development of broodstock is of primary importance. The rapid growth and development of Largemouth Bass makes raising adult fish relatively simple and there is little excuse for a hatchery not having young, vigorous brood fish for spawning each year. The experience of a number of hatcheries throughout the southeast indicates that excellent results can be obtained from 1-yr-old fish that have reached a size of 320 to 454 g (Figure 16.1.1). Where 1-yr-old fish do not reach this size range, 2 years should be allowed for production of brood fish.

A key factor in developing Largemouth Bass brooders that will spawn at one year of age is supplying an ample quantity of forage fish of a consumable size during the time water temperature is warmer than 15°C. Table 16.1.2 illustrates the relationship between Largemouth Bass and some prey species. Amounts of live food equal to as much as 20% of the body weight of the bass/day may be consumed under optimum conditions, resulting in extremely rapid growth. Under ideal pond conditions, it is reasonable to expect bass to grow to a size of 180 to 320 g in 120 days at water temperatures of 24 to 30°C. Where a growing season of 225 to 250 days (frost to frost) exists, spawning size bass can be reared on a formulated feed in one year when stocked fingerlings at 40,000/ha. Periodic grading will be necessary to minimize cannibalism, removing fish that have not been aggressively feeding on the formulated feed as well as any apparent cannibals that are much larger than their siblings.

How the bass fingerlings are to be used is an important question to answer when developing a broodstock management plan. Suitable brooders can be developed on the hatchery to meet the demand for fingerlings, particularly when producing fingerlings for stocking farm ponds. However, if the fingerlings are produced to be stocked into a natural setting they should have a similar genetic profile to the largemouth bass in the locale where the hatchery fish are going to be stocked. The exception would be where the Florida-strain of Largemouth

Table 16.1.2. Forage sizes able to be consumed by various size classes of Largemouth Bass (from Lawrence 1958).

Largemouth Bass		Forage Fish (total length cm)				
Total Length (cm)	Mouth Width (mm)	Gizzard Shad	Bluegill Shiner	Golden	Goldfish	Largemouth Bass*
3.8						2.4
6.4	6.8	3.9	3.6	4.5	4.0	3.4
8.9	8.8	4.5	4.0	5.2	4.5	4.4
11.4	10.8	5.2	4.5	6.0	5.0	5.4
14.0	13.7	6.2	5.2	7.2	5.8	6.8
16.5	16.5	7.2	5.8	8.2	6.5	8.2
19.0	19.3	8.1	6.5	9.3	7.2	9.6
21.6	22.7	9.3	7.3	10.7	8.1	12.0
24.1	25.9	10.4	8.1	11.9	8.9	13.2
26.7	29.3	11.6	8.9	13.2	9.8	14.5
29.2	32.5	12.6	9.7	14.6	10.6	15.8
31.8	35.5	13.6	10.4	15.6	11.4	16.9
34.3	39.1	14.9	11.3	17.0	12.4	18.3
36.8	42.5	16.0	12.1	18.3	13.3	19.6
39.4	46.1	17.2	12.9	19.7	14.2	22.9
41.9	52.8	19.5	14.5	22.4	15.9	24.9
44.5	57.7	21.2	14.7	24.3	17.2	26.3
47.0	62.8	23.0	16.9	26.2	18.5	27.8
49.5	67.7	24.6	18.5	28.1	19.8	29.2
52.1	72.7	26.3	19.3	30.1		29.8
54.6	78.9	28.4	20.8			34.2
57.2	85.1	30.5	22.3			35.5
60.0	91.5	33.7				36.9

*Cannibalism

Bass are stocked in locales where the Northern-strain of Largemouth Bass exists in an effort to enhance the sport fishery.

A hatchery manager is often caught in a dilemma: broodstock developed on the hatchery over several generations become domesticated and are easier to produce, but may not be as genetically fit for stocking back into the wild as a less domesticated fish. Dependence on wild fish is generally not desirable in an aquaculture program as little is known about their previous history and such fish may not be available on a short notice to meet a hatchery's schedule. In addition, wild fish may not be resistant to diseases present in the hatchery, or may introduce a disease that was not previously present at the hatchery. Where fish are being produced for stocking in the wild, a periodic reintroduction of wild fish should be done to avoid excessive domestication. Wild bass should be collected from the same watershed that their progeny are to be stocked back into. To best represent the genetic characteristics of the wild population, at least 100 potential brooders should be collected from the wild. However, adverse public opin-

Figure 16.1.1. Largemouth brooder can be sexually mature at one year of age and are often suitable as brooders.

ion among sport fishermen may also develop from this practice unless a good public relations program is carried out to explain why wild fish are being collected as brooders.

Largemouth Bass brooders can be produced by stocking a brood pond with 250 to 370 selected bass fingerlings/ha, along with an adequate supply of forage fish. A minimum of 5 kg of forage fish is needed to be consumed/kg gain by the brooders. Young brooders should gain 100% of their body weight from one spawning season to the next while a 50% weight gain by older brooders would be acceptable. As an example, based on a 50% weight gain, a 5:1 FCR and a 350 day growing season, the quantity of forage needed to be consumed from one year to the next by a 1 kg fish is as follows:

Initial weight = 1 kg
Final weight = Initial wt. (1 kg) + 50% weight gain (0.5 kg) = 1.5 kg
Forage fish requirements at a 5:1 feed conversion
Maintenance, at 0.5% body wt./d = 1.5 kg × 0.5% × 350 days = 2.62 kg forage
Forage for wt. gain = 0.5 kg gain × 5: 1 FCR = 2.5 kg forage
Total forage to be consumed = 5.12 kg

The time of year when food is available is important in bass nutrition. Bass form eggs in the late summer and fall of the year preceding spawning. Food intake and rate of digestion slow to a marked degree when water temperatures drop below 10°C. Metabolic activity also decreases and growth rate probably declines if water temperatures rise above 32°C.

A loss in weight of 10 to 20% can be expected for brood fish during the spawning period. Lost weight should be regained and some additional growth made before egg formation begins. To do this, a suitable food supply should be made available to the fish as soon as they are placed in the summer holding pond.

Attention should be given to quantity, quality, and availability of forage fish provided. The quantity must be adequate to meet maintenance requirements and desired weight gain. Quality refers to the forage being: a suitable species of fish that can be produced on the hatchery, in good condition and not a disease carrier, and is small enough for capture and consumption by the bass. Availability pertains to the amount of cover and depth of edges in the pond where the bass brooders are held. The pond should be free of filamentous algae or rooted vegetation that would provide cover shelter for the forage fish. Pond edges should have a minimum depth of 60 cm at 1 to 2 m from shore so that the adult bass can range over the entire pond and readily capture the forage, even those close to the shoreline.

The density at which broodstock are held from one spawning season to the next is important in conditioning the fish for optimum performance. If broodstock are maintained on a diet of forage fish, the maximum standing crop of bass and forage should not exceed 1,200 kg/ha. This figure includes provision for a weight gain of 50%. When a formulated diet is given, the density of brood bass should not exceed 1,000 kg/ha to minimize the risks of poor water quality associated with feeding higher standing crops. Separate ponds should be used for holding brood males and females, holding each sex separately by year class. Older brooders will spawn earlier in the spawning season than younger ones and if both are held in the same pond seining out the older brooders may stress the younger fish. If enough ponds are not available, brooders can be tagged by year class and held in a communal pond.

Frequent inspection of the brood holding pond is recommended during the summer, fall and spring months. The abundance of appropriate-size forage fish should be determined every 2 to 4 weeks during these months by seining. The forage supply should be replenished when a quadrant seine haul with a 4.6 m seine yields less than 15 to 25 forage fish of a size which can be swallowed by the average size bass being fed, The practice of feeding tadpoles, scrap Bluegill, crayfish, and odd lots of miscellaneous fishes that may occur at a hatchery cannot be depended on as a satisfactorily source of forage for the hatchery broodstock. Instead, ponds should be set aside for producing an adequate quantity of a suitable forage species of a usable size to provide the calculated amounts of forage needed for both maintenance and growth of the bass broodstock being held for future use.

Species of forage fish that appear to be best for propagation as bass food include Goldfish *Carassius auratus*, Common Carp *Cyprinus carpio*, Fathead Minnow *Pimephales promelas*, and Golden Shiner *Notemigonus crysoleucas*. Goldfish are preferred because of ease in spawning, high production, and rapid growth. Common Carp have many similar characteristics as Goldfish, however, they tend to grow too fast and become too large to be consumed by brood bass. Another disadvantage is that Common Carp do not spawn over as long a spawning period as the Goldfish, thus limiting the availability of small-size forage. Fathead Minnows and the Golden Shiner seldom get too large for bass to swallow, but the yield/ha from the forage production ponds is not as much as Goldfish and Common Carp. Moreover, when adult Fathead Minnows and Golden Shiner are stocked into ponds designated to bass brood holding ponds before the bass are stocked, their progeny do not grow rapidly enough to be of a desirable size by the time bass are returned to the brood holding pond. Adult Fathead females range from 6 to 7.5 cm and males 7.5 to 8.5 cm, but larger size forage are preferred for 2.5 to 3.0 kg and larger bass brooders. Additional fish from a forage production pond should be added over the course of the summer.

Bluegill can be employed instead of one of the above species, but they do not seem to be as preferred by bass as much as some of the cyprinids, and cannot be produced in as large a quantity/ha. They also require a higher water temperature for spawning and that influences the time when the young-of-the-year are large enough for forage.

Tilapia are excellent forage species, easy to propagate, prolific, fast growing, disease resistant, and extremely hardy for transferring in water warmer than 21°C. However, they will not survive at water temperatures much below 12°C. Broodfish must be overwintered under controlled conditions at 21°C or higher and stocked in the spring when water temperatures are >21°C. Their progeny will become forage for the bass over the course of the summer. This frequent spawning over the summer is an advantage over using Goldfish or carp as forage for Largemouth Bass.

Goldfish, carp, and tilapia at rates of 225 kg/ha or more have an added advantage in that their feeding habits help inhibit the development of noxious species of vegetation, making vegetation control unnecessary in ponds.

When bass brooders are moved to spawning ponds, the ponds that were holding the brooders should be refilled and stocked with mature forage fish so they can spawn in that pond while the bass brooders are spawning in another pond. This will allow more forage to be produced in the pond before the bass are returned. During this period, the pond should be fertilized and the forage fish given a supplemental feed. This should result in 335 to 675 kg/ha of forage fish large enough for bass food being present in the brood holding pond when the bass are returned. In addition, other ponds will need to be set aside to produce additional forage that can be added to the brood holding pond to ensure that adequate quantities of forage are maintained in the bass brood pond. These should be available beginning 6 to 8 weeks after bass have been returned to the brood holding pond. An abundant supply of consumable forage fish should also be available in the spring before the brooders are transferred to the spawning pond. Broodfish usually are not fed when they are in the spawning ponds.

Where pond space is limited, replacement bass can be grown in Bluegill or Redear spawning-rearing ponds, sacrificing the small sunfish consumed by the bass in return for the savings in pond space. A stocking rate of 12 to 25 bass fingerling/ha can be employed without seriously reducing bluegill fingerling production. Table 16.1.3 illustrates production of Bluegill when this method was followed. In the southern part of their range, brood Largemouth Bass can be grown to maturity in a year, while in the northern part of the range they may not be mature until two years old.

Largemouth Bass can be trained to consume pelleted feed and grown to be suitable broodstock without having to provide live forage. Brood bass have been successfully maintained for as many as seven spawning seasons on the Oregon Moist Pellet ration. Feed costs are comparable to the cost of feeding live fish. More labor is required when feeding a pelleted feed since feeding should be done daily during the growing season. Feeding a pelleted formulated feed helps to reduce the possibility of bass tapeworm infestation that can be transmitted through the forage fish, and allows controlling bacterial diseases by feeding medicated feed.

Table 16.1.3. Results of growing brood bass in Bluegill *Lepomis macrochirus* spawning-rearing ponds.

Pond (ha)	Bass Harvested (no./ha)	Av. wt. Bass (g)	No. Bluegill harvested/ha	Estimated Bluegill loss (no./ha)
0.69	10	454	247,000	60,000
0.73	15	345	355,000	45,000
0.52	22	454	270,000	93,000
0.52	47	358	250,000	150,000
0.45	47	195	116,000	74,000
0.32	99	340	34,000	559,000
1.30	141	318	37,000	286,000

Formulated feeds can be used to produce brooders but growth will be slower and egg quality the first year may not be as good as from fish given a natural diet. Spawns were obtained from 135 g female bass that had been fed an artificial diet. The small eggs they produced had a low hatching percentage. This was attributed to the small egg size and not to the diet as two-year-old fish fed the same diet produced large eggs with a high percentage hatch.

After the initial broodstock have been obtained, annual replacement of 25 to 33% of the population should be planned for. This would include: bass culled at the start and end of the spawning season; fish lost in the spawning pond, particularly males as they are more vulnerable to predation as they guard nests along the shoreline; fish that did not survive through the winter; and fish that have become too large for efficient use. Some culling at the beginning and end of each spawning season should be used to eliminate malformed, diseased, extremely thin, or otherwise undesirable individuals. If this practice is conscientiously followed, the overall health and quality of the remaining brooders will help to ensure a high survival of the remaining fish. Bass can be expected to spawn satisfactorily for three or four seasons or longer. Excellent results have been obtained following a practice of replacing broodstock after they have been used for three spawning seasons. Thus the stock would be replaced at the end of 3 years of age in the southern USA and 4 years in the north.

There are several advantages associated with replacing brood fish when they are 3–4 years old. If growth is normal, the fish will have reached between 1.4 to 1.8 kg in size by the end of this time. Fish larger than 2 kg are more difficult to handle, require more food for maintenance and have a greater risk of mortality during succeeding holding periods. At hatcheries where bass tapeworm *Proteocephalus ambloplitis* is a chronic problem, this parasite will damage the ovaries of the bass and after several years cause enough damage to greatly limit egg production (Figure 16.1.2). The major disadvantage of disposing of older broodstock, aside from the psychological urge to retain a large fish, is the greater production of young/brooder possible from the larger, older fish. It is often easier to meet production goals by spawning a few large female bass rather than a number of smaller brooders but such approach limits genetic diversity.

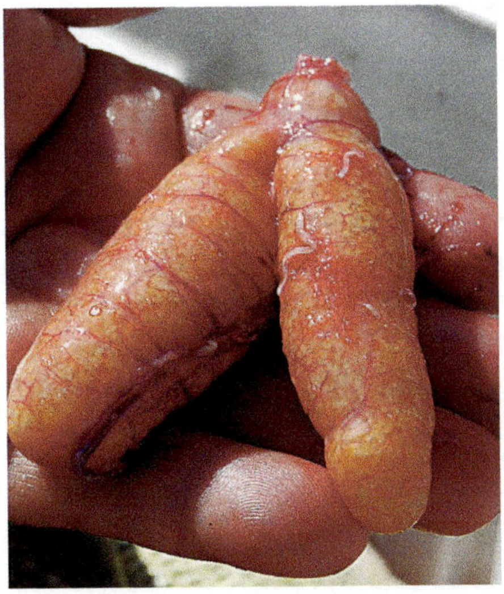

Figure 16.1.2. Largemouth Bass tapeworm *Proteocephalus ambloplites* capable of causing significant damage to the ovaries of Largemouth Bass.

Brooders that are older than 4 years old can still be effectively used as long as they do not weigh more than 2 kg. In an experiment where brood fish were maintained on an artificial diet of Oregon Moist Pellet, growth was slower than for those receiving forage fish. At 7 years of age, artificially fed bass weighed 1.4 to 1.8 kg. These fish spawned satisfactorily for seven spawning seasons but brood mortality over the 7-year period was about 88% of the initial number.

Broodstock replacement should be done in a systematic basis, with each year-class being identified so that replacement can be made with certainty. Keeping the age groups separate in different ponds is the most effective way to keep track of the brooders by year-class. Fin clipping can be used as a temporary way to identify the year-class but fin regrowth can confuse the identification of fish that were previously clipped. Tagging, particularly with PIT tags, can be effective. Generally, the 1-year-old fish can be separated from the 3-year-olds on the basis of size.

Bass brooders handle better when the water temperature is 15–26°C and when being moved they should held in water containing 0.25–0.50% salt. When moving them from one set of water quality conditions to another they should be acclimated, paying particular attention to temperature and pH. The fish should be handled with a soft-mesh net avoiding knotted netting whenever possible. Keep in mind that any stress to the brooders when collecting them from brood ponds and stocking them into spawning ponds may delay spawning and in turn disrupt the production schedule.

16.1.4 Fry Production

Three general methods of producing largemouth bass fry have been successfully used: 1) Spawning-rearing pond method, 2) Fry transfer from ponds, and 3) Egg transfer from ponds and spawning tanks.

16.1.4.1 Spawning-rearing

With the spawning-rearing method, adult fish are stocked in a suitable pond at low density for natural spawning. The young bass that are hatched in the spawning pond remain in the pond a month or more until they have grown to a size large enough to be distributed. This approach can be appropriate for hatcheries where ponds are unsuitable for use as rearing ponds or where the workload is such that fry transfer is not feasible. Spawning-rearing has the advantage of lower manpower requirements, since no fry transfer is involved. Handling damage to small fry is eliminated, and the adult fish guards the developing young against predator attacks. Clarity of the pond water is not necessary for harvesting the fish. Spawning-rearing has the disadvantage in that the yield of fingerlings can be unpredictable both in the number of fish obtained and the size of fish harvested. Cannibalism becomes a major cause of fish loss if harvesting is delayed.

In the spawning-rearing pond method, a low stocking rate of 50 to 75 adults/ha is generally used. Fertilizer application is similar to that recommended for a rearing pond. When the schools of small fish disperse, prompt harvesting by draining the pond is advisable unless an effective trapping method can be employed to remove 50 to 75% of the fingerlings in a short period of time. Failure to separate the young and adult fish after the young reach a size of about 3 cm can result in a loss of fingerlings from adult predation. About the same time, the small bass also begin to consume one another, bringing about further losses. An added disadvantage of this method is the greater unpredictability of the spawning-rearing pond in producing a known number of bass on a predetermined date as needed to meet distribution schedules. Necessity of harvesting small bass quite early makes it difficult on occasion to meet any minimum production size requirements that may exist. These factors render the spawning-rearing pond method less desirable than the fry transfer method for producing fingerling bass, although it may still have a place on some hatcheries or in certain situations.

16.1.4.2 Fry Transfer

Fry transfer is the traditional approach for largemouth bass production at many hatcheries. Adult fish spawn naturally in ponds and the fry are collected, typically by seining, after they leave the nest but before the schools break up. The fry are transferred to prepared rearing ponds and stocked at specified densities for growth to distributable size.

Pond selection is a key step in the successful production of bass fry using a fry transfer system. As mentioned before, ponds have personalities and this particularly applies to selecting ponds for bass spawning using fry transfer. Keeping good records is important in learning a pond's personality. The fact that a given pond has produced well in several previous years is the best assurance available that it should again be used. Conversely, a pond that was a failure or a mediocre producer in former years should be rejected unless the cause of failure has been determined and corrected. A desirable spawning pond for fry transfer is one of moderate depth, protected from wind action, and range in size from 0.3 to 0.6 ha. The pond should not have chronic weed problems that would interfere with seining or provide shelter for the bass fry, or develops dense phytoplankton blooms that would make it difficult to see schools of bass fry as they swim along the shoreline. A pre-flooding treatment of the pond bottom with an acceptable herbicide is suggested if undesirable vegetation has previously been a problem.

Drying the pond thoroughly prior to flooding and stocking is recommended where possible. It is desirable to flood the pond 10 days to 3 weeks prior to the time bass fry are expected to begin feeding unless a residual supply of food from a previous production cycle is present, as would be the case when a pond has been drained and immediately refilled for the bass crop. The 3-week period provides enough time in the early spring for natural food organisms to develop in the pond to provide live food suitable for the bass fry. By late spring or early summer, 10 days are probably adequate for this purpose. Also, if the water supply is surface water (e.g. from a reservoir pond or streams), it typically carries food organisms that will speed up the development of a suitable food chain in the nursery pond.

A majority of bass culturists prefer to leave the spawning pond unfertilized to reduce the possibility that a bloom of phytoplankton will develop and make observation of fry difficult. However, zooplankton is desirable as a food for the bass fry. A conservative rate of organic fertilizer may be added to stimulate a zooplankton bloom. Superphosphate at a rate of 9 kg P_2O_5/ha/application is generally combined with the organic material. If the organic fertilizer, such as grass hay, is low in nitrogen, then a source of inorganic nitrogen, at a rate of 4.5 to 9.0 kg N/ha, should also be added. Development of a phytoplankton bloom can interfere with observation of spawning activity and the collection of small fry. However, it does not seriously hamper the removal of larger fry of a size of 15 to 20 mm, as this size fish can be harvested by trapping or seining as they begin to cruise along the pond shoreline in search of food.

The best time to stock brood fish is thought to be about the time that the main spawn has been observed at the hatchery in past years. This can be determined by checking hatchery records for previous years. If adequate hatchery records are not available, a guide can be based on the average date in the spring of the last killing frost recorded by the U.S. National Weather Service or equivalent agency for your area. At this time the brood fish are examined and the ripe fish sorted and stocked. Gravid females are selected on the basis of an obviously distended, soft, pendulous ovarian region, and a swollen, red and protruding vent. When viewing the ventral side of a female, the two lobes of the ovary should be evident with a slight depression between them (Figure 16.1.3). Mature males will flow a small volume of a watery milt when stripped that may be hard to see if the fish is stripped more than once. The ventral side of a male should not show the two lobes as seen in a female but should have a rounded center ridge extending the length of the abdomen.

Lower quality fish should be returned to the holding pond for a week or two and then examined again to see if their quality has improved. Often a week or two of additional holding is enough for young brooders to develop into good quality brooders. Often less than optimum quality brooders will spawn but not yield as many eggs. Such fish can be transferred to a separate spawning pond and fry obtained if the demand for fry is very high that season to justify the effort.

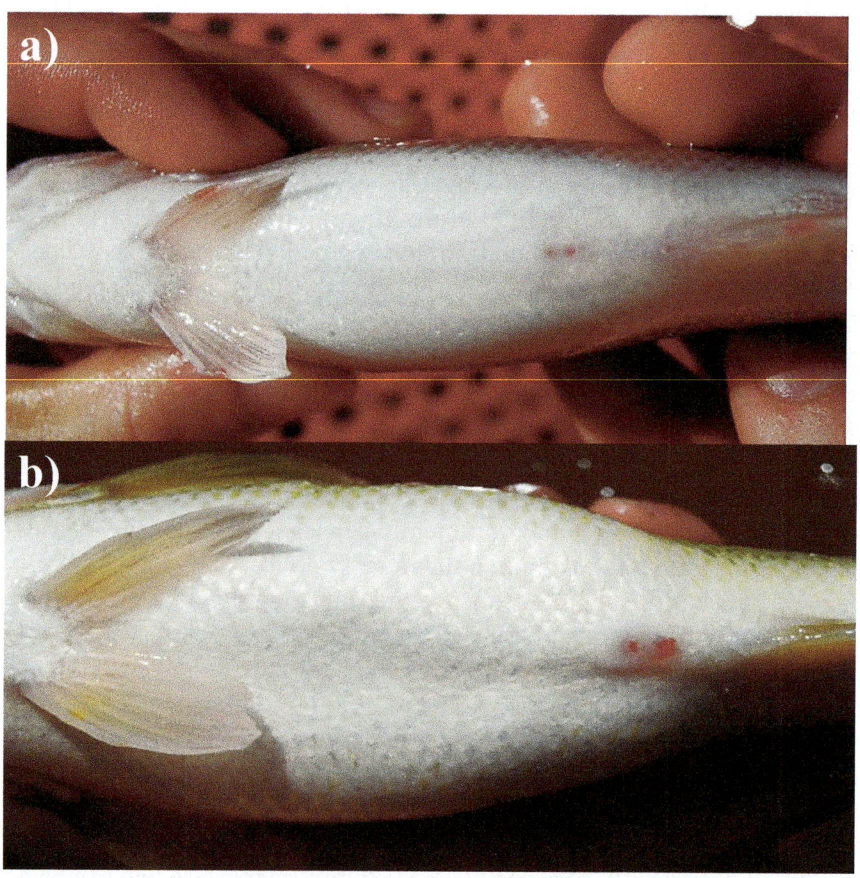

Figure 16.1.3. Male (a) and female (b) Largemouth brooders.

A key aspect of a fry transfer plan is to get as much fry as needed as quickly as possible to stock available nursery ponds at a specific point in time. That means having the brooders spawn successfully in a short period of time and produce enough fry for completely stocking a pond with similar age fish. In temperate climates, close attention needs to be given to the weather before stocking the brooders. It is common to have 1–2 bands of warm weather to come in the spring that last only 4–5 days. If brood males and females are stocked into a spawning pond during one of these short periods of warm weather where the water warms to about 18°C, the fish may spawn. However, another cool band of weather may arrive that slows down the development of any bass eggs in the nests, giving the opportunity for fungus to develop over the eggs and the male may quit guarding the nest site. A few fry may survive in the spawning pond and may become cannibals on fry from later spawns. The survivors from any early spawn can get mixed in when fry from later spawns are collected for transfer to nursery ponds. This can result in a severe cannibalism problem in the nursery pond.

Ideally, male and female brooders are held in separate ponds through the winter until water temperatures have stabilized at 18°C or above before they are stocked into spawning ponds. If male and female brooders have to be held together over the winter they should be separated and held separately a month before the start of the spawning season to avoid any spawning in the brood holding pond. After reviewing water temperature data from past years as to when the water temperature should remain ≥18°C. the brooders can be stocked on the appropriate date when active spawning can be anticipated along with good hatch rates.

The number of brood fish to stock in the spawning pond is dependent upon the following factors: 1) number of fry desired, 2) size of spawners available, 3) condition of spawning fish, and 4) productivity of pond being used. Federal warm-water hatcheries usually stock brooders at 100 to 250 adults/ha. Considering the reproductive potential of the females and the amount of space utilized by a nesting male, 85 to 100 fish should be stocked/ha when using 1.4 to 1.8 kg brood fish. The stocking rate can be increased 175 to 200/ha when using 350 to 450 g brooders. Overstocking will reduce the percentage of brooders that spawn successfully.

Use of more males than females is recommended. A sex ratio of 2:1 or 3:1 is considered to be optimum. A female fish will not release all her eggs at a time and one female may spawn in more than one nest during a short period. In order to have nests for receiving all eggs potentially available, plenty of males are needed.

When gravid females and mature males are stocked into clean ponds containing water of 18–19°C or higher, nesting activity often begins within 24 hours and spawning will normally occur within 72 hours post-brood stocking. Fry generally will hatch 48 to 96 hours after spawning. The transparent larvae are about 4 mm long at hatch and are almost invisible in the nest for 5 to 7 days. They come up off the nest as swim-up fry at 8 to 10 days after spawning. They can be transferred to rearing ponds any time after they swim up from the nest. Table 16.1.4 shows the time sequence involved in spawning and rearing Largemouth Bass of an assumed spring temperature regime.

Spawning ponds should be monitored daily to observe the brood bass behavior. The presence of nest sites along the shoreline should be noted along with the male behavior. Males often do not aggressively guard recently built nests if there are no eggs in the nest and will swim off as a person walks up. Males guarding eggs in a nest will guard the nest more actively. Males guarding recently laid eggs will often swim off several meters when someone walks up and may not return until the person leaves. Males guarding nests having eggs close to hatching or that contain sac fry will guard the nest more aggressively and only swim away a

Table 16.1.4. Time schedule for rearing 40 mm bass fingerlings at a water temperature of 19 °C[1].

Day 1	Fill spawning pond when fish are ripe and water warms to 18°C
Day 2	Stock brooders
Day 4	First Spawns
Day 7.5	Hatching
Day 12	Mouth forms
Day 14	Free swimming fry (swim-up)
Day 16	50% of fry are feeding
Day 17	Yolk sac absorbed
Day 28	Advanced fry stage (15 mm)
Day 42	40 mm fingerlings

[1]For each 1° deviation from 19 °C adjust time 0.1 day (2.4 hr.)/24 hours at the different temperatures within a range of 15 to 30 °C.

meter or so then quickly return to the nest. It is not uncommon for such males to bite at a stick or a person's hand when placed near the nest. Observing male behavior will give you an idea of how soon there might be fry that need collecting from a specific nest site. A small sample of eggs can be taken from the nest to get a closer idea of how old the eggs are in a nest, with younger eggs being somewhat clear to a light gold/brown in color, while older eggs will be a dark brown/gold with dark eyed embryos.

Fry that have just begun to swim-up will be in a tight school in the immediate area of the nest. They can be harvested using a dip net frame where mosquito netting is tightly stretched across the frame. The net, approximately 40 cm in diameter, is placed under a school of swim-up fry and gently lifted up to the surface. The net is lifted out of the water allowing the water to drain off, stranding the fry on the netting. The net is then inverted into a tub of water to rinse the fry off into the water.

When the bass fry are 10–15 mm they can be seined out of the spawning pond and transferred to a nursery pond. Before harvesting, a staff member should walk the pond bank to find schools of bass fry appropriate for collection by first looking for the male and his movement. This should be done in the morning before the wind starts making waves that make it more difficult to see the fry. Males guarding schools of fry typically are swimming on the outside of the school. Observing the male will make it easier to find the school of fry. Once the school is seen, the size of fish should be noted as well as how the school is dispersed along the pond bank. The fish can be harvested using a one-person push net (0.5–1 m wide with a 25 cm deep bag, with a 1 mm mesh). They can also be seined by a two-person crew using a 3 × 1.2 m seine made of a 0.75–1.5 mm soft mesh material. The trouble with seining or using a push net is that walking in the water will make the water turbid and difficult to see any fry that may have escaped, or to find other nearby schools. To avoid any turbidity issues the fry can also be collected using a push net from a boat. Once the fry are crowded together in the net they should be moved in water to a transport container. Once captured, the fry can be enumerated by visual estimation (Figure 16.1.4).

As the fry grow larger they are stronger swimmers and will swim along the shoreline not necessarily near the nest site. Such fish need to be seined or trapped out of the spawning pond as soon as possible. By the time the fish reach 20 mm the school starts to break up and the fish be-

Figure 16.1.4. Visual estimation of the quantity of swim-up Largemouth Bass collected.

come more difficult to harvest. At the same time the male brooder will quit guarding the school making the fingerlings susceptible to be eaten by the adult brooders. Any reproduction from the first spawns that escaped capture can cannibalize on the fry produced from later spawns.

Disease, inadequate food, cold windy weather, or predation by either adult or fingerling bass may cause a school of fry to break up. At Marion National Fish Hatchery fry were trapped or seined from the pond edge even if the schools broke up. Little difficulty was experienced in collecting 250,000 to 375,000 fry/ha or more in this manner, and ample fish can be obtained to stock rearing ponds where fry schools have broken up as long as a disease epidemic or cannibalism has not occurred.

If a spawning pond is harvested frequently enough it can provide fry for 2–3 weeks. After that point the brooders should be seined out and the pond drained to harvest any remaining fry.

16.1.4.3 Egg Transfer

With egg transfer, the male bass are not allowed to build nests on the natural pond bottom. Instead an artificial substrate is provided as a nest site. Spawning takes place over the provided artificial substrates and the substrates with eggs attached are collected and taken to the hatchery for incubation. This is a much more intensive approach that the two earlier methods described but yields are high, less pond space is needed for spawning, production goals are met quickly, and the output is more predictable. Egg transfer is also appropriate for indoor spawning under controlled water temperatures allowing an advancement of the natural spawning season.

Feed-trained brooders are preferred when using the egg transfer approach. Brooders fed formulated feeds are accustomed to being held at high densities in brood ponds and adapt well to being stocked at high densities in spawning ponds. They are also accustomed to hatchery staff being daily at the brood holding pond and will be less disturbed by human presence around the spawning pond.

Small ponds (0.04 to 0.10 ha) with an average depth of 90 cm are preferred as spawning ponds when using the egg transfer approach. Small ponds have more shoreline relative to their total area and are easier to monitor. Walled ponds are preferred where the water depth next to the wall is 20–25 cm deep. This will allow spawning mats to be placed close to the wall so that the mat can be placed and removed from the pond bank without a staff member having to enter the water. Mats being placed close to the wall makes it easier for the male to guard the mat. However, small ponds are more subject to water temperature change, so it is important to not stock brooders too early before the temperature stabilizes. The spawning pond should not have a history of aquatic weed problems or dense algal blooms. The pond should be filled and not stocked until any turbidity has settled out. Gypsum can be added at 1,000 kg/ha to settle any clay turbidity.

Spawning mats can be made of several types of materials. Some hatcheries have used nylon-felt pads or doormats; others prefer rubber-coated fiber mats such as Spawntex®. The size of the mat may be as large as 45 × 76 cm for use with larger brooders or as small as 25 × 25 cm for use with 1-year old brooders. The mats should be placed in the spawning pond the same day the brooders are stocked. If they are placed too soon they may get covered with sediment and may be less attractive to the males. The mats should be pinned securely to the pond bottom. A large 10-cm nail going through the mat pinning it to the bottom is adequate to keep the mat in place. The nail should be placed in the center of the mat as the nail head is often the focal point of egg depositing. Mats should not be held in place with a rock or other object that offers a higher profile than the mat itself. If a rock is placed on the mat the brooders may spawn on the rock making egg collection and incubation more complicated. Mats should be spaced about 3 m apart in a single row around the pond. Spacing the mats much closer may result in the males fighting to establish their territory instead of protecting the eggs (Figure 16.1.5).

Figure 16.1.5. Placing spawning mats around the shoreline of the Largemouth Bass spawning ponds.

The number of brooders to stock depends on the age of the brooders available. A stocking rate of 5,500 fish/ha is appropriate for 1-year-old fish and 2,000 to 5,000 fish/ha for older fish at a sex ratio of 2 females to 1 male. Nesting begins shortly after male fish are added, usually within 72 hours. Stocking brooders on a Friday lets the brooders settle in over the weekend and spawns may be found the next Monday. Brooders should not be fed while in the spawning pond. If spawning mats are properly placed, up to 90% or more of the mats will be used.

Once spawning begins, each spawning mat should be observed daily (Figure 16.1.6). If the water is clear, the mats can be observed from the pond bank and any eggs present can be seen. Where ponds are turbid, it is good to mark each mat location when the mat is placed in the pond using a numbered stake to help locate it later. When the sky is overcast or visibility limited by wave action or turbidity, an underwater viewing device enables the fish culturist to determine presence or absence of eggs. This can be a specially constructed viewer or short section of 15.24 cm (6") diameter PVC pipe. If a male starts to build a nest on the pond bottom, placing a mat on that site often results in the male using the mat, and the resultant eggs can be transferred on the mat to the hatchery for incubation.

When a spawn is noted, the mat should be removed and taken to the hatchery for incubation. A new mat should be placed back in the same location where any egg-bearing mat was removed. Egg-bearing mats can be left in the pond longer before being collected but the risk of egg loss is greater.

Largemouth Bass are adaptable to indoor spawning if a clean water supply of the right temperature is available and the photoperiod is appropriate (Figure 16.1.7). Brooders can be collected from outdoor holding ponds once the water temperature is ≥15°C and brought indoors and acclimated over a 3–4 week period to a temperature of >18°C and a 14 h light: 10 h dark photoperiod. During the initial acclimation period broodfish can be easily startled by sudden noises or bright lights and try to jump out of the holding tank or raceway. They should be held in tanks or raceways that provide enough freeboard to minimize the risk of fish jumping out. The fish should be held in an area of the hatchery where there is a minimum of foot or vehicular traffic. Ideally the lighting system in the portion of the hatchery where the

Figure 16.1.6. Largemouth Bass eggs on mat.

Figure 16.1.7. Indoor raceway for spawning of Largemouth Bass.

brooders are being held can be controlled to daily dim-up and dim-down as part of the daily photoperiod acclimation cycle. An abrupt turning of the lights on or off can startle the fish and cause some to try and jump out of the acclimation tank or raceway. Lights in other portions of the building holding the fish being acclimated should not be turned on until the lights in the acclimation area have reached full illumination.

During both the acclimation period and the spawning period the water exchange should be around 0.25 and 0.5 times/h. If well water is being used for acclimation or the spawning raceways, the source water may be too cool and must be warmed up before use. However, warming up well water reduces its gas solubility. Well water has an overburden pressure that increases its gas solubility. That solubility is reduced when the groundwater is brought to the surface and the pressure relieved. These two factors can result in well water being supersaturated with carbon dioxide and other gases that need to be driven off before use. Well water is often low in oxygen and may contain dissolved minerals such as iron and manganese that may precipitate out once the water is aerated. Such water may need to be filtered and aerated before it is added to the brood acclimation and spawning tanks. The goal during acclimation should be to bring gradually bring the water temperature up to at least 15°C. If starting with a 12 h light: 12 h dark photoperiod, the light phase should be increased 30 min./week over a 4 week period. The goal is to have the fish acclimated to a 14 h light:10 h dark photoperiod and a temperature of >15°C with the spawning raceway having similar conditions.

The ability to stimulate Largemouth Bass to spawn under indoor conditions by controlling the temperature and photoperiod offers several management alternatives. Instead of waiting until the water temperature in the ponds has come up enough for bass to start spawning, brooders can be spawned indoors several weeks ahead of the outdoor spawning season. This can make outdoor ponds available for bass nursery that otherwise would be used for bass spawning. Using such ponds for nursery of fry obtained indoors can result in bass fingerlings being available for distribution several weeks ahead of the normal distribution schedule.

Similarly, where cool water is available brooders can be held indoors and their spawning delayed so fry are available for stocking nursery ponds in the late summer and fingerlings being available for distribution in the fall. The adaptability of Largemouth bass brooders to indoor hatchery conditions allows fry to be available over most of the year if there is the demand for such fish.

Once brooders have adapted to hatchery conditions that duplicate the temperature and photoperiod that would be found at the start of the outdoor season outdoor they can be moved to spawning raceways where such conditions are also provided. Spawning raceways are stocked at a 1:1 to a 1:1.5 male: female ratio, providing one spawning mat for every male stocked. Spawntex® spawning mats (approximately 60 × 45 cm) are set approximately every 2 m along the sides of the raceway, staggering them so that no two mats are directly across from each other. Attention should be given when placing the mats to be sure that there are no air pockets in the mats that would prevent them from sinking to the bottom. Once broodstock and mats are in the raceways, the mats should be checked daily with the first spawns anticipated within 72 h of brood stocking (Figure 16.1.8).

16.1.4.4 Egg Transfer: Mat and Egg management

Mats containing eggs should be moved to the hatchery the same day they are found. Largemouth Bass eggs can hatch in two days at ≥27°C, and the adhesiveness of the eggs to the mat is reduced the closer the eggs are to hatching. Delays in collecting mats can result in egg loss. Mats should be carefully removed from the spawning pond or raceway allowing the water to drain off slowly. Felt mats can be rolled up with the eggs inside and transported to the hatchery. Less flexible mats such as a Spawntex® mat should be transported in a tub of water. Several mats at a time can be placed in the tub as long as the eggs remain wet.

When mats are collected from ponds, eggs are exposed to fungus that may hinder successful egg incubation. The mats should be given a prophylactic treatment of formalin at 150 ppm for 30 min. This can be done while the mats are being transported to the hatchery or

Figure 16.1.8. Indoor spawning with brooder on nest.

placed in tanks for incubation. Fungus problems are not as common when Largemouth Bass are spawned indoors in aerated well water, but prophylactic treatments can be beneficial. Hydrogen peroxide can be given as a flush treatment at 100 mg/L twice a day during incubation.

Another approach to minimize the effects of fungus is to raise the water temperature where the eggs are being incubated. Raising the temperature a few degrees can speed up the development rate of the bass embryos so they are exposed to potential fungal problems for a shorter period of time. For example, Largemouth Bass eggs incubated at 18–19°C might hatch in 84–96 h but if the eggs are incubated at 22–23°C, they should hatch in 40–48 h with the hatch rate being about 30% higher.

Egg-bearing mats can be placed in clean well-aerated tanks or large aquaria for incubation (Figure 16.1.9). Water exchange rates of once/h are acceptable. Mats should be suspended vertically in the tank allowing a space of at least 5 cm between mats at a density of 10–12 mats/m^3 of incubation tank. A water current should be provided between mats using an overhead spray bar or by providing aeration from along the tank bottom. Mats collected the same day should be incubated in a common tank to help ensure similarity in fry age for pond stocking. Mats collected on different days should not be incubated together in the same tank. Hatching may occur within 48 hours depending on the age of the eggs when collected and the water temperature at which they are incubated.

If the mats are held in aquaria the sac fry can be siphoned off daily from the bottom of the tank, counted volumetrically and pooled together with other same-age fry into other aquaria or troughs. The fry are held there until they swim up and are ready to be stocked into nursery ponds. Sac fry can be held at densities of 5,000 to 8,000/L where the water flow assures ample DO. Depending upon water temperature during yolk absorption, a period of 3 to 7 days is normally required before swim-up and first feeding. Sac fry can also be shipped to other locations using plastic bags with water and oxygen before they develop enough to be able to accept exogenous feeds. Loading rates of 3,500 to 4,000 sac fry/L can be used for transporting sac fry for periods of up to 72 hours.

Figure 16.1.9. Incubation of mats containing Largemouth Bass eggs.

When mats are incubated in larger tanks it is not practical to siphon sac fry off the bottom. There the mats are suspended vertically in the water column and so the sac fry can drop to the tank bottom where they remain until they swim up. The mats are removed once all the eggs have hatched. The swim-up fry are captured daily with fine-mesh dip nets. Increasing the light intensity will attract the swim-up fry to the surface making them easier to collect. The swim-up fry (Figure 16.1.10) can be counted by number/unit volume or weight, or by visual estimation, and stocked into nursery ponds.

Largemouth Bass eggs can be incubated in more traditional hatchery incubators such as MacDonald jars, or Heath trays. Bass eggs are highly adhesive when spawned but become progressively less firmly attached as the eggs develop and can be separated from the spawning mat. One or two dozen vigorous swishing motions of the spawning mat in a bucket of water will dislodge the eggs that are slightly heavier than water. After separation from the spawning mat, the eggs are cleaned using two different mesh screen boxes. One screen retains the eggs but allows the fine material to pass through. The second screen allows the eggs to pass out but retains the coarse material. Final cleaning is done by decanting to eliminate floating material that was not removed by screening, and siphoning off any detritus that was not separated from the eggs by screening.

Clean eggs are measured by settled volume and a smaller volume counted to determine the number per unit of volume. A prophylactic treatment against fungus, such as formalin at 150 ppm for 15 minutes, should be given. Handling eggs during the cleaning, enumeration and prophylaxis is done in water with occasional stirring to ensure good aeration.

Each tray of a Heath vertical incubator (Figure 5.1) can be loaded with 150,000 to 200,000 eggs/tray. A water flow of 5.7 to 7.7 L/minute will maintain 6 to 7 trays of eggs. After hatching, there is some downstream migration of larvae from tray to tray and it is best to keep the older lots of eggs and fry below the younger ones. Daily cleaning to remove dead eggs, debris, and fungus is required.

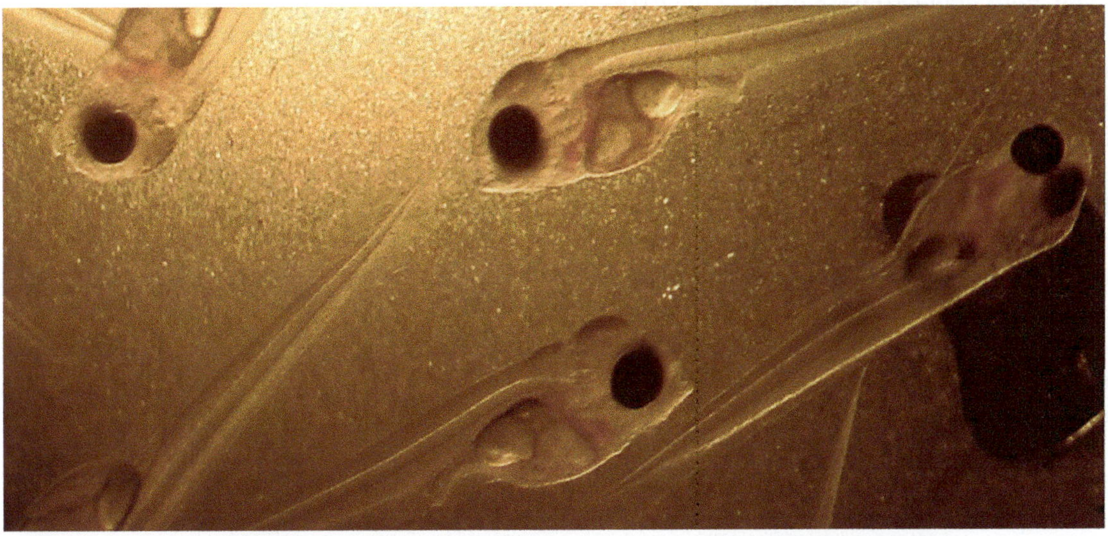

Figure 16.1.10. Largemouth Bass with functional mouthparts, ready for stocking into a primary nursery pond.

16.1.5 Fingerling Production

16.1.5.1 Rearing Small Fingerlings

Largemouth Bass fingerlings are produced in one or more phases, typically starting with a primary nursery pond that is prepared to provide an abundance of zooplankton. It may take 7–14 days, depending on water temperature and fertilization protocol, to produce an adequate initial zooplankton density of the proper size. Lead-time in preparing a nursery pond should take into consideration the source of the fry to be stocked, and the type and size of zooplankton needed for such fry. When collecting egg-bearing mats and incubating them in the hatchery, this will provide 6–7 mm ± fry, whose mouthparts have recently formed and are ready to feed. Such fry should be stocked into nursery ponds where copepod nauplii are abundant. This might be a pond that has been filled for only 7–10 days. If fry are produced using a spawning pond with fry transfer approach, the fry may be 14–25 mm when they are collected for transfer to the nursery pond. At that point the fry are large enough to consume cladocerans, adult copepods, and insect larvae. For an adequate abundance of such food organism to develop the nursery pond should be prepared and filled 10–14 d prior to the anticipated stocking date. An adequate density of cladocerans at the time of stocking bass fry should be around 200 /L.

Preparation of nursery ponds should begin at about the time brood bass are stocked into spawning ponds with new nursery ponds prepared weekly depending on the anticipated length of the spawning season. Preparation consists of filling the ponds with clean, fish-free water, or sterilizing the filled ponds if they are not free of unwanted life. Fertilization with an organic nitrogen-superphosphate combination at a rate of 9 kg N and 9 kg P_2O_2/ha/application at weekly intervals until an abundant zooplankton supply develops is generally an effective protocol. Ordinarily, this will require three to four applications. Organic fertilizers, such as rice bran, that are lower in nitrogen, tend to favor copepod and cladocerans production compared to organic fertilizers such as cottonseed meal that are higher in nitrogen that favors rotifer production. After sufficient plankton develops, biweekly applications of fertilizer should be made until about 10 days prior to the expected harvest of the fingerling bass.

Stocking rates of fry depend on how the fish are to be used. If the fry are being produced for stocking new or renovated farm ponds then a stocking rate of 125,000–250,000 fry/ha should yield small 2.5–5.0 cm fingerlings that are suitable for such use (Figure 16.1.11). If fingerlings are being stocked into populations where predators are already present then fry should be stocked at 75,000–100,000/ha to give fish ≥5 cm. To have larger bass fingerlings for use in a corrective restocking program, the first step is to stock fry at 150,000/ha into nursery ponds to produce 4–5 cm fingerlings that are suitable for feed training and then grown to larger sizes using formulated feeds.

The length of the nursery period may range from 14 to 28 days in the southeastern USA in April and May when the water temperature range is 21 to 27°C. The longer culture period is for fry stocked at lower densities to produce larger fingerlings. Yield from nursery ponds are based on the availability of appropriate natural food organisms and may be as high as 168 kg/ha. If ponds were prepared properly, and the fry counted and handled correctly, then a nursery pond survival should range from 75 to 90%. If a higher survival percentage is obtained, one can suspect inaccurate estimates of the initial number of fry stocked. Less than 75% survival suggests a need to improve techniques related to enumeration and handling, food production, disease control, predator or competitor control, or eliminating other interfering factors.

Figure 16.1.11. Largemouth Bass fingerlings harvested from a primary nursery pond.

The quantity of fry stocked is estimated by visually comparing pans containing unknown numbers of fry with a counted standard, use of a number-weight relationship, or use of number-volume relationship. Each method has its advantages and disadvantages. All are thought to be sufficiently accurate for adequate estimation of the numbers of fish on hand, provided they are properly employed. Size of the fry handled and personal preference are probably the most important factors influencing the method selected. Use of the visual comparison method of estimating fry numbers is probably employed more widely for Largemouth Bass than either gravimetric or volumetric measurement.

The age of the fry that are stocked into a nursery pond is not as much of an issue as is the uniformity of the age of the fish being stocked. At the Marion Fish Hatchery, Marion, Alabama, stocking nursery ponds with swim-up fry obtained by egg transfer has given survival rates as good or better than nursery ponds stocked with more advanced fry collected from spawning ponds. Both sets of fry grew to a distributable size in about the same number of nursery days (21 to 25 d). Working with recent swim-up fry hatched in the hatchery has the advantages of ease of handling, predictability in the quantity of fry available, and ease in enumeration. Pond-collected fry can be variable in size and age as it is not uncommon for two or more schools of fry to merge as they swim along the shoreline.

After nursery ponds have been stocked, the brooders from the spawning pond can be seined out and returned to a brood holding pond. Any fry remaining in the spawning pond can be left for growth to harvestable size. Otherwise, the spawning pond should be drained, all fish removed, and the pond prepared for a new crop of fish. If further bass fry production is needed, the brood bass can be examined before being returned to the brood holding pond and any unspent fish transferred to a clean, freshly filled pond. Additional eggs can usually be obtained from partially spawned or un-spawned fish, which often makes this practice quite productive for late-hatched fry.

If one of the original spawning ponds has failed to give the expected quantity of eggs or fry within four weeks of being stocked, the pond should be drained and the brooders transferred to a freshly prepared pond for spawning. Females may begin to reabsorb their eggs four weeks after the peak spawning period.

16.1.5.2 Large Fingerling Production—Extensive Culture

If pond space is available, Largemouth Bass harvested for primary nursery ponds can be stocked at low densities (30,000/ha) to produce 8 cm fish in about 40 days. This approach depends on having enough natural food organisms to be available, particularly midge larvae. Ponds will need to be fertilized as discussed earlier and allowed 21–28 days to develop an abundance of adult insects and their larvae. Yields that have been obtained at the Marion Hatchery, Marion, Alabama are illustrated in Table 16.1.5. Careful attention needs to be given to the size uniformity of the fish being stocked. A centimeter variation in initial length will quickly expand and result in significant cannibalism. Periodic seining inspections should be made of rearing ponds. When a size disparity is evident to the extent that cannibalism is possible, the pond should be drained, and the fish shipped or graded to separate the large and small sizes into different ponds. This will minimize losses that if unchecked can drastically reduce production.

16.1.5.3 Large Fingerling Production—Intensive Culture

Largemouth Bass can be trained to consume a formulated feed once they are >2.5 cm, and remain on formulated feeds throughout their life. Suitable fish for training can be obtained from primary nursery ponds stocked with a common age fry obtained using an egg transfer approach. Stocking fry at 150,000/ha should yield 4–5 cm fingerlings with a 75–80% survival in 21–28 days at water temperatures of 22–25°C.

Feed training can best be done in a rectangular or circular tank of suitable size that is at least 60 cm deep. The tank should be well aerated and have a water exchange of at least once an hour. Largemouth Bass fingerlings will not pick up feed off the bottom of the tank but will consume formulated feed as it drops through the water column. A 60 cm deep tank will allow the fingerlings the opportunity to strike at the feed as it drops through the water column, particularly if the feed is given slowly. Holding the fish in a deep hapa placed in the tank facilitates examining the fish periodically for any cannibals that need to be removed, or disease problems that need to be treated. Water temperature should be in the 24 to 30°C range. Fish should be stocked at about 3 to 9/L of tank with as similar size fish as possible. *Columnaris* is a common bacterial problem seen when feed training Largemouth Bass. Fish should be treated prior to stocking and then every 2–3 days during training with 5 mg/L potassium permanganate to keep disease problems to a minimum.

The same day that the fish are stocked they should be offered freeze-dried krill (*Euphausia* sp.). The dry krill should be crushed by hand at the water surface and the particles allowed to drop through the water column. The krill should be fed a minimum of 5 times/day for two days to train the fish to accept a feed. Starting the 3rd day the fish are weaned to a 2 mm pelleted ration (75% krill meal and 25% a commercial trout meal) for 2 to 3 days. That is followed by a pelleted feed (50% krill and 50% commercial trout meal) for 2 to 3 days, then a

Table 16.1.5. Extensive production of Largemouth Bass fingerlings in ponds.

Stocking Rate no./ha	Culture Period (days)	Survival (%)	Weight Gain (kg/ha/day)	Production/ha Number	Production/ha kg	Mean TL (cm)
14,000	39	92.8	1.78	12,992	69.4	7.5
23,900	37	84.3	2.15	20,148	79.6	6.8
30,000	70	70.3	1.82	21,090	127.6	7.8
34,500	37	80.7	2.92	27,842	108.1	6.8
36,000	26	73.8	3.23	26,568	84.0	6.3
36,600	65	65.0	2.72	23,790	176.7	8.3
39,600	37	62.8	2.91	24,869	107.6	7.0

pelleted feed (25% krill and 75% commercial trout meal) for 2 to 3 days. Finally the fish are fed a pelleted ration made from 100% trout meal for another 2 to 3 days. After that point the fish should remain in the treatment unit and be fed a 1.5 mm floating commercial trout pellet for an addition 3 to 5 days before being harvested and stocked into growout units. At the end of the training period more than 75% of the fish should be actively feeding on the dry ration. The non-feeders will be smaller and can be graded out, size-sorted, restocked, and retrained on the above protocol. Any fish that is much larger than its siblings should be considered as a potential cannibal and removed from the population.

There are several modifications of the above feed training protocol that are also effective. Dried shrimp (*Penaeus* sp.) can be substituted for krill, particularly as a component of the transition pellets. Softening commercial pellets in water before feeding or coating the pellets in fish oil has also been effective in feed training. Another approach has been to feed live Artemia nauplii every two hours for the first three days of feed training, then feed a high protein 620–920 μm commercial starter feed for three days. After this, start with a 75% starter feed/ 25% 1 mm commercial pellet mixture, then reduce the percent starter feed and increase the percentage of the pelleted feed over a 4 to 10 day period (depending on the feeding response) until only the pelleted feed is given (Table 16.1.6). Feed is given every 2 h using automatic feeders with the hatchery staff periodically inspecting the feeding response, presence of unconsumed feed, and proper functioning of the automatic feeder.

After the feed-trained fish have been harvested, they can be stocked into raceways or earthen ponds for further growth. If stocking ponds, the fish should be held in a 6 cm mesh hapa in the pond and fed for 3–4 days before being released into the pond. This is to help ensure that the fish continue to accept the floating feed once they are released into the pond. The stocking rate is dependent upon the size to which the bass is to be reared. A stocking rate of about five bass/m^2, 2 g each, can result in 18 cm TL fish in 100 to 120 days, with a yield of >4,000 kg/ha.

Standing crops of around 16 kg/m^2 are possible in raceways. For the most efficient use of the raceway, an initial stocking rate should be about 8 kg/m^2 although a lower density would reduce the need for fish handling as growth occurred.

16.1.6 Harvest and Distribution

Methods of harvesting bass fingerlings include draining, trapping, and seining. When large numbers of bass are required to meet specific distribution schedules, draining is the har-

Table 16.1.6. Production of Largemouth Bass in ponds using Oregon Moist Pellets.

Pond No.	Stocking rate. No./ha	Days fed	Yield (kg/ha)	Av. Wt. (g)	Survival (%)	FCR
1	27,417	140	4,216	157.5	97.7	1.60
2	38,532	98	3,080	81.6	98.0	1.43
3	49,400	105	3,051	88.6	69.7	1.65
4	49,400	112	3,576	87.0	83.1	1.52
5	49,400	114	4,274	125.0	69.2	1.43
6	49,400	114	4,249	119.0	72.3	1.45
7	52,873	117	3,992	94.0	80.3	1.52

vesting method of choice. The pond water level is lowered as rapidly as possible, and the fish seined from the pond's catch basin or other areas of concentration as soon as possible. In some ponds, removal can begin long before the water is shallow enough to wade with hip boots. In other instances, the water level must be lowered so that 1/3 to 1/2 of the bottom is exposed before harvest can begin. The catch basin should be cleaned prior to fish removal. Doing so keeps suspended silt to the minimum, helps improve water quality during harvest and helps in reducing disease troubles. Prompt completion of the harvest is desirable once the water level is reduced. In addition to cannibalism and disease, other causes of losses at this time include snakes, birds, raccoons, and oxygen depletion.

Trapping is a widely employed means of harvesting both fry and fingerling Largemouth Bass. Fish of a size of 1.5 to 5.0 cm move about the edges of the pond in search of suitable food and can be taken in several types of traps with the glass-V trap, or modifications of it. Advanced fry and small fingerling bass are attracted to a stream of inflowing water and will swim into a trap placed in the current. Such a combination of water flow and trapping can harvest a major portion of the small bass in the pond. If Largemouth Bass is one of the principle species being produced at the hatchery, adding a permanent seining basin at the shallow end of the pond in front of the water inlet will facilitate harvests.

Traps when set under the correct conditions enables the harvest of bass with less labor than by seining; helps to provide clean fish which suffer little stress during harvesting; minimizes the impact of vegetation interfering with the harvest; and allows grading during the process of trapping. Unfortunately, use of traps cannot always be relied upon to provide a given number of fish on a specified date. Even though fish are abundant, they sometimes will not enter the trap. Under such conditions, seining and draining must be done.

Seining small bass without lowering the pond water level is an effective technique for obtaining advanced fry or fingerlings up to 5 cm TL. Particularly when the smaller sizes of fish are still in schools, they can be collected by seining along the edges of the pond. After the schools have broken up and the fish are more dispersed, they are not difficult to harvest by seining as long as there is little to no aquatic weeds in the pond.

Harvesting and distribution of Largemouth Bass presents problems not encountered with several other species. They are more cannibalistic than most species produced at the warmwater fish hatcheries in the southeastern USA. Cannibalism can result in a loss of fingerlings in the rearing pond, while being held in the hatchery, and as well as when being transported. Trying to keep the fish as uniform in size as possible is key in trying to minimize cannibalism. This begins with stocking nursery ponds with as similar age and size fry as possible, not over stocking to the point of creating a shortage of food and an increased competition that results in enough size variation among the fingerlings that allows cannibalism to occur. The longer the culture period typically results in more size variation and greater cannibalism. Sampling during the culture period to determine the magnitude of size variation is an important part of controlling cannibalism. The length of the culture period should short enough that the degree of size variation is not so large that aggression and cannibalism become issues.

Every time Largemouth Bass are harvested they should be graded. The bass should be graded into size categories such that the initial size variation within a given size category will not have increased toward the end of the next culture phase to the point where cannibalism becomes a problem. Largemouth Bass fingerlings are aggressive and often will attempt to swallow their cohorts even though the size difference between fish is not great enough for one to swallow the other. Such attempts often injure both fish resulting in diseases such as columnaris developing and mortality. If Largemouth Bass fingerlings are handled and graded carefully, and given prophylactic treatments for external parasites and bacterial infections, they can be held 7 to 10 days without harm.

Largemouth Bass are different than most of the hatchery-produced fish in the southeastern USA, in that fingerlings are commonly harvested and distributed during the spring and summer when the water temperatures may be high. To minimize the temperature effects, the fish should be harvested early in the day and procedures need to be implemented during handling and transport to prevent the water temperature from rising to stressful levels. Using insulated distribution units is the first step in controlling temperature fluctuations. When it is necessary to transport heavy loads of fish, particularly for long periods, the judicious use of ice can be effective in minimizing increases of water temperature in the transport container. Adding ice will lower the water temperature, and in turn, reduce metabolic waste production, and lower the oxygen uptake of fish, while increasing the solubility of oxygen added to the water. The best practice is to stabilize water temperature at about 16°C by the periodic addition of chipped or crushed ice. Adding ice in lumps on a less frequent schedule than that employed with crushed ice may save time but may result in a greater temperature fluctuation. Attention should be given to the water source used to make the ice, avoiding the use of large quantities of ice made from potable water sources treated with chlorine.

Loading rates for Largemouth Bass fingerlings may range from 60 up to 120 g/L of water depending on water quality, temperature, condition of the bass, and size of fish involved. The demands of a given distribution trip, such as the number of locations to receive fish, the number of fish to be provided per location, as well as the total distance traveled must be considered as to the loading rate, and the type of equipment used to transport the fish.

Toward the end of the trip, a gradual rise in temperature is usually allowed so that upon arrival the water temperature in the distribution tank is within 6 to 7°C of where the fish are to be stocked. An adequate acclimation period should be allowed for so that the water temperatures of where the fish are held and where they are to be released should be within 1–2°C before being released.

16.1.7 Cultural Problems

Several problems in Largemouth Bass culture are important enough to receive special emphasis. As discussed earlier, it is not uncommon to have significant changes in the weather during the spring in North America. One week the temperature may be high enough to stimulate bass to spawn, and then the next week the temperature drops again, causing the adult bass to abandon the nest site, eggs taking longer to hatch and becoming infested with fungus. Such temperature fluctuations make zooplankton development in nursery pond more difficult to predict.

The influence of cold weather at spawning time can be minimized by delaying brood fish stocking until after the average date of the last killing frost in the spring. Prior to the start of the spawning season broodfish should be sorted and held by sex according to their apparent degree of gonadal development. Less developed brooders can be held by sex for possible later use that spring. Spawning activity in the brood holding pond can be inhibited by drastically reducing the water level if nest building is observed. Often the water temperature of the water supply source (reservoir or well) is cooler than that of the brood ponds. Using one of these water sources to provide a continuous flow into the brood pond may be enough to suppress brood spawning activities until water temperatures have stabilized in the spawning ponds into acceptable ranges for bass spawning. Chemical treatment of brood fish immediately prior to premature spawning may be a useful deterrent. It has been observed that applying 0.5 ppm $CuS0_4$ to the spawning pond suspended spawning for about 7 days.

The larval stage of the bass tapeworm *Proteocephalus ambloplites* (Figure 16.1.2) can over time cause significant damage to bass ovaries and can only be eliminated by rearing bass in an environment free of the immature parasites. Use of forage fish free of tapeworms and parasite-free broodstock is necessary before the immature tapeworms can be controlled. Feeding a formulated diet to the brooders will break the life cycle of the parasite.

A common disease problem affecting Largemouth Bass brooders fish, especially at spawning time, is hemorrhagic septicemia caused by *Aeromonas hydrophila*. Males often have eroded caudal fins resulting from sweeping out nest sites; the handling of female brooders and the act of spawning can be stressful. Several control methods for hemorrhagic septicemia have been tried. A prophylactic approach for brooders that accept formulated feeds is to feed oxytetracycline for 10 days prior to handling and pond stocking at a rate of 50 to 75 mg of active antibiotic/kg of fish/day. At the end of the spawning season, before brooders are returned to their holding ponds, they should be held in the hatchery and given a potassium permanganate bath at 8 mg/L for an hour.

Columnaris disease caused by the bacteria *Flavobacterium columnare* is a problem in bass fingerling production. It is most troublesome during warm weather when fingerlings are being harvested and handled. A bath treatment using potassium permanganate bath at 8 mg/L for an hour is helpful to control it while the fish are being held in the hatchery and if necessary a pond treatment of 2 to 3 mg/L can be applied.

Another cause of mortality is the Largemouth Bass Virus (LBV), which is a virus in the Iridoviridae family. It has been isolated from adult bass with mortalities occurring under warmwater conditions during the summer. LBV can infect the swim bladder making it difficult for the fish to swim upright. Fish can also carry the virus without showing any symptoms. It can be detected from tissue samples using quantitative polymerase chain reaction (see Getchell et al. 2007). Good records should be kept as to the origin of hatchery broodstock and whether they may have been exposed to LBV. As with many other virus problems LBV can be a chronic problem that only causes mortality under certain conditions. Good brood records are needed as to any possible exposure to this disease as well as to the distribution of the progeny that could act as disease carriers.

Cannibalism among bass fingerlings in rearing ponds can be a problem as the fish grow and there is more diversity in sizes. Overstocking rearing ponds, stocking mixed sizes of fry, having inadequate food production, and delays in harvesting, all contribute to such losses. Attention is needed during any primary or secondary nursery period as to the extent of size variation seen and if it appears that cannibalism is occurring. Seine samples are needed to confirm these observations. If cannibalism is occurring then the pond should be completely harvested, the fish graded into closely grouped size categories and where possible the fish restocked by size category.

16.1.8 Recommended Reading

Benz, G. W., and R. P. Jacobs. 1986. Practical field methods of sexing Largemouth Bass. The Progressive Fish-Culturist 48:221–225.

Brown, T. G., B. Runciman, S. Pollard, and A. D. A. Grant. 2009. Biological synopsis of Largemouth Bass (*Micropterus salmoides*). Canadian Manuscript Report of Fisheries and Aquatic Sciences 2884. Available: http://www.dfo-mpo.gc.ca/Library/337843.pdf. (May 2020).

Davis, J. T., and T. Lock. 1997. Culture of Largemouth Bass fingerlings. SRAC Publication No. 201, Southern Regional Aquaculture Center, Stoneville, Mississippi. Available: http://agrilife.org/fisheries2/files/2013/09/SRAC-Publication-No.-201-Culture-of-Largemouth-Bass-Fingerlings.pdf. (May 2020).

Engle, C., and B. Southworth. 2013. Costs of raising Largemouth Bass fingerlings. ETB260 Aquaculture/Fisheries Center, University of Arkansas at Pine Bluff, Pine Bluff. Available: https://www.uaex.edu/publications/PDF/ETB-260.pdf. (May 2020).

Getchell, R. G., G. H. Groocock, V. L. Schumacher, S. G. Grimmett, G. A. Wooster, and P. R. Bowser. 2007. Quantitative polymerase chain reaction assay for Largemouth Bass virus. Journal of Aquatic Animal Health 19(4):226–233.

Glenewinkel, H., A. Barkoh, T. Engeling, L. Hall, J. Paret, and T. Owens. 2011. Guidelines for the culture of Black Bass. Management Data Series No. 267. Texas Parks and Wildlife Department Inland Fisheries Division, Austin. Available: https://agrilifecdn.tamu.edu/fisheries/files/2013/09/Guidelines-for-the-Culture-of-Black-Bass.pdf. (May 2020).

Grizzle, J. M., and C. J. Brunner. 2003. Review of Largemouth Bass virus. Fisheries 28:10–14.

Heidinger, R. C. 1975. Life history and biology of the Largemouth Bass. Pages 11–20 *in* R. H. Stroud and H. E. Clepper, editors. Black bass biology and management. Sport Fishing Institute, Washington, D.C.

Isaac, J., Jr., T. M. Kimmel, R. W. Bagley, V. H. Staats, and A. Barkoh. 1998. Spawning behavior of Florida Largemouth Bass in an indoor raceway. The Progressive Fish-Culturist 60(1):59–62.

Matthews, M. D., and R. B. Stout. 2013. Out-of-season spawning method for Florida Largemouth Bass to produce advanced-sized fingerlings by early spring. North American Journal of Aquaculture 75:524–531.

Matthews, M. D., J. C. Sakmar, and N. Trippel. 2012. Evaluation of hydrogen peroxide and temperature to control mortality caused by Saprolegniasis and to increase hatching success of Largemouth Bass eggs. North American Journal of Aquaculture 74:(4): 463–467.

Mayes, K. B., P. M. Rosenbloom, and T. M. Brandt. 1993. Raceway spawning of Florida Largemouth Bass: effects of acclimation time and hormone treatment on spawning success. The Progressive Fish-Culturist 55:1–8.

16.2 Sunfish *Lepomis* sp. and Their Hybrids

16.2.1 Introduction

Members of the sunfish genus are probably the most widely sought-after group of game fishes in the USA. Although relatively small in size, they are widely distributed in North America, easy to catch, grow rapidly, and are relished as food. Their suitability as a component of the combination of species used in stocking farm ponds has created a sustained demand for seed to stock new ponds and to restock old ones. The most important of the hatchery produced sunfishes is the Bluegill *Lepomis macrochirus*. Next in importance is the Redear Sunfish, *L. microlophus*. The other ten species are not widely propagated. Of the numerous hybrid combinations possible, the Redear Sunfish male × Green Sunfish female (*L. cyanellus*) cross appears to be most promising.

The life span of Bluegill is relatively short. Only 75 to 80% of the fingerlings stocked into a managed fish pond can be expected to survive the first year. The second year 15% of the remainder can be expected to die. Most do not live for more than four years but older fish are found in the northern part of their range with an 11-year-old fish recorded.

Bluegill are one of the faster growing sunfish. They should grow to a size of 50 to 75 g in the first year, and reach 150 to 225 g during the second year. Beyond the second year, growth is slower and fish larger than 350 to 450 g are not common. However, the maximum size recorded is 39 cm in length, weighing 2.1 kg.

Sexual dimorphism is distinct in mature Bluegill in the breeding season and most all other species of *Lepomis*. Adult male Bluegill are generally larger and darker in color than females (Figure 16.2.1). The opercula flap is dark blue or black in the male with more blue on the head and upper body being characteristic. The genital orifice of gravid females has a distinctive appearance as compared to the male. Gravid males can be stripped to yield a flow of milt. Sexual maturity in the Bluegill will occur in the first season of growth in the southern part of their range. Young Bluegill fry spawned in May can reproduce in September of the same year in Alabama. Up to three years may be required to reach sexual maturity in the most northern part of their range.

Bluegill and other sunfish are generally classified as intermittent spawners. The Bluegill is probably one of the most fecund of the species in this group. Well-conditioned Bluegill females will spawn every 30 to 45 days as long as water temperature is favorable. In the southern half of the range, spawning commences at a surface water temperature of about 25°C and continues intermittently as long as it is at this level or higher. In the northern part of the range, spawning takes place at a water temperature of 21 to 22°C. At the lower temperature there is less frequent spawning during the growing season.

Sunfish are nest builders, selecting sites in relatively shallow water over firm or gravelly bottoms when available. Usually a number of males occupy a general area, each making a bowl-shaped nest in close proximity to one another (Figure 16.2.2). There may be twenty or more nests in a nesting area. Each male guards his nest and limited territory, but no parental care is given to the young. There are three types of males that may participate in the spawning. Parental males grow to a large size before maturing and are the dominant males when it

Figure 16.2.1. Male Bluegill *Lepomis macrochirus* (above) and female Bluegill (below).

Figure 16.2.2. Several sunfishes such as Bluegill *Lepomis macrochirus* illustrated here, are nest builders where a male will build and guard a nest in close vicinity to other males and their nests.

comes time to reproduce. Such males build the nests and guard the nest site. These males take on a distinct breeding coloration and appearance. Males will have a large dark opercular flap, the upper body may appear purplish and the area below the pectoral fin may appear orange to red-brown. A second set of males, often called cuckold males, also try to participate in the spawning process but do not establish territories as do the parental males. These cuckold males come in two types, "sneakers" that try to sneak in when the female is releasing eggs and fertilize some eggs before it is chased away by the parental male. Another type of cuckolder, "female mimics," are males that mimic the color pattern of the female and will swim close to the ovulating female to fertilize eggs as they are released. By appearing like a female, they are less likely to be chased off by the parental male.

Once nests are established and males are defending their territories, females will swim over the nesting area, entering a male's nest multiple times, releasing 20+ eggs each time. A male may end up guarding a nest with several thousand eggs. Depending upon the size and age of the female, she may release 40 to 60 eggs/g body weight in a single spawn. Fertilized eggs are 1.0 to 1.2 mm in diameter and will hatch 23 to 26 hours at a water temperature of about 27°C. A drop in water temperature of 5.5°C can almost double the length of the incubation period. At hatching, the fry are almost colorless and are about 3.3 mm TL. The fry remain in the nest subsisting on the yolk sac for 3 to 4 days. They leave the nest to begin searching for exogenous food when they are 5.0 to 5.5 mm TL. There is no definite pattern of schooling in fry, but concentrations can be found in the vicinity of the nesting area. They gradually disperse to the shallow marginal waters in a matter of hours or days after leaving the nest to feed on zooplankton. After first feeding the young are likely to become more selective but continue on a zooplankton diet until a size of 30 to 50 mm is reached. At this size they become insectivorous and continue this food habit throughout their adult life.

16.2.2 Production

16.2.2.1 Spawning-Rearing

Sunfish will naturally spawn in static water ponds, and as such, are produced in most hatchery situations using a Spawning-Rearing approach. Such an approach is straightforward requiring only a basic input of resources. However, this apparent simplicity of cultural methods can give unpredictable results if attention is not given to proper pond preparation and pond management. This includes eliminating any possible contamination of the pond with other fish species, proper pond fertilization, and control of aquatic weeds. Bluegill will spawn numerous times in a spawning-rearing pond, resulting in unknown qualities of fish of various size categories. In most cases, to meet production goals, Bluegill are produced in excess to compensate for any pond to pond variation in production.

Brood Management.—Best spawning success is obtained using mature 135 to 270 g brood fish. Fish larger or smaller can be used but with less consistent or predictable results. However, good production has been obtained on occasions from 1-year old Bluegill that weigh 50 to 75 g at spawning. When 50 to 75 g brooders are used, an increase in the number per unit of area is needed to adequately stock the rearing pond. Yearling brood fish display less uniformity as to time of spawning. This results in a greater size variation among the fingerlings produced.

Broodfish are generally obtained by grading or hand-selecting larger fingerlings from the previous year's crop. These replacements may either be mixed with adults stocked in spawning-rearing ponds or used alone to stock production ponds. The recommended procedure is to keep year classes of brood fish separate so that systematic replacement can be carried out. Use of brood fish for three spawning seasons was practiced at the Marion Hatchery in Alabama for a number of years with satisfactory results. Since Bluegill and Redear spawn at 1 year of age at this latitude, the fish were replaced at an age of 3 or 4 years. In the southern half of the USA, brood fish average weights might be approximately the following for the respective ages: 0–1 year, 50–75 g; 1–2 year, 150–225 g; 3–4 year, 225–400 g. In more northern portions of the USA the growing season is shorter and it will take longer to produce suitable size brooders.

A special holding pond is normally established to maintain brood fish from the time of harvest of the previous crop to the time when current production ponds are stocked. Supplemental feeding is practiced by most hatcheries although specific diets for Bluegill have not been developed. A commercial floating catfish feed containing 30 to 35% protein is recommended where the weight of brood fish held exceeds 700 kg/ha. When the standing crop in the holding pond is between 250 and 700 kg of brood Bluegills/ha, a lower percent protein commercial pellet is adequate, and at standing crops below 250 kg/ha, no supplemental feeding should be needed in well-fertilized ponds. Feeding at 2 to 3% of body weight/day is appropriate during the spring, summer and fall months. During the late fall and winter, the feeding rate is reduced to 1.0 to 1.5% and when water temperatures fall below 10°C feeding can be suspended. Feed is given daily during the warm months and on alternate days during the cool period. The dry feed should be pelleted to a size small enough for the fish to swallow. A gain of 100% in weight from year to year should result in acceptable spawning success unless other factors override the influence of growth.

Many hatcheries produce a crop of bass or other species in the same pond that is later used for Bluegill production. Once the bass are produced and harvested, the pond is prepared for use as a spawning-rearing pond for Bluegill or other sunfish. The challenge to this approach is that in temperate climates, the water temperature in the spring some years rises sooner than other years reaching temperatures suitable for Bluegill and Redear spawning. However, such pond designated for Bluegill and Redear spawning are still being used for bass production for several more weeks.

In the meantime, Bluegill brooders, still in the brood holding pond, may have spawned, but there is not enough time for any resultant fry to reach a suitable size for harvest before the brooders must be harvested. This premature spawning in the holding pond results in some brood females having released their initial batch of eggs in the holding pond. When such females are stocked into the spawning pond, they may require additional time until a new batch of eggs is mature enough to be spawned. This results in more variation among brooders as to when they spawn once they are stocked into the spawning-rearing ponds, and in turn more size variation in the progeny.

Ideally, cool water can be added to a Bluegill brood holding pond to maintain the pond water temperature below 21°C and delay their spawning until a spawning-rearing pond is available. In temperate climates, depending on the depth of the well, the water temperature of well water is often less than 21°C and can be added to the brood holding pond. In the spring, stream water is often cooler than pond water and can be added, but both well water and stream water have issues as discussed earlier in Chapter 2 that need to be considered. Redear and Green Sunfish will spawn at temperature several degrees cooler than Bluegill, necessitating a temperature of less than 20°C to suppress spawning of these species.

Selecting brood holding ponds that have deep shorelines will limit the amount of suitable area for nest building and in turn limit spawning, but most hatchery ponds are going to have shallow enough shorelines to allow nest building.

Separating fish by sex serves to prevent or minimize wild spawning. Unfortunately, this approach is limited by the fact that sunfish cannot be sexed with 100% accuracy. This is especially true for younger brood fish.

Spawning-Rearing Pond Management.—When using a spawning-rearing approach with a species such as Bluegill that will spawn multiple times, the goal is to get a strong first wave of spawning that will provide the majority of fingerlings to be produced in the pond. Enough brood fish should be stocked to produce enough fry during one short spawning period to utilize all available food, thus suppressing further spawning. However, the number should not be great enough to limit growth to the point where the young Bluegills are too small for distribution. The optimum fingerling size to be harvested from the spawning-rearing pond is 3 to 6 cm TL or 2,200 to 3,500 fish/kg if they are to be for stocking farm ponds.

For spawning-rearing ponds of average productivity (i.e. carrying capacity about 750 kg/ha) a stocking rate of about 200 adults/ha weighing 150 to 300 g each is suggested. Larger fish or less productive ponds may justify reducing the stocking rate to 150/ha while the maximum number would be about 250/ha for highly productive ponds. Use of yearling or stunted adult fish requires an increase of 50 to 200% in the number of brood fish stocked, depending on the expected egg output. A 1:1 sex ratio is suggested since no research is available to indicate otherwise.

Spawning-rearing ponds for sunfish can be stocked in the winter, spring, or early summer prior to the distribution season for which fingerlings are produced. About 60 days are required to produce a harvestable size fingerling under average conditions. Density of small fish, food supply available, and water temperature are factors thought to be most important in obtaining the desired size.

Both zooplankton and midge larvae are important in the diet of small Bluegill. Since a size of 2.5–5.0 cm is adequate for sunfish fingerlings stocked in farm ponds, small crustaceans and small midges appear to be needed in abundance, with only enough larger food organisms to feed the brood fish. Unless the water supply already supports an abundant supply of natural food, the spawning pond should be filled 2 to 3 weeks before the first Bluegill hatch to allow a food supply to develop. Pond fertilization using an organic source of nitrogen can speed up the development of food organisms in the pond. Inorganic fertilizer should be added to provide 9.1 kg/ha nitrogen, 9.1 kg phosphorus as P_2O_5, and 2.25 kg K_2O at weekly intervals until phytoplankton develops sufficiently to prevent light penetration to deeper depths than 45 cm is recommended. Later applications at the above rate should be made when the water clears enough to permit light penetration to a depth of 60 cm. Fertilization can be discontinued when water temperature drops below 10°C.

Attempts to pond-rear Bluegill fingerlings to distributable size on artificial food can result in increasing the size variation. While the small fingerlings readily accept pelleted fish feed, the largest fish tend to consume more than the smaller ones, and size difference becomes greater. In two growth trials on a production hatchery, Bluegill were fed at a rate of 11.3 kg/ha/day for about 60 days. The results were an increased size variation with the largest fish being too large for efficient distribution.

16.2.3 Intensive Fry Production

16.2.3.1 Fry Transfer

To produce a more uniform size fingerling than seen with a spawning-rearing method the spawning activity needs to be separated from the rearing phase. Bluegill fry can be produced similar to intensive fry production of tilapia in earthen ponds. In the case of Bluegill, brooders are stocked into spawning ponds having clean catch basins. Appropriate stocking rates are 75–90 kg of brooders using a 1:1 sex ratio, using the lower weight of brooders when the average weight is closer to 50 g and the higher rate when the average weight is >100 g. While in the spawning pond the brooders should be fed at 1% body weight/d a floating pelleted feed. Brooders are allowed to spawn for 28 days during which time the majority of the eggs will have hatched and the fry developed to the point that they are swimming up in the water column. At that point, the pond is slowly drained into the catch basin so that the fish can be captured in the first few hours after dawn. The fry are more delicate than tilapia fry and should be harvested by crowding the fry using a large fine mesh dip net then transferring the fry in water to a transport tank that has a similar water temperature as the water from where the fry are being harvested. A fry harvest of 7,500–9,500 10–11 mm fry/kg of female stocked should be expected. Once most of the fry have been harvested, the brooders should be harvested, held in the hatchery and given a prophylactic treatment with potassium permanganate at 10 mg/L for 30 min. Brooders should be examined and any brooders in good condition can be restocked into other spawning ponds. Anticipate a 10–20% loss of brooders in the spawning pond and an additional 10% discarded after being examined in the hatchery. Four to five spawning cycles may be possible during the summer in a temperate climate as long as well-conditioned brooders are available and two or more spawning ponds are available. After a spawning cycle is complete the pond should be dried and prepared for a new cycle. At the same time another pond has been already stocked with brooders and they are actively reproducing.

Fry Transfer is more labor intensive than Spawning-Rearing but is more productive, resulting in two times the amount of fingerlings/kg brood female, and yields larger fingerlings with less size variation than the Spawning-Rearing approach. However, allowing only a 28 day spawning period results in some egg and fry loss at harvest.

16.2.3.2 Egg Transfer

Bluegill can be spawned indoors or outdoors under controlled conditions. This allows more control of fish age, egg hatching success and fry harvest. In outdoor concrete tanks filled to a depth of 45 cm with clean water, brooders are stocked at one male and two females/m^2. Artificial nests, 25 cm plastic bowls with a layer of gravel in the bottom, are added at 1/0.8–1 m^2 of tank. Males will select a nesting site within a day after artificial nests are added and spawning will begin soon thereafter. By having the bowls set in clear water, the male activity can be observed and eggs when deposited in the nest can be seen. If needed, a bowl can be lifted up for closer examination and returned to the same site if no eggs are found. Each bowl should be observed twice daily and removed if containing eggs. A new bowl should be added where the previous bowl was. The eggs will be attached to the gravel and sides of the bowl. The eggs are best incubated in place, immersing the bowl in an aerated aquarium with a 5 L/min. flow of water into the bowl. Eggs should hatch in 24 h after being collected at 26–28°C. Hatch rates of >85% should be anticipated. Sac fry are allowed to remain in the nests until they swim-up about 5 days after hatch. A yield of 30,000–40,000 swim-up fry/kg female should be expected.

Bluegill can be induced to spawn naturally under indoor conditions where temperature and photoperiod are controlled. Brooders can be acclimated over a two-week period to a temperature of 15°C and a 8 h light: 16 h dark photoperiod and held for four weeks. The temperature and photoperiod are then gradually increased over a two-week period to 25°C and a 14 h light: 10 h dark cycle. Spawning is expected within a month after reaching the 25°C, 14:10 h conditions. Fry production is similar to that of outdoor tank spawning but with the advantage that fry can be available year-around for laboratory studies and other applications where fry are needed throughout the year.

A consideration with egg transfer is how to feed recent swim-up fry. Such fry have a mouth gape too small to initially feed on some strains of artemia, a widely used first food in the hatchery for many freshwater fish larvae. If first-feeding Bluegill fry are to be held in the hatchery for several days they need to be provided rotifers, copepod nauplii or 1st instar artemia of a strain known to yield small 1st instar nauplii (see 17.14. Live Food Production). After a day or so they can be fed just artemia nauplii and by day 14 weaned over several days to a high quality fine particle commercial feed. With careful management, a 40–50% survival can be obtained at 24 d post-hatch.

Recent swim-up bluegill fry are delicate and care needs to be given to their handling whether produced indoors or outdoors. Fry transfer from spawning ponds using a 28 d cycle provides a larger fry (11–12 mm) than the 6–7 mm recent swim-up fry obtained with egg transfer procedures. Fry obtained by egg transfer must be fed in the hatchery for several weeks to obtain a size similar to those collected from fry-transfer ponds before being transferred to outdoor nursery ponds. Transferring recent swim-up fry directly to a nursery pond can result in a reduced survival. In a series of trials conducted at Auburn University stocking recent swim-up fry directly into nursery ponds resulted in a survival of 30–35% compared to a 60–70% survival obtained by stocking the larger fry collected as part of the Fry Transfer approach. However, the quantity of fry produced by Egg Transfer was about three times that obtained/kilo of female using the Fry Transfer approach. Even after taking into account the lower nursery pond survival when fry from the Egg Transfer approach were stocked soon after swim-up, the quantity of fingerlings produced/kg female brooder was twice that of Fry Transfer, and four times the quantity of fingerlings obtained/kg female using the Spawning-Rearing approach. The Spawning-Rearing approach is the simplest but the results can be variable from pond to pond and often less than 30% of the fingerlings harvested are ≥6 cm in length. But if pond space is available and considering its low labor requirements, Spawning-Rearing is a practical approach for many hatcheries. Using fry obtained by Fry or Egg Transfer for stocking nursery ponds will produce a greater quantity of fingerlings, with less size variation, and ≥70% of the fingerlings can be expected to be ≥6 cm in length. Egg or fry transfer is preferred when fingerlings of similar age with less size variation is desired.

16.2.4 Issues during nursery production

16.2.4.1 Vegetation

Vegetation control is necessary in Bluegill culture, with rooted plants and filamentous algae being controlled by appropriate herbicidal treatment. Ponds should be adequately fertilized to maintain a phytoplankton bloom that can shade out many of the aquatic weeds. Not only is the fish crop harvested more readily from a weed-free pond, but also the food chain is more productive. Moreover, small fish can have greater access to the natural food organisms present without the interference of aquatic weeds.

16.2.4.2 Feeding

Fertilization alone will produce enough natural food organisms to support about 500–750 kg/ha of Bluegill fingerlings. Bluegill will accept formulated feed when cultured in nursery ponds, offering the feed twice a day, beginning with a high percent protein powdered feed for several weeks followed by a No. 2 crumbled trout feed. Depending on the density at which the fry were stocked, by 12 weeks the fish should be large enough to be weaned to a 1.6 mm floating feed. Depending on the length of the culture period, yields of fingerlings of >1000 kg/ha can be obtained by providing formulated feeds. Feed rates should be adjusted based to feeding activity, and that is temperature dependent with little to no feeding activity expected at ≤12°C.

16.2.4.3 Harvesting

Harvesting the crop of Bluegill fingerlings can be done as soon as the major portion has reached a size large enough for distribution. Fingerlings withstand handling stress much better at water temperatures of less than 20°C than at higher temperatures. Both large and small Bluegill can be easily damaged by poor handling. The large numbers of bluegill produced at a hatchery can lead to hatchery staff becoming complacent and use more rapid, and more casual handling methods than appropriate. This results in skin and scale loss leading to development of columnaris and fungal diseases and other conditions that reduce fish survival. To minimize losses at harvest, a clean, silt-free catch basin is needed for collection, cool water for collection and transportation, and careful handling during capture.

After removal from the production pond, it is desirable to retain small sunfish in a holding tank for about 18 to 24 hours to allow them to recover from the stress of being harvested before further processing. Grading should be done if as much as 2.5 cm of length separates the largest fish from the smallest. Hardware cloth baskets of different size mesh are ordinarily employed for this purpose, although the variable spaced metal rod grader also works satisfactorily. Grading enables a more accurate count of small fish produced from the spawning-rearing pond to be made, assures a more attractive and uniform lot of fish for distribution, and increases the accuracy of the size and count of the fish provided to the clients.

16.2.4.4 Distribution

Sunfish are best transported at rates of 60 to 120 g of fish/L of water. Distribution units are supplied with bottled oxygen and electrical agitators to provide aeration. The agitators provide ample oxygenation during transport, but are supplemented with pure oxygen while the truck is stopped or during emergencies where power failure is encountered. Loads above 120 g/L have been hauled for time intervals of less than 6 hours. For larger fish, a more efficient distribution tank, and water of better quality may permit a heavier loading rate.

16.2.5 Other Sunfish Species

16.2.5.1 Introduction

The foregoing discussion deals more specifically with Bluegill sunfish fingerling production than with other sunfish. Another commonly hatchery produced sunfish is the Redear Sunfish *Lepomis microlophus*. It is often stocked along with bluegill in farm ponds using a 4 Bluegill: 1 redear fingerling ratio. An advantage of Redear is that they will feed on snails and other mollusks not reality consumed by Bluegill. They also grow to a larger size than Bluegill.

Redear Sunfish do not accept formulated feed as readily as Bluegill making it more difficult to hold high densities of brooders or increase the fingerling production. Redear Sunfish are propagated using a Spawning-Rearing approach similar to that used with Bluegill. Redear spawn heavily in late spring at a water temperature of about 18–21°C. Following the spring spawn, a limited amount of spawning will take place during the warmer months. Another large wave of spawning will take place in late summer or early fall. To obtain highest output from broodstock, the early spawn must occur in the spawning-rearing pond and not in the broodstock holding pond. Redear Sunfish begin spawning about the same time as Largemouth Bass so spawning ponds must be stocked before bluegill spawning ponds are. In north temperate regions, the late summer or early fall spawn may come too late to permit the fry to reach distributable size in that growing season. Redear Sunfish will freely hybridize with Bluegill. Extra care must be given in preparing holding and spawning ponds to make sure no fish were still alive in the ponds prior to filling, and the ponds are filled with filtered water to avoid contamination with other fishes. Ideally, Redear and Bluegill are held in ponds that are some distance apart on the hatchery to minimize the chance of a bird picking up a fish and dropping it into another pond.

16.2.5.2 Sunfish Hybrids

One issue with Bluegill in farm ponds is that they can reproduce too much, resulting in too many fish for the available food supply, and the population becoming stunted and not reaching an acceptable size. One approach is after the initial stocking of Bluegill has had some time to grow, is to add a predator, such as Largemouth Bass, that will consume most of the Bluegill reproduction, leaving enough food for the remaining Bluegill to grow to an acceptable size. Another approach to reduce excess reproduction is to produce a hybrid that is less reproductively successfully than its parents. Depending on species of sunfish used and the cross made, the 1st generation hybrid, often referred to as the F1 (first filial) generation, may grow faster than either parent species. In most cases, the 1st generation hybrid is reproductively viable but many crosses give less progeny than would pure crosses of either parent species. In some crosses, the sex ratio of the 1st generation hybrid is often skewed toward male, which further reduces the potential for over-reproduction. However, the progeny (the F2 generation) from the 1st generation hybrid is reproductively viable and an individual fish may have inherited genes similar to that of the male grandparent species or the female grandmother species. When they (the F2 generation) reproduce with other progeny from that generation, the genes are further remixed. This re-mixture of genes continues in later generations and any genetic advantage expressed in the 1st generation becomes lost. The advantage of stocking a 1st generation hybrid is lost if the pond is to remain in production for several years. Hybrid sunfish are best used where there is intensive fishing pressure and the subsequent generations are not counted on to provide quality fishing.

The two most common sunfish hybrids produced by hatcheries are: female Redear X male Bluegill, and female Green Sunfish × male Redear Sunfish. As long as brooders are carefully selected, the Spawning-Rearing method can be used to produce hybrid sunfish fingerlings. But no errors can be made in species identification or sexing of the broodfish. If there is a mistake, those fish may mate with the opposite sex of the same species, resulting in purebred fingerlings of that species, while the other brooders have mated with the other species producing hybrid fingerlings. This will result in a population of fingerlings where the vast majority are hybrids, but some purebred fingerlings are mixed in and are difficult to sort out.

To minimize errors in sexing, only mature fish that are in good condition should be selected, preferably using fish that are two years old or older. Sex identification of gravid fishes can be done by use of secondary sex characteristics such as coloration, opercula flap, abdominal distention, and vent appearance. The best way to be sure of the sex is that mature males should release milt when stripped and gravid (ripe) females should release some eggs when light pressure is applied to the abdomen. If there are problems in being confident of the sex of the fish, possible males can be given an injection of hCG at 500 IU/kg and reexamined 24 h later for milt flow. Possible females can be confirmed by trying to take an egg sample from the ovary using a 1.1–1.2 mm clear plastic catheter. The catheter is gently inserted through the urogenital opening, up the oviduct, and into the ovary. If in the ovary, a little pressure on the catheter will allow eggs to enter the catheter. When the catheter is withdrawn and examined the presence of eggs will confirm that the fish is a female.

Spawning ponds should be carefully prepared, making sure that there are no fish remaining from previous crops. Extra care is needed when using surface water to fill spawning ponds as often there may be other sunfish living in the water supply that can enter the pond and contaminate the crop being produced. Otherwise, the pond needs to be limed and fertilized as if it were to be prepared for Spawning-Rearing of Bluegill. Brooders are stocked at a 1:1 sex ratio, stocking 75 to 100 pair/ha. Brooders can be given a floating feed and the progeny offered a powdered feed and later a crumbled feed. Yields of 5 cm hybrid fingerlings may be approximately 250,000/ha, while similarly managed redear spawning ponds would yield 300,000–375,000 5 cm fish/ha or 375,000–450,000 3–5 cm Bluegill fingerlings/ha.

Management of the spawning rearing pond, its harvest, as well as the distribution of the fingerlings follows the same procedures as described earlier for Bluegill.

16.2.6 Recommended Reading

Arslan, T., and R. P. Phelps. 2011. Egg transfer to increase productivity and predictability in Bluegill fry production. North American Journal of Aquaculture 73(2):135–140.

Brunson, M. W., and J. E. Morris. 2000. Species profile: sunfish. SRAC Publication 724, Southern Regional Aquaculture Center, Stoneville, Mississippi. Available: https://agrilifecdn.tamu.edu/fisheries/files/2013/09/SRAC-Publication-No.-724-Species-Profile-Sunfish.pdf. (May 2020).

Dupree, H. K., and J. V. Huner. 1984. Propagation of black bass, sunfishes, tilapias, eels and hobby fishes. Pages 119–135 *in* H. K. Dupree and J. V. Huner, editors. Third report to fish farmers. U.S. Fish and Wildlife Service, Washington, D.C.

Heidinger, R. C., and W. M. Lewis.1972. Potentials of the Redear Sunfish X Green Sunfish hybrid in pond management, The Progressive Fish-Culturist 34(2):107–109,

Lovshin, L. L, and M. Matthews. 2003. Intensive culture of Bluegill *Lepomis macrochirus* in earthen ponds. Journal of the World Aquaculture Society 34:106–112.

Mischke, C. C., and J. E. Morris. 1997. Out-of-season spawning of sunfish *Lepomis* spp. in the laboratory. Progressive-Fish Culturist 59:297–302.

Morris, J. E., C. C. Mischke, and D. L. Garling, editors. 2002. Pages 2–3, 12–18 *in* Sunfish culture guide. NCRAC Culture Series 102, North Central Regional Aquaculture Center Michigan State University, East Lansing, Michigan. Available: https://www.ncrac.org/files/biblio/Sunfish_Culture.pdf. (May 2020).

Pierce II, R. A., and C. E. Hicks. 2012. Bluegill sunfish production in Missouri. University of Missouri Extension Publication G9473. Available: http://extension.missouri.edu/explorepdf/agguides/wildlife/g09473.pdf. (May 2020).

Williamson, J. H., G. J. Carmichael, K. G. Graves, B. A. Simco, and J. R. Tomasso. 1993. Centrarchids. Pages 145–199 *in* R. R. Stickney, editor. Culture of nonsalmonid freshwater fishes, 2nd edition. CRC Press, Boca Raton, Florida.

16.3 Striped Bass *Morone saxatilis* (Walbaum) and White Bass *Morone chrysops*

16.3.1 Introduction

Striped Bass is an important game and commercial species native to the Atlantic coast of North America from the St. Lawrence River to the St. Johns River in Florida, and the Gulf of Mexico from the Suwannee River in Florida westward to several coastal rivers of Texas. Striped Bass or its hybrids have been introduced to the Pacific coast of the USA as well as many inland rivers and reservoirs. Currently they are found in 48 of the 50 States of the USA. In addition, Striped Bass or its hybrids have been introduced into ≥16 countries around the world for culture or as a sportfish.

Over the wide natural distribution of Striped Bass there are population differences in spawning habitat, egg and larval characteristics, adult morphometrics such as lateral line scale count and fin ray count, and environmental requirements. Most of the Atlantic strains are migratory, spending much of their life in the ocean. The strains found along the Atlantic Coast south of North Carolina and along the Gulf Coast of the Gulf of Mexico are more riverine and seek cool thermal refuges within rivers during the summer heat. These differences must be considered in which hatchery methods to use and how will any fish produced be used thereafter.

Hatchery propagation of Striped Bass in the USA dates back to the latter part of the nineteenth century (1882), providing a substantial base of production experience on which to draw in establishing cultural procedures. The Santee River diversion project in South Carolina resulted in a dam-locked reproducing population of striped bass in Lakes Marion and Moultrie. The South Carolina Department of Natural Resources established the Moncks Corner Striped Bass Hatchery beside the tailrace of the Cooper River. There procedures were developed for induced spawning, egg incubation and larval management of Striped Bass (Bayless 1972, Bishop 1976). These efforts resulted in a dependable source of larvae for distribution throughout much of the USA and internationally to establish Striped Bass in many reservoirs and rivers where they previously did not exist.

Another cultured species in the Moronidae family is the White Bass *Morone chrysops*. It is native to the Great Lakes and Mississippi River drainages but has been introduced throughout much of the USA. It is used in hatcheries primarily to make a hybrid with Striped Bass. The hybrid is suitable for some reservoirs in the southern USA too warm for pure–line Striped Bass. The hybrid is also more suitable for culture as a food fish than pure-line Striped Bass. Pure-line White Bass are also a popular sportfish and are produced in hatcheries to provide a sport fisheries in the USA.

16.3.2 Biology

Striped Bass have a relatively long lifespan, up to 30 years in some cases. The maximum recorded size is a 57 kg fish caught by a commercial fisherman in the late 1800s. The growth rate of Striped Bass is rapid under optimum conditions. Fry will grow from a length of 5 mm to 5 cm on a diet of natural food in a period of 50 days at a density of 100,000/ha. Fish 5 cm long have been grown to a size of 18 cm in a period of 120 days on a formulated diet. A 4-year-old fish is expected to reach a length of 45 to 60 cm and a weight of 1.8 to 2.7 kg.

There is little sexual dimorphism exhibited for this species, although female fish in a given year class tend to be larger than males. Typically, fish weighing >10 kg are females. In the spawning season, the abdominal development and vent appearance are distinctive in females while males will show a flow of milt when stripped. Male fish attain sexual maturity at an age of 2 years. Females begin to be sexually mature at 4 years of age. Breeding age for females is reported to be from 4 to 14 years.

Under natural conditions, Striped Bass make a spawning migration from salt water or brackish water environments in to freshwater, or in some cases from one freshwater location to another. In temperate climates in the early spring males start to migrate towards the spawning area as water temperatures rise to about 12°C. Females migrate later as the water temperature approaches 15°C and above. Spawning activities often peak at about 18°C. Fish may remain in the spawning area for 4–5 weeks before the temperature goes too high.

Spawning grounds are often river rapids or behind dams where water currents are strong enough to keep the striped bass eggs up in the water column until they hatch and the larvae are able to swim on their own. As a female swims into the spawning grounds, she will be followed by a group of males already at the spawning grounds. The males will swim close to the female, often nudging the vent area of the female as time of ovulation approaches. As the female begins to ovulate she will become stationary, often with the head downward, begins to quiver, and release eggs into the water column. Numerous males will try to get close to the female and release their sperm into the water. This process will be repeated several times until the female has released the ovulated eggs. No parental care is given to the eggs or young. Most females will spawn only once a year. Average fecundities are approximately 100,000 eggs/kg female and a large female may produce as many as 5 million eggs.

At time of ovulation, eggs are 1.1 to 1.4 mm in diameter. After water hardening they swell to a diameter of about 3.6 mm or more and have a large perivitelline space. The eggs are slightly heavier that water with a specific gravity of about 1.0005. Egg dimensions can vary by strain of fish, and water hardness of the water into which the eggs were released. Eggs from Gulf Coast strain Striped Bass are often smaller and less buoyant than that of Atlantic Coast females and require a greater current to keep the eggs up in the water column.

Enough stream flow is needed to keep the eggs and larvae in suspension for a long enough period until the eggs have hatched and the larvae are able to swim on their own. In the absence of sufficient current, eggs and young larvae will settle out of the water column and die. The time to hatch for Striped Bass eggs is temperature dependent and may range from 80 h at 12°C to 29 h at 24°C. First feeding commences 5 to 7 days after hatching. Striped Bass larvae may be 2.0–3.7 mm TL at hatching and average about 5 mm TL at first feeding.

Preferred natural foods for first feeding Striped Bass include copepod nauplii and small cladocerans. Rotifers will be consumed in the absence of preferred foods. As the larvae grow there is a progression to larger zooplankton with immature insects being consumed when the fish are larger than 10 mm TL. Adults feed mainly on fish but the transition size is not well defined. Under hatchery conditions cannibalism has been noted in fry less than 10 mm TL. It is probable that change to a fish diet does not ordinarily take place until a size of 6 to 10 cm is reached.

White Bass reach a weight of 2–3 kg and are smaller, stockier in appearance, and proportionally, have a smaller head than Striped Bass. The hybrid commonly weighs between 2.5–5 kg and can be larger. Optimum temperatures of growth of White Bass are between 25– 27°C compared to approximately 20°C for Striped Bass.

White Bass are found in open waters of large lakes and reservoirs as well as slow-moving low turbidity rivers. They reach sexual maturity when 2 years old in the southern part of their range. In the early spring once the water temperature is around 14–18°C mature adults migrate up smaller streams to areas of rocks and gravel having a good water flow to spawn. They also congregate on the down-stream side of dams that present a similar habitat. Males often arrive at the spawning area weeks before the females. Once the females arrive they spawn similar to Striped Bass with one female being chased by several males. The female releases her eggs in the water where they are fertilized by multiple males. The eggs are heavier than Striped Bass eggs and soon sink to the bottom. They become adhesive soon after being in contact with water and will stick to the rocks and gravel as they settle to the bottom. The time to hatch is approximately 48–72 h, depending on the water temperature.

Striped Bass and White Bass rarely hybridize naturally as the two species typically spawn at different times of the year, at different temperatures, and in different types of habitat. However, hybrids can be produced in a hatchery setting. Hybrids between the two species can be made in either direction: Striped Bass female × White Bass male commonly called Palmetto bass; and a White Bass female × Striped Bass male, commonly called Sunshine bass. The hybrid is more suitable for warmer conditions than Striped Bass. They grow faster than Striped Bass during the first few years post-hatch. Hybrids are easier to train to a pelleted feed, and are more disease resistant than striped bass, making them more suitable for use in aquaculture. Private commercial hatcheries primarily produce the sunshine bass (White Bass female × Striped Bass male hybrid) as White Bass females are more easily acquired from the wild and are easier to domesticate and maintain under hatchery conditions. However, White Bass females are smaller and give a smaller quantity of eggs/female than a Striped Bass female. Providing quality female Striped Bass to make the female Striped Bass × male White Bass hybrid is more a problem. In most cases Striped Bass females have to be collected from the wild, but fewer females are needed as a much greater quantity of eggs/female is obtained.

Larval *Morone* have a yolk sac that contains a large oil globule that helps maintain buoyancy in the water column. Mouthparts are typically developed by 3–5 days post-hatch and they begin feeding on small zooplankton. The size of the larvae depends on the species of female being spawned. At 12 h post-hatch, larval Striped Bass (prolarvae) average about 4.7 mm in total length (TL), while White Bass larvae are considerably smaller (ca. 2.9 mm TL). Striped Bass × White Bass larvae (4.5 mm) are slightly smaller than Striped Bass × Striped Bass. Larvae of the White × Striped cross are slightly larger (3.0 mm) than that of a pure White Bass larvae. The cross using a Striped Bass female has an advantage as the eggs are non-adhesive, are easy to incubate, and the larger larvae are able to take 1st instar artemia nauplii as a 1st food. Using a White Bass female results in adhesive eggs that must be treated before being incubated and the smaller larvae must be fed rotifers (which are more complicated to produce) as a 1st food.

16.3.3 Broodstock

The source of broodstock must take into consideration how the progeny will be used. If the progeny are to be stocked into the wild then brooders should be collected from the same watershed that the progeny are to be released. Where the progeny are going to be used in aquaculture or be hybrids then brooders can be collected from a wide range of locations and in many cases be domesticated and several generations developed for hatchery use. However, it should be kept in mind that the hybrid if stocked into a natural setting can backcross to their parent species if there is a reproducing population of the parent species.

Striped Bass brooders are typically collected from the wild, particularly when used as part of a government program because getting official permission for brood collection is easier. They can be obtained from a commercial capture fisheries but often such fish have been stressed too much during capture to be suitable brooders. Best quality brooders are collected once they have migrated to spawning grounds or into areas where their migration is blocked and are congregating in strong current below a dam. They are best collected through electrofishing using pulsed AC voltage. If done properly, electrofishing can be one of the least stressful methods to capture brooders. Stationary and drift gill nets with a 5–13 cm bar mesh can be used if the nets can be monitored frequently enough. Leaving a fish tangled in a gill net for over 30 min can be very stressful and damage the fish. Hook and line capture of brooders is an alternative particularly for private hatcheries when complying with the sportfishing regulations in place at that time. Government regulations often do not permit private hatcheries to use electrofishing equipment or set gill nets in areas where striped bass may be gathering to spawn. Hook and line fishing and the associated fight can be particularly stressful on female Striped Bass to the point where they will not spawn after arrival at the hatchery. Hook and line captured males typically will still produce sperm after arrival at the hatchery.

All females are not equally mature when collected from a natural spawning area. Despite appearing very similar in gross appearance, there may be some mature females that could ovulate within a few hours while others may be 15 h or more away from ovulating. A rough idea of maturity can be determined in the field or at the hatchery taking an egg sample (Figure 16.3.1) as described in Chapter 4. Samples of eggs that appear very small (<600 mm), often not very uniform in size, yellowish in color, difficult to see through, and with no visible oil drops, suggests that the female is immature and should be released back to the wild as she

Figure 16.3.1. Sampling female Striped Bass *Morone saxatilis* with a catheter to collect an egg sample to determine egg maturity.

may be a year away from spawning. A female within 15 h of spawning will have greenish/opaque-appearing eggs with small but visible oil droplets dispersed throughout the egg. The response of such females to hormone injection can be variable, and if females are available that are more advanced in their egg development, they would be preferred. Any females with translucent eggs where some of the oil droplets have merged together to make larger droplets would be considered an eligible fish for induced spawning.

At some hatcheries, the staff prefer to collect and examine the females in the field, select eligible females based on egg development, inject selected fish while still in the field, and then transport the injected females to the hatchery. At other hatcheries, the preference is to take all females of an appropriate size to the hatchery, and there the eggs are examined and eligible fish injected. Doing these activities at the hatchery can be an effective way to reduce stress on the brooders and increase the odds that a fish will spawn. Egg sampling in the more controlled hatchery conditions helps to reduce the chance of damaging the oviduct when taking an egg sample, and in turn, the chance of a blood clot that could block the oviduct and prevent the flow of ovulated eggs.

Striped Bass are easily stressed no matter how captured, particularly large gravid females. Poor handling of such fish may result in such females not responding to being hormone-induced to spawn. Minimizing stress begins with selecting a collection site that is easily accessible and is within a few hours' drive of the hatchery. A plan should be developed ahead of time as to how to reduce brood handling. Brooders should be handled using a soft-mesh netting (not knotted netting that can cause scale and mucous loss and hemorrhaging). To move brooders from one tank to another, a stretcher made of knotless netting should be used. This allows a fish to be guided into the stretcher and the stretcher rolled up around the brooder to control the fish's movement and reduce handling damage. When electrofishing, a second boat should be available to transport any brooders from the electrofishing boat back to the dock so the potential brooders can be examined and loaded on the transport truck or into a holding net where they can swim freely. Holding brooders too long in a small holding tank on the electrofishing boat can be enough stress to prevent a female that is otherwise in good condition from spawning.

To move any captured brooders from the collecting site to the hatchery, a truck with a large tank is needed. The transport tank should be an insulated circular tank about 1.8 m in diameter and 60–90 cm deep, provided with bottled oxygen and set up to have a continuous circular flow. Appropriate loading rates may range from 40 to 60 kg/m^3 of tank depending on how long the fish are to be transported. Striped Bass will swim around into the current in a circular tank and be less damaged than if a rectangular tank is used where fish could swim into a corner and be damaged. The tank should have tight sealing lids large enough to easily allow stocking and harvesting of large fish. The water level in the tank should be filled up to the brim of the tank and kept there so little to no wave action will occur when the truck is underway. A chimney is needed so excess oxygen can escape but no water is lost.

The transport tank water should be filled with water from the site where the fish were collected and salt (NaCl) added to give a salinity of 5–10 g/L. Tricaine methanesulfonate (MS-222) is added as an anesthetic to slow the fish's metabolic rate, but not enough to put the fish to sleep. The dose rate is fish size dependent. At temperatures of 13–18°C a recommended dose is 5–8 mg/L for fish with a total length of $\leq$75 cm, and 2.5–5 mg/L for fish >75 cm.

Female Striped Bass are generally induced to spawn soon after their arrival at the hatchery. Male Striped Bass may have been collected a week or more ahead of the females and held at the hatchery in circular tanks until needed. Similarly, White Bass males and females are often collected and can be held for several weeks before being used.

Both Striped Bass and White Bass can be reared from egg to mature adult under hatchery conditions if the proper water temperatures and foods can be provided. Striped Bass can be domesticated and quality brooders obtained under hatchery conditions in locations where the water temperature can be maintained below 26°C. At higher temperatures females may not produce good quality eggs but males can produce sperm successfully when reared in outdoor ponds at temperatures >26°C. Fish can be cultured to adulthood in either indoor or outdoor circular tanks on recirculating or flow-through systems using tanks ranging size from 3.7 m diameter × 0.9 m deep to 6.1 m diameter × 1.5 m deep at standing crops of approximately 5 kg/m^3. Water quality should be monitored frequently to ensure safe conditions that do not stress the fish. Fish should be fed a high-quality feed, increasing the pellet size as the fish get larger, offering a 1/4" to 3/8" (6.35 to 9.52 mm) pellet to brooders. Wild-caught brooders often do not learn to take a pelleted feed and must be fed live fish. Keep in mind that any forage fish is a potential disease carrier and such fish should be given prophylactic treatments with potassium permanganate to control bacteria and parasites before being fed to brooders.

Development of domesticated Striped Bass brooders is a long-term commitment of time, space and record keeping as it may take 3 years for the majority of the males to become mature and 4 years or more for the majority of the females to be mature. Manual stripping of a hormone-induced female is very stressful with a high likelihood that she will die in the process. A broodstock replacement plan is needed that includes replacing each induced-spawned female every year. Inbreeding can develop rapidly, because spawning only 1–2 Striped Bass females may provide all the eggs needed that year. This limited gene pool can be a major issue if the progeny are to be stocked into the wild. After the first year of growth the fish are large enough so they can be PIT tagged and fish of different origin or year-class can be communally reared. Keeping track of the brood history is essential in trying to minimize any negative genetic effect.

Wild-caught White Bass can also be successfully reared under hatchery conditions in both outdoor ponds and in tanks. Domestic broodstock can also be developed from egg to mature adult in two years. Domesticated brooders will have been fed formulated feeds from <45 d post-hatch to mature adult, so accepting a formulated diet is not an issue. However, optimal diet formulations for broodfish are still being studied. Wild-caught adult white bass can be feed trained beginning with chopped fish or a pelleted moist salmon diet then weaned to a dry pellet over a several week period.

Whenever introducing wild fish to the hatchery, they should be quarantined and given prophylactic treatments for bacteria and parasites. Where brooders can be observed in tanks, signs of possible parasite problems can include: flashing, where the brooder will scrape its body along the tank wall or bottom; coughing, where the normal pattern of opercular movement is interrupted by a flaring of the opercula followed by strong contraction and expulsion of water; and visible mucous buildup and fin erosion. When such signs are seen, prophylactic treatments should be given. Keep in mind that such treatments may harm the effectiveness of the biological filter if the brooders are being held on a recirculating system. The fish should be on a flow-through water supply while being treated.

The Great Lakes basin has been a common source of White Bass brooders but with Viral Hemorrhagic Septicemia (VHS) causing problems in the basin, fish transportation out of the basin is being controlled. Hatchery managers need to know the origin and history of any White Bass being introduced to the hatchery, both from a genetic perspective and disease history.

Photoperiod and temperature regulation is an important part of hatchery-held broodstock management. Brooders are best introduced to tanks during the winter when the day length is short and the water temperature is low. Female White Bass should be stocked at temperatures of ≤12°C. Male Striped Bass should be stocked at temperatures of ≤15°C. Brooders can be held at conditions that represent the typical 12 month natural cycle of temperature and photoperiod of the hatchery locale. The cycle can also be condensed to a 9 month cycle if there is a need to start the spawning season sooner. A more common cycling pattern is "cold banking," that is using a cycle where the brooders are exposed to a changing temperature and photoperiod but as the cycle starts to approach spawning conditions, the cycle is not advanced. The goal is to keep the brooders for a longer period in a pre-spawning mode, then advance the cycle to that of spawning conditions. This will extend the time period as to when brooders can be available for spawning. Domesticated White Bass females and Striped Bass males held under cold-bank conditions can be capable of reproducing months after their natural spawning time. Having fry available for a more extended period can allow two or more 30-day primary nursery crops, reusing the same pond.

16.3.4 Spawning

Striped Bass are sensitive fish and upon arrival at the hatchery they should be held in circular tanks (ca. 1.8 m diameter) in a quiet area where they will not be disturbed by other hatchery activities. The receiving tanks should be prepared to have a temperature and salinity as similar as possible as the fish were transported in. If not already done in the field, females should be weighed, eggs sampled, and selected fish injected. Each female needs to be identifiable. Ideally, enough tanks would be available so each female is held separately. Depending on the size of the fish, two females, obviously different in size, could be stocked per tank. Fish can be tagged with an external tag if necessary, but keep in mind that tagging can be stressful. Data should be recorded for each female as to its ID and tank number, its origin/genetic strain, its weight, hormone quantity given and time when given, the water temperature at the time when the female was stocked into the holding tank, and the stage of egg development at that time. Similar data is needed for the males. Males can be given external tags, and several held together in the same tank.

Once the fish are successfully stocked into their holding tanks, they should be acclimated to the water quality of the hatchery water supply providing a continuous flow of water, keeping in mind that the fish do best when held at a stable temperature within the 17–20°C range. The final tank water temperature should also be similar to that of the water where the eggs are to be incubated. Dissolved oxygen levels should be kept near saturation throughout the holding period.

16.3.4.1 Induced Spawning

Striped Bass are spawned either by hand-stripping or allowing them to spawn naturally after hormone injection. On occasions, females are collected on the spawning grounds that are judged to be within 8 h of ovulation. Such fish may spawn naturally when stocked with males in a circular tank but most females will need an exogenous source of hormones to stimulate a spawn. Human Chorionic Gonadotropin (hCG) is the most widely used hormone to stimulate ovulation in striped bass and white bass females and to increase milt production in the males. Use of hCG is discussed in Chapter 3. When using wild-caught striped bass females, the female may be injected at the collection site or upon arrival at the hatchery. In either case, the female is given a single injection at a rate of 275 to 330 IU/kg of body weight. The male fish can be given a lesser dose, e.g. to 110 to 165 IU/kg. White Bass females are injected with hCG at a dose between 1,100–2,200 IU/kg and males at half those dose rates. Both Striped Bass and White Bass are injected intramuscularly about midway above the lateral line and below the dorsal fin.

The response time following HCh injection is dependent on the stage of egg development at the time of injection, the water temperature, how much the female has been stressed, and if and how the female was anesthetized. Some hatcheries will take a second egg sample from the ovary at 10–12 h post-injection to see if the eggs are continuing to advance toward ovulation. Others will wait until 20 to 28 h to handle the fish to see if she is releasing eggs. Careful record keeping will help to develop guidelines for a specific hatchery as to when the females should be checked for ovulation. Excessive handling of the females will delay or prevent ovulation. Anesthetics such as MS-222, can slow the response to the hormone contributing to the unpredictability of ovulation.

If the females can be quietly observed, the female's swimming pattern may help suggest if a female is approaching ovulation. The closer the female is to ovulation, the more her eggs have taken on water, resulting in that portion of the body having a lower specific gravity than other portions of the body. This difference in weight distribution may result in a female nearing ovulation to have problems staying horizontal in the water column and the posterior portion of the fish may float up higher than the head.

Without experience and good records it is hard to anticipate ovulation. The time to ovulation can range from 15 to 60 h post-hCG injection. Once the potential time to ovulation as suggested by the egg samples collected from the ovary is reached, the female should be handled to see if she is releasing eggs. A light pressure should be applied to the abdomen to determine its firmness, starting just below the pectoral fins going toward the oviduct. If the entire area is firm and no eggs come out the oviduct, then the female has not started to ovulate. Ovulation is not instantaneous and there can be more than a 10 minute difference in when a female releases her first eggs and the last eggs are ovulated. This is particularly true when working with large females. If the abdominal area near the oviduct is soft but more anterior portions of the abdomen are firm, then the female may have just begun to ovulate. A few eggs may flow from the oviduct with most of the eggs appearing round in shape. This would suggest that the female has just started to ovulate and should be released back into the tank and examined 20–30 min later. Fish having more of the abdomen being soft with eggs flowing freely from the oviduct suggests that the female has adequately ovulated and is ready to be stripped. As the lumen of the ovary becomes full of eggs, the eggs press against each other, changing their shape. Eggs coming out of the oviduct should be hexagonal in shape. With a large female, the first eggs to exit the oviduct may appear to be over-ripe, but that is because they were the first eggs to be ovulated, and the remaining eggs should be of good quality.

Once a female starts to ovulate, the ovulated eggs are not receiving oxygen from the blood stream but are loose in the lumen of the ovary, and egg quality begins to drop within 30–60 min depending on the water temperature. In over-ripe eggs, the yolk membrane will start to become detached from the egg shell and may appear wrinkled. If eggs collected near the end of the stripping appear over-ripe, then the fish may have been stripped too late and all the eggs may be bad.

White Bass females typically respond to hCG and ovulate 25–50 h post-injection. Egg samples are not generally taken from white bass brooders for staging egg development, although their eggs follow a similar pattern to ovulation as Striped Bass. The fish are smaller than Striped Bass and as such are easier to handle and are also hardier. The onset of ovulation is determined by applying a slight pressure to the abdomen near the urogenital opening to determine if there is an egg flow. If there is no flow then slightly more pressure is applied to see if any ovarian fluid is released. If there was no flow of ovarian fluid then the female is returned to the holding tank and re-examined about 3–4 h later. If some fluid and eggs were released the fish should be re-examined 1–2 h later. If both single eggs and some egg clumps are seen the female should be ready to strip within an hour. A free flow of eggs with no clumps in response a slight pressure to the abdomen suggests that the female has adequately ovulated and should be stripped.

Keep in mind that a few degrees difference in temperature will impact the timing of when ovulation will occur, as well as just as how mature the female was when injected, and how she had been handled. Response time also is influenced by brood history. Female White Bass brooders held under "Cold-Banking" conditions for a month and brought up to 18°C prior to injection may ovulate 20–24 h sooner than recently captured females that are injected soon after capture.

Striped Bass and White Bass need to be anesthetized for stripping. Both have sharp spines and opercula that can cut the person handling the fish. MS-222 at 1,000 mg/L can be sprayed over the gills or added to the holding tank water at 100 mg/L. To avoid egg loss, the urogenital opening is covered or plugged with a finger and the tail elevated. Three workers make up a spawning team: one person holds the head, a second the tail, while the third applies pressure to the abdomen to strip the eggs. Wet fertilization is most commonly used with Striped Bass. Males freely run large volumes of sperm that can be stripped directly into the container of eggs (Figure 16.3.2). Sperm also can be collected ahead of time and if not exposed to water can be held at room temperature for about 30 min and then used. White Bass eggs start to become adhesive once in contact with water. They can be dry fertilized or stripped into a fertilizing solution, and once sperm is added, water is added to activate the sperm. If the dry method is used care should be taken to keep the eggs clean and free of water during stripping. Striped Bass give so many eggs that to maximize genetic diversity the eggs should be divided into 4–5 different bowls and each bowl of eggs fertilized with the sperm of a different male. Once the sperm is mixed in with the eggs, water is added to activate the sperm. After a minute of gently stirring, Striped Bass eggs should be fertilized and can be added to the incubator.

In the case of White Bass eggs, they need to be treated to control their adhesiveness. The most widely used method with white bass eggs is using a 150–300 mg/L tannic acid solution. Techniques to do so are given in Chapter 17.4. A MacDonald jar is filled about half full with the tannic acid solution, the fertilized eggs added, and the eggs and tannic acid solution vigorously agitated for 8–12 min. At the end of that time period, the eggs are rinsed several times in clean water and then incubated in the jar.

1.3.4.2 Egg Incubation & Larval Management

Incubation is usually done in MacDonald jars, but can be done in cylindrical cones. The loading rate in MacDonald jars can be up to 100,000–200,000 eggs/jar. The initial water flow rate is slow for the first hour or so, providing just enough to gently roll the eggs

Figure 16.3.2. Wet fertilization of Striped Bass eggs.

but not pushing them up into the water column. The eggs are water hardening during this period, decreasing in specific weight, increasing in diameter and taking up more space in the incubator. Striped Bass eggs may expand their diameter three times or more than their initial diameter over more than an hour period. Diameters of water–hardened eggs may range from 1.5 to 4.2 mm, varying by strain of Striped Bass and the water hardness where the eggs are incubated.

White Bass eggs do not expand as much during water hardening, only about 1.2 times their diameter at ovulation. The water flow rate needs to be slowly increased to help meet the oxygen demand of the embryos and wash away the increased quantity of waste, while being careful to not flush any eggs out of the jar.

Water supply lines should be inspected and flushed prior to loading eggs into the jars. No air bubbles should appear in the incubation jars as they may push eggs out of the jar. Quality of the hatchery water supply is of major importance in the hatching success of Striped Bass eggs. Table 16.3.1 lists the physical and chemical water quality criteria for incubation of Striped Bass eggs. Extra effort is needed in providing the best quality water for egg incubation. The total hardness should be at least 50 mg/L with ≥100 mg/L being preferred. Appropriate water temperatures are in the range of 16–20°C and should be similar to that where the brooders were held. At 16 to 18°C, the incubation period for Striped Bass eggs is 40 to 48 h. For White Bass the incubation period is around 43 h at 18–20°C. Stability of water temperature is key to a predictable hatching time. A 4°C variation in water temperature can accelerate or delay embryo development by >12 h.

Table 16.3.1. Water quality parameters for Striped Bass egg incubation[1].

Parameter	Range	Optimum
Temperature	14.5–24.0 °C	17–18.5 °C
pH	7.0–8.5	7.5–8.5
Dissolved Oxygen	4–10 ppm	saturation
Carbon Dioxide		<10 ppm
Ammonia		<0.5 ppm
Copper		<0.1 ppm
Hydrogen Sulfide		<0.3 ppm
Nitrates		<2.0 ppm
Nitrites		<2.0 ppm
Phosphates		<6.0 ppm
Potassium		<2.0 ppm
Sodium		<5.0 ppm
Zinc		<0.1 ppm
Iron		<0.5 ppm

[1]From Bonn et al. 1976.

Percent fertilization can be determined at approximately 3 h post-fertilization and by 12–18 h. dead opaque eggs will appear. These dead eggs represent both unfertilized eggs and poor quality eggs that failed to develop past that point in time. A 50% fertilization rate or more should be expected. In MacDonald jars, the lighter dead eggs will float higher in the water column that the viable eggs and by careful control of the water flow most of the dead eggs can be washed out of the jar before hatch or can be siphoned out.

On occasions, there is a need to ship eggs from one hatchery to another but there are windows of time when the eggs are more delicate than others. If eggs must be transported, they should not be shipped within two hours post-fertilization, as this is a period of early cell division where excess movement can result in abnormal cell divisions. Another period is as the embryo approaches gastrulation. The two time-windows when eggs can be best moved are 2–12, and 24–36 h post-fertilization. Transit times must be planned to avoid eggs hatching during transit and water quality problems resulting from egg shell decomposition. From 12,000 to 15,000 eggs/L of water can be packaged in a plastic bag containing 7 to 8 L of water and 15 L of oxygen. Transit time at this rate is 3 or 4 hours. Reducing the loading rate to 5,000 eggs/L will allow a transit time of up to 24 h.

During incubation, the effluent from the MacDonald jars can be allowed to flow into a common drain line and discharged from the hatchery building. About an hour before the time of hatching is anticipated, the effluent from the incubators should be diverted into aquaria or tanks where the larvae can be collected. Aquaria (114–284 L) are convenient to use as it is easier to see the larvae and easier to see areas that need to be cleaned. Each aquarium should have one or more 2.54 cm diameter drain pipes. Each drainpipe is covered with a 300 μm mesh screen mounted on 15.2 cm diameter PVC pipe where most of the PVC has been cut out leaving only enough to provide a frame for the screen. Even within a common set of eggs, the time to hatch may range from 6 to 12 h from the time the first larvae appear until 95% of the eggs have hatched. Staff should be available during this period of hatching to ensure that the drain screens do not clog or that the aquaria do not overflow. A hatch rate of >50% should be expected.

Recently hatched Striped Bass larvae are not strong swimmers. At hatching they will float out of the incubator into the aquarium along with any shell debris. Poor quality or deformed larvae often are not strong enough to float out of the incubator and should be discarded. In the aquarium the healthy larvae will swim up into the water column for a few millimeters then drift back down. They are dependent on the water current to keep them up in the water column. To prevent the effluent water from impinging the larvae on the screen, a pressurized perforated air line or diffuser hose is placed around the base of screened cylinder to give a screen of air bubbles rising along the mesh screen. This will help prevent larvae from being trapped on the drain screen. The water supply line should be placed to keep the water column mixed and no eddies of larvae to form. Mild supplemental aeration may be needed to help keep the aquaria well mixed, however, too strong a current may weaken the larvae. By day 4 Striped Bass larvae are starting to swim horizontally. Striped Bass larvae and hybrids based on female Striped Bass have developed enough to begin feeding by day 5. Larvae of White Bass and hybrids, based on female White Bass, are about one-third smaller in size and are ready to feed by day 4 post-hatch.

Larval densities up to 10,000 to 15,000 larvae/L can be held in an aquarium with 2 water changes per hour and aeration to keep water circulating until mouthparts are formed. If the larvae are to be held in the aquaria and fed the density must be reduced drastically to around 250–500/L by the time of first feeding.

16.3.4.3 Induced Spawning and Natural Fertilization in Tanks

Striped Bass can be stimulated with hCG to spawn naturally in a tank. Tank spawning has three major advantages compared to hand-stripping: 1) manpower requirements are reduced by an estimated 75%; 2) stress on the spawning fish, especially the female, is greatly reduced; and 3) there is less danger that the eggs will become overripe following ovulation. Some disadvantages include numerical estimates of eggs and fry are less precise. Fungus on the eggs is more difficult to treat. A female may release her eggs but sometimes they do not get fertilized. Having adequate hatchery space can be a limitation as each set of spawners occupies tank space for 3 to 7 days, requiring a large number of tanks if a significant quantity of fry are to be produced.

16.3.4.4 Hormone-induced Natural Spawning

A 1.5–1.9 m dia. circular tank that is about 0.8 or 0.9 m deep with a center drain can be used for natural spawning. A screen made of a 45 cm diameter cylindrical frame covered with 500 μm mesh material should be made to fit tightly around the base of the standpipe. The other end should be open and extend about 10 cm above the top of standpipe. A pressurized perforated air line should surround the base of the screen giving a curtain of air bubbles from the base of the screen up the sides of the screen to keep eggs and debris from clogging the screen. A flow of water is introduced at the surface at a 15 degree angle to create a circular current of 10 to 15 cm/second. The flow should be approximate 30 to 38 L/minute for a 2 m^3 tank. The spawning tanks should set indoors out of direct sunlight in a quite area so the brooders are not disturbed, using a water supply where the temperature can be kept stable.

The female may be given a single injection of hCG at 330 IU/kg body weight or an implant containing 150 mg LH-RH. Spawning should occur 20–48 h after a hCG injection. One or two males are injected about 24 hours before spawning is expected at a rate of 110 to 165 IU of hCG/kg of body weight and added to the tank with the female. After injection the fish are not disturbed until spawning takes place and the eggs have water-hardened. Movement of the brooders helps stir the eggs to keep them in suspension. In addition, air lines with air stones extending to the tank bottom may be needed to provide additional turbulence to keep the eggs suspended in the water column.

16.3.4.5 Egg and Larval Management

After the eggs have water hardened, the brooders and eggs can be removed. Eggs can be concentrated using a fine mesh dip net and scooped up in water from the net for loading in incubators. They can also be carefully siphoned into a container half filled with water. Once concentrated the eggs can be either siphoned or transferred in water to an incubator. Care must be used in handling the eggs to minimize damage.

The eggs can be left in the tank to hatch but fungus can become a problem at water temperatures below 17.5°C. Fungus will develop on the dead eggs and at the lower water temperatures the live eggs are slower to develop. This gives the opportunity for fungus to smother live eggs and kill them. Eggs can be cleaned manually when hatching is slow but this requires more labor. Egg treatment with standard fungicides can be substituted for manually removing dead and fungus-covered eggs.

At hatch, the larvae drift in the current and are not able to swim horizontally. The fry disperse themselves more evenly about the tank by the second or third day after hatching. Harvesting of fry is best done by waiting 4 to 5 days after hatching as the fry move to the tank perimeter. The water level is lowered using a siphon inside the cylindrical outlet screen. After reducing the water volume, the fry can be dipped or siphoned for transfer, keeping them in water at all times.

16.3.4 Larval Management Considerations

16.3.4.1 Swim Bladder Inflation

Whether the larvae were hatched in MacDonald jars and held in aquaria or were allowed to hatch in a large tank, a common issue is swim bladder inflation. Striped Bass larvae inflate their swim bladder within the first 4–8 days post-hatch by gulping surface air and forcing it through their pneumatic duct. However, some larvae are not successful in inflating their swim bladder and cannot float. They must swim continuously to remain in the water column otherwise they sink to the bottom of the culture tank. They spend energy trying to remain in the water column and have difficulty capturing food, and generally do not survive. One of the reasons the larvae cannot inflate their swim bladder is an oil film resulting from shell debris, can develop on the water surface. This film can block the larvae from being able to come to the surface to get the gulp of air needed to inflate the swim bladder. Oil film control begins with siphoning out any shell debris that settles to the bottom of the larva holding tank. The surface film can be adsorbed on to oil absorbent polypropylene cloths. A gentle spray of water on the surface can help break up the oil film. Placing an air stone just below the water surface will give a fine spray of water droplets that will break up the surface film and push it to the sides of the tank.

16.3.4.2 Fry Transport

Many hatcheries do not have access to quality Striped Bass brooders and are dependent on other hatcheries to spawn the brooders and ship the larvae to them. Striped Bass larvae are usually shipped in oxygenated plastic bags during the yolk sac absorption period. Highest survival is obtained when 1 or 2 day old larvae are used. A preferred shipping bag is a 3-mil thick square bottom bag 56 × 38 × 38 cm with a capacity of about 80 liters. Depending upon the time and distance of the shipment, each bag can carry 50,000 to 100,000 fry. The bag containing fry is shipped in an insulated Styrofoam shipping box with a tightly fitted cover. That box is often placed inside a corrugated cardboard box for protection.

Before packing, the holding tank should be cleaned to eliminate any dead larvae, egg shells, or debris that might be transferred with the live larvae into the shipping bag. The number of fry are estimated volumetrically or visually, added up to 10 L of fish and water to the bag. The remaining volume of the bag is filled with oxygen, the bag sealed and the box closed.

Properly packed larvae can be transported by airplane, automobile, or other means for 24 to 48 h with little mortality. Preferably, the water temperature during travel should remain stable in a range of 16 to 18°C. A small amount of ice in the corners of the box can be added to hold the temperature in this range. Upon arrival at the destination, the boxes are unpacked and the bags floated in the receiving water for 30 min to 1 h to equalize temperature. The bag is then opened and water from where the fish are to be stocked is gradually added over a 30 to 60 min period to permit the larvae to adjust to differences in water quality. Provision for supplemental aeration should be made after the bag is open. If the larvae are ready for food they can be stocked directly into outdoor nursery ponds. If the larvae need to be held until mouthparts are functional then the larvae can be held in indoor aquaria or holding tanks.

16.3.4.3 Larval Feeding

Once the larvae have inflated their swim bladders and are ready to feed they can be fed in the hatchery for several days and then stocked into nursery ponds or they can be stocked directly into nursery ponds. Striped Bass larvae or hybrid larvae from a female Striped Bass are larger than White Bass larvae or hybrid larvae from a female White Bass and this influences what is an acceptable first food. Larvae from a Striped Bass female are able to accept 1st instar artemia nauplii as a 1st food beginning on day 5. In many cases, pond survival of Striped Bass larvae can be improved by feeding the larvae in the hatchery artemia for three days. Artemia should be offered at least every 3 h at a density of 5/mL while the fish are held in the hatchery at a density of 250– 500/L of holding tank. As Artemia nauplii will not survive in freshwater for more than a few hours, new nauplii need to be added frequently and dead nauplii and other debris need to be removed from the tank bottom and off the drain screens.

Larvae obtained from White Bass females are smaller than Striped Bass larvae at the time of first feeding and artemia nauplii are too large as a 1st food. Rotifers are a more appropriate size and can be produced in the hatchery for feeding such larvae but in most cases, it is more practical to stock the larvae as soon as they have functional mouthparts into outdoor nursery ponds prepared to have an abundance of rotifers.

16.3.5 Producing Small Fingerlings

Pond production of fingerlings of Striped Bass and its hybrids is typically divided into a primary nursery phase (Phase-1 fingerling production) and a secondary nursery phase (Phase-2 fingerling production). In Striped Bass stocking programs, fish of at least 3.8 cm TL are acceptable. This size fish, often referred to as a Phase-1 fingerling, can be harvested from a primary nursery pond and stocked into reservoirs and streams. When Striped Bass are produced for aquaculture, a larger size fingerling is desired and a secondary nursery is used to produce the larger fingerling, a Phase-2 fish.

16.3.5.1 Pond Preparation

Success in producing small fingerlings begins with the care given in selecting and preparing the primary nursery pond. Ponds built according to specifications outlined in Chapter 2 are best. The size should not exceed 0.6 to 0.8 ha. Larger ponds can be used successfully but the challenge is to have enough similar age and size fry available for stocking a large pond at one time. The pond should have an outlet structure equipped with a seining basin. Preferably, a supply of fresh water is located at the deep end of the pond. The bottom of the pond should be graded to slope to the drain and be free of seepage from adjoining ponds. This makes it possible to dry the pond bottom completely between crops of fish. The preferred water supply is a good quality spring or well. Surface water from a reservoir or stream can be used, but use of such water complicates pond preparation and may be a factor in reduced survival.

The pond water should have an alkalinity of at least 50 mg/L with an alkalinity closer to 100 mg/L preferred. Agricultural lime ($CaCO_3$) added to the pond bottom will help to increase the alkalinity in low hardness waters, but higher alkalinities can be obtained using gypsum (calcium chloride, $CaCL_2$). Nursery pond survival of striped bass is much better in the higher alkalinity settings.

About 6 months before starting the crop, a soil sample to determine the lime requirement should be taken from the ponds to be used. Lime applications based on the recommendation of the soils testing laboratory should be applied to the pond bottom when the pond is dry or filled with water.

The pond bottom should be dry for 3 weeks or longer prior to filling for a Striped Bass crop. If the pond has a problem with rooted weeds or filamentous algae, a preflooding treatment with an approved herbicide should be given. If there are puddles of water in the pond or in the catch basin, fish or other unwanted organisms from a previous crop might still be alive in such waters. A fish toxicant and/or insecticide should be added to any puddle of water left in the pond before it is refilled. Once these potential problems have been taken care of, the first application of fertilizer should be applied. Composition of the fertilizer mixture varies according to pond bottom soil and water quality. An effective fertilization protocol for Striped Bass used at the Marion, Alabama hatchery has been:

Cottonseed meal
400–500 kg/ha as an initial application to a dry pond
200–250 kg/ha, weekly to a filled pond
Liquid inorganic fertilizer (10-34-0)
9.3–18.7 L/ha as an initial application to a full pond
4.7–9.3 L/ha, weekly to a filled pond
Fish meal or meat & bone meal at 5 kg/ha if bloom fading

A primary nursery pond should be filled 5 to 21 days before the fry are to be stocked, depending on the type of fish larvae to be stocked, and how quickly the appropriate zooplankton population builds. Rotifers should be the dominant zooplankton at the time ponds are stocked with larvae obtained from a White Bass female. This should occur at approximately 6–8 days post-filling, at densities of rotifers of 500–1,000/L. Ponds prepared for larvae obtained from a Striped Bass female should be a mixture of rotifers, cladocerans, and copepods at the time of larval stocking, approximately 10–14 days post-filling. Target densities of cladocerans are 100–200/L.

16.3.5.2 Pond Management

Striped Bass larvae can be stocked into the rearing pond as soon as the majority are swimming horizontally and have a functional digestive system. Delays for more than 24 hours after yolk sac is absorbed will require that the larvae be fed Artemia nauplii in the hatchery before stocking. The pond stocking rate for producing small fingerling Striped Bass should be based upon expected survival and productivity of the individual pond, its history, and how well it was prepared. If a survival of 40% is expected and the pond productivity is 85 kg/ha of 1 g fingerlings, the stocking rate would be 212,500 fry/ha (e.g., 85 kg averaging 1 g = 85,000 fish which = 40% of the number stocked; 100% = 212,500). The range of stocking rates for small fingerling rearing ponds is generally 125,000 to 250,000/ha.

To estimate numbers, volumetric or visual comparison methods can be used. Visual comparison is thought to give a more precise estimate but may be more time consuming. In order to make best use of the supply of fry and to make most efficient use of pond space, it is desirable to know within about 10% ± the number of fry stocked.

Before the larvae are prepared for transportation to the nursery pond the pond water quality should be monitored to be sure that the nursery pond is suitable for receiving the larvae. Ideally the nursery pond would be monitored for several days prior to the anticipated stocking date to be sure that the water quality conditions were stable without wide daily variations. Wide daily variations in parameters such as pH or DO are stressful to the larvae and zooplankton production, and such ponds should be avoided. Pond DO levels at the time of stocking should be at least 5.0 mg/L. Ideally the pH should not exceed 8.5. Zooplankton densities should be in the target ranges with no indication of sexual reproduction within the rotifers or cladocerans.

Larval stockings should be done early in the morning to avoid high light intensities. They can be transported in plastic bags as discussed earlier or in small tanks or other containers of enough volume so one container can hold all the larvae needed to stock one nursery pond and have enough volume so that pond water can be added to acclimate the fry to pond water conditions. Tempering over a period of an hour or more should be done if appreciable temperature and/or water quality differences exist between the holding water and the rearing pond. Bags should be floated in the receiving water for 30 min to 1 h to equalize temperature. The bag is then opened and the receiving water gradually added to the shipping water over a 30 to 60-minute period to permit the larvae to adjust to differences in water quality. Provisions for supplemental aeration should be made after the bag is open.

Many hatcheries are designed so the larvae can be acclimated to pond water quality the night before the larvae are to be stocked into the ponds. The larval holding tank in the hatchery is supplied with both the high-quality water used to hold the larvae during their early development and a pond water supply line to acclimate the larvae to pond water conditions. Starting in the evening, over approximately a 12 h period, the flow rate of the high quality water is reduced while the flow rate of the pond water is increased. The next morning the holding tank should be receiving only pond water. At that point the larvae can be counted and transported to the pond for stocking while minimizing the acclimation time out at the pond.

Striped Bass larvae are delicate and may not survive the transport and stocking of the nursery pond if not handled carefully. The nursery pond should be checked within a few days post-stocking to verify survival. This can be done beginning the night after the larvae were stocked using a flashlight to illuminate the area of the pond where the larvae were stocked. The larvae should be attracted to the light within a few minutes and are visible as the light reflects off their eyes. This technique can be repeated at several points along the shoreline to get an idea of the larval distribution. The larvae are not strong swimmers for the first few days post-stocking and can be collected in a 1 mm mesh tow net having a 0.5 m^2 mouth opening. The net is towed about 20–25 m across the pond to determine the relative abundance of the larvae. Both techniques do not give an exact estimate of survival but can indicate if there is survival and whether it is justified to continue managing the pond or to drain it and start again.

Once stocked, the ponds should receive addition inputs of fertilizer as described earlier. By 14–21 d post-stocking the fish may have grown large enough to start taking a formulated feed. A high percent protein (>40%) powdered feed can be offered at 5–10 kg/ha/d given in 3–4 feedings/d. The feed rate and particle size are increased as the fish become larger. A safe feeding rate/d should not exceed 30–40 kg/ha and emergency aeration also should be available.

At a mean water temperature of about 26°C, a period of 50 to 60 days will be needed for the Striped Bass larvae to reach an average size of 1 g. After about 4 weeks of age, survival of the fry stocked can be verified by taking samples with a fine-mesh seine. Where survival of fry is low, it is advisable to terminate the crop and restock either with the same species or another species on the production schedule to make efficient use of the rearing space.

16.3.5.3 Harvesting and Distribution

Striped Bass fingerlings can be harvested and transported with less problems once reaching a size of 0.4 g (2,500/kg). This might be 30–60 days post-stocking depending on fish survival, the availability of natural food, whether they accepted formulated feed, and the water temperature. Such fingerlings are often referred to as Phase-1 fish. Striped Bass fingerlings can be delicate to handle and may go into shock, flaring the gills and go into tetany. This is more common when the fish are crowded together and lifted out of the water, particularly when handled at higher water temperatures and low hardness conditions. Harvest procedures should be reviewed to reduce fish handling to a minimum, moving the fish in water whenever possible, while minimizing loud or sudden noises. Harvesting should start before daylight, ideally on cloudy or cool days.

16.3.5.4 Harvesting

Striped Bass can be effectively harvested by trapping, seining, and pond draining. A glass-V-trap is effective at trapping small 2.5 to 5.0 cm Striped Bass in ponds that have drain structures. The trap is floated in front of the outlet screen with spacers which maintain a distance of about 10 to 12 cm between the trap and the outlet screen. As water is released, the small fish congregate at the outlet and enter the trap. It is then a simple matter to remove the fish with a dip net as they accumulate in the trap. Up to 80% of the crop can be removed in this way. The trap should not be left unattended and high densities should not be allowed to accumulate. At night a light may be used to attract the fish to the trapping area. Traps can also be used around the margin of the rearing pond if the small fish are moving around the pond parallel to the shore using wings to divert the fish into the trap. Traps are an effective alternative for ponds that are difficult to drain or which infested with vegetation that interferes with seining.

Seining can be used to remove part of the crop. The pond water level should be lowered to reduce the volume to about 50% before seining. The seine should be used to crowd the fish toward the catch basin but not to harvest all the fish in one attempt. A smaller seine with knotless netting should be used to harvest a portion of the fish concentrated by the larger seine, not crowding the fish too much in the net and moving them in a bucket of water from the seine to the transport truck. Freshwater should be available to improve pond water quality and keep the temperature down in the area where the fish are being harvested. Seining up too many fish and crowding them can damage the fish and cause them to go into shock.

Striped Bass are a species that can be effectively harvested by draining the pond. The fish will follow the water current into the catch basin as the pond is drained if the pond is properly constructed. The pond bottom should be relatively free from rooted weeds or filamentous algae that might trap the fish during the draining. Harvesting from the catch basin is an efficient use of hatchery labor while not unduly stressing the fish. Attention should be given to not draining the pond too fast and stranding the fish out in the pond. After the fish begin to enter the harvest basin they should be removed periodically. It is best to limit the number of fish in nets and containers to reduce handling stress. A flow of fresh water at the drain helps to prevent stress from silt turbidity, high temperature, and low oxygen. If ponds are being drained when air temperatures are above 29°C work should commence shortly before daybreak and terminate by mid-morning.

16.3.5.5 Holding

After harvest, the fish should be handled a minimum number of times before distribution. Fish that learned to accept a formulated feed in the primary nursery pond may have outgrown those that did not by the time the pond is harvested. If there is too much size variation the fish should be graded once they are in holding tanks using a tank grader. Using a tank grader will avoid having to lift the fish out of the water with a net to put them into a floating grader.

A recovery/hardening period of 24 to 48 hours following the pond harvest is advisable. During this time the digestive tract can be emptied and prophylactic treatments given. Prophylactic treatments include formalin at 125–150 mg/L for an hour for external parasites and a 1% salt bath for 4 to 5 hours as a general tonic and restoration of osmotic balance after the stress of harvesting.

16.3.5.6 Distribution

Striped Bass fingerlings are transported in conventional distribution equipment but should be given special care during hauling. Carrying the fish in a cushion of water to load the distribution tanks has already been mentioned. Loading rate for small fingerlings in distribution tanks should not exceed 60 g/L of water for trips of 1 to 4 hours. This should be reduced to 40 g/L for 4 to 8 hour trips and 30 g/L if the trip is longer than 8 hours. Total hardness should be increased to >75 mg/L using calcium chloride. Salt at a rate of 0.5 to 1.0% is highly beneficial, and some hatcheries add an anesthetic such as MS 222 or quinaldine at low rates to keep the fish calm during the trip.

A stable temperature of 20 to 22°C is desirable especially for long trips. Ice can be used during the summer months to maintain suitable temperatures. If the receiving waters are considerably warmer than 22°C it may be advisable to allow slow warming to take place during transport but this should not be more than 1°C/hour during the last part of the journey. An aeration system, combining bottled oxygen released into the tank water through carbon-rod air stones and 12-volt electric agitators operating at the surface of the tank, does an effective job of maintaining water quality. The mesh size of the screen around the agitator blade should be fine enough to keep small fingerlings from coming in contact with the agitator blade. If foaming results from running an agitator, an anti-foam solution can be added at a rate of 3 mL/100 L of tank water.

All of the work of producing and distributing a load of Striped Bass fingerlings can be nullified by failing to take proper precautions when releasing the shipment. Since most introductions go into waters containing an established fish population, it is essential that small fish be liberated in the best condition possible to escape predation and to be able to compete for food and space with the resident fish population.

Tempering to equalize water quality differences should be done for a period of at least 1 hour. This is best done by gradually replacing water in the hauling tank with receiving water. After this is done, a bucketful of fish should be released and observed as to reaction to the new environment. If normal movement occurs with no evidence of stress, the load can be unloaded in water, preferably through a quick release valve and out via a chute or discharge pipe. If it is not possible to discharge the fish in water directly to the receiving water then the fish can be carefully netted up, added to a bucket of transport truck water, and carried to the receiving water.

16.3.6 Culture of Large Fingerlings

When Striped Bass or their hybrids are to be stocked into reservoirs, their chance of survival can be increased by stocking larger fingerlings that were produced during a

Phase-2 production cycle. When Striped Bass or their hybrids are produced for aquaculture, the production cycle is broken up into three phases: Phase-1, small fingerling production; Phase-2, large fingerling production; and Phase-3, food fish production. As part of Phase-1 production, the fish can be trained to take a formulated feed. Once harvested from the primary nursery pond, they can be graded and restocked for Phase-2 production in ponds, raceways or cages to produce an advanced fingerling to a larger size using commercial feeds. The choice of using ponds, tanks or raceways for a secondary nursery depends upon the facilities and production inputs available. There are advantages and disadvantages associated with each setting that may be inherent in the particular situation being considered.

16.3.6.1 Pond Production

If ponds are to be used, then weed control must be considered. To control unwanted vegetation, the rearing pond is drained and the exposed bottom treated with a herbicide. After the pre-flooding treatment, the pond is fertilized as for small fingerling rearing but allowing a period of 14 to 21 days to elapse between flooding and stocking whenever possible. This allows an insect population to develop as a food source for the fingerlings. There is no need to treat the pond with toxicants, unless wild fish are likely to be present or if disease has been a problem earlier in the pond.

The portion of the fish that were accepting a formulated feed in a primary nursery pond may need to be retrained as part of stocking a Phase-2 production pond. If the fish are to be stocked directly into the pond after being graded, it may be advisable to stock the fish into a small mesh cage in the pond where they can be fed a formulated feed for 3 to 5 days until they are actively feeding before being released to swim freely in the pond. An alternative is to stock the fish into a rearing pond that is less than half full that has a catch basin and a water supply by the basin. The fish can be fed in the basin and surrounding area and water added slowly to help with the aeration as the pond is being filled. However, this approach can make the fish vulnerable to predation by fish-eating birds and other animals. Both day and nighttime surveillance would be needed.

Stocking rates for the Phase-2 production ponds depend on the desired size of fish to be produced, the standing crop that can be supported, the length of the growing season and water quality. Common stocking densities range from 20,000–50,000/ha. Ideally, previous records of yield, growth rate and survival, feeding rates and feed conversion, as well as water quality information would be available to help guide the selection of a stocking rate. At a pond stocking rate of 50,000 fingerlings with a 80% survival where the daily feeding rate does not exceed 40 kg/ha/d, a yield of 2000 kg/ha might be expected after 120 days of culture with the fish averaging 150 g each. Dissolved oxygen concentrations should be maintained as near saturation as possible with significant reductions in growth expected at saturation level of 40% and less (Green et al 2016).

Before releasing the fish in the pond, another formalin treatment should be given. Careful handling to minimize stress is necessary to hold down post-stocking mortality. If mortality following stocking continues after 3 days, pond treatment should be given with a suitable therapeutic agent, or medicated feed can be given for a period of about 7 days.

For the first 10 to 14 days after stocking the pond, the feeding rate should be 10% of the body weight of fish/day with the daily ration given in 4 to 5 feedings. After 14 days, the number of feedings can be reduced to 3 times/day and the amount of feed increased 50 to 100% depending on the fish appetite. Feed quantities are adjusted at 7 to 14-day intervals until the level reaches 40 kg/ha/day. The feed rate is held at that level to minimize water quality issues until the fish reaches the desired size. Higher feeding rates can be used if supplemental aeration and/or continuous water exchange is possible. Rates as high as 56 kg/ha/day might be fed for 4 to 6 weeks if the water is of good quality initially and some water exchange is done. Although not essential, use of a floating pellet is preferred for feeding Striped Bass in ponds as less time is required to feed the fish, there is less waste, and it is possible to determine the feed intake more exactly if the pellets are floating.

Striped bass are highly susceptible to handling stress, particularly when held in low hardness waters. Sampling to determine fish growth rate or to adjust feed rate should be kept to a minimum in an effort to reduce fish disease outbreaks and mortality. As an alternative, the feed rate may be increased at 2-week intervals based upon observed consumption. Usually, the amount should be increased 50 to 100% each time until maximum safe level for the pond is reached. If after an increase, and it is apparent that the amount provided is not being consumed, a corresponding reduction should be made. Three feedings/day, 7 days/week is recommended. See Table 16.3.2 for recommended feed sizes and rates.

In the right setting, Striped Bass respond well to being feed trained and produced in 600 to 2,000 L circular or rectangular tanks. The tanks should have a good quality water supply with a temperature about 24°C. Uniform size fish of 3 to 5 cm are placed in the tank at a rate of 3.5 to 5 fish/L. If it has not been done previously, a formalin treatment for external parasites should be given the first day after stocking the rearing tanks. A pelleted formulated feed of suitable particle size should be offered 5 to 6 times daily at an initial rate of 10% of the biomass/day. If the feeding response is poor after 48 hours, ground fresh, cooked, canned or frozen fish can be mixed with the dry feed at a 1:1 ratio to increase acceptance of the artificial feed. If the ration includes ground fish, the quantity should be increased to as much as 15% body weight/day. A period of 7 to 10 days is ample to train the Striped Bass fingerlings to take formulated feeds. As soon as they are feeding readily, the fish component can gradually be reduced to a low level (e.g., 1:9 fish dry feed ratio), or eliminated completely.

Table 16.3.2. Recommended feed sizes and feed rates for various sizes of Phase-2 Striped Bass and hybrid Striped Bass[1].

Fish wt (g)	Total Length (mm)	Feed Size	Feed rate (% B Wt)
≥ 0.4	30	# 2 crumble	25
≥ 0.8	44	# 3 crumble	15
≥ 1.5	51	# 4 crumble	10
≥ 9.0	90	3/32" pellet	7.5
≥ 25	127	1/8" pellet	5
≥ 75	192	5/32" pellet	3
≥ 150	238	3/16" pellet	3

[1]Modified from Smith et al. 1990.

16.3.6.2 Tank and Cage Culture

Striped Bass can be reared intensively in cages, tanks and raceways. The procedure described earlier for training the small fingerlings to accept artificial food should be followed until the small fingerlings are feeding. The container is then stocked for rearing to the desired size. Depending upon the situation, 1 or 2 water changes/hour should be provided. The maximum density for rearing in a tank or raceway has not been specifically established. However, based upon experience in training small fingerlings to feed in 1.4 m^3 tanks, the best starting density of 1 to 2 g fish is 3 to 5 fish/L. After a weight increase of 100 to 200% the lot should be divided equally and half stocked in another tank of similar size. This will involve dividing the stock at 2 to 4 week intervals to make best use of rearing space, maintaining the standing crop between 50 and 100% of the maximum capacity of the tank. As size of fingerlings increases, their environmental tolerances are greater and the weight of fish/volume of water can be increased. The main concern is minimizing environmental stress to maintain the fish in a healthy condition. For optimum growth dissolved oxygen concentrations should be near saturation at all times.

With the above approach, the number of fish/m3 will need to be reduced 5 to 6 times to grow fingerlings from a size of 1 to 2 g to 54 g or 15 cm TL. While this appears to be a considerable amount of fish handling, the alternative is to stock at a much lower density when the fish are small, resulting in underutilization of expensive tank space until the last 2 or 3 weeks of the production period. Since the fish are in an intensive or controlled environment, transfer can be made without too much trouble but a prophylactic treatment of $KMnO_4$ may be necessary after each handling.

The growth of Striped Bass to a length of 15 cm and an average size of 54 g will result in a fish density of 1 fish/3 to 6 L. The above loading rates for intensive culture are approximate since the optimum density in a given situation will be dependent upon water quality, number of water exchanges/hour, characteristics of the container, care given by the culturist, and other influences.

Growth rate in cages suspended in a pond or reservoir can be faster than in a tank or raceway. However, the difficulty of managing the fish density, disease control, and other factors make cage culture more of a challenge. Cages have the advantage in they can be suspended in under-utilized bodies of water and a continuous source of flowing water is not necessary.

Some of the advantages of intensive culture in cages, raceways or tanks over pond culture are better inventory management and ease of harvesting. Shipping can be done on short notice, only requiring clearing the digestive tract, administering needed prophylactic treatment, and making a final inventory.

16.3.6.3 Transportation of Large Striped Bass Fingerlings

Transportation of large Striped Bass fingerlings is done keeping in mind the same precautions as for small fingerlings. However, precautions may be intensified because of the increased value of the larger fish. The loading rate can be increased up to 120 g/L of tank capacity. The fish are more tolerant of temperature and water quality differences at the release point, but care to avoid handling stress should be taken at each step of the operation.

16.3.7 References

Bonn, E. M., W. M. Bailey, J. D. Bayless, K. E. Erickson, and R. E. Stevens, editors. 1976. Guidelines for Striped Bass culture. Striped Bass Committee, Southern Division, American Fisheries Society, Bethesda, Maryland.

Green, B. W., S. D. Rawles, S. A. Fuller, B. H. Beck, and M. E. McEntire. 2016. Hypoxia affects performance traits and body composition of juvenile hybrid Striped Bass (*Morone chrysops* × *M. saxatilis*). Aquaculture Research 47:2266–2275.

Smith, T. I. J., W. E. Jenkins, and R. V. Minton. 1990. Chapter 9: Production of advanced fingerling and subadult Striped Bass and Striped Bass hybrids in earthen ponds. Pages 121–139 *in* R. M. Harrell, J. H. Kerby, and R. V. Minton, editors. Culture and propagation of Striped Bass and its hybrids. Striped Bass Committee, Southern Division, American Fisheries Society, Bethesda, Maryland.

16.3.8 Recommended Reading

Anderson, R. O. 1994. Apparent problems and potential solutions for production of fingerling Striped Bass, *Morone saxatilis*. Journal of Applied Aquaculture 2(3/4):101–118.

Bayless, J. D. 1972. Artificial propagation and hybridization of Striped Bass, *Morone saxatilis* (Walbaum). South Carolina Wildlife and Marine Resources Department, Columbia, South Carolina. Available: https://www.nrc.gov/docs/ML0734/ML073400432.pdf. (May 2020).

Bosworth, B. G., G. S. Libey, and D. R. Notter. 1997. Egg, larval, and fingerling traits of crosses among Striped Bass (*Morone saxatilis*), White Bass (*M. chrysops*), and their F 1 hybrids. Aquaculture 154:201–217.

Bergerhouse, D. 1993. Lethal effects of elevated pH and ammonia on early life stages of hybrid Striped Bass. Journal of Applied Aquaculture 2(3):81–100.

Denson, M. R., and J. Smith. 1996. Larval rearing and weaning techniques for White Bass *Morone chrysops*. Journal of the World Aquaculture Society. 27(2):194–201.

Garber A. F., and C. V. Sullivan. 2006. Selective breeding for the hybrid Striped Bass (*Morone chrysops*, Rafinesque X *M. saxatilis*, Walbaum) industry: status and perspectives. Aquaculture Research 37(4):319–338.

Geiger, J. G. 1983. Zooplankton production and manipulation in striped bass rearing ponds. Aquaculture 35:331–351.

Geiger, J. G., and N. C. Parker. 1985. Survey of Striped Bass hatchery management in the Southeastern United States. North American Journal of Aquaculture 47(1):1–13.

Harrell, R. M., J. H. Kerby, and R. V. Minton, editors. 1990. Culture and propagation of Striped Bass and its hybrids. Striped Bass Committee, Southern Division, American Fisheries Society, Bethesda, Maryland.

Harrell, R. M., editor. 1997. Striped Bass and other *Morone* culture. Developments in Aquaculture and Fisheries Science 30. Elsevier Science, Amsterdam, The Netherlands.

Hodson, R. G., R. W. Clark, M. S. Hopper, A. S. McGinty, G. M. Weber, and C. V. Sullivan. 2000. Reproduction of domesticated striped bass: commercial mass production of fingerlings. Pages. 23–32 *in* C. C.-T. Tamaru, C. S. Tamaru, J. P. McVey, and K. Ikuta, editors. Spawning and maturation of aquaculture species. Proceedings of the 28th US-Japan Natural Resources Aquaculture Panel, US-Japan Cooperative Program in Natural Resources (UNJR) Technical Report No. 28, University of Hawaii Sea Grant College Program, Honolulu.

Jenkins, W. E., and T. I. Smith. 1987. Natural and induced production of Striped Bass hybrids in tanks. Proceedings of the Annual Conference Southeastern Association of Fish and Wildlife Agencies 39:255–260.

Kohler C. C. 1997. White Bass production and broodstock development. Chapter 6. Pages 217–234 *in* R. M. Harrell, editor. Striped Bass and other *Morone* culture. Elsevier Science Press, Amsterdam, the Netherlands.

Lane, R. L., and C. C. Kohler. 2006. Effects of dietary lipid and fatty acids on White Bass reproductive performance, egg hatchability, and overall quality of progeny. North American Journal of Aquaculture 68:141–150.

Ludwig, G. M., and D. L. Tackett. 1991. Relation between bar grader size and size of Striped Bass. Progressive Fish-Culturist 53:128–129.

Lyon, D. D., A. Barkob, D. Smith, and D. Begley. 2006. Guidelines for the culture of Striped Bass and Palmetto Bass. Management Data Series No.243. Texas Parks and Wildlife Department, Inland Fisheries Division, Austin. Available: https://tpwd.texas.gov/publications/pwdpubs/media/mds_inland/mds-243.pdf. (May 2020).

McEntire, M., S. Snyder, and D. Freeman. 2015. Comparison of growth between *Morone* hybrids (Palmetto and Sunshine) in earthen ponds. Journal of the World Aquaculture Society 46(5):557–563.

McGinty, A. S., and R. G. Hodson. 2008. Hybrid Striped Bass: hatchery phase. SRAC Publication No. 301, Southeastern Regional Aquaculture Center, Stoneville, Mississippi. Available: http://fisheries.tamu.edu/files/2013/09/SRAC-Publication-No.-301-Hybrid-Striped-Bass-Hatchery-Phase.pdf. (May 2020).

Mylonas, C. C., Y. Magnus, A. Gissis, Y. Klebanov, and Y. Zohar. 1997. Reproductive biology and endocrine regulation of final oocyte maturation of captive White Bass. Journal of Fish Biology 51:234–250.

Rulifson, R., and L. Wilson.1999. Striped Bass stocking programs in the United States: Ecological and Resource Management Issues. Canadian Stock Assessment Secretariat Research Document 99/007. Fisheries and Oceans Canada, Ottawa, Canada.

Setzler, E. M., W. R. Boynton, K. V. Wood, H. H. Zion, L. Lubbers, N. K. Mountford, P. Frere, L. Tucker, and J. A. Mihursky. 1980. Synopsis of biological data on Striped Bass, *Morone saxatilis* (Walbaum). NOAA Technical Report NMFS Circular 433.

Smith, S. A. 2010. Common diseases of cultured Striped Bass, *Morone saxatilis*, and its hybrid (*M. saxitilis* X *M. chrysops*). VCE Publications/600/600–080 Virginia Cooperative Extension, Virginia Tech, Blacksburg. Available: https://www.pubs.ext.vt.edu/600/600-080/600-080.html. (May 2020).

Woods, L. C. 2005. Striped Bass and hybrid Striped Bass culture. Pages 339–353 *in* A Kelly and J. Silverstein, editors. Aquaculture in the 21st Century. American Fisheries Society, Symposium 46, Bethesda, Maryland.

Section III

Toolbox with More Details and Other Hatchery Topics

Chapter 17

Toolbox with Recommended References

17.1 Aquaculture/Fisheries Management Planning

Whether planning for next year's activities or planning to establish a new hatchery, a lot of detail needs to be considered (as discussed in Chapter 1). Goals need to be clearly defined and a strategy to accomplish them spelled out. Over the years, the goals of a fish hatchery have evolved as we have gained a better understanding of what fisheries management must consider when using hatchery-produced fish. A traditional role of a fish hatchery has been to provide fish to stock back into natural settings. Many of the initial efforts often gave little to no consideration of whether the location to be stocked will support additional biomass, whether the stocked fish can successfully compete in the new setting, or whether they will dominate and result in a reduction of the native stock.

Another issue is that the hatchery-produced fish to be stocked in a given body of water may taxonomically be the same species as the native stock occurring in that watershed but be of a different genetic profile. Consequences on the genetic profile of the native stock should be taken into consideration when hatchery-produced fish are to be stocked into natural waters. As genetic tools have continued to be refined and applied to fisheries management, it appears in many cases that the "species" in one watershed may be genetically different than the same "species" in an adjacent watershed. A hatchery manager should be aware of the range of opinions regarding stocking of natural settings with hatchery-produced fish and take into consideration the arguments for and against stocking hatchery-produced fish into a particular setting and adjust the production plans, taking those issues into consideration. Final decisions on what are stocked where are often made higher up in the administrative chain. However, hatchery managers need to stay up-to-date on the genetic characteristics of the fish they are producing as well as the profiles of the fish where the hatchery-produced fish are to be stocked, and keep administrators informed of possible issues.

Aquaculture is continuing to provide a greater portion of the world's seafood, and fish hatcheries are a key component in that growth. Aquaculture goals include providing a quality product acceptable by the public that is produced in a profitable socially acceptable manner while minimizing environmental consequences.

Goals need to be clearly defined and a strategy to accomplish them spelled out. Decisions need to be based on up-to-date data. This information can be used to build an enterprise budget based on the data collected on-site from previous production cycles or from information available from other sources that use similar methodologies under similar circumstances. New enterprises, whether the focus is on natural resource management or providing fish for human consumption or enjoyment, should consider a broad set of questions when establishing goals and procedures, and be aware of the range of perceptions regarding fish hatcheries as illustrated in several of the following publications:

17.1.1 Planning and Policy

Brown, C., and R. L. Day, 2002. The future of stock enhancements: lessons for hatchery practice from conservation biology. Fish and Fisheries 3(2):79–94.

Brugère, C., N. Ridler, G. Haylor, G. Macfadyen, and N. Hishamunda, 2010. Aquaculture planning: policy formulation and implementation for sustainable development. FAO Fisheries and Aquaculture Technical Paper. No. 542. Rome, FAO. Available: http://www.fao.org/3/i1601e/i1601e00.pdf. (May 2020).

Helfrich, L. A., and D. L. Garling, 2009. Planning for Commercial Aquaculture. Virginia Cooperative Extension Publication 420-012. Virginia Cooperative Extension, Virginia Tech, Blacksburg; Virginia State University, Petersburg.

Meade, J. W. 1989. Aquaculture Management. Springer, New York.

Oregon Department of Fish and Wildlife, 2010. Fish Hatchery Management Policy. Available: https://www.dfw.state.or.us/fish/hatchery/docs/hatchery_mgmt.pdf. (May 2020).

Patrick, W. S., O. Bin, K. A. Schwabe, and P. W. Schuhmann. 2006. Hatchery programs, stock enhancement, and cost effectiveness: a case study of the Albemarle Sound/Roanoke River stocking program 1981–1996. Marine Policy 30:299–307.

Pomeroy, R. 2013. Decision-Making Factors for Investment in Aquaculture, AQUACULTURE FACT SHEET, Connecticut Sea Grant Extension Publication Number CTSG-03-13. Available: http://media.ctseagrant.uconn.edu/publications/aquaculture/decision.pdf. (May 2020).

Science for Environment Policy. 2015. *Sustainable Aquaculture*. Future Brief 11. Brief produced for the European Commission DG Environment by the Science Communication Unit, UWE, Bristol, UK. Available: https://ec.europa.eu/environment/integration/research/newsalert/pdf/sustainable_aquaculture_FB11_en.pdf. (May 2020).

Tett, P., and I. Galparsoro, 2018. Discussion of spatial planning for aquaculture. AquaSpace project (H2020 no 633476), SAMS, Oban, Scotland. Available: http://www.aquaspace-h2020.eu/wp-content/uploads/2018/09/AquaSpaceMM-Discussion-MSP-28Aug18.pdf. (May 2020).

Webster, D., J. Buttner, and G. Flimlin. 2008. Planning for success in your aquaculture business. NRAC Publication No. 101-2008. Northeastern Regional Aquaculture Center. University of Massachusetts Dartmouth, North Dartmouth. Available: https://studylib.net/doc/14807108/nrac-publication-no.-101-2008. (May 2020).

17.1.2 Record Keeping and Budget Development

Good record keeping is an essential part of a successful hatchery operation. Information needs to be recorded on a wide range of topics. Periodic reviews of activities need to be made based good record keeping. This can be done as part of the weekly work plan, at the end of specific production phases, as well as during annual reviews. Budgets need to be prepared, reviewed and adjusted, and be justifiable based on sound record keeping. Details on production practices need to be recorded as to: their effectiveness and reliability; labor, physical resources, and facility requirements; and the associated costs. Such records need to be easily recorded and accessible. As discussed earlier we often have a variety of ways to produce a crop of fish and make it available to the end-user. A good record keeping system and procedures to review that information must be a part of a successful hatchery in its selection of the most appropriate approaches to produce and provide a crop of fish.

An example of the types of day-to-day data that has been recorded as part of the Texas Fish Hatcheries program included:

Trip Sheets

Trip sheets are required for every stocking event. Trip sheets have been standardized and can be printed with pre-trip information from the Fish Hatchery Database (FHD). All relevant trip sheet information, including the pre- and post-trip check lists should be completed by the driver within 24 h after fish delivery. The trip sheet should be given to the individual responsible for data entry into the FHD, and the information should be entered into the database within 48 h of receipt. Another individual should verify the data entered into the database no later than 7 d after the information has been added to the database.

Harvest Data

Harvest data should include, at least, all routine production statistics for each production unit (e.g., pond). The data may include: number of fish harvested, average TL of fish, number of fish/kg, and weight (kg) of fish harvested. This information is used to evaluate production methods and outputs (i.e., pond management success). Therefore, adequate and accurate data are essential. If a pond has multiple partial harvests, sampling each partial harvest is necessary, and the weighted mean values for the production variables reported. Further, whenever multiple ponds or multiple harvests (i.e. multiple datasets) are involved in the estimation of production statistics, weighted mean values must be reported. In these cases, avoid reporting "pond averages," which may obscure deficiencies in production.

Water Quality Data

Water quality data are used to make pond (or culture unit) management decisions or to refine management strategies. Thus, accurate recording and timely reporting of water quality data are important. Water quality data should be reported to production managers in a timely manner so that water quality management decisions can be timely made. If water quality data are collected using a multi-parameter data logger, the data should be downloaded daily into the FHD. Data entry into the FHD is required weekly. Meters must be calibrated daily or according to manufacturer's instructions to ensure proper performance.

Spawning and Incubation Data

Spawning and incubation data are necessary for proper allocation of pond space for fingerling production and therefore must be comprehensive. Information recorded on spawning and incubation data sheets should include: fish weight and number, number of spawns produced, number of fry produced, fry produced per kg of female, and the destination of the fry.

Pond Production Data

Pond production data including, but not limited to, filling dates, water quality data, zooplankton data, feeding and fertilization records, stocking records, sampling records, and harvest data, must be entered into the FHD in a timely manner. Production data should be entered into the database within 7 d of the occurrence.

Status Reports

Data for monthly status reports are entered into tables in the Hatchery Production Status Report by the 10th of each month. Annual reports for spring production are due by September 1. These reports are prepared by hatchery biologists.

Program Summary Report

Program summaries are prepared by the respective program directors and disseminated to staff before the annual programs evaluation meeting. This meeting normally convenes in January.

Examples of record keeping and budget development are as follows:

Carolina Farm Stewardship Association. 2016. Record Keeping. Carolina Farm Stewardship Association, Pittsboro, North Carolina. Available: https://www.carolinafarmstewards.org/record-keeping/. (May 2020).

Cornell University. 2016. Northeastern Beginning Farmers Project, #15 Record Keeping.15A Plant Science. Cornell University. Ithaca, New York. Available: http://smallfarms.cornell.edu/2017/05/01/15-record-keeping/. (May 2020).

Daily, S., D. Selock, and S. Kohler. 2002. Fish-Farm Business Plan Workbook. Technical Bulletin Series, Number 117. North Central Regional Aquaculture Center. University of Iowa, Ames. Available: https://www.ncrac.org/files/biblio/TB117.pdf. (May 2020).

Engle, C. E. 2012. Determining the profitability of an aquaculture business: using income statements and enterprise budgets. SRAC Publication No. 4402. Southern Regional Aquaculture Center, Stoneville, Mississippi. Available: http://fisheries.tamu.edu/files/2013/09/SRAC-Publication-No.-4402-Determining-the-Profitability-of-an-Aquaculture-Business-Using-Income-Statements-and-Enterprise-Budgets.pdf. (May 2020).

Engle, C. R., K. K. Quagrainie, and M. M. Dey. 2016. Seafood and aquaculture marketing handbook. Wiley, New York.

FAO Fisheries Department, Inland Water Resources and Aquaculture Service of the Fishery Resources Division. 2003. THE FAO AQUACULTURE TRAINING MANUALS, Simple methods for aquaculture. Chapter 16. MONITORING, RECORD KEEPING, ACCOUNTING AND MARKETING. FAO, Rome. Available: http://www.fao.org/tempref/FI/CDrom/FAO_Training/FAO_Training/General/x6709e/x6709e16.htm. (May 2020).

Gerloff, D. C., and R. W. Holland, Jr., 2012. Establishing and using a farm financial record-keeping system. University of Tennessee Agricultural Extension Service. Available: https://trace.tennessee.edu/cgi/viewcontent.cgi?article=1000&context=utk_agexfinman. (May 2020).

Pomeroy, R. 2015. Aquaculture Record Keeping. Publication Number CTSG-03-15. Connecticut Sea Grant College Program, The University of Connecticut, Groton. Available: http://media.ctseagrant.uconn.edu/publications/aquaculture/records.pdf. (May 2020).

Saini, V. P. 2016. Record keeping at ornamental fish production and marketing establishments. Pages 153–164 in A. K. Jain, V. P. Saini, and I. K. Vaneet, editors. Best management practices for freshwater ornamental fish production. National Fisheries Development Board, Hyderabad, India. Available: https://www.oftri.org/record-keeping-at-ornamental-fish-production-and-marketing-establishments/. (May 2020).

17.1.3 Cost-Benefit Analysis

Another approach to planning that is particularly appropriate for government programs is Cost-Benefit analysis. This approach is based on reviewing the strengths and weaknesses of the alternatives available for accomplishing a project's goals. At the hatchery level this might be as simple as the consideration of various production alternatives to produce a given size fish. Cost-Benefit analysis can also be used to help define the role of a hatchery. For example, if the objective is to improve sportfishing, one approach might be for a hatchery to stock more fish. If so what species, what size, and how many? Other approaches to meeting the goal of improving sportfishing might not involve a hatchery. It may be possible to improve the fishery by improving the fertility of the lake where the fishing is taking place, or improve access to the lake so more people can reach the lake to fish. What are the likely results and the probability of each approach meeting the goal of providing more sportfishing? What are the costs of each alternative? Who will benefit from the selected action? What are the negative aspects of the selected action? Is the benefit sustainable? What types of actions at what cost will be needed to keep it sustainable? In other words, there are tradeoffs and an approach most appropriate for meeting the objective must be selected taking these tradeoffs into consideration. In such cases, a hatchery staff must be prepared to offer information to the decision makers as to what the hatchery can or cannot do.

The following web sites offer some more guidance on how a Cost-Benefit analysis can be applied to various management situations.

https://www.fema.gov/benefit-cost-analysis
http://www.sjsu.edu/faculty/watkins/cba.htm
http://asafm.army.mil/Documents/OfficeDocuments/CostEconomics/guidances/cba-gd.pdf

17.2 Aquatic Weed Control

A variety of techniques are available for controlling aquatic weeds with specific techniques most effective on specific types of aquatic weeds. Mechanical control can be a short-term answer if applied as soon as the weed begins to appear. Once the weeds are established it is like cutting your grass—it must be repeated frequently to keep it under control. Mechanical control will not get rid of the problem after aquatic weeds have become established.

Biological control can be effective if enough time is available to graze down the density of the aquatic weed to an acceptable level. However, the use of fish as a biological weed control method is not practical in primary nursery ponds, as the culture period is short and having larger herbivorous fish mixed in with the young fingerlings makes a safe efficient harvest difficult.

Dyes can be added to hatchery ponds in an attempt to limit light penetration, in turn limiting emergent aquatic weed growth. This is appropriate in ponds where the fish are not highly dependent on natural food organisms and the fish are being fed a formulated feed. Aquashade® and other similar commercial products can be effective in shading the bottom enough to limit the establishment of aquatic weeds in areas of a pond that are 0.61 m (2 ft.) deep or more.

Chemical herbicides are the most commonly used approach to control aquatic weeds. The starting point is to identify the aquatic weed causing the problem and apply an appropriate herbicide for that weed. In addition, consideration of how water quality (particularly pH and alkalinity) and soil characteristics impact a herbicide's efficacy, and its impact on non-target organisms needs to be known. The effective dose of an herbicide may differ from one hatchery to another. It is often best to treat a limited area to determine if the dose rate applied was efficacious and does not produce any unforeseen consequences. Prompt treatment of aquatic weeds when they are first observed and are just being to grow offers the best chance for their control. Waiting until the weeds are mature permits the plants to store nutrients in the roots, which allows new growth to occur post-treatment, or to produce seed resulting in the weed spreading. The older the weed is, the more resistant the leaves and stems may be to the herbicide. Prompt treatments will help to give a better control while reducing the total quantity of herbicide needed.

Localized patches of submergent aquatic weeds can be treated using a granular herbicide distributed by hand or with a small agricultural seed spreader. Water-soluble herbicides should be diluted in water before application. This will help to ensure a more uniform distribution of the herbicide at the desired dose rate, reducing the chance of "hot spots" of high herbicide concentrations that might kill non-target organisms. Water-soluble herbicides can be applied to small ponds using a sprayer. Larger ponds can be treated by adding the diluted herbicide into the wake of a motorized boat. Deep ponds may be thermally stratified and herbicides may not reach any aquatic weed rooted in the pond bottom below the thermocline. In this case, the diluted herbicide can be pumped or siphoned from the holding container through a hose to below the thermocline and dispersed throughout the water column by moving the hose up and down in the water column.

Localized patches of emergent or floating aquatic weeds as well as marginal vegetation are often treated using a backpack pressure sprayer to spray the herbicide directly on the plants. A drenching spray of the herbicide mixed with a carrier should be applied. The diluent/carrier used depends on the nature of the herbicide and its solubility/emulsification. When water is used as the carrier, the emulsification of the herbicide can be improved by adding a liquid household detergent at 1 mL/35 L of carrier. It also will act as surfactant, which facilitates the herbicide to enter the plant. There are a number of commercial surfactants designed for use with specific herbicides that increase the effectiveness of the herbicide. For harder to kill weeds or those with waxy leaves, an oil is a better carrier than water. An oil carrier makes it easier to see what has been sprayed and less volume of carrier is needed to obtain a drenching coverage.

When spraying emergent or floating aquatic weeds, or marginal vegetation, the spray nozzle should be adjusted to give a droplet size fine enough to give an optimum coverage without excess runoff or drift. To treat a dense stand of 1.2 to 1.5 m tall emergent weeds using water as the carrier, a total volume of herbicide and water of 0.38–0.47 L/m^2 of area being treated should be used. However, when treating the same stand of vegetation using oil as a carrier, the total volume of herbicide and oil needed would be only 0.09 L/m^2. As with any chemical being applied to a pond, the percent active ingredient and the area or volume of the pond to be treated must be known to calculate the quantity of herbicide needed. See Section 17.3 of this chapter for more on dose calculations.

Always keep in mind that herbicides are meant to kill. This not only includes aquatic plants but also at the right dose will kill non-target organisms, including humans. Several safety precautions need to be kept in mind.

1) Study the label on the herbicide container and follow the manufacturer's recommendations.
2) Know the toxicity of the product to both cold- and warm-blooded animals.
3) Do not guess at the concentrations. Know the ppm concentration and strength of the treatment solution.

4) Know the exact area or volume of the location to be treated beforehand, and prepare only enough herbicide for that specific area or volume. This will reduce the quantity of herbicide to dispose of later.
5) Use protective clothing if specified by the herbicide manufacturer.
6) Guard against herbicide drift onto susceptible plants and animals in adjacent areas.
7) Combine the herbicide and carrier outside the spray tank to minimize the possibility of an insoluble residue or gummy precipitate clogging the sprayer or distribution lines.
8) Thoroughly clean the equipment after use.
9) Dispose of the empty containers in a safe way.
10) Do not use the weed spraying equipment to apply insecticides on desirable plants.

The following publications provide detail on aquatic plant identification and their control. Keep in mind that regulations on herbicide use are subject to change, and new products come into the market while others are withdrawn.

Avery, J. L. 2003. Aquatic weed management herbicide safety, technology and application techniques, SRAC Publication No. 3601. Southern Regional Aquaculture Center, Stoneville, Mississippi. Available: https://www.ncrac.org/files/biblio/SRAC3601.pdf. (May 2020).

Durborow, R. M., C. S. Tucker, B. I. Gomelsky, R. J. Onders, and S. D. Mims. 2008. Aquatic weed control in ponds. Kentucky State University Aquaculture Program, Kentucky State University, Frankfort. Available: http://www.ksuaquaculture.org/Pubs.htm/AquaticWeedControlInPonds12-6-07Comprs.pdf. (May 2020).

Haller, W. T. 2009. Appendix B: aquatic herbicide application methods. Pages 152–156 in L. A. Gettys, W. T. Haller, and M. Bellaud, editors. Biology and control of aquatic plants: a best management practices handbook. Aquatic Ecosystem Restoration Foundation, Marietta, Georgia. Available: http://www.aquatics.org/bmp.html. (May 2020).

Lovell, R. G. 2007. Field guide to aquatic plants of Alabama, Alabama Department of Conservation and Natural Resources. Available: https://www.outdooralabama.com/sites/default/files/fishing/Freshwater%20Fishing/Aquatic%20Plants/Field%20Guide%20to%20Aquatic%20Plants%20of%20Alabama%20(revised%202011).pdf. (May 2020).

Masser, M. P., T. R. Murphy, and J. L. Shelton. 2013. Aquatic weed management: herbicides. SRAC Publication No. 361. Southern Regional Aquaculture Center, Stoneville, Mississippi. Available: http://pods.dasnr.okstate.edu/docushare/dsweb/Get/Document-1836/SRAC-361web.pdf. (May 2020).

Selden, G. 2015. Aquatic herbicide mode of action and use implications. SRAC Publication No. 3602. Southern Regional Aquaculture Center, Stoneville, Mississippi.

State of Connecticut. 2014. Nuisance aquatic vegetation management, a guidebook. State of Connecticut, Department of Environmental Protection, Pesticides Management Program, Hartford.

U.S. Army Corps of Engineers. GLMRIS.ANL.GOV. 2012. Aquatic herbicides. Available: http://glmris.anl.gov/documents/docs/anscontrol/AquaticHerbicides.pdf. (May 2020).

U.S. Army Corps of Engineers. GLMRIS.ANL.GOV. 2012. Light attenuating dyes. Available: http://glmris.anl.gov/documents/docs/anscontrol/LightAttenuatingDyes.pdf. (May 2020).

UF/IFAS Center for Aquatic and Invasive Plants. 2015. Background on the aquatic herbicides registered for use in Florida. Available: https://plants.ifas.ufl.edu/manage/control-methods/chemical-control/details-about-the-aquatic-herbicides-used-in-florida/. (May 2020).

UF/IFAS Center for Aquatic and Invasive Plants. 2018. Background on the aquatic herbicides registered for use in Florida. In Plant Management in Florida Waters—An Integrated Approach. Available: http://plants.ifas.ufl.edu/manage/developing-management-plans/chemical-control-considerations/herbicides-registered-for-use-in-florida-waters/. (May 2020).

17.3 Chemical Use and Dose Calculations

17.3.1 Chemical Use

A wide variety of chemicals are used on a hatchery for a wide range of activities. Their effects are not specific to just the one species of plant or animal being targeted but there will be effects on non-target plants and/or animals as well. Such effects on non-targeted organisms must be considered. Some chemicals are slow to break down and their effects may carry over to subsequent crops. Careful records need to be kept as to what chemical was used, at what dose, and when, as this may affect how a pond is used in the future, as well as how the fish produced can be used.

Aquaculture is not as well established in some countries as are other forms of animal production, and as a result such countries may have a smaller portfolio of drugs and other chemicals that have been approved for use with fish destined for human consumption. In the USA, drug approval is the responsibility of the Food and Drug Administration (FDA). This agency has established a series of items that must be addressed in the drug approval process. Currently there are only a limited number of products that have FDA approval. Efforts are being made by the U.S. Fish and Wildlife Service through its Aquatic Animal Drug Approval Partnership (AADAP) program to obtain more drugs approved for use in aquaculture and fisheries. This is a cooperative effort by federal and state agencies as well as the private sector. When there is a drug/chemical with great potential benefit to aquaculture or fisheries, which has been proven not to have any negative effects on human health, such drugs/chemicals may be used as part of an effort by AADAP to obtain FDA approval. Such materials are eligible for an Investigational New Animal Drug (INAD) exemption. Participation in an AADAP-managed INAD allows drug/chemical users to use such materials while collecting needed data for FDA review and approval. Some materials used in hatchery management are considered by FDA to be low regulatory priority drugs and can be used in U.S. aquaculture and fisheries management. However, regulations and policies can change, and a hatchery manager needs to keep current as to what can be used.

It is the responsibility of hatchery staff in the USA to use only approved drugs, or participate in INAD programs specific for a given drug or chemical, or use materials that are of low regulatory priority. What drugs/chemicals that can be used changes, and a hatchery manager must remain up to date as to what products can be used and how. Such details are available for the USA at: https://www.fda.gov/animal-veterinary/aquaculture/approved-aquaculture-drugs and https://www.fws.gov/fisheries/aadap/PDF/approveddrugs.pdf.

In the USA, the sale and use of many of the herbicides and other chemicals used at a hatchery are controlled, and the applicator must obtain a Certified Pesticide Applicator's license before such controlled chemicals can be purchased or used. For details on obtaining a license, contact your State extension service.

It is the obligation of the hatchery manager know how the fish produced on the hatchery will be used by the end user and whether any previous exposure to a specific drug or chemical while on the hatchery would make the fish un-usable for that purpose.

17.3.2 Dose Rate

It is very important to know what you are treating and apply the chemical or drug at the correct concentration. This means correctly knowing the area or volume of the unit to be treated, correctly calculating the quantity of chemical to be used, and then accurately measuring that quantity. Whether the objective is to control aquatic weeds, control fish disease organisms, or fish predators, the material used is meant to be toxic to the target organism and, in most cases, at too high a dose can be toxic to humans and other non-target organisms. For a chemical to be effective it also must be stored properly or it may degrade and lose its potency. The label on the container in which the chemical is sold will give information on how it should be stored, the percent active ingredient, toxicity information, and recommendations on how to handle it safely.

When chemicals are added to a volume of water, whether that might be a pond, a holding tank, or a transport tank, the dose is often expressed as parts per million (ppm). A part per million is one unit of one material in a million parts of another. In a hatchery, water quality parameters may be expressed as parts per million or a weight/volume ratio. For example, an oxygen concentration of 5 ppm would be 5 mg/L as a weight/volume ratio. When adding chemicals to water to treat fish diseases, eliminate gill-breathing predators, or to control aquatic weeds, dose rates are often expressed as ppm or mg/L. Mineral content of pond soils may be expressed as mg/kg of soil.

A part per million can be expressed in English or metric units of measurement as the volume of the container. Metric units are much more convenient to use for most hatchery applications. In the metric system one part per million (ppm) can be expressed as 1 mg/L or 1 g/m^3. Most containers used on a hatchery are either square, rectangular, cylindrical, spherical or a combination of shapes and their dimensions can be measured in millimeter, centimeters or meters keeping in mind that 1 cm^3 of water $\approx$ 1 mL.

Ideally the tank volume or the pond area or volume is already known for every tank or pond at the hatchery. The volume of each pond used at a hatchery should be measured when the pond is dry, including the volume of the catch basin. The volume of a holding tank may vary depending on how deep is the tank and how the water level is regulated. In tanks with a standpipe, an approach is to record the volume of the tank when using a standpipe of a specific height and marking that standpipe as to the volume obtained, and what 1 ppm is for that volume when that pipe is used. It is helpful to have a wall-mounted display board that identifies each pond and tank, giving its volume, what 1 ppm is, when the pond/tank was stocked with what, when it is scheduled to be harvested, and what staff member is responsible for that pond or tank. Having the volume already measured and what a ppm already calculated for that container will save time and help avoid mistakes that may occur when the staff is in a hurry to get a job done.

When calculating how much chemical to add, the percent active ingredient must be known. Some materials such as potassium permanganate are considered to be 100% active, while some materials such as Dylox may be sold having a percent active ingredient ranging from 6.2 to 98% depending on the product form and manufacturer. Always read the label to determine the percent active ingredient and take that information into consideration when calculating how much of the commercial product is needed for an effective dose. Also keep in mind the cost of the effective dose is dependent on the percent active ingredient. One manufacturer's product may appear cheaper/total weight of product, but when considering the percent active ingredient, it may be more expensive than that of another manufacturer's product per gram or volume of active ingredient.

When preparing a chemical to add to a small volume of water it may be advisable to prepare a stock solution beforehand. For example, to treat a series of aquaria, each holding 20 L of water at 2 mg/L with potassium permanganate ($KMn0_4$), which is 100% active, a total of 40 mg of $KMnO_4$ would be needed/aquarium. This small amount/aquarium may be difficult to weigh accurately as the hatchery may not have a balance suitable for weighing out this small a quantity. Similarly, it may not be convenient to weigh out such small quantities for each aquarium when a number of aquaria need to be treated.

In the above example, a stock solution can be prepared so a easily-measured volume of stock solution containing 40 mg of $KmnO_4$ can be added to each aquaria. To prepare enough to treat 20 aquaria a total of 800 mg would be needed. To make a suitable stock solution 1 g of $KmnO_4$ would be dissolved in 100 mL of distilled water to give a 10 mg/mL solution. From there, 4 mL of stock solution containing a total of 40 mg of KmnO4 would be pipetted into the aquaria.

Some chemicals such as formalin are liquids, which would be added as a liquid to a container of water holding the fish to be treated. When using very dense liquids, the relationship that 1 mL = 1 g does not hold true, and adjustments in what volume to use must take into consideration the specific gravity of the material. In cases such as when using formalin, the specific gravity of formalin is 1.08 at 25°C. For practical purposes it is assumed that 1 g of formalin = 1 mL of water, so the dose is calculated as to the volume of formalin needed. One ppm of formalin would be 1 mL formalin/m^3 of water being treated. When treating smaller volumes of water, one ppm formalin would be 1 μL /L of water being treated. Depending on the size of the container and the dose rate, dilution of the formalin may not be necessary. For example, to treat a 40 L aquarium with 100 ppm (100 mg/L) of formalin, a total of 4 mL of formalin would be needed (40 L × 100 mg = 4,000 mg = 4 g = 4 mL). This value then is considered for dose calculations as being 100% active. Dose rates of other chemicals are often calculated based on the % active ingredient of the brand being used and if a liquid measured taking its specific gravity into consideration.

In the case of the herbicide 2,4-D, it is prepared based on the % active ingredient in the product. For example, the milliliters of commercial 2,4-D needed/L of carrier to give a 0.1% active spray solution in a water carrier is calculated as follows:

$$= \frac{1000\text{ (mL)} \times 0.1\%\text{ (desired conc.)}}{37\%\text{ (\% active)}} = \frac{1000 \times 0.001}{0.37} = 2.7\text{ mL/L}$$

If the sprayer tank has a total volume of 475 L, the total volume of the commercial product needed to be added to the tank = 475 × 2.7 = 1283 mL = 1.283 L, plus enough water and surfactant to give a total volume of 475 L to make a 0.1% solution of 2,4-D.

To treat a 1000 m^2 pond that has an average depth of 0.7 m with 1 mg/L of copper sulfate (which is considered 100% active) the pond volume first needs to be calculated.

$$\text{Pond volume} = \text{surface area} \times \text{average depth} = 1000\text{ m}^2 \times 0.7\text{ m} = 700\text{ m}^3$$

Keep in mind that 1 ppm = 1 mg/L or 1 g/m^3, then the quantity of copper sulfate needed = pond volume × dose rate of active ingredient = 700 m^3 volume × 1 g/m^3 dose rate = 700 g.

Copper sulfate is a soluble mineral (32 g/100 mL at 20°C) but its solubility is impacted by the alkalinity of the water. Dose rates may need to be adjusted taking alkalinity into consideration. When the target dose is 1 mg/L the dose rate should be adjusted when the alkalinity is ≥50 mg/L and ≤250 mg/L. It is much more toxic at alkalinities of <50 mg/L and precipitates out at >250 mg/L. The formula for adjusting a 1 mg/L dose rate as a function of alkalinity is as follows:

Corrected dose = theoretical dose × (alkalinity/100)
If the alkalinity is 50 mg/L the quantity of copper sulfate is reduced
= 1 × (50/100) = 0.5 mg/L copper sulfate.

If the alkalinity is 150 mg/L the quantity of copper sulfate needed is increased
= 1× (150/100) = 1.5 mg/L of copper sulfate.

The efficacy and toxicity of several chemicals used at a hatchery can be impacted by water quality characteristics. Any time a new treatment protocol is being used at the hatchery, a bioassay should be conducted, treating the same size fish, held in the same water quality as where the fish would be treated on a larger scale at the hatchery. Dose rates tested should include the intended dose rate to be used at the hatchery, 2X of that dose, as well as 5X and 10X of the intended dose rate.

More details on some of the commonly used chemicals at a hatchery are as follows:

Bowker, J. D., and J. T. Trushenski, editors. 2019. Guide to using drugs, biologics, and other chemicals in aquaculture. American Fisheries Society Fish Culture Section. Fish Culture Section, American Fisheries Society. Available: https://drive.google.com/file/d/1zINB0VMIMplnCaOkeAxIH_Oz-vOBcHPZt/view. (May 2020).

Coyle, S. D., R. M. Durborow, and J. H. Tidwell. 2004. Anesthetics in aquaculture. SRAC Publication No. 3900. Southern Regional Aquaculture Center, Stoneville, Mississippi. Available: http://fisheries.tamu.edu/files/2013/09/SRAC-Publication-No.-3900-Anesthetics-in-Aquaculture.pdf. (May 2020).

Dorman, L. 2007. Aquaculture producer's quick reference handbook. MP435. Cooperative Extension Program University of Arkansas at Pine Bluff, Pine Bluff. Available: https://www.uaex.edu/publications/PDF/MP435.pdf. (May 2020).

Finlayson, B. J., R. A. Schnick, R. L. Cailteux, L. DeMong, W. D. Horton, W. McClay, C. W. Thompson, and G. J. Tichacek. 2000. Rotenone use in fisheries management: administrative and technical guidelines manual. American Fisheries Society, Bethesda, Maryland. Available: https://rotenone.fisheries.org/wp-content/uploads/2014/11/Planning-and-Standard-Operating-Procedures-for-the-Use-of-Rotenone-in-Fish-Management.pdf. (May 2020).

Francis-Floyd, R. 1995. The use of salt in aquaculture. Fact Sheet VM 86. Institute of Food and Agricultural Service, Cooperative Extension Service, University of Florida, Gainesville. Available: http://fisheries.tamu.edu/files/2013/09/The-Use-of-Salt-in-Aquaculture.pdf. (May 2020).

Mischke, C., and J. Avery. 2013. Toxicities of agricultural pesticides to selected aquatic organisms, SRAC Publication No. 4600. Southern Regional Aquaculture Center, Stoneville, Mississippi. Available: http://fisheries.tamu.edu/files/2013/09/SRAC-Publication-No.-4600-Toxicities-of-Agricultural-Pesticides-to-Selected-Aquatic-Organisms1.pdf. (May 2020).

Swann, L., and S. Fitzgerald, 1992. Use and application of salt in aquaculture. North Central Regional Aquaculture Center. Publication No. 105. Available: http://www.ncrac.org/files/biblio/Use%20and%20Application%20of%20Salt%20in%20Aquaculture.pdf. (May 2020).

U.S. Fish and Wildlife Service. 2015. Quick Desk Reference Guide to: Approved Drugs for Use in Aquaculture. Available: https://www.fws.gov/fisheries/aadap/PDF/approveddrugs.pdf. (November 2020).

Watson, C., and R. P. E. Yanong, 1989. Use of copper in freshwater aquaculture and farm ponds. Fact Sheet FA-13, Department of Fisheries and Aquatic Sciences, Florida Cooperative Extension Service, Institute of Food and Agricultural Sciences, University of Florida, Gainesville. Available: http://edis.ifas.ufl.edu/pdffiles/FA/FA00800.pdf.

Whetstone, J. M., D. C. Smith, and M. Watson. 2001. Use of rotenone for management of fish populations, HGIC 1713. Clemson University Cooperative Extension Service, Clemson University, Clemson, South Carolina. Available: https://hgic.clemson.edu/factsheet/use-of-rotenone-for-management-of-fish-populations/. (May 2020).

World Health Organization. Calcium hypochlorite, Fact Sheet 2.19. Available: http://www.who.int/water_sanitation_health/hygiene/emergencies/fs2_19.pdf. (May 2020).

17.3.3 Area and Volume Calculations

The following references provide detail on how to calculate areas and volumes, and convert from one unit of measure to another.

Creswell, R. L. 1993. Aquaculture desk reference. Van Nostrand Reinhold, New York. https://fau.digital.flvc.org/islandora/object/fau%3A6552. (May 2020).

Masser, M. P., and J. W. Jensen. 1991. Calculating area and volume of ponds and tanks. SRAC Publication No. 103. Southern Regional Aquaculture Center, Stoneville, Mississippi. Available: http://counties.agrilife.org/valverde/files/2014/11/Calculating-Area-and-Volume-of-Ponds-and-Tanks..pdf. (May 2020).

Math2.org Math Tables: Areas, Volumes, Surface Areas. Available: http://math2.org/math/geometry/areasvols.htm. (May 2020).

17.4 Control of Egg Adhesiveness

There are a number of hatchery-produced fish species that have adhesive eggs. Depending on the species, the eggs may be deposited individually on a substrate, deposited individually over each other to form an egg mass, or scattered over a substrate. For improved control over the egg incubation, a female may be manually stripped and the egg adhesiveness controlled while it is forming, or for natural spawns after the eggs have become adhesive, the eggs can be separating from the matrix in which they are held.

17.4.1 Egg Coatings

Eggs of fish that become adhesive following contact with water can be treated during this initial period by coating the eggs with Fuller's earth, a kaolinite or montmorillonite clay, or with whole-fat powdered milk. A slurry of one of these materials in water is prepared and added to a bowl of fertilized eggs a few minutes after fertilization. The eggs are gently stirred for 10–20 minutes or longer in the slurry, during which time the eggs become adhesive and the particles stick to the eggs. The objective is to coat each egg with enough particles so there are not any large patches of the egg's surface exposed that could stick to other eggs or the sides of the incubator. Excess particles are rinsed off and the eggs are ready for incubation. This approach will work for Channel Catfish, Clarias, Paddlefish, Sturgeon, Walleye, and other species with moderately adhesive eggs, allowing the eggs to be incubated in MacDonald jars or similar incubators.

17.4.2 Tannic Acid

Tannic acid is effective in reducing egg adhesiveness of eggs of White Bass, Walleye, and several other species that are manually spawned and treated soon after being fertilized. In the case of White Bass, fertilized eggs are added to a MacDonald jar half full of aggressively aerated water. Tannic acid is added at 150–300 mg/L and the eggs treated for 7–12 min. The aeration will keep the eggs stirred during treatment. After the treatment period the aeration is withdrawn and fresh water is added to flush the tannic acid out of the incubator and gently roll the eggs during incubation. Higher rates for shorter periods have been effective with Tench *Tinca tinca*, and Walleye *Sander vitreus*.

17.4.3 Enzymes to Prevent Adhesiveness

17.4.3.1 Alkaline Proteases

Alcalase® is a commercially available enzyme, produced by the bacteria *Bacillus licheniformis*, often added to laundry detergents to remove protein-based stains. It has been shown to be effective in reducing the adhesiveness of Common Carp, European Catfish *Silurus glanis*, Tench, and Walleye eggs. Concentrations of 1 to 2 mL diluted in 1 L of water are added to fertilized eggs. The eggs are gently mixed in the solution for 2 to up to 30 minutes, species dependent. After treatment the eggs are rinsed and added to MacDonald or similar incubators. The use of Alcalase® has given hatch rates equal or better to eggs treated with a milk powder and talc suspension while shortening the time to hatch. Maxatase, another alkaline protease similar to Alcalase, was effective in reducing the adhesiveness of *Clarias gariepinus* following a 90 s exposure.

17.4.3.2 Cysteine Proteases

To incubate Channel Catfish or Blue × Channel Catfish eggs in a MacDonald, the matrix holding the egg mass together must be dissolved after it forms, or the matrix be prevented from forming. The proteolytic enzyme, bromelain, has been used to prevent the matrix to form, resulting in individual eggs suitable for incubation in MacDonald jars. This enzyme derived from the stem and juice of pineapples is sold as a health food item to aid in human digestion and control inflammation, and is also used for meat tenderizing. It is sold as the quantity of gel dissolving units (GDU)/g of tablet or powder. The GDU may range from 600–2400 GDU/g depending on the commercial brand used. For treating catfish eggs, a solution of 4000 GDU/L of water is prepared, allowing the bromelain-containing material to dissolve over several hours before use. The solution is added to an equal volume of catfish eggs, about 3 to 5 minutes after having fertilized the eggs, and then gently stirring the eggs for several minutes. After treatment the eggs are rinsed and loaded into a MacDonald jar for incubation. For more detail regarding bromelain and catfish, see p. 11 of Bosworth et al. 2011.

17.4.4 Salt/Urea treatment

For highly adhesive eggs, such as that of Common Carp, a combination of salt (NaCl) and urea is effective to eliminate their adhesiveness. A solution of salts (4 g salt and 3 g urea/

L of water) is added to the eggs at the time of fertilization. The eggs are stirred continuously, keeping the eggs completely immersed in the solution at all times. As the solution becomes cloudy, fresh solution is added. This method requires about an hour, during which time the eggs become non-adhesive and water-hardened. The duration of the treatment period depends on the water hardness and may be as short as 45 min at a hardness of $>$100 mg/L and 1.5 h at a hardness of $<$ 30 mg/L. To be sure that the eggs were treated long enough, a sample should be taken at what is the assumed appropriate time and placed in freshwater and observed 5 min later to see if the eggs had become adhesive. If the eggs prove to not be adhesive, the eggs should be rinsed in freshwater and then treated with a tannic acid solution (0.5 g tannin/L of water) for 10 s and rinsed several times in freshwater. The tannic acid solution will toughen the egg shells, reducing the chance of their rupture. At that point, the eggs can be loaded into MacDonald jars or similar incubators and incubated.

17.4.5 Post-Adhesiveness Onset Treatments

Where there is a desire to separate eggs held in a matrix, or that are adhering to a substrate, such eggs can be separated from the matrix or substrate using a 1.5% sodium sulfite solution. In the case of Channel Catfish, the eggs are naturally deposited in layers making an egg mass. However, the matrix holding the egg mass together can be dissolved allowing the eggs to be incubated in MacDonald jars using approximately 1 L of the 1.5% sodium sulfite solution/ 500 g of egg mass. The egg mass is placed in a pan of the sodium sulfite solution and the egg mass is gently pulled apart, exposing more of the egg mass to the solution. Protective gloves should be used as the sodium sulfite solution may irritate human skin. Within 5–6 min of gentle mixing, the eggs should be free from any matrix and at that point the eggs should be rinsed in freshwater several times. The eggs can be enumerated based on settled volume and then loaded into MacDonald jars for incubation.

17.4.6 What Technique to Use?

The effectiveness and consequences of reducing the adhesiveness with the above techniques vary depending on the fish species, the nature of the egg, and water quality conditions. If the treatment selected does not result in individually separate eggs, some eggs will stick together, increasing the likelihood of fungus problems. When using clay to coat the eggs, it is possible to apply such a thick coating around an egg that gas exchange is reduced. Over-exposure to tannic acid can toughen the egg shell, making it difficult for the embryo to break out of the egg and hatch. Over-exposure to protease can weaken the egg shells, making the eggs more delicate to handle and increase the risk of rupturing eggs if the water flow in the incubator is too strong. Using an enzyme to reduce the adhesiveness of the eggs can be accomplished in a lesser period of time that using a salt/urea treatment, saving on labor but the cost and availability of such enzymes may be an issue. Keep in mind that with some treatments, such as coating, the eggs often can be loaded in the incubator before water hardening is complete, but when using a salt/urea treatment, the treatment must be continued until water hardening is complete. How long water-hardening takes will vary depending on the total hardness of the water that the eggs are being held. In soft water it may take 30–45 min longer than in hard water.

As with most techniques used on a hatchery, some small-scale trials need to be done to learn how the local hatchery conditions, particularly, water quality, might affect the implementation of a given technique. That also applies to selecting a technique to manage egg adhesiveness. Does the technique improve hatching rate, larval quality, reduce labor spent? Is it appropriate for the facilities available, the water quality conditions, and labor available? Is it cost effective?

17.4.7 References

Bosworth, B., N. Chatakondi, and J. Avery. 2011. Page 11 *in* Producing Hybrid Catfish Fry: Workshop Manual. USDA-ARS Catfish Genetic Research Unit and Mississippi State University National Warmwater Aquaculture Center.

Demska-Zakęś, K., Z. Zakęś, and J. Roszuk. 2005. The use of tannic acid to remove adhesiveness from Pikeperch, *Sander lucioperca*, eggs. Aquaculture Research 36:1458–1464.

Dumas, R. F., and J. S. Brand. 1972. Use of tannin solution in walleye and carp culture. The Progressive Fish-Culturist 34:7.

Horvath, L. 1980. Use of a proteolytic enzyme to improve incubation of eggs of the European Catfish. The Progressive Fish-Culturist 42:110–111.

Khan, H. A., S. D. Gupta, P. V. G. K. Reddy, and S. K. Sahoo. 1986. Use of milk, urea, sodium sulphite and human urine for degumming fertilized eggs of Common Carp, *Cyprinus carpio* L. Aquacultura Hungarica. Szarvas 5:47–54.

Krise, W. F. 1988. Optimum protease exposure time for removing adhesiveness of walleye eggs. The Progressive Fish-Culturist 50:126–127.

Krise, W. F., L. Bulkowski-Cummings, A. D. Shellman, K. A. Kraus, and R. W. Gould. 1986. Increased Walleye egg hatch and larval survival after protease treatment of eggs. The Progressive Fish-Culturist 48(2):95–100.

Kristan, J., M. Blecha, and T. Policar. 2016. Alcalase treatment for elimination of stickiness in Pikeperch (*Sander lucioperca* L.) eggs under controlled conditions. Aquaculture Research 47(12):3998–4003.

Linhart, O., M. Rodina, D. Gela, M. Kocour, and M. Rodriguez. 2003. Improvement of Common Carp artificial reproduction using enzyme for elimination of egg stickiness. Aquatic Living Resources 16:450–456.

Linhart, O., M. Rodina, D. Gela, M. Flajšhans, and M. Kocour, 2003. Enzyme treatment for elimination of egg stickiness in Tench (*Tinca tinca* L.), European Catfish (*Silurus glanis* L.) and Common Carp (*Cyprinus carpio* L.). Fish Physiology and Biochemistry 28:507–508.

Rottmann, R. W., J. V. Shireman, C. C. Starling, and W. H. Revels.1988. Eliminating adhesiveness of White Bass eggs for the hatchery production of hybrid Striped Bass. The Progressive Fish-Culturist 50(1):55–57.

Verreth, J., and A. G. Meyling. 1989. The use of an industrial enzyme, Maxatase, to eliminate the adhesiveness of *Clarias gariepinus* (Burchell) eggs (Abstract). *In* N. De Pau, editor. Aquaculture: a biotechnology in progress: Volume 1. European Aquaculture Society, Bredene, Belgium.

Woynarovich, E. 1962. Hatching of carp eggs in Zuger-glasses and breeding of carp larvae until an age of 10 days. Bamidgeh 14:38–46.

Woynarovich, E., and A. Woynarovich. 1980. Modified technology for elimination of stickiness of Common Carp *Cyprinus carpio* eggs. Aquacultura Hungarica. Szarvas 2:19–21.

Zarski, D., S. Krejszeff, D. Kucharczyk, K. Palinska-Zarska, K. Targonska, K. Kupren. P. Fontaine, and P. Kestmont. 2013. The application of tannic acid to the elimination of egg stickiness at varied moments of the egg swelling process in Pikeperch, *Sander lucioperca* (L.). Aquaculture Research 46(2):324–334.

17.5 Counting Techniques

Fish and their eggs are measured as to their number, length, weight, or condition. Algae and zooplankton are counted to determine their number/unit volume of water. The populations we work with are often very large and it is not practical to count or weigh every individual. We rely on sampling to determine the number of organisms in a population, their density/unit area or volume, or their average for a specific parameter such as length or weight. The issue then becomes how many samples are needed to describe the population and the nature of the tools we use to do the sampling. If we measure the weight of the population, and take enough samples to accurately determine the mean number of organisms/unit weight, then we can extrapolate the number of organisms in the total population. Sampling accuracy refers to how closely the samples represent the true value for the population. More samples will help account for the variability in a population making the mean of the samples a more accurate estimate for the population mean. Details regarding how to calculate an appropriate sample size for a yet to be determined population size are given in the following publications and websites:

Engineering Statistical Handbook, 3.3.3.3. Selecting sample sizes. NIST/SEMATECH e-Handbook of statistical methods. Available: https://www.itl.nist.gov/div898/handbook/. (May 2020).

Henderson, H. F. 1980. Chapter 12: Some statistical considerations in relation to sampling populations of fishes. *In* T. Backiel and R. L. Welcomme, editors. Guidelines for sampling fish in inland waters. Fisheries Department, FAO, Rome. EIFAC Technical Paper 33:176. Available: http://www.fao.org/3/AA044E13.htm#ch12.

Miranda, L. 2007. Approximate sample sizes required to estimate length distributions. Transactions of the American Fisheries Society 136:409–415.

Lockwood, R. N., and D. B. Hayes. 2000. Chapter 6: Sample size for biological studies. *In* J. C. Schneider, editor. Manual of Fisheries Survey Methods II. January 2000. Michigan Department of Natural Resources, Fisheries Division. Fisheries Special Report 25. Available: https://www.michigan.gov/documents/dnr/SMII_Assembled_Doc_2017_final_552610_7.pdf.

Smith, S., 2013. Determining sample size: how to ensure you get the correct sample size. E-Book (c) Qualtrics Online Sample. Available: https://uncw.edu/irp/ie/resources/documents/qualtrics/determining-sample-size-2.pdf.

It can be hard to ensure that the sample adequately represents the population as a whole. What we can do is to use sampling techniques that have an adequate level of precision. That is a sampling method that, in itself, did not contribute to the variation in measurements recorded. Sample variation can come from a number of sources depending on how the population is sampled and the type of data being collected. Basic techniques to describe populations commonly encountered at a fish hatchery are expressed as the number or weight/unit area, volume or weight. Algae or zooplankton typically are expressed as the number/unit volume; fish eggs as the number/unit area, weight or volume; fish larvae as the number/unit weight or volume; and larger fish most commonly as the number or weight/unit area or volume. The techniques to collect this data vary, each with its advantages and disadvantages.

17.5.1 Counting algae

For cultured algae, samples can be taken from the culture unit and counted directly using a hemocytometer. Details on how to use a hemocytometer are available at:

Anonymous. Cell counting using a Haemacytometer, based on: S. Schoen "Cell Counting" *in* C. S. Lobban et. al. (1988) Experimental phycology—a laboratory manual; R.R.L. Guillard "Counting Slides" in A. Sournia (1978) Phytoplankton Manual. UNESCO. Available: http://www.marine.csiro.au/microalgae/methods/haemacytometer%20counting.htm. (May 2020).

Anonymous. 2014. Algae—Botany 330. Quantitative determinations of algal density and growth. University of Wisconsin, Madison. Available: https://courses.botany.wisc.edu/botany_330/Quantitative_Determinations.pdf. (May 2020).

Creighton University. Cell Counts using a Hemacytometer. Available: http://medschool.creighton.edu/fileadmin/user/medicine/images/TeachResearchPatient/Reearch/Flow_Cytometry_Core_Files/Protocols/Using_a_Hemacytometer.pdf. (May 2020).

17.5.2 Counting Zooplankton

Zooplankton are larger than single-cell algae and are commonly counted in a Sedgewick-Rafter counting chamber. If properly filled, a Sedgewick-Rafter will have a volume of 1 mL. *Artemia nauplii* and rotifers are produced at high enough densities/mL that formalin can be added to a sample to give a 5 % solution, and the sample counted directly, if the concentration is very high it can be diluted, fixed in formalin and then counted. Keep in mind that zooplankton once fixed in formalin will settle to the bottom of the container and must be resuspended uniformly within the water column of the container holding the sample when the sample for counting is taken. A minimum of two samples of the preserved sample should be counted.

Pond densities of zooplankton are typically <5/ mL of a given type of zooplankton and to get an adequate quantity (25–50/mL) of the target zooplankton to count on Sedgewick-Rafter counting chamber, the zooplankton must be concentrated.

For example:
10 L of pond water is filtered through a plankton net and concentrated to 500 mL
The concentrated pond sample (500 mL) preserved with formalin at 5% of total volume (95% + 5%) = 500 mL sample + 26.3 mL formalin = 526.3 mL
One mL preserved sample counted on Sedgewick-Rafter counting chamber = 60/ mL zooplankton/mL
Actual sample count/mL of concentrated pond water = 60 /0.95 = 63.2 /mL
Total zooplankton collected = 63.2/mL * 500 mL of concentrated sample = 31,6008
Total zooplankton/ L pond water = 31,600/10 L pond water filtered
= 3,160/L = 3.16/mL

Key things when counting zooplankton are to remember: the initial volume of the sample and to what extent it was concentrated, the quantity of formalin added to fix the sample, and the volume actually counted. This information is needed so the number counted can be back-calculated to express the number counted as the number of zooplankton/mL or /L.

Details on how to use a Sedgewick-Rafter counting chamber are given at the following web sites:

Britton, L. J., and P. E. Greeson, editors. 1987. Pages 105–108. *In* Chapter A4 methods for collection and analysis of aquatic biological and microbiological samples. Book 5, Laboratory analysis, techniques of water-resources investigations of the United States Geological Survey. USGS. Available: https://pubs.usgs.gov/twri/twri5a4/pdf/twri_5-A4_e.pdf. (May 2020).

Gittleson, S. M., and M. Ganapathy. 2011. Cell counting with the Sedgewick-Rafter Chamber and Whipple Micrometer Disc. Protocol Online. Available: http://www.protocol-online.org/prot/Protocols/Cell-Counting-with-the-Sedgewick-Rafter-Chamber-and-Whipple-Micrometer-Disc-4315.html. (November 2020).

How to use a Sedgewick Rafter Counting Chamber. Available: https://www.youtube.com/watch?v=PMvzK5G-G7M. (May 2020).

HAMP How to use the Sedgewick Rafter slide. Available: https://www.youtube.com/watch?v=yQRYagnkW-0. (May 2020).

17.5.3 Counting Fish Eggs

Fish are counted at various stages in their life history when at a hatchery. For fish induced to spawn and manually stripped, the eggs can be counted at the time of stripping, prior to fertilization. The total weight of eggs collected can be recorded and three to four samples containing at least 25 eggs/sample weighed and counted. Based on the number of eggs in the sample and weight of the sample the number of eggs/unit weight of eggs can be calculated. Multiplying the average number of eggs/g of sample times the total weight of eggs collected (g) will give the total number of eggs collected.

A gravimetric method can also be used to estimate the number of eggs post-fertilization if the eggs are tolerant to handling. An egg sample of known number is drained of excess moisture and weighed. Then, the entire lot is drained and weighed with the total number being determined by extrapolation. A gravimetric or direct count is suitable for eggs with relatively tough shells after water hardening but a volumetric count is better for eggs with a more delicate shell.

Errors associated with counting eggs by weight are less when using the same balance (ideally with a precision of $\pm$ 0.1 g) to weigh the samples and the overall quantity of eggs. Using a less sensitive balance to weigh the total quantity of eggs can result in over or underestimating the total weight. A common error is not draining the total quantity of eggs to be weighed of water to the same extent that the eggs in the sample were dried. A few extra drops of water in the total quantity of eggs being weighed will contribute to overestimating the total quantity of eggs. The egg size varies with the female size. A larger female will commonly give a larger egg (less eggs/g of eggs) than a smaller female. It is best to determine the number of eggs/g of eggs for each female rather that assuming a mean number/g for the species. Brood diet will also affect egg weight. Sample counts need to be taken each spawning season to determine the number of eggs/g of eggs to account for any differences that may be due to brood age or diet.

Egg counts after water hardening are commonly made volumetrically. After fertilization and water hardening, non-adhesive, sinking eggs can be enumerated based on settled volume. Eggs can be added to a MacDonald jar for incubation where the jar has gradations marked on the side to allow its volume to be measured. The total volume of eggs can be measured by stopping the water flow in the jar, allowing the eggs to settle in the jar and then recording the settled volume of eggs in the jar. Once the settled volume is recorded, the water flow is gently returned to the jar, and the eggs continue to be incubated. Egg samples are collected from the incubator and are added to a 10 mL graduated cylinder. The settled volume (at least 1 mL of eggs) is recorded for three or more samples and the number of eggs in that settled volume counted and expressed as the number of eggs/mL.

Total number of eggs in the jar = the mean no. eggs/ mL $\times$ total mL of settled eggs.

Eggs that have been treated to reduce their adhesiveness and have become water hardened can be counted using a graduated cylinder to measure the total quantity of settled eggs. Non-adhesive eggs can be enumerated right after being fertilized if it can be done quickly. Samples need to have their volumes recorded and set aside in water to be counted later. In the meantime, known volumes of fertilized eggs should be added quickly to the incubators because the eggs will continue to become larger as they absorb water during water hardening, which will reduce the number of eggs/mL count. If large quantities of eggs need to be loaded, then samples taken at different points in time should be taken to represent the counts/mL stocked during a particular period of time.

Errors associated with egg counts based on the settled volume can come from not letting the eggs in the jar or cylinder completely settle before estimating the total settled volume. Errors associated with counts of eggs still in the water -hardening phase include basing the counts on samples that over-estimate the number stocked in the incubator. If the samples to determine the number of eggs /mL are taken soon after the eggs are fertilized, they may have just begun to water harden. Any eggs to be stocked later will have hydrated more by the time they are stocked and have a larger diameter resulting in a lower number of eggs/mL To get a more accurate estimate of the number of eggs stocked, samples to determine the number of eggs/mL should be taken over the course of stocking and recording the total quantity of eggs stocked represented by each of the samples. Another issue is the precision of the measurements. The samples may be measured within ± 0.1 mL, but the jar or cylinder used for incubation may be only marked with graduations to the nearest 50–100 mL.

For fish that have adhesive eggs scattered over a flat artificial substrate, the quantity of eggs can be determined by measuring the area over which the eggs are scattered and then counting the number of eggs in samples of known areas. The drawback is that the eggs are often not uniformly distributed over the spawning mat. Samples need to be distributed over the mat so they reflect the egg distribution over the mat. In addition, depending on the texture of the spawning mat, some eggs will drop into the mat and not be visible and will not be included in the count. Where a more accurate count of the number of eggs on a spawning mat is needed, the mat can be held in a 1.5% sodium sulfite solution for approximately 5 minutes to reduce the adhesiveness of the eggs. Keep in mind that the sodium sulfite solution is anoxic and a prolonged exposure may damage the eggs. Once the eggs are loose, the mat needs to be rinsed in a pan of fresh water, exchanging the water several times. The eggs will settle to the bottom of the pan along with any debris that was rinsed off the mat. Once the debris is removed from the eggs, the eggs can be counted volumetrically and incubated in in a MacDonald-like incubator.

When eggs that have already developed past the blastula stage are counted, the percent live eggs can be determined at the same time as the overall egg count is being made. This percentage of live eggs can be considered as the fertilization rate. The cells in a dead egg will have started to decompose and will appear white in color. In each sample, if the number of dead eggs and total eggs in a sample are counted then the percentage of live eggs can be calculated. During incubation, additional mortality may occur, particularly during the gastrulation and neurulation phases. Counts of live eggs made after these phases will better reflect the number of larvae that can be expected at hatching.

References

Stone, N., and G. M. Ludwig. 1993. Estimating numbers of golden shiner eggs on spawning mats. Progressive Fish-Culturist 55(1):53–54.

17.5.4 Counting Larvae

The hatch rate can be determined in several ways depending on the nature of the larvae. The recently hatched larvae of several species have large yolk sacs and cannot swim up in the water column. Such larvae can be quantified by the number/ unit volume, and if hardy enough by number/unit weight. Issues associated with getting counts of such larvae include larval size, which may vary by brood age, brood size, and diet. Ideally, the fry count should be based on the mean fry weight and the total weight of fry for each female, but in many cases, this is not practical. Counts based on counts from common age females may be more practical. Another key factor affecting counts of fry is that the fry develop rapidly and their size and weight from one day to another has changed. Fry of different ages should not be pooled for enumeration as it makes it more difficult to obtain representative samples.

Some larvae, such as Common Carp, are difficult to count soon after hatching as many of the larvae are attached by their adhesive head pads to whatever substrate available. Such larvae are better to count after they have developed more and are swimming up in the water column.

Larvae that are up in the water column can be sampled based on the number of larvae/unit volume. To facilitate counting, the volume of the container holding the larvae can be reduced to concentrate the larvae. After the sample counts have been made, the volume of the holding container is recorded, and the number of larvae calculated. Once that is done, the volume of the container can be restored to full volume and the larvae held until the next phase of activities.

Depending on the larval size, an appropriate concentration for counting should be about 1 larvae/2–4 mL of sample. Careful measurement of the larval holding container volume is needed, subtracting the volume taken up by the screen, the drainpipe or any other structure in the container from the gross volume of the container. Samples for counting can be taken using 50 mL beakers, recording the actual volume of the sample. Counting can be made by pouring a portion of the beaker volume into a Petri plate and counting the number of larvae present, emptying the Petri plate, and repeating the process until the entire volume has been counted. The larval holding container should be mixed well before sampling. A minimum of five samples should be taken and counted.

References

Chatain, B., L. Debas, and A. Bourdillon. 1996. A photographic larval fish counting technique: comparison with other methods, statistical appraisal of the procedure and practical use. Aquaculture 141(1–2):83–96.

Lemarie, D. P., D. A. Weller, D. D. Theisen, and L. L. C. Woods III. 1997. Technical notes: evaluation of an electronic fry counter with Striped Bass embryos and larvae. The Progressive Fish-Culturist 59(1):71–74.

17.5.5 Counting Fingerlings and Larger Fish

When the fish are larger and can be handled safely they can be enumerated based on sampling to determine the average weight, and determining the total weight of fish on hand. However, there are several sources of error that can result in a misleading estimate of average weight and total weight. When harvesting a pond, there is often a bias towards harvesting the larger fish first and the smaller fish later. An average fish weight, based on the early part of a harvest, may be higher than the true population mean. Using this mean to divide into the total weight of fish harvested to calculate the total number of fish harvested will result in an underestimate of the number of fish harvested and a misleading estimate of survival. Using a sample taken later may give a smaller average weight and could lead to an overestimation of the number of fish harvested. Samples need to be taken throughout the harvest to get a better estimate of the average weight of the fish in the population.

Whether harvesting a pond or a tank of fish in the hatchery, a question needs to be asked. Did the equipment used to take the sample equally select for all the sizes of fish in the population being sampled? Using the wrong mesh size may allow some size classes of fish to escape. Grading the fish into several size classes will reduce the variance in each size class and give a better estimate of the average weight for that size class. That average weight, when divided into the total weight of that size class, will give a better estimate of the number of fish in that size class compared to using a general average weight determined from the entire population.

When fish are held in holding tanks they may sort themselves out relative to the water current or depth of the container. Depending on where the sample was taken in the tank, it may not represent the population. Multiple samples over the course of harvesting the tank need to be taken to get a better estimate of the population mean for that tank when there is a difference in fish behavior by size class.

Another common error seen with estimating the number of fish present based on average weight and total weight, is the impact of contamination. This contamination may be inorganic or organic debris, algae or other aquatic plants, and unwanted animals such as other species of fish, tadpoles or frogs. It is easy to clean the samples taken to determine the average weight but it is more difficult to eliminate contaminates when the larger population is being weighed. Such contamination, if included in the total weight measured, will result in a total weigh greater that the true fish population weight and an overestimation of the total number of fish on hand. The total population must be cleaned before being weighed to avoid overestimating the quantity of fish on hand.

When weighing small fish, a fine mesh net may be used to collect the fish, but a fine mesh will slow water from draining off the net, resulting in water possibly being added along with the fish when the fish are weighed. A few drops of water added along with the fish being weighed to determine the average weight can result in overestimating the average weight of the fish. This effect can be reduced by allowing enough time for any excess water to drain off the net and touching the net holding the fish to a towel to absorb excess water on the net before the fish are transferred to the weighing container. A related error common when adding small fish to a tared container already holding water, is touching the net into the water, and in doing so, removing water from the weighing container. This can lead to underestimating the average fish weight. This effect can be avoided by not letting the net touch the surface of the water in the tared container. If touching the water surface is unavoidable such effects can be minimized by carefully shaking the net over the weighing container after the fish have been added so any excess water on the net will fall back into the container.

Small fish, 10–25 mm in length, can be enumerated by visual estimation with $\leq$5% error if carefully done. The key to success is keeping conditions as standard as possible. This begins with using round pans of the same size and color, filled to the same depth with clean water. A known number of fish are added to one bowl at a high density, depending on the fish size, approximately 50–100 fish/L. This bowl is the standard to compare other bowls to. Another bowl, having the same volume of water, has progressively more fish added to it until the fish density appears to be the same as the standard bowl. The water in both bowls is gently swirled to give a uniform distribution of the fish within the bowl. When the second bowl appears to have the same number of fish as the standard, the bowl is recorded as having the same number of fish as the standard, and then the bowl is emptied into a larger container to hold the counted fish. A new bowl is prepared and the process continued until all the fish have been processed. New standard bowls may need to be prepared periodically once the fish in the earlier standard appear to be stressed.

The total number of fish counted =
(n of bowls estimated × n fish in standard) + the fish in the standard.

The key to success with this technique is to have healthy fish of uniform size, and using bowls similar in size, color and water volume. Factors that hinder the accuracy of this technique include having mixed sizes of fish, insects, tadpoles or plant material in the bowls. Such material distracts the estimator's eye and makes it harder to make a good estimate.

17.6 Disease Management

Fish at most fish hatcheries are not disease-free and carry some type of parasite, bacteria or virus. In most cases, these disease organisms do not cause a problem unless the fish are stressed, often as a result of poor water quality or poor handing. Good management can minimize the effects of such stressors. Prophylactic treatment of the fish any time they are held in holding tanks at the hatchery prior to being transferred elsewhere can reduce the chance of a parasite problem. Prophylactic treatment should always be a part of the protocol in preparing fish for transport off-station. You do not want to have the recipients of the fish you transported coming back to you saying they never had a particular disease problem until they got fish from your hatchery, or all the fish you shipped got sick and died within a few days after you delivered them.

When there is a risk of spreading a disease problem within the hatchery it may be necessary to sterilize production ponds, holding tanks, transportation equipment as well as all fish handling equipment that may have become in contact with such fish and the water in which the fish were held.

Chlorine at 10 ppm active ingredient for an hour is an effective method for sterilizing most hatchery settings. Household bleach (5 % chlorine) or HTH (65 % chlorine) are commonly used products for this purpose. All chlorine residues need to be eliminated before any treated materials come in contact with fish again. Flushing or rinsing with clean water followed by sun drying for a day is generally effective in removing any chlorine residues. When the facility or equipment needs to be used sooner, chlorine can be removed using 7.4 ppm sodium thiosulfate for each ppm of chlorine.

References

The following references provide details on specific disease issues and how to manage them as well as other hatchery biosecurity issues.

Bartholomew, J., G. Kurath, and E. Emmenegger. 2011. About Viral Hemorrhagic Septicemia (VHS) virus potential threat of Great Lakes VHS virus in the Western United States. Available: http://depts.washington.edu/wracuw/publications/pdfs/VHSV%20Extension%20Pub.pdf. (May 2020).

Borrego, J. J., E. J. Valverde, A. M. Labella, and D. Castro. 2017. Lymphocystis disease virus: its importance in aquaculture. Reviews in Aquaculture 9:179–193.

Bowker, J. D., and J. T. Trushenski, editors. 2016. Guide to using drugs, biologics, and other chemicals in aquaculture. American Fisheries Society, Fish Culture Section. Available: https://drive.google.com/file/d/1zINB0VMIMplnCaOkeAxIH_OzvOBcHPZt/view. (May 2020).

Dadar M., K. Dhama, V. N. Vakhariac, S. H. Hoseinifard, K. Karthik, R. Tiwarif, R. Khandia, A. Munjal, C. Salgado-Mirandah, and S. K. Joshii. 2017. Advances in aquaculture vaccines against fish pathogens: global status and current trends. Reviews in Fisheries Science and Aquaculture. 25(3):184–217.

De Swaef, E., W. Van den Broeck, K. Dierckens, and A. Decostere. 2006. Disinfection of teleost eggs: a review. Reviews in Aquaculture. 8:321–341.

Durborow, R. M., R. L. Thune, J. P. Hawke, and A. C. Camus. 1998. Columnaris disease a bacterial infection caused by Flavobacterium columnare, SRAC Publication No. 479. Southern Regional Aquaculture Center, Stoneville, Mississippi. Available: https://appliedecology.cals.ncsu.edu/wp-content/uploads/479.pdf. (May 2020).

Francis-Floyd Francis-Floyd, R. 1995. The use of salt in aquaculture. Fact Sheet VM 86. Institute of Food and Agricultural Service, Cooperative Extension Service, University of Florida, Gainesville. Available: http://fisheries.tamu.edu/files/2013/09/The-Use-of-Salt-in-Aquaculture.pdf. (May 2020).

Goldthwaite, D. B., and T. G. Carey. 1983. Planning A Fish Health Program for Hatchery Management, Chapter 3 in F. P. Meyer, J. W. Warren, and T. G. Carey, editors. 1983. A guide to integrated fish health management in the Great Lakes basin, Special Publication 83-2. Great Lakes Fishery Commission, Ann Arbor, Michigan. Available: http://www.glfc.org/pubs/SpecialPubs/sp83_2/pdf/chap3.pdf. (May 2020).

Kelley, A. N. 2013. Medicated feed for food fish. SRAC Publication No. 473. Southern Regional Aquaculture Center, Stoneville, Mississippi. Available: http://fisheries.tamu.edu/files/2013/09/SRAC-Publication-No.-473-Medicated-Feed-for-Food-Fish.pdf. (May 2020).

Kent, M. L., S. W. Feist, C. Harper, S. Hoogstraten-Miller, J. M Law, J. M. Sánchez-Morgado, R. L. Tanguay, G. E. Sanders, J M. Spitsbergen, and C. M. Whipps. 2009. Recommendations for control of pathogens and infectious diseases in fish research facilities. Comparative Biochemistry and Physiology Part C: Toxicology & Pharmacology. 149(2):240–248.

Kittel, C., R. Schmid, R. Sparrow, J. Mathews, D. Lyon, and D. Smith. 2013. A biosecurity manual for inland fisheries division hatcheries. Management Data Series No. 276, 2013. Texas Parks and Wildlife Department Inland Fisheries Division, Austin. Available: https://tpwd.texas.gov/publications/pwdpubs/media/pwd_rp_t3200_1776.pdf. (May 2020).

Kumar, G., T. S. Byars, T. E. Greenway, S. Aarattuhodiyil, L. H. Koo, M. J. Griffin, and D. J. Wise. 2019. Economic assessment of commercial-scale Edwardsiella ictaluri vaccine trials in U.S. catfish industry. Aquaculture Economics and Management. 23(3):254–275.

LaPatra, S., E. Misk, L. Al-Hussinee, and J. S. Lunsden. 2016. Rhadoviruses of Fish. Chapter 16 in Kibenge, F. S. B. and M. G. Godey, editors. Aquaculture Virology. Academic Press, New York.

Oidtmann, B., P. Dixon, K. Way, C. Joiner, and A. E. Bayley. 2018. Risk of waterborne virus spread—review of survival of relevant fish and crustacean viruses in the aquatic environment and implications for control measures. Reviews in Aquaculture 10:641–669.

Meyer, F. P., J. W. Warren, and T. G. Carey, editors. 1983. A guide to integrated fish health management in the Great Lakes basin, Special Publication 83-2. Great Lakes Fishery Commission, Ann Arbor, Michigan. Available: http://www.glfc.org/pubs/SpecialPubs/sp83_2/home.html. (May 2020).

Ministry of Natural Resources, Ontario. 2011. Fish culture technical bulletin: best management practices, egg disinfection procedures for Muskellunge and Walleye, Bulletin 2011–01. Available: https://docs.ontario.ca/documents/2546/268426.pdf. (May 2020).

Sadler, J., and A. Goodwin. 2007. Disease prevention on fish farms. SRAC Publication No. 4703. Southern Regional Aquaculture Center, Stoneville, Mississippi. Available: http://fisheries.tamu.edu/files/2013/09/SRAC-Publication-No.-4703-Disease-Prevention-on-Fish-Farms.pdf. (May 2020).

U.S. Fish and Wildlife Service. 2015. Quick Desk Reference Guide to: Approved Drugs for Use in Aquaculture. Available: https://www.fws.gov/fisheries/aadap/PDF/approveddrugs.pdf. (November 2020)

Yanong, R. P. E. 2003. Fish health management considerations in recirculating aquaculture systems—Part 1: Introduction and General Principles. UF/IFAS Circular 120. Available: http://edis.ifas.ufl.edu/FA099. (May 2020).

Yanong, R. P. E. 2003. Fish health management considerations in recirculating aquaculture systems—Part 2: Pathogens. UF/IFAS Circular 121. Available: http://edis.ifas.ufl.edu/pdffiles/FA/FA10000.pdf. (May 2020).

Yanong, R. P. E., 2003. Fish health management considerations in recirculating aquaculture systems—Part 2: General recommendations and problem-solving approaches. UF/IFAS Circular 122. Available: http://edis.ifas.ufl.edu/pdffiles/FA/FA10100.pdf . (May 2020).

17.7 Effluent Management

Most warmwater fish hatcheries produce their fish in ponds, and when it comes time to harvest the fish the ponds must be drained. When in use, the hatchery holding tanks or raceways have a continuous discharge of water that must be dealt with. How to manage hatchery effluents is becoming more of an issue and a hatchery manager must keep the hatchery in compliance with the regulations regulating hatchery discharges. Regulations on the management of hatchery effluents vary from location to location and are subject to change. A hatchery manager must remain current as to what are the appropriate regulations and acceptable techniques to minimize discharge impacts.

Issues to be considered include: the volume of water being discharged; the release of organic matter and its BOD, and other sediments when a pond is drained; chemical discharges following aquatic weed control efforts or fish disease treatments; the escape of non-native fish, plants, diseases or other organisms in the hatchery discharge; and any temperature or other water quality changes in the receiving stream that might occur as a result of the hatchery discharge.

The following references provide more background on effluent management.

References

Boyd, C. E. 2003. Guidelines for aquaculture effluent management at the farm-level. Aquaculture 226:101–112.

Boyd, C . E., and J. F. Queiroz. 2001. Feasibility of retention structures, settling basins, and best management practices in effluent regulation for Alabama Channel Catfish farming. Reviews in Fisheries Science 9(2):43–67.

Department of Primary Industries. 2008. Best practice environmental guidelines for recirculating aquaculture systems. Fisheries Victoria Report Series No. 37. Department of Primary Industries. Available: https://vfa.vic.gov.au/aquaculture/publications/best-pratice-environmental-management-guidelines-for-recirculating-aquaculture-systems. (November 2020).

Environmental Protection Agency. 2004. Federal Register 40, CFR Part 451 Effluent limitations guidelines and new source performance standards for the concentrated aquatic animal production point source category; Final Rule. Available: https://www.epa.gov/eg/concentrated-aquatic-animal-production-effluent-guidelines. (May 2020).

Environment Protection Authority. 2003. Aquaculture management and the Environment Protection (Water Quality) Policy. EPA 488/03. Adelaide, Australia.

Fornshell, G. 2001. Settling basin design. WRAC-106. Western Regional Aquaculture Center, University of Washington, School of Aquatic & Fishery Science. Seattle, Washington. Available: https://www.uidaho.edu/-/media/UIdaho-Responsive/Files/Extension/county/twin-falls/WRAC-106.pdf. (May 2020).

Mugg, J., A. Serrano, A. Libeti, and M. A. Rice. 2000. Aquaculture effluents, a guide for water quality regulators and aquaculturists. NRAC Publication No. 00-0003. Northeastern Regional Aquaculture Center. University of Massachusetts Dartmouth, North Dartmouth, Massachusetts.

Powell, A., and J. W. S. Scolding. 2018. Direct application of ozone in aquaculture systems. Reviews in Aquaculture; Burwood. 10(2):424–438.

Siddiqui, S. A. 2003. Wastewater treatment technology in aquaculture. World Aquaculture, Sept. 2003. 49–52. Available: https://www.was.org/magazine/ArticleContent.aspx?Id=132. (May 2020).

Summerfelt, R. C. 2003. "Introduction to" Aquaculture Effluents: overview of EPA guidelines and standards and BMPs for ponds, raceways, and recycle systems". Natural Resource Ecology and Management Conference Papers, Posters and Presentations 17. Available: https://lib.dr.iastate.edu/nrem_conf/17/. (May 2020).

Tucker, C. S. 2005. Quality of effluents from Channel Catfish hatcheries in northwestern Mississippi. North American Journal of Aquaculture. 67:354–358.

Tucker, C. S., and J. A. Hargreaves. 2008. Environmental best management practices for aquaculture. Blackwell Publishing, Ames, Iowa.

Tuckett, Q. M., C. V. Martinez, J. L. Ritch, K. M. Lawson, and J. E. Hill. 2016. Preventing escape of non-native species from aquaculture facilities in Florida, Part 4: operational strategies. FA198. University of Florida Institute of Food and Agricultural Sciences, Gainesville. Available: https://edis.ifas.ufl.edu/fa198. (November 2020).

Verdegem, M. C. J. 2013. Nutrient discharge from aquaculture operations in function of system design and production environment. Reviews in Aquaculture 5:158–171.

Yeo, S. E., F. Binowski, and J. E Morris. 2004. Aquaculture effluents and waste by-product characteristics, potential recovery, and beneficial reuse. NCRC Publication Office, North Central Regional Aquaculture Center, Iowa State University, Ames. Available: https://lib.dr.iastate.edu/cgi/viewcontent.cgi?referer=https://www.google.com/&httpsredir=1&article=1014&context=ncrac_techbulletins. (May 2020).

17.8 Facilities and Equipment

A hatchery manager needs to have a broad range of knowledge, including the infrastructure needed for a hatchery. Often a hatchery manager will be involved in the design and construction of a new facility or the renovation of an existing one. A hatchery manager needs to be able to communicate with architects and engineers and explain how a hatchery functions, what construction materials are appropriate, what materials or construction techniques are not and why, how ponds are used and should be constructed, why one type of equipment is acceptable and another is not. New equipment or new production systems are always being designed. A hatchery manager needs to be able to evaluate such alternatives as to whether such equipment or production systems will significantly improve hatchery production and whether such changes are economically justifiable.

An overview of the components of a warmwater fish hatchery and construction criteria are given as excerpts from:

United States Department of the Interior, Fish and Wildlife Service, Bureau of Sport Fisheries and Wildlife: Standards for Fish Hatchery Development; Warmwater Hatcheries. August 1969.

The details provided in the following excerpts from this publication have been incorporated into the construction of many of the U.S. Fish and Wildlife Service hatcheries built around the USA. These guidelines should be reviewed and updated based on the availability and appropriateness of new materials and construction practices.

Land

Area required

Adequate acreage for facilities, protection of the water supply, and allowances for future expansion. Include sufficient land area for treatment of effluent, and possible water reuse and recirculation systems.

Topography

Sufficient elevation differential between water source and production facilities for desired aeration and gravity flow. Area should be gently sloping or moderate elevation that can be reduced to provide adequate fall for drainage. Avoid areas subject to flooding.

Type of Soil

Avoid rocky terrain or unstable soil conditions such as swamps, bogs, etc. Obtain subsoil information during site investigation. Conduct test borings prior to selecting pond sites. Seepage tests mandatory. Avoid rocky soil, limestone substrata or old streambeds. Seek soils reasonably impervious to water losses.

Right-of way

By purchase or permanent easement.

Boundary Markers

Permanent monument type, easily visible, placed when land acquired.

Fences

Appropriate fences to be used for protection of facilities and fish where necessary. Areas hazardous to the public will be fenced or covered. Appropriate hedge planting may be used. Boundary fences will be placed on the property lines.

Water Supply

Source

Ground water, preferred: springs or wells. Surface water can be suitable with lakes and reservoirs preferred over streams.

Quantity

Four to ten cubic ft/s, dependent on acreage planned and future requirements.

Quality

Long term chemical analysis, and biological or live fish tests, with emphasis on periods of turnover where reservoir or lake supply is contemplated. Refer to established tolerances, applicable for chemical analysis. Studies should include examination of watershed for potential sources of pollution including turbidity. Consider equipment to filter and sterilize water.

Temperature

70°–90° F preferred during growing season.

Availability

Gravity flow preferred. Water shut-off alarm systems mandatory.

Turbidity

Low turbidity.

Water Supply Lines

Type

Pressure pipe: polyvinyl chloride, concrete, steel, unless size or soil conditions make other materials desirable. Teflon, nylon, or other proven, durable inert substances acceptable. Under no conditions should copper, brass, or zinc galvanized pipe be used.

Size

Adequate to carry 1.5 times the quantity required, at velocities not to exceed 10 ft/s. Main supply line adequate to fill a one-acre pond in two days and all ponds in 14 days or less. Maintenance lines sized to maintain pond levels. Consider possible future enlargement of the hatchery in sizing supply lines.

Rearing Facilities: Raceways for Intensive Culture

Size

Rectangular raceways 8’ × 80’ × 30”, or 6’ × 50’ × 8”.
Circular tanks 36’ diameter.
Concrete construction.

Floor slope

1” in 100’

Intake Control

Headbox with concrete overflow wall and adjustable metal weir plate control for individual raceways, or by pipe discharging above the water surface. Inlet should be full width of raceway.

Wall and Floor

Double slots at drain end of raceway or holding tank, either 2” double angle or 2” channel of non-corrosive metal. Single wall slots spaced to permit dividing a raceway into three equal sections. In fabrication of slots, removable spacers should be provided, allowing a tolerance of no more than 1/8” out of plumb in 5 ft of span.

Outlet control

Overflow full width of raceway, with standpipe or valve that is tamper proof.

Screens

Perforated non-corrosive metal, with supporting braces when used with large fish.

Walkways

14”–16” concrete walkways, broom finish or swinging aluminum skid-proof grating. “Gript- Strut” or equivalent crosswalks. For safety all open flumes, control structures, etc. should be covered with non-slip grating.

Raceway Freeboard

6”–12”

Water exchange rate

Minimum of three times/h

Raceway arrangement

Double in series or in banks

Water collection and dispersal for series

Flume or pipe, oversize where possible

Electrical Lines

To be laid at the time of construction either in raceway walls or alongside with outlets spaced to satisfy operational requirements. Consider requirements for automatic feeder installations, floodlights etc.

Use non-corrosive outlets with both 110 volt and 220 volt lines for flexibility. All electrical work to meet code.

Rearing and Holding facilities: Earthen Ponds

Total acreage

As desired or feasible, 20 to 30 acres normally preferred, depends on demand.

Size

0.75 to 1 acre preferred, 0.5 to 4 acres allowable, 0.1–0.5 acres for special purposes.

Depth

Minimum of 2.5 ft shallow end, 5 ft at deep in for rearing ponds; broodstock, somewhat deeper, depending on latitude. Deeper ponds (10'–12') may be desirable in northern areas, and for Channel Catfish rearing regardless of climate.

Dikes

A 2:1 slope is standard with riprap or 3:1 slope without riprap. The latter used where soil, wind and other conditions permit stabilization with vegetation. 12' dike tops with gravel surface; core wall mandatory, seed banks with low growing grass.

Freeboard

18" sufficient. Ponds oriented to limit sweep of prevailing winds.

Outlet structure

Standard plans available. Provide steps and walk around catch basin. Outside kettles should be considered.

Pond water supply

Valved water supply line at catch basin. Where desirable, two water supplies, the main supply at the pond outlet and a supplemental supply on the opposite side of the pond from the drain structure. Water supply should enter the pond above the water surface not lower than top of the drain structure.

Other pond considerations

Consider ponds with concrete side walls and natural bottoms for holding broodstock or finishing large fish. Approximate size, 8'–12' wide, 50'–120' long and 3'–5' deep.

Effluent Treatment

Provide retention or settling basin of size and design that will effectively settle out solids from used water prior to release from the hatchery proper.

Rearing and Holding Facilities: Troughs and Tanks

Troughs

Type	Fiberglass, metal, wood.
Size	Rectangular 14" × 14" × 8", or 16' × 16" × 16".
Screens	Perforated metal or synthetic.
Water exchange	9–10 exchanges/h

Arrangement	Double with individual supply and drains. If used in series, allow fall between tanks of at least 12", four ft aisle between sets.
<u>Tanks</u>	
Type and size	Circular 4–8' diameter, sloping bottom 1/4" per foot of radius. Rectangular 3' × 3' × 30' in pairs, concrete.
Screens	Perforated metal or synthetic.
Water exchange	Five exchanges/h
Arrangement	Set for convenience with sufficient aisle space for handling and removing fish.

Hatchery Buildings, General Considerations

<u>General Layout</u>

Arranged to expedite work; placement of facilities to present pleasing appearance; compatible with topography and approach routes. Consideration of local architecture style desirable. Provide adequate spacing between buildings for fire control. 50-yr lifespan with minimum maintenance, constructed as fireproof as possible. Modern simple design avoiding dormers, ells, and projections. Buildings requiring heat should be designed for economical heating. Avoid condensation problems in tank room by providing adequate insulation, ventilation, and heating. Selection of building exteriors (colors and materials) should harmonize with existing buildings or surroundings.

Hatchery Building Components

<u>Arrangement</u>

To include hatchery room, incubation area, feed storage room (outside bulk storage tanks), material storage, crew room, toilet facilities, and a small office area. Freezer room, cool room, feed preparation room may be needed in some cases. All doors and door frames should be of rust resistant construction, aluminum suggested. Administrative offices and visitor facilities are not recommended for inclusion in the hatchery building.

<u>Tank Room</u>

Adequate size for anticipated number of troughs or tanks required, allowing 2 ½' aisles between tanks and 4'–6" around ends. Floor: concrete with broom or similar finish, slope 1" per 10' for drainage; Walls: spray glaze or glazed tile; Ceiling: a water-proof material. Water supply and drain systems designed toward flexibility and alteration. Buried lines kept to a minimum. No windows, artificial lighting preferred.

<u>Incubation Area</u>

Separate room or designated area in tank room for egg incubation. Floor: concrete with drain trenches around perimeter or aluminum "Gript Strut" floor with a common drain to permit flexibility in arranging incubators, small tanks, or troughs within the room. Walls: spray glaze or glazed tile. Ceiling: a water-proof material.

Feed Storage

Separate dry feed storage area recommended because of possible undesirable odors but located near the use area. A low-humidity, vermin-free storage area large enough to hold ¼ of the annual dry feed requirements should be provided. Proper ventilation and temperature control is needed. The storage area should be accessible with enough room so that large trucks have access to an elevated loading dock so that bags of feed can be unloaded directly into the feed storage area. If moist pellets are used, a cold storage area large enough to hold a 60-day supply at 10°F should be included. If a hatchery plans to use 50 tons or more of feed a year then metal feed silos should be added for bulk storage.

General Storage

A room containing shelves and storage lockers is needed for general storage near the tank room.

Office

Approximately 150 ft^2 is needed for daily record keeping. Interior construction and décor should be the same as used for the main offices located in a separate Administration Building.

Laboratory

May be included in the hatchery building or Administration Building, depending on the individual setting. For more details see Administration Building-Laboratory.

Crew Area

Minimum of 150 ft^2, large enough to meet staff needs. One or more rooms should contain storage lockers, one for each employee. Shower facilities should be provided. One adequately equipped area of pleasant decor should be set aside as a lunch room.

Employee Restroom

Tile or poured epoxy type flooring with floor drain for easy cleaning; glazed tile walls; ceiling of acoustical tile.

Administrative Building Components

To include office space, laboratory, employee restrooms and visitor facilities with public restrooms. A multi-purpose room that can easily be converted into a conference room should be considered. At other times this area might serve as an employee's lunch room, auxiliary office space, etc.

Offices

Minimum of 300 ft^2. Two or more separate offices with a formal closet in each. Vinyl floor; walls of wood-panel, plaster, or dry wall construction; ceiling of acoustical tile.

Hallways

Same as offices

Laboratory

150' minimum; floors vinyl tile; walls glazed tile; ceiling acoustical tile preferred. Equipped and sized in accordance with anticipated needs.

Visitor's Facilities

Type and size to be determined by anticipate visitor load. Public restrooms should be included within the structure if not otherwise provided. Consider modest aquarium facilities for live fish display, separate lecture room or conference room optional. Floor vinyl or indoor-outdoor carpeting; wood-paneled or glazed tile walls; ceiling acoustical tile.

Public Restrooms

Area adequate for anticipated visitor load, designed for use by the handicapped. Tile or poured type epoxy floors with floor drains for easy cleaning. Glazed tile walls; ceiling acoustic tile preferred.

Other Buildings

Holding House

Average size 38' × 52'. Walls masonry; Fully-enclosed or open at window height depending on climatic conditions. Walk-in doors. Center drive with overhead doors at each end to provide 9' clearance. Floor concrete broom finish, sloped toward center with common drain for holding tanks. Type, number and size of tank consistent with needs, 10 to 16 tanks normal for 20 to 50 acres of ponds. Consider use of monorail system for loading out fish. Provide adequate space in building for storage of equipment.

Garage and Storage Building

Size of building or buildings dependent upon the number of truck stalls required and the amount of material to be stored. Concrete floors, broom finish, 1" to 10' slope to doors. Walls unfinished, overhead fiberglass doors 10' wide by 9' high. Open ceiling. No windows. Provide stalls of sufficient length for shelves or other storage in front of vehicles.

Shop

Minimum of 300 ft^2, floor 1" to 10' slope to door or center drain. Provide heating and electrical systems to satisfy requirements including 220 volt outlets. Overhead door minimum of 10' wide by 9' high. Build-in cabinets for tool storage and adequate work bench area.

Oil and Paint Storage

Provide separate building, or materials may be stored in another building if special 2-hour fire rated room with outside access is provided. Square footage 100–200 ft^2 depending on requirements. Explosion-proof electrical installation. Provide heat to room if water-based paints are to be stored.

Fertilizer and Chemical Storage

Separate building 100–300 ft^2, size determined by fertilizer needs. Explosion-proof electrical fittings. Provide unloading ramp at truck bed height.

Visitor Center

May be included as part of Administration Building or as separate facility. See Administration building for more details.

Hatchery Grounds

Display Pools

Use of display pond in many instances can serve in place of more expensive public use structures. Display ponds require a minimal capital investment and less maintenance than aquaria.

Landscaping

Limited lawn areas. Include the planting of easily maintained trees and shrubs of native varieties. Re-forestation should be accomplished where desirable.

Roads and Drives

Roadways and drives located to expedite hatchery operations in a safe manner for both personnel and visitors. Provide as much privacy to hatchery residents as possible. Allow parking area of a size adequate to handle needs including buses and other large vehicles.

Signs

Include entrance, approach, and direction signs placed at appropriate locations, clearly visible from roadways. Public should be aware of hazards through the proper location of signs. Informational materials should include items to promote safety.

Other Infrastructure Considerations

Maintain up to date maps showing water supply lines, drain lines, underground electrical and gas distribution, and any other buried infrastructure. Mail boxes placed to be inconspicuous. Metal flagpoles of a respectable height in relation to buildings and trees.

Safety

Hatchery areas that are particularly hazardous should be excluded to visitors in the interest of public safety. Screens, fences, or other protection must be provided around dangerous water areas or exposed high voltage electrical equipment.

References

The following publications provide some additional background regarding facilities and equipment used at a fish hatchery.

Bankston, J. D., Jr., and F. E. Baker. 1994. Selecting the proper pump, SRAC Publication No. 372. Southern Regional Aquaculture Center, Stoneville, Mississippi. Available: http://fisheries.tamu.edu/files/2013/09/SRAC-Publication-No.-372-Selecting-the-Proper-Pump.pdf. (May 2020).

Bankston, J. D., Jr., and F. E. Baker. 1995. Powering aquaculture equipment. SRAC Publication No. 375. Southern Regional Aquaculture Center, Stoneville, Mississippi. Available: http://fisheries.tamu.edu/files/2013/09/SRAC-Publication-No.-375-Powering-Aquaculture-Equipment.pdf. (May 2020).

Bankston, J. D., Jr., and F. E. Baker. 2013. Piping systems. SRAC Publication No. 373. Southern Regional Aquaculture Center, Stoneville, Mississippi. Available: http://fisheries.tamu.edu/files/2013/09/SRAC-Publication-No.-373-Piping-Systems.pdf. (May 2020).

Brune, D. E., G. Schwartz, A. G. Eversole, J. A. Collier, and T. E. Schwedler. 2004. Partitioned aquaculture systems. SRAC Publication No. 4500. Southern Regional Aquaculture Center, Stoneville, Mississippi. Available: http://fisheries.tamu.edu/files/2013/09/SRAC-Publication-No.-4500-Partitioned-Aquaculture-Systems.pdf. (May 2020).

Coche, A. G., J. F. Muir, and T. Laughlin. 1995. Pond construction for freshwater fish culture: building earthen ponds (FAO Training Series). FAO, Rome.

Hargreaves, J. A., and C. S. Tucker. 1999. Design and construction of degassing units for catfish hatcheries. SRAC Publication No. 191. Southern Regional Aquaculture Center, Stoneville, Mississippi. Available: https://agrilifecdn.tamu.edu/fisheries/files/2013/09/SRAC-Publication-No.-191-Design-and-Construction-of-Degassing-Units-for-Catfish-Hatcheries.pdf. (May 2020).

Hill, J. E., and Q. M. Tuckett. 2018. Preventing Escape from Aquaculture Operations. SRAC Publication No. 4312. Southern Regional Aquaculture Center, Stoneville, Mississippi. Available: https://tal.ifas.ufl.edu/extension-and-outreach/extension-publications/#Non-Native%20Species. (May 2020).

Huguenin, E., and J. Colt. 2002. Design and operating guide for aquaculture seawater systems, 2nd edition. Developments in aquaculture and fisheries science 33:1–328. Elsevier Science, Amsterdam, The Netherlands.

Jensen, G. 1998. Handbook for common calculations in finfish aquaculture. Louisiana State University Agricultural Center, Louisiana Cooperative Extension Service, Pub 8903 (500) 6/98. Available: https://www.lsuagcenter.com/~/media/system/2/6/1/a/261a0177f4653e91262ea3dc65beb3fc/pub-8903finfishcalculations.pdf. (May 2020).

Jensen, G. J., J. D. Bankston, and J. W. Jensen. 1989. Pond aeration types and uses of aeration equipment. SRAC Publication No. 371. Southern Regional Aquaculture Center, Stoneville, Mississippi. Available: https://appliedecology.cals.ncsu.edu/wp-content/uploads/371FS.pdf. (May 2020).

Losordo, T. M., M. P. Masser, and J. Rakocy. 1998. Recirculating aquaculture tank production systems: an overview of critical considerations, SRAC Publication No. 451. Southern Regional Aquaculture Center, Stoneville, Mississippi. Available: http://fisheries.tamu.edu/files/2013/09/SRAC-Publication-No.-451-Recirculating-Aquaculture-Tank-Production-Systems-An-Overview-of-Critical-Considerations.pdf. (May 2020).

Malone, R. 2013. Recirculating aquaculture tank production systems: a review of current design practice. SRAC Publication No. 453. Southern Regional Aquaculture Center, Stoneville, Mississippi. Available: http://fisheries.tamu.edu/files/2013/09/SRAC-Publication-No.-453-Recirculating-Aquaculture-Tank-Production-Systems-A-Review-of-Current-Design-Practice.pdf. (May 2020).

Masser, M. P., and A. Lazur. 1987. In-pond raceways. SRAC Publication No. 170. Southern Regional Aquaculture Center, Stoneville, Mississippi. Available: http://www.aces.edu/dept/fisheries/aquaculture/pdf/170fs.pdf. (May 2020).

Masser, M. P., J. Rakocy, and T. M. Losordo. 1999. Recirculating aquaculture tank production systems management of recirculating systems. SRAC Publication No. 452. Southern Regional Aquaculture Center, Stoneville, Mississippi. Available: http://fisheries.tamu.edu/files/2013/09/SRAC-Publication-No.-452-Recirculating-Aquaculture-Tank-Production-Systems-Management-of-Recirculating-Systems.pdf. (May 2020).

Sta-Rite Industries, Inc. 2003. Sta-Rite Industries basic training manual, third edition. Available: https://docplayer.net/17163027-Sta-rite-industries-basic-training-manual.html. (May 2020).

United Nations Development Program, Food and Agriculture Organization of the United Nations, Rome. 1984. Inland Aquaculture Engineering, ADCP/REP/84/21. FAO, Rome. Available: http://www.fao.org/docrep/x5744e/x5744e00.htm#Contents. (May 2020).

Westers, H., 1983. Considerations in hatchery design for the prevention of diseases. Chapter 4 *in* F. P. Meyer, J. W. Warren, and T. G. Carey, editors. A guide to integrated fish health management in the Great Lakes basin, Special Publication 83-2. Great Lakes Fishery Commission, Ann Arbor, Michigan. Available: http://www.glfc.org/pubs/SpecialPubs/sp83_2/home.html. (May 2020).

17.9 Feeds and Fertilizers

Feeds and fertilizers are one of the greatest production costs at a hatchery. Issues include selecting the proper size feed of the appropriate nutrient content for specific life stages of a given species of fish, and then offering such feeds in a proper manner at an appropriate frequency, while obtaining an acceptable feed conversion and not creating unmanageable water quality conditions. Proper feed storage is also an issue to avoid spoilage, contamination and lost to insects and other animals.

The following references cover the above topics as well as others.

References

Chiba, L. I. 2014. Animal nutrition handbook section 18: diet formulation and feed ingredients. Available: http://www.ag.auburn.edu/~chibale/an18dietformingredients.pdf. (May 2020).

Davis, D. A., editor. 2015. Feed and feeding practices in aquaculture. Woodhead Publishing Series in Food Science, Technology and Nutrition. Elsevier, Amsterdam, Netherlands.

Hasan, M. R., T. Hecht, S. S. De Silva, and A. G. J. Tacon, editors. 2007. Study and analysis of feeds and fertilizers for sustainable aquaculture development. FAO Fisheries Technical Paper No. 497. FAO, Rome. Available: http://www.fao.org/docrep/011/a1444e/a1444e00.htm. (May 2020).

Hinshaw, J. M. 1999. Trout production feeds and feeding methods. SRAC Publication No. 223. Southern Regional Aquaculture Center, Stoneville, Mississippi. Available: http://fisheries.tamu.edu/files/2013/09/SRAC-Publication-No.-223-Trout- Production-Feeds-and-Feeding-Methods.pdf. (May 2020).

Li, M. H., and E. H. Robinson. 2013. Feed ingredients and feeds for Channel Catfish. SRAC Publication No.1806. Southern Regional Aquaculture Center, Stoneville, Mississippi. Available: http://fisheries.tamu.edu/files/2013/09/SRAC-Publication-No.-1806-Feed-Ingredients-and-Feeds-for-Channel-Catfish.pdf. (May 2020).

Majoros, F. 1984. Chapter 18, Fish feed production, storage and transportation *in* C. H. Clay, editor. Inland aquaculture engineering, ADCP/REP/84/21. FAO, Rome. Available: http://www.fao.org/docrep/x5744e/x5744e0j.htm#chapter%2018%20fish%20feed%20production,%20storage%20and%20transportation. (May 2020).

O'Keefe, T., and C. A. Campabadal. 2015. Chapter 12, Storage and handling of feeds for fish and shrimp. Pages 299–313 *in* D. A. Davis, editor. Feed and feeding practices in aquaculture. Woodhead Publishing Series in Food Science, Technology and Nutrition. Elsevier, Amsterdam, Netherlands.

Rice, M. A., D. A. Bengtson, and C. Jaworski. 1994. Evaluation of artificial diets for cultured fish. NRAC Fact Sheet No. 222-1994. Northeastern Regional Aquaculture Center. University of Massachusetts Dartmouth, North Dartmouth, Massachusetts. Available: https://www.ncrac.org/files/biblio/NRAC0222.pdf. (May 2020).

Royes, J.-A. B., and F. Chapman. Preparing your own fish feeds, UF/IFAS Circular 97. Available: http://edis.ifas.ufl.edu/pdffiles/FA/FA09700.pdf. (May 2020).

U.S. Soybean Export Council (USSEC). Storage and handling of feeds for fish and shrimp. Available: http://indianasoybean.com/images/stories/Workshops/FeedManagement/asafeedhandstorepub.pdf. (May 2020).

Watson, C. A., and C. E. Cichra. 2013. Fertilization of fresh water fish ponds. UF/IFAS Publication #FA17. Available: http://edis.ifas.ufl.edu/fa003. (May 2020).

17.10 Fish Transport

A major activity at a fish hatchery is the on-station and off-station transport of fish. This includes all life stages from eggs to broodstock. Methods of transport depend on the life stage of the fish and its water quality tolerances, the equipment available, and distance to be covered. Modes of transport can be grouped into two basic groups: closed systems and open systems. Closed systems are typically small containers used to transport eggs, larvae, or small fish, but can be used when a limited number of high value larger fish need to be transported. Such systems are considered as closed because once the fish are added, typically with oxygen added, the container is closed. A closed container will prevent oxygen from being lost but also does not allow volatilization of metabolites that could potentially reach toxic levels. Such containers do not require a source of electricity to power supporting equipment. This allows a closed container to be transported in a variety of ways including air shipment. Open systems allow gas exchange. Volatile metabolites can be driven off if the water can be agitated mechanically or aerated. Open systems are typically larger, are used for land transportation, and require mechanical aeration and/or compressed oxygen.

Specifics regarding the transport and distribution are given in the earlier chapters for the fishes covered in this book. The following references provide additional information on open and closed systems, loading rates for other species and life stages as well as information on equipment and chemicals used with fish transport.

References

Cole, B., C. S. Tamaru, R. Bailey, C. Brown, and H. Ako. 1999. Shipping practices in the ornamental fish industry. Center for Tropical and Subtropical Aquaculture Publication Number 131. Center for Tropical and Subtropical Aquaculture, University of Hawaii at Manoa, Honolulu. Available: http://www.ctsa.org/files/publications/CTSA_1316316728603111649761.pdf. (May 2020).

Crosby, T. C., J. E. Hill, C. V. Martinez, C. A. Watson, D. B. Pouder, and R. P. E. Yanong. 2014. Preparation of ornamental fish for shipping, UF/IFAS FA-120. Available: http://edis.ifas.ufl.edu/pdffiles/FA/FA12000.pdf. (May 2020).

Harmon, T. S. 2009. Methods for reducing stressors and maintaining water quality associated with live fish transport in tanks: a review of the basics. Reviews in Aquaculture. 1:58–66. Available: http://citeseerx.ist.psu.edu/viewdoc/download?doi=10.1.1.502.4939&rep=rep1&type=pdf. (May 2020).

Jensen, G. L. 1990. Transportation of warmwater fish procedures and loading rates. Southern Regional Aquaculture Center. SRAC Publication No. 392. Southern Regional Aquaculture Center, Stoneville, Mississippi. Available: https://agrilifecdn.tamu.edu/fisheries/files/2013/09/SRAC-Publication-No.-392-Transportation-of-Warmwater-Fish-Procedures-and-Loading-Rates.pdf. (May 2020).

Jensen, G. L. 1990. Transportation of warmwater fish, loading rates and tips by species. SRAC Publication No. 393. Southern Regional Aquaculture Center, Stoneville, Mississippi. Available: http://fisheries.tamu.edu/files/2013/09/SRAC-Publication-No.-393-Transportation-of-Warmwater-Fish-Loading-Rates-and-Tips-by-Species.pdf. (May 2020).

Lim, L. C., P. Dhert, and P. Sorgeloos. 2007. Preparation for transport: fish packaging technology: Pages 25–45 *in* International transport of live fish in the ornamental aquatic industry. S. A. Fossa,

G. M. O. Bassleer, C. Lim, and A. Ploeg, editors. Ornamental Fish International OFI Educational Publication 2, OFI, Maarssen, The Netherlands.

Watson, C., K. Heym Kilore, and C. Martinez. 2010. Shipping fish in boxes. 2010. SRAC Publication 3903. Available: https://agrilifecdn.tamu.edu/fisheries/files/2013/09/SRAC-Publication-No.-3903-Shipping-Fish-in-Boxes.pdf. (May 2020).

Wurts, W. A. 1990. Pure oxygen and live fish transport, WKY-214, Kentucky State University Cooperative Extension Program. Available: http://agrilife.org/fisheries2/files/2013/09/Pure-Oxygen-and-Live-Fish-Transport.pdf. (May 2020).

Wurts, W. A. 1995. Using salt to transport live fish, WKY-216, Kentucky State University Cooperative Extension Program. Available: https://freshwater-aquaculture.extension.org/wp-content/uploads/2019/08/UsingSalt.pdf. (May 2020).

Wynne, F. S., and W. Wurts. 2011. Transportation of warmwater fish, equipment and guidelines. SRAC Publication No. 390. Southern Regional Aquaculture Center, Stoneville, Mississippi. Available: http://fisheries.tamu.edu/files/2013/09/SRAC-Publication-No.-390-Transportation-of-Warmwater-Fish-Equipment-and-Guidelines.pdf. (May 2020).

Wynne, F. S., and W. Wurts. 2015. Live-hauling Channel Catfish. Kentucky State University Cooperative Extension Program. Available: http://aquaculture.ca.uky.edu/publication/live-hauling-channel-catfish-manual. (May 2020).

17.11 Genetics

Genetic management is core to a successful hatchery program. Goals may vary among hatcheries. When hatcheries are used to support natural populations, the goal is often to produce fish with a diverse genetic profile that reflects that of the natural stocks where the hatchery fish are to be stocked, and not selecting for any specific trait. Where a hatchery is to support an aquaculture operation, the goal is to produce fish genetically selected to perform well under a narrow set of conditions. In either case, good record keeping is key to success. Inbreeding is a basic concept that a hatchery manager must have a clear understanding of. As our understanding of genetics has improved so has the tools to manipulate it also has improved. The following references provide background on fish genetics and how it can be manipulated.

References

Beaumont, A., P. Boundry, and K. Hoare. 2003. Biotechnology and genetics in fisheries and aquaculture, second edition. Wiley, New York.

CIHEAM/CSIC-IATS, 2009. Protocol H—Induction of gynogenesis using UV-light and cold shock [Practical guide of protocols: chromosome set manipulation] *in* A. Felip, M. Carrillo, M. P. Herráez, S. Zanuy and B. Basurco, editors. Advances in fish reproduction and their application to broodstock management: a practical manual for sea bass. Zaragoza: CIHEAM/CSIC-IATS. (Options Méditerranéennes: Série B. Etudes et Recherches; n. 63). Available: http://om.ciheam.org/option.php?IDOM=566. (May 2020).

Dunham, R. A., K. Majumdar, E. Hallerman, D. Bartley, G. Mair, G. Hulata, Z. Liu, N. Pongthana, J. Bakos, D. Penman, M. Gupta, P. Rothlisberg, and G. Hoerstgen-Schwark, 2001. Review of the status of aquaculture genetics. Pages 137–166 *in* R. P. Subasinghe, P. Bueno, M. J. Phillips, C. Hough, S. E. McGladdery, and J. R. Arthur, editors. Aquaculture in the Third Millennium. Technical Proceedings of the Conference on Aquaculture in the Third Millennium, Bangkok, Thailand, 20-25 February 2000. NACA, Bangkok and FAO, Rome. Available: http://www.fao.org/docrep/003/AB412E/ab412e03.htm. (May 2020).

Fisch, K. M., C. C. Kozfkay, J. A. Ivy, O. A. Ryder, and R. S. Waples. 2015. Fish hatchery genetic management techniques: integrating theory with implementation. North American Journal of Aquaculture 77:343–357.

Flagg, T. A., and C. F. Nash, editors. 1999. A conceptual framework for conservation hatchery strategies for Pacific salmonids. U.S. Department of Commerce. NOAA Technical Memo NMFS-NWFSC-38.

Gjedrem, T., editor, 2005. Selection and breeding programs in aquaculture. Springer, Netherlands.

Gjedrem, T., N. Robinson, and M. Rye. 2012. The importance of selective breeding in aquaculture to meet future demands for animal protein: a review. Aquaculture 350–353:117–129.

Hulata, G. 2001. Genetic manipulations in aquaculture: a review of stock improvement by classical and modern technologies. Genetics 111:155–173.

Rothbard, S. 2006. A review of ploidy manipulations in aquaculture: the Israeli experience. Israeli Journal of Aquaculture/Bamidgeh 58(4):266–279.

Rottmann, R. W., J. V. Shireman, and F. A. Chapman. 1991. Induction and verification of triploidy in fish. SRAC Publication No. 427. Southern Regional Aquaculture Center, Stoneville, Mississippi. Available: http://agrilife.org/fisheries/files/2013/09/SRAC-Publication-No.-427-Induction-and-Verification-of-Triploidy-in-Fish.pdf. (May 2020).

Tave, D., editor, 1993. Genetics for fish hatchery managers. Springer, New York.

Tave, D. 1999. Inbreeding and broodstock management. Fisheries Technical Paper. No. 392. FAO, Rome. Available: http://www.fao.org/3/a-x3840e.pdf. (May 2020).

17.12 Grading

Grading is a very useful tool at a hatchery as discussed in detail in Chapter 11.4. It allows selection of similar size fish and reduces competition. It can be used to separate the sexes when there is sexual dimorphism in growth. Grading can begin in the hatchery to reduce cannibalism by making the population to be similar in size so one fish cannot swallow another. Fingerlings that have learned to take formulated feed can be separated from those that have not, and those that were not feed-trained can be retrained. Grading can take place in ponds to make a partial harvest using a smaller mesh seine to crowd the fish into a larger mesh grading sock where the smaller fish can escape back into the pond and the larger fish be harvested for marketing or restocking. Details on graders and their use are available in the following references.

References

Crosby, T. C., J. E. Hill, C. V. Martinez, C. A. Watson, D. B. Pouder, and Roy P. E. Yanong. 2017. Grading ornamental fish, UF/IFAS Circular 118. Available: http://edis.ifas.ufl.edu/pdffiles/FA/FA11800.pdf. (May 2020).

FAO, Management for Freshwater Fish Culture. No date. Chapter 12: Grading and sorting fish. Available: http://www.fao.org/fishery/static/FAO_Training/FAO_Training/General/x6709e/x6709e12.htm#top. (May, 2020).

Kelly, A. M., and D. Heikes. 2013. Sorting and grading warmwater fish. SRAC Publication. 391. Southern Regional Aquaculture Center, Stoneville, Mississippi. Available: http://agrilife.org/fisheries/files/2013/09/SRAC-Publication-No.-391-Sorting-and-Grading-Warmwater-Fish.pdf. (May 2020).

Leitritz, E., and R. C. Lewis. 1976. Grading. Pages 117–123 *in* Trout and salmon culture (hatchery methods). Fish Bulletin 164. State of California Department of Fish & Game. Available: https://oac.cdlib.org/view?docId=kt5q2nb139&brand=oac4&doc.view=entire_text. (May 2020).

17.13 Induced Spawning

Many species of fish will mature under hatchery conditions but lack the proper environmental stimuli to trigger the cascade of hormone release needed for final maturation and ovulation including several species discussed in the earlier chapters of this book. The lack of natural endogenous hormones can be overcome by providing appropriate exogenous hormones at a suitable dose rate and if properly administered can induce final maturation and ovulation. Induced spawning is also a useful tool for use with species that would otherwise spawn naturally in a hatchery setting. Induced spawning allows better scheduling of hatchery activities. Knowing on what days eggs will be available will allow better planning of the rest of the production cycle. Being able to obtain viable unfertilized eggs through induced spawning allows a variety of genetic manipulations. Induced spawning is a technique more appropriate for group synchronous spawners that will release a large quantity of eggs at one time. It is less practical for asynchronous spawners that release only a few eggs at a time. Always keep in mind that successful induced spawning requires quality mature broodfish.

The following series of references provide a good introduction to the various steps involved in induced spawning, the hormones that might be used and other related aspects.

References

Chapman, F. A. 2016. A semen extender for the short-term storage of fish, FA193. Available: https://edis.ifas.ufl.edu/pdffiles/FA/FA19300.pdf. (May 2020).

Chatakondi, N. G., I. A. E. Butts, R. A. Dunham. 2018. Hormone preparation, dosage calculation, and injection techniques for induced spawning of foodfish. SRAC Publication No. 425. Southern Regional Aquaculture Center, Stoneville, Mississippi. Available: https://fisheries.tamu.edu/files/2019/01/SRAC_0425.pdf. (May 2020).

Gallego V., and J. F. Asturiano Nemesio. 2018. Fish sperm motility assessment as a tool for aquaculture research, a historical approach. Reviews in Aquaculture (Online) 1–28. doi:10.1111/raq.12253. Available: https://riunet.upv.es/bitstream/handle/10251/104903/Gallego%20and%20Asturiano%20%28RAq%29.pdf?sequence=3&isAllowed=y. (May 2020).

Lee, C. S., C. S. Tamaru, and C. D. Kelley. 1986. Technique for making chronic-release LHRH-a and 17α-methyltestosterone pellets for intramuscular implantation in fishes. Aquaculture 59:161–168.

Mylonas, C. C., N. J. Duncan, and J. F. Asturiano. 2016. Hormonal manipulations for the enhancement of sperm production in cultured fish and evaluation of sperm quality. Aquaculture 472(1):21–44.

Rottmann, R. W., J. V. Shireman, and F. A. Chapman. 1991. Introduction to hormone-induced spawning of fish. SRAC Publication No. 421. Southern Regional Aquaculture Center, Stoneville, Mississippi. Available: https://www.ncrac.org/files/biblio/SRAC0421.pdf. (November 2020).

Rottmann, R. W., J. V. Shireman, and F. A. Chapman. 1991. Capturing, holding, handling, transporting, injecting and brood fish for induced spawning. SRAC Publication No. 422. Southern Regional Aquaculture Center, Stoneville, Mississippi. Available: http://agrilife.org/fisheries/files/2013/09/SRAC-Publication-No.-422-Capturing-Handling-Transporting-Injecting-and-Holding-Brood-Fish-for-Induced-Spawning.pdf. (May 2020).

Rottmann, R. W., J. V. Shireman, and F. A. Chapman. 1991. Determining sexual maturity of broodstock for induced spawning of fish. SRAC Publication No. 423. Southern Regional Aquaculture Center, Stoneville, Mississippi. Available: http://agrilife.org/fisheries/files/2013/09/SRAC-Publication-No.-423-Determining-Sexual-Maturity-of-Broodstock-for-Induced-Spawning-of-Fish.pdf. (May 2020).

Rottmann, R. W., J. V. Shireman, and F. A. Chapman. 1991. Hormonal control of reproduction in fish for induced spawning. SRAC Publication No. 424. Southern Regional Aquaculture Center, Stoneville, Mississippi. Available: http://agrilife.org/fisheries/files/2013/09/SRAC-Publication-No.-424-Hormonal-Control-of-Reproduction-in-Fish-for-Induced-Spawning.pdf. (May 2020).

Rottmann, R. W., J. V. Shireman, and F. A. Chapman. 1991. Techniques for taking and fertilizing the spawn of fish. SRAC Publication No. 426. Southern Regional Aquaculture Center, Stoneville, Mississippi. Available: https://www.ncrac.org/files/biblio/SRAC0426.pdff. (November 2020).

Rurangwa, D. E., E. Kimeb, F. Olleviera, and J. P. Nash. 2004. The measurement of sperm motility and factors affecting sperm quality in cultured fish. Aquaculture 234:1–28.

Woynarovich, E., and L. Horváth, 1980. Chapter 8. Collection and preservation of fish pituitary glands *in* The artificial propagation of warm-water finfishes—a manual for extension. FAO Fisheries Technical Paper 201. Available: http://www.fao.org/docrep/005/AC742E/AC742E09.htm. (May 2020).

17.14 Live Food Production

The larvae of only a few of the commonly cultured fishes readily accept a formulated feed as a first feed. Therefore, some type of live food needs to be provided during these early developmental stages. This can be done by preparing ponds to provide appropriate types of live foods at appropriate densities and then the fish larvae are added to the pond. The other approach is to hold the fish larvae in the hatchery and bring live food to them until they can be trained to accept a formulated feed or the larvae have grown large enough so they can better survive in outdoor ponds. Brine shrimp (*Artemia*) are the most widely used live food in freshwater fish hatcheries because its nauplii are small enough to be consumed by the first feeding larvae of the majority of the cultured freshwater fishes. The encysted artemia eggs resist desiccation and can be stored until needed, providing a suitable size first food within 24 h before it is needed.

Rotifers are a smaller live first food, but its production is more complicated and requires a greater lead-time to have them available. However, they are an appropriate first food for some freshwater fish such as White Bass *Morone chrysops* and are widely used with many marine fish. Several crustaceans including *Daphnia* and *Moina* are also cultured for use in freshwater hatcheries to feed fish larvae.

Details on how to produce and use live feeds are given in the following references:

References

Baptist, G., D. Meritt, and D. Webster. 1993. Growing microalgae to feed bivalve larvae. NRAC Fact Sheet No. 160–1993. Northeastern Regional Aquaculture Center. University of Massachusetts Dartmouth, North Dartmouth, Massachusetts. Available: https://shellfish.ifas.ufl.edu/wp-content/uploads/Growing-Microalgae-to-Feed-Bivalve-Larvae.pdf. (May 2020).

Conceicao, L. E. C., M. Yufera, P. Makridis, S. Morais, and M. T. Dinis. 2010. Live feeds for early stages of fish rearing. Aquaculture Research 41:613–640.

Davis, D. A., T. J. Derbes II, and M. E. Head. 2018. Culture of small zooplankton for the feeding of larval fish. SRAC Publication No. 0701. Southern Regional Aquaculture Center, Stoneville, Mississippi. Available: https://fisheries.tamu.edu/files/2019/01/SRAC_0701.pdf. (May 2020).

Delbos, B., and M. H. Schwarz. 2008. Rotifer production (as a first feed item) for intensive finfish larviculture. Virginia Cooperative Extension. Publication 600-105. College of Agriculture and Life Sciences, Virginia Polytechnic Institute and State University, Blacksburg, Virginia. Available: https://www.pubs.ext.vt.edu/600/600-105/600-105.html. (May 2020).

Dhert, P., and P. Sorgeloos.1994. Live feeds in aquaculture. Pages 99–109 *in* K. P. P. Nambiar and T. Singh, editors. Aquaculture towards the 21st Century: proceedings of INFOFISH-AQUATECH '94, International Conference on Aquaculture, Colombo, Sri Lanka, 29–31 August 1994.

Dhert, P., G. Rombaut, G. Suantika, and P. Sorgeloos. 2001. Advancement of rotifer culture and manipulation techniques in Europe. Aquaculture 200:129–146.

Lavens, P., and P. Sorgeloos, editors. 1996. Manual on the production and use of live food for aquaculture. FAO Fisheries Technical Paper. No. 361. FAO, Rome Available: http://www.fao.org/3/a-w3732e.pdf. (May 2020).

Ohs, C. L., E. J. Cassiano, and A. Rhodes. 2016. Choosing an appropriate live feed for larviculture of marine fish. UF/IFAS Publication #FA167. Fisheries and Aquatic Sciences Department, Florida Cooperative Extension Service, Institute of Food and Agricultural Sciences, University of Florida, Gainesville. Available: http://citeseerx.ist.psu.edu/viewdoc/download;jsessionid=5B422C56C320A318D5D5DF9345D6D141?doi=10.1.1.652.8763&rep=rep1&type=pdf. (May 2020).

Rottmann, R. W., J. S. Graves, C. Watson, and R. P. E. Yanong. 2017. Culture techniques of *Moina*: the ideal daphnia for feeding freshwater fish fry. UF/IFAS CIR1054. Fisheries and Aquatic Sciences Department, Florida Cooperative Extension Service, Institute of Food and Agricultural Sciences, University of Florida, Gainesville. Available: http://edis.ifas.ufl.edu/pdffiles/FA/FA02400.pdf. (May 2020).

Tamaru, C. S., H. Ako, R. Paguirigan, Jr., and L. Pand. 1999. Enrichment of artemia for use in freshwater ornamental fish production. CTSA no. 133. Center for Tropical and Subtropical Aquaculture. University of Hawaii at Manoa, Honolula. Available: http://www.ctsa.org/files/publications/CTSA_1336316728593134906001.pdf. (May 2020).

Treece, G. D., and D. A. Davis. 2000. Culture of small zooplankters for the feeding of larval fish. SRAC Publication No. 701. Southern Regional Aquaculture Center, Stoneville, Mississippi. Available: https://www.ncrac.org/files/biblio/SRAC0701.pdf. (May 2020).

Treece, G. D. 2000. Artemia production for marine larval fish culture. SRAC Publication No. 702. Southern Regional Aquaculture Center, Stoneville, Mississippi. Available: http://agrilife.org/fisheries/files/2013/09/SRAC-Publication-No.-702-Artemia-Production-for-Marine-Larval-Fish-Culture.pdf. (May 2020).

Vos, J., and N. L. de la Rosa. 1980. Manual on Artemia production in salt ponds in the Philippines. FAO/UNDP-BFAR Brackishwater Aquaculture Demonstration and Training Project PHI/75/005. Available: http://www.fao.org/3/AC062E/AC062E00.htm. (May 2020).

17.15 Nursery Pond Preparation

The role of nursery ponds, their preparation, and management are discussed in detail in Chapter 7. Additional detail on pond preparation and management is given in the following references.

Ludwig, G. M. 1999. Zooplankton Succession and larval fish culture in freshwater ponds. SRAC Publication No. 700. Southern Regional Aquaculture Center, Stoneville, Mississippi. Available: http://agrilife.org/fisheries/files/2013/09/SRAC-Publication-No.-700-Zooplankton-Succession-and-Larval-Fish-Culture-in-Freshwater-Ponds.pdf. (May 2020).

Ludwig, G., N. M. Stone, and C. Collins. 1998. Fertilization of fish fry ponds. SRAC Publication No. 469. Southern Regional Aquaculture Center, Stoneville, Mississippi. Available: http://agrilife.

org/fisheries/files/2013/09/SRAC-Publication-No.-469-Fertilization-of-Fish-Fry-Ponds.pdf. (May 2020).

Morris, J. E., and C. C. Mischke. 1999. Plankton management for fish culture ponds. Technical Bulletin Series #114. Iowa State University Agricultural Experiment Station. Iowa State University, Ames. Available: https://lib.dr.iastate.edu/cgi/viewcontent.cgi?article=1010&context=ncrac_techbulletins. (May 2020).

17.16 Predator Control

There is a wide range of animals found on a hatchery that can prey on one or more life stages of fish. Several types of predators and their control are discussed in Chapter 8.2. What tools to use to control predators depends on where the hatchery is located and what are the government regulations affecting their use. Such regulations change frequently and it is up to the hatchery manager to keep up to date as to the current regulations.

The following references provide more detail on specific groups of vertebrate predators, the problems they can cause, and how they have been managed in some locations.

References

Avery, M. L., and M. Lowney. 2016. Vultures. Wildlife Damage Management Technical Series 5. USDA, APHIS, WS National Wildlife Research Center. Fort Collins, Colorado. Available: https://www.aphis.usda.gov/wildlife_damage/reports/Wildlife%20Damage%20Management%20Technical%20Series/Vultures.pdf. (May 2020).

Barras, S. C. 2007. Avian predators at aquaculture facilities in the southern states. SRAC Publication No. 400. Southern Regional Aquaculture Center, Stoneville, Mississippi. Available: https://appliedecology.cals.ncsu.edu/wp-content/uploads/400.pdf. (May 2020).

Falker, S. T., and M. C. Brittingham. 2005. Controlling birds at aquaculture facilities. College of Agricultural Sciences, The Pennsylvania State University, University Park. Available: https://extension.psu.edu/controlling-birds-at-aquaculture-facilities. (May, 2020).

Fowler, J. F. and Avery, J. L., "TURTLES" (1994). The handbook: prevention and control of wildlife damage. Available: https://digitalcommons.unl.edu/icwdmhandbook/82/. (May 2020)

Hill, E. P. 1994. River Otter (*Lutra canadensis*). The handbook: prevention and control of wildlife damage. Available: https://digitalcommons.unl.edu/cgi/viewcontent.cgi?article=1039&context=icwdmhandbook. (May 2020).

Hygnstrom, S. E., R. M. Timm, and G. E. Larson. editors. 1994. Prevention and control of wildlife damage. University of Nebraska-Lincoln. Available: http://digitalcommons.unl.edu/icwdmhandbook/. (May 2020).

Miller, J. E. 2018. Wildlife Damage Management Technical Series: Muskrats. U.S. Department of Agriculture Animal & Plant Health Inspection Service Wildlife Services. Available: https://www.aphis.usda.gov/wildlife_damage/reports/Wildlife%20Damage%20Management%20Technical%20Series/Muskrat-WDM-Technical-Series.pdf. (May 2020).

Moler, P. E. 1994. Frogs and Toads. The Handbook: Prevention and Control of Wildlife Management Technical Series 78. Digital Commons @ University of Nebraska—Lincoln. Available: https://digitalcommons.unl.edu/cgi/viewcontent.cgi?article=1076&context=icwdmhandbook. (May 2020).

Roy, L. A., N. M. Stone, and A. M. Kelly. 2015. Control of bullfrogs and their tadpoles on commercial fish farms. University of Arkansas at Pine Bluff, Aquaculture/Fisheries, FSA9619-PD-12-2015N. Available: https://www.uaex.edu/publications/pdf/FSA-9619.pdf. (May 2020).

Seamans, T. W., and A. L. Gosser. 2016. Bird dispersal techniques. Wildlife Damage Management Technical Series 2. Digital Commons @ University of Nebraska—Lincoln. Available: http://digitalcommons.unl.edu/cgi/viewcontent.cgi?article=1001&context=nwrcwdmts. (May 2020).

Smith, A. E., S. R. Craven, and P. D. Curtis. 2000. Managing Canada geese in urban environments: a technical guide. Jack Berryman Institute Publication 16, and Cornell University Cooperative Extension, Ithaca, New York. Available: https://ecommons.cornell.edu/handle/1813/66. (May 2020).

Taylor, J. D., G. K. Yarrow, and J. E. Miller. 2017. Beaver. Wildlife Damage Management Technical Series 11. Digital Commons @ University of Nebraska—Lincoln. Available: http://digitalcommons.unl.edu/cgi/viewcontent.cgi?article=1011&context=nwrcwdmts. (May 2020).

United States Department of Agriculture. Animal and Plant Health Inspection Service. 1997. Bird Predation and Its Control at Aquaculture Facilities in the Northeastern United States. USDA-APHIS Fact Sheets on Wildlife Damage Management. Paper 33. Digital Commons @ University of Nebraska—Lincoln. Available: http://fisheries.tamu.edu/files/2013/09/Bird-Predation-and-Its-Control-at-Aquaculture-Facilities-in-the-Northeastern-United-States.pdf. May 2020).

Wade, D. A., and C. W. Ramsey. 1986. Identifying and managing aquatic rodents in Texas: beaver, nutria and muskrats. B-1556, Texas Agricultural Extension Service. Texas A&M University, College Station. Available: http://wildlife.tamu.edu/files/2010/04/6_Idetifying_and_managing_Aquatic_Rodents.pdf. (May 2020).

17.17 Water Filtration

The purity required of a hatchery water supply depends on how it is to be used on the hatchery. Ground water is typically free of predators or competitive organisms, possibly due to the ground water being anaerobic. Many minerals may be in a reduced state and are soluble under such conditions. Minerals such as iron and manganese will precipitate out of solution when such anaerobic water is aerated. The precipitation of minerals is not much of an issue when ground water is used to fill ponds, but precipitation of such minerals in the hatchery can block valves, and coat fish eggs and larvae causing them to suffocate. In the hatchery, such precipitants must be allowed to settle out or be filtered out before such water can be used for egg incubation or larval rearing.

When surface waters are used, there are a variety of undesirable organisms that must be dealt with. Material such as fish eggs and larger items can be filtered out when filling a fish pond as discussed in Chapter 8.3.1. In the hatchery there is often a need for as clean of water as possible as young delicate life stages are being managed, as well as to keep diseases under control. A first step obtaining clean water is the use of a pressurized sand filter. Sand filters as used with swimming pools are commonly used in fish hatcheries.

Sand filters filled with pool grade silica sand (0.45–0.55 mm diameter) are capable of filtering out particles of 20–40 μm and larger. The size of the filter is dependent on the volume of clean water needed and how frequently it is acceptable to clean the filter. It is best to use a sand filter that is larger than the theoretical needed to filter a specific volume of water. As the filter gets clogged, the water flow rate through the filter will become reduced. Reading on the pressure gauge should be monitored and recorded at least twice a day, once at the start of the workday and again at the end of the day. Normal pressure levels depend on pump size, pipe size, and piping configurations. The pressure reading of a recently backwashed filter should be recorded as the baseline or initial level (typically about 10 psi). As the filter becomes clogged and the pressure builds to be about 10 psi above the initial level then it is time to backwash the filter. As the pressure builds in the sand filter, backpressure builds causing the pump to work harder, use more electricity and may cause the motor to overheat, while the water flow becomes reduced.

A recently washed sand filter will remove particles of about 20–40 μm, but a number of disease organisms are smaller and can pass through a sand filter. To remove smaller particles a diatomaceous earth (D. E.) filter may be added on the downstream side of the sand filter. D.E. filters use porous fossilized diatom shells to coat a filter grid further restricting the diameter of materials that can pass through the filter. D. E. filters can remove particles as small as 1–3 μm. D. E. filters are of particular value in marine recirculating systems where the parasite *Oodinium* can be a problem. This parasite has a small free-living life-stage (dinospore) that can be filtered out of the water column when a D. E. filter is used, effectively breaking the life cycle of this parasite.

As the D. E. filter gets clogged, pressure will build and a pressure gauge may show pressures of about 8–10 psi above that seen with a clean filter. At that point the filter needs to be backwashed and new D. E. material added to recoat the filter grid.

In an effort to eliminate or reduce the risk of bacterial or viral diseases, the hatchery water supply can be passed through an Ultraviolet (UV) filter after passing through a sand and/or D. E. filter. UV light is classified into three wavelengths, UVA, UVB, and UVC. The UVC wavelength (100–290 nm) is the most potent. It can alter the DNA material of bacteria, viruses, and other microorganisms preventing them from successfully reproducing. Dosages for a 90% kill of most bacteria and viruses range from 2,000 to 8,000 $\mu W \cdot s/cm^2$. Efficacy of UV exposure depends on: water clarity (pre-filter to improve clarity); water flow rate (slower the better, recirculate to increase exposure); width of water column between the UV lamp and the containment wall (the narrower the width the greater the efficacy); age and cleanliness of the UV bulb (older the bulb reduces transmission, fouling reduces transmission); and water temperature (reduced efficacy at lower temperatures).

Most UV filter designs can provide effective bacterial control at a water flow rate of 75–95 liters per hour. UV filter bulbs are often rated for 10,000 h of use, but in many hatchery settings, it is convenient to replace the bulb at the start of each spawning season. For a UV filter to be effective, the water supply concentration of iron should be <0.3 mg/L, manganese <0.05 mg/L, the total hardness <120 mg/L and have a turbidity of <1 NTU.

Further information on water filtration is available at the following sites:

An example on the setup and use of a sand filter is available at:
http://www.pentairpoolme.com/uploads/document/file/533/Manual_FRP_Series.pdf. (May 2020).

An example of the use of a D. E. filter is available at:
https://hayward-pool-assets.com/assets/documents/pools/pdf/manuals/Pro-grid-DE2420.pdf. (May 2020).

For more information regarding UV filters see:
http://www.water-research.net/Waterlibrary/privatewell/UVradiation.pdf. (May 2020).
https://ww2.health.wa.gov.au/Articles/U_Z/Ultraviolet-disinfection-of-drinking-water. (May 2020).

17.18 Water Quality

A successful hatchery needs to be able to monitor its water quality and appreciate the significance of the values obtained for the parameters monitored. A hatchery must be ready to control a given water quality condition before it becomes out of control. It is important to understand how the water quality parameters interact, and how trying to manage one may negatively affect another.

References

Boyd, C. E. 1998. Pond water aeration systems. Aquacultural Engineering 18(1):9–40.

Boyd, C. E. 1998. Water quality for ponds aquaculture. Research and Development Series No. 43, International Center for Aquaculture and Aquatic Environment, Alabama Agricultural Experiment Station, Auburn University, Auburn. Available: http://soiltesting.tamu.edu/publications/AU43.pdf. (May 2020).

Boyd, C.E. 2018. Ammonia nitrogen dynamics in aquaculture. Understanding the fate of aquafeeds and environmental waste loads. Global Aquaculture Advocate. 10/29/2018. Available: https://www.aquaculturealliance.org/advocate/ammonia-nitrogen-dynamics-in-aquaculture/?headlessPrint=AAAAAPIA9c8r7gs82oWZBA. (May 2020).

Boyd, C. E., C. S. Tucker, and R. Viriyatum. 2011. Interpretation of pH, acidity, and alkalinity in aquaculture and fisheries. North American Journal of Aquaculture 73(4):43–408.

Cowan, P. A., D. B. Porcella, V. D. Adams, and L. A. Gardner. 1978. Water quality analysis laboratory procedures syllabus. USU, UWRL Reports Paper 520. The Utah Water Research Laboratory at DigitalCommons@USU, Utah State University, Logan. Available: https://digitalcommons.usu.edu/cgi/viewcontent.cgi?article=1519&context=water_rep. (May 2020).

Durborow, R. M., D. M. Crosby, and M. W. Brunson. 1997. Nitrite in fish ponds. SRAC Publication No. 462. Southern Regional Aquaculture Center, Stoneville, Mississippi. Available: http://agrilife.org/fisheries/files/2013/09/SRAC-Publication-No.-462-Nitrite-in-Fish-Ponds.pdf. (May 2020).

Francis-Floyd, R., C. Watson, D. Petty, and D. B. Pouder. 2015. Ammonia in aquatic systems. Publication #FA16. Florida Cooperative Extension Service, Institute of Food and Agricultural Sciences, University of Florida, Gainesville. Available: https://edis.ifas.ufl.edu/fa031. (May 2020).

Hargreaves, J. A. 1999. Control of clay turbidity in ponds. SRAC Publication No. 460. Southern Regional Aquaculture Center, Stoneville, Mississippi. Available: http://agrilife.org/fisheries2/files/2013/09/SRAC-Publication-No.-460-Control-of-Clay-Turbidity-in-Ponds.pdf. (May 2020).

Hargreaves, J. A., and C. S. Tucker. 2002. Measuring dissolved oxygen concentration in aquaculture. SRAC Publication No. 4601. Southern Regional Aquaculture Center, Stoneville, Mississippi. Available: http://agrilife.org/fisheries/files/2013/09/SRAC-Publication-No.-4601-Measuring-Dissolved-Oxygen-Concentration-in-Aquaculture.pdf. (May 2020).

Hargreaves, J. A., and C. S. Tucker. 2004. Managing ammonia in fish ponds. SRAC Publication No. 4603. Southern Regional Aquaculture Center, Stoneville, Mississippi. Available: https://fisheries.tamu.edu/files/2013/09/SRAC-Publication-No.-4603-Managing-Ammonia-in-Fish-Ponds.pdf. (May 2020).

Jensen, G. L., J. D. Bankston, and J. W. Jensen. 1989. Pond aeration: types and use of aeration equipment. SRAC Publication No. 371. Southern Regional Aquaculture Center, Stoneville, Mississippi. Available: https://appliedecology.cals.ncsu.edu/wp-content/uploads/371FS.pdf. (May 2020).

Kepenyes, J., and L. Váradi. 1984. Chapter 21: Aeration and oxygenation in aquaculture in Inland aquaculture engineering, ADCP/REP/84/21, FAO, Rome. Available: http://www.fao.org/docrep/x5744e/x5744e0m.htm. (May 2020).

Lazur, A. M., C. E. Cichra, and C. Watson. The use of lime in fish ponds. FA38. Florida Cooperative Extension Service, Institute of Food and Agricultural Sciences, University of Florida, Gainesville. Available: http://edis.ifas.ufl.edu/fa028. (May 2020).

Tucker, C. 2005. Pond aeration. SRAC Publication No. 3700. Southern Regional Aquaculture Center, Stoneville, Mississippi. Available: http://fisheries.tamu.edu/files/2013/09/SRAC-Publication-No.-3700-Pond-Aeration.pdf. (May 2020).

Tucker, C. S., and L. R. D'Abramo. 2008. Managing high pH in freshwater ponds. SRAC Publication No. 4604. Southern Regional Aquaculture Center, Stoneville, Mississippi. Available: http://agrilife.org/fisheries/files/2013/09/SRAC-Publication-No.-4604-Managing-High-pH-in-Freshwater-Ponds.pdf. (May 2020).

Wurts, W. A., and R. M. Durborow. 1992. Interactions of pH, carbon dioxide, alkalinity and hardness in fish ponds. SRAC Publication No. 464. Southern Regional Aquaculture Center, Stoneville, Mississippi. Available: http://fisheries.tamu.edu/files/2013/09/SRAC-Publication-No.-464-Interactions-of-pH-Carbon-Dioxide-Alkalinity-and-Hardness-in-Fish-Ponds.pdf. (November 2020).